TELECOMMUNICATIONS TOPICS

*Applications of
Functions and Probabilities in
Electronic Communications*

E. Bryan Carne

Prentice Hall PTR
Upper Saddle River, NJ 07458
www.phptr.com

ISBN 0-13-645565-4

90000

9 780136 455653

Library of Congress Cataloging-in-Publication Data

Carne E. Bryan
Telecommunications topics : applications of functions and probabilities in electronic communications / E. Bryan Carne.
  p. cm.
Includes index.
ISBN 0-13-645565-4 (case : alk. paper)
1. Signal theory (Telecommunication)--Mathematics. I. Title.
TK5102.9.C35 1998
621.382'2--dc21
                                         98-4540
                                         CIP

Editorial/production supervision: *Patti Guerrieri*
Cover design director: *Jerry Votta*
Cover designer: *Design Source*
Manufacturing manager: *Alan Fischer*
Marketing manager: *Kaylie Smith*
Acquisitions editor: *Bernard Goodwin*
Editorial assistant: *Diane Spina*

 © 1999 Prentice Hall PTR
Prentice-Hall, Inc.
A Simon & Schuster Company
Upper Saddle River, NJ 07458

Prentice Hall books are widely used by corporations and government agencies for training, marketing, and resale.

The publisher offers discounts on this book when ordered in bulk quantities. For more information, contact: Corporate Sales Department, Phone: 800-382-3419; Fax: 201-236-7141; E-mail: corpsales@prenhall.com; or write: Prentice Hall PTR, Corp. Sales Dept., One Lake Street, Upper Saddle River, NJ 07458.

All products or services mentioned in this book are the trademarks or service marks of their respective companies or organizations.

Printed in the United States of America
10 9 8 7 6 5 4 3 2

ISBN   0-13-645565-4

Prentice-Hall International (UK) Limited, *London*
Prentice-Hall of Australia Pty. Limited, *Sydney*
Prentice-Hall Canada Inc., *Toronto*
Prentice-Hall Hispanoamericana, S.A., *Mexico*
Prentice-Hall of India Private Limited, *New Delhi*
Prentice-Hall of Japan, Inc., *Tokyo*
Simon & Schuster Asia Pte. Ltd., *Singapore*
Editora Prentice-Hall do Brasil, Ltda., *Rio de Janeiro*

To Joan

Contents

PROBABILITIES

APPLICATIONS

Preface

This book is written for those of us who came first to computer engineering and then discovered that digital communications were vital to the continuing expansion of the information age. As a result, we picked up bits and pieces of lore—enough to make a go of communicating, but not enough to fully appreciate so demanding a discipline. Also, it is for those who, through a better understanding of the applications of functions and probabilities to digital communications, wish to further their comprehension of what may be the most important technical development of the last quarter of the twentieth century. In many ways, *Telecommunications Topics* complements my previous book *Telecommunications Primer* (Prentice Hall; 1995), which provides an overview of modern, mostly digital communications without supporting analysis.

The 22 topics relate to one another, and progress from techniques *for* analysis to the analysis *of* operating techniques. Taken from the current telecommunications environment, they reflect some of the problems that have been investigated and solved over the last 30 years of intense development in predominantly digital communications. They are divided into three sections—Functions, Probabilities and Applications—which can be summarized as follows.

Functions   The notion of surrogate functions is fundamental to the analysis of communication systems. Chapter 1, *Signals*, describes the advantages and limitations of mathematical functions that mimic signals. With them, models are built to explore the behavior of systems. As a framework for future results, a chart is included that maps the relationships among time and frequency domains and the techniques employed. In Chapter 2, *Periodic Functions*; Chapter 3, *Transient Functions*; Chapter 4, *LTI Systems*; and Chapter 5, *Autocorrelation and Spectral Density*; the techniques of deterministic analysis are reviewed and applied to periodic and transient signals. On a limited understanding of signal processing, the foundation is laid for the analysis of digital signals.

Probabilities To come closer to real-world situations, ideas of randomness and measures of probabilities are introduced in Chapter 6, *Probability*. Chapter 7, *Random Variables*, develops the concepts of Binomial-, Poisson-, and Normal-distributions and introduces the properties of uniform-, exponential-, and Erlang-distributed random variables. The second section closes with Chapter 8, *Random Processes*. It provides a limited look at random processes and introduces the all-important Wiener-Khinchine connection between the autocorrelation function and the power spectral density of power-type functions.

Applications Chapters 9 through 22 apply the results of the first eight chapters to the modern communications environment. In Chapter 9, *Queues*, steady-state results are developed for three basic queueing systems, and Erlang's formulas for telephone traffic are introduced. In Chapter 10, *Noise*, signal-to-noise ratio is defined, white noise is characterized, and the effect of noise on binary decisions is analyzed. Chapter 11, *Information Theory*, provides an elementary view of Shannon's theory including self-information and entropy, describes Shannon-Fano coding, and discusses Shannon's capacity theorem and the Hartley-Shannon law.

Chapter 12, *Digital Voice Signals*, describes the processes (sampling, quantizing, companding, and reconstruction) required to produce a digital signal from an analog voice signal. The effect of signal-to-noise ratio on the performance of PCM signals is discussed.

Chapters 13 through 17 are concerned with topics in data communication. In Chapter 13, *Data Signals*, several signal formats are described, and the frequency spectrums of random data signals are estimated. Coding and scrambling are discussed. In Chapter 14, *Intersymbol Interference*, pulse shaping, equalization, and regeneration of data signals are considered. Chapters 15 and 16 are devoted to error correction. In Chapter 15, *Error Detection*, parity checking, the use of checksums and cyclic redundancy checking are discussed. In Chapter 16, *Error Correction*, ARQ and forward error correction techniques are described. The throughput of various correction strategies is calculated for a range of data rates and other conditions. Chapter 17, *Access to Shared Media*, describes and analyzes popular techniques for providing shared access to common facilities.

Chapters 18 through 21 examine the characteristics of various types of radio signals. In Chapter 18, *Amplitude Modulation*, five methods of producing AM are analyzed. In Chapter 19, *Angle Modulation*, narrowband and wideband angle modulation is discussed, and the spectral characteristics of frequency modulation are developed. In Chapter 20, *Digital Modulation*, techniques for amplitude, phase, and frequency keying are discussed. Chapter 21, *Spread Spectrum Modulation*, describes direct spreading and frequency-hopping techniques and CDMA.

Finally, Chapter 22, *Transmission Media*, describes some of the properties of wire cables, optical fibers, cellular radio, and communication satellites.

Since very little happens in a vacuum, I would like to thank the many students, particularly those involved in continuing education, who asked questions of me and stimulated this work. The contents should be comfortable for those who have received courses in technical analysis. Hopefully, the result will be an increased understanding of some of the practices of digital communications.

E. Bryan Carne
Peterborough, NH

1

Signals

Signals can be measured; in mathematical terms they are real and finite. However, they are not easy to describe and they challenge the ingenuity of engineers who would analyze them. To perform this task, we build models of the environment that contain mathematical functions whose values we can calculate, and whose behavior mimics that of the signals we wish to analyze. These functions are divided into non-exclusive categories on the basis of:

- variation of their values with time
- degree to which their behavior can be described and predicted
- time for which they exist
- whether they are power-type or energy-type functions.

1.1 VARIATION OF VALUES WITH TIME

Functions are divided by the way their values vary over time.

- **Analog**: a continuous function that assumes positive, zero, or negative values. Changes occur smoothly and rates of change are finite.

Analog signals are generated by many physical processes; they play a large part in the *transmission* of messages by electronic media. Indeed, over many *electrical* transmission systems, data signals are coded in analog form. Information is carried in the variations of amplitude, frequency, and/or phase of the signal.

- **Digital**: a function that assumes a limited set of positive, zero, or negative values. Changes of value are instantaneous, and the rate-of-change at that instant is infinite—at all other times, it is zero. A common class of digital functions is called *binary;* they exist in two states only.

Binary signals are generated by computers and similar electronic devices. Binary techniques dominate the *processing* of messages, and, as explained in § 22.2, digital signals are used exclusively in the transmission of messages by optical fibers. With binary signals, information is carried in the sequence of states (1s and 0s).

The separation into analog or digital is probably the most important of all the divisions I describe. As noted, it implies an important difference in the way information is carried. Digital functions can be converted to analog functions, and vice-versa.

1.2 CERTAINTY WITH WHICH BEHAVIOR IS KNOWN

Functions are divided by the degree of certainty with which their behavior is known.

- **Deterministic**: at every instant, a deterministic function exhibits a value (including zero) that is related to values at neighboring times in a way that can be expressed exactly.

Thus, we can determine the time at which a deterministic function will achieve a particular value, and the value the function will have at a particular time. As a consequence, deterministic signals can be described by familiar mathematical functions.

- **Probabilistic**: a function whose future values are described in statistical terms

Although we may have a set of historical values, neither the value of a probabilistic function at a specific future time, nor the future time at which the function will have a specific value, can be known for sure. Future values are estimated from the statistics associated with past values and the assumption that future behavior is somehow connected to it. An important sub-class of probabilistic functions is random functions.

- **Random**: a probabilistic function whose values are limited to a given range. Over a long time, each value within the range will occur as frequently as any other value.

1.3 TIME FOR WHICH THEY EXIST

Functions can be divided on the basis of the time for which they exist.

- **Transient**: a function that exists for a limited time only

It represents the class of real-world signals that are turned on, and then turned off. If the function is deterministic, it can be analyzed with the help of *Fourier transforms*. (Determinism is necessary, because evaluating these transforms requires integrating $\pm\infty$ with respect to time, that is, from way in the past to way in the future.)

- **Eternal**: a function that exists for all time

It describes the performance of communication systems that are operating in the steady state (i.e., any behavior solely due to turn on has long since disappeared so that it is as if the systems have been operating forever). An important sub-class of eternal functions is periodic functions.

- **Periodic**: an eternal function whose values repeat at regular intervals

To know everything about a periodic function, only a single period need be studied. Periodic functions can be decomposed into a related set (fundamental plus harmonics) of sinusoidal functions through the use of *Fourier series*.

1.4 ENERGY-TYPE OR POWER-TYPE

- **Energy**: the capacity for doing work
- **Power**: the rate of doing work, i.e., energy per unit time

According to Ohm's Law, when a signal of $e(t)$ volts is applied to a resistor of R ohms, a current of $i(t) = e(t)/R$ amps will flow. To standardize comparisons between signals, it is customary to put $R = 1$ ohm so that $e(t) = i(t)$, and to substitute the quantity $s(t)$ for $e(t)$ and $i(t)$. To reflect this arrangement, $s(t)$ is expressed in *signal units*, power is expressed in *signal watts*, and energy is expressed in *signal joules*.

(1) Signal Power

The average signal power in signal watts developed in a 1-ohm resistor in T seconds due to a signal $s(t)$ signal units is

$$\overline{P}[T] = \frac{1}{T} \int_{\alpha}^{\alpha+T} s^2(t)\,dt \tag{1.1}$$

where α is an arbitrary instant in time. As $T \to \infty$ we obtain the signal power averaged over all time

$$\overline{P}[\infty] = \lim_{T \to \infty} \frac{1}{T} \int_{\alpha}^{\alpha+T} s^2(t)\,dt \tag{1.2}$$

For a signal of limited duration, i.e., a transient signal, the integral is finite, so that $\overline{P}[\infty]$ is zero. For an eternal signal, the integral is infinite, so that $\overline{P}[\infty]$ is finite.

For a periodic signal $s_p(t)$, averaging over one period (T_1) is the same as averaging over all time, so that

$$\overline{P}[T_1] = \frac{1}{T_1} \int_{\alpha}^{\alpha+T_1} s_p^2(t)dt = \left\langle s_p^2(t) \right\rangle = \overline{P}[\infty] \tag{1.3}$$

where, as before, α is a point in time, integration occurs over one period, and the angular brackets $\langle \ \rangle$ stand for time-averaging, i.e., the function $\frac{1}{T_1} \int_{\alpha}^{\alpha+T_1}(\ )dt$.

(2) Signal Energy

In T seconds, the signal energy (in signal joules) is

$$E[T] = \overline{P}[T]T = \int_{\alpha}^{\alpha+T} s^2(t)dt \tag{1.4}$$

so that, over all time, the total signal energy $E[\infty]$ is

$$E[\infty] = \int_{-\infty}^{\infty} s^2(t)dt \tag{1.5}$$

For a transient signal, the integral is finite so that $E[\infty]$ is finite. For an eternal signal, the integral is infinite so that $E[\infty]$ is infinite.

(3) Parameters for This Classification

Thus, we distinguish between functions on the basis that they are *energy*-type or *power*-type functions.

- **Energy-type function**: evaluating $E[\infty]$ yields a finite value, and evaluating $\overline{P}[\infty]$ yields zero. Transient functions fulfill this condition.
- **Power-type function**: evaluating $E[\infty]$ yields an infinite value, and evaluating $\overline{P}[\infty]$ yields a finite value. Eternal functions fulfill this condition.

Treating power-type functions as energy-type functions can lead to indeterminant results, and treating energy-type functions as power-type functions can result in zeros. Because of their differences, you need to be aware of the nature of the functions used in specific situations. For example, in Chapter 5, I employ distinctive notation for power-type and energy-type correlation and spectral density functions.

EXAMPLE 1.1   Regular Pulse Train

Given   A regular pulse train consists of an unlimited sequence of identical square pulses. Each pulse is of amplitude A signal units, occupies the center of a timeslot of duration T seconds, and is of duration t seconds ($t < T$).

Problem

a) What is the energy in a pulse? In what units is it measured?

b) Show how your expression for pulse energy can be converted into an expression for the power averaged over all time (i.e., $\bar{P}[\infty]$) of the train. In what units is it measured?

Solution

a) The pulse energy (i.e., $E[t]$) is $A^2 t$ signal joules.

b) Now $E[t] = E[T]$ so that $\bar{P}[T] = \frac{1}{T}E[T] = A^2 t/T$. Because we are dealing

with an eternal periodic function, $\bar{P}[\infty] = \bar{P}[T]$, and $\bar{P}[\infty] = A^2 t/T$ signal watts.

EXAMPLE 1.2   Unit Step Function

Given   A unit step function

$$s(t) = \begin{cases} 0, t < 0 \\ 1, t \geq 0 \end{cases}$$

To assist in visualization, the step function is drawn in **Figure 1.1**.

Problem   Is it an energy-type or power-type function?

Solution   Calculate $E[\infty]$ and $\bar{P}[\infty]$.

$$E[\infty] = \int_{-\infty}^{\infty} |s(t)|^2 dt = \int_{-\infty}^{0} 0^2 dt + \int_{0}^{\infty} 1^2 dt = t\big|_0^{\infty} = \infty \text{ signal joules}$$

$$\bar{P}[\infty] = \lim_{T \to \infty} \frac{1}{T} \int_{0}^{T} |s(t)|^2 dt = \lim_{T \to \infty} \frac{1}{T} t\big|_0^{T} = 1 \text{ signal watt}$$

Because $\bar{P}[\infty]$ is finite and $E[\infty]$ is infinite, the unit step is a power function.

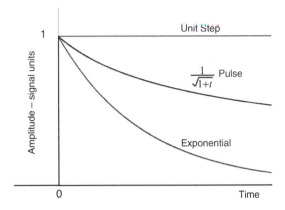

Figure 1.1 Three Signals Evaluated in Examples 1.2, 1.3, and 1.4

EXAMPLE 1.3 Exponential Function

Given An exponential function

$$s(t)=\begin{cases}0, t<0 \\ \exp(-t), t\geq 0\end{cases}$$

To assist in visualization, the exponential function is drawn in Figure 1.1.

Problem Is it an energy-type or power-type function?

Solution Calculate $E[\infty]$ and $\overline{P}[\infty]$.

$$E[\infty]=\int_0^\infty \exp(-2t)dt=\tfrac{1}{-2}\exp(\text{-2t})\Big|_0^\infty=\tfrac{1}{2}\ \text{signal joules}$$

$$\overline{P}[\infty]=\lim_{T\to\infty}\tfrac{1}{T}\int_0^T \exp(-2t)dt=\lim_{T\to\infty}\tfrac{1}{-2T}\big[\exp(-2t)-1\big]=0\ \text{signal watts}$$

Because $\overline{P}[\infty]$ is zero and $E[\infty]$ is finite, the exponential function is an energy function.

EXAMPLE 1.4 $1/\sqrt{1+t}$ Function

Given A $1/\sqrt{1+t}$ function, i.e.,

$$s(t)=\begin{cases}0, t<0 \\ 1/\sqrt{1+t}, t\geq 0\end{cases}$$

To assist in visualization, a $1/\sqrt{1+t}$ function is drawn in Figure 1.1.

Problem Determine whether it is power-type or energy-type.

Solution Calculate $E[\infty]$ and $\overline{P}[\infty]$.

$$E[\infty]=\int_0^\infty \tfrac{1}{1+t}dt = \log_e(1+t)\big|_0^\infty = \infty \text{ signal joules}$$

$$\overline{P}[\infty]= \lim_{T\to\infty} \tfrac{1}{T}\int_0^T \tfrac{1}{1+t}dt = \lim_{T\to\infty} \tfrac{1}{T}\log_e(1+T)$$

As $T \to \infty$, $\log_e(1+T)$ grows larger at a much slower rate than T; hence, in the limit, $\overline{P}[\infty]$ becomes zero[1]. Because $\overline{P}[\infty]$ is zero and $E[\infty]$ is infinite, the $1/\sqrt{1+t}$ function is neither a power function nor an energy function.

Comment That such a function exists should be no surprise. It is at the interface between power and energy functions.

1.5 REAL-WORLD SIGNALS

The goals of signal analysis are to determine the nature of the frequency spectrums of signals and to achieve an outcome that closely matches observed behavior with as little analytical complexity as possible. In descending order of simplicity (i.e., ascending order of complexity), we employ

- **Periodic functions**: eternal deterministic functions that can be analyzed with the help of Fourier series
- **Transient deterministic functions**: functions of limited duration that can be analyzed with the help of Fourier transforms
- **Probabilistic eternal functions**: functions that can be analyzed with the help of random variables and processes.

These methods, and their extensions, are explained in the next seven chapters. In the balance of the book, they are applied to the analysis of telecommunications situations.

---

[1] This assertion can be confirmed by the application of l'Hôpital's rule for evaluating indeterminate forms (after Guillaume de l'Hôpital, a late 17th century French mathematician).

In Chapters 2 through 5, I describe: the application of Fourier series techniques to determine amplitude spectrums of periodic functions; the application of Fourier transform techniques to determine magnitude density spectrums of transient deterministic functions; and the use of autocorrelation functions to determine energy spectral densities and power spectral densities. In Chapters 6 and 7, I describe probability and random variables; among other uses, they are the basis for Chapter 8, in which I describe the application of autocorrelation functions and power spectral densities to the analysis of random processes. To give meaning to these chapters, I list some of the ways in which periodic, transient, and probabilistic functions are described.

(1) Periodic Functions

For periodic functions, transformations from time domain to frequency domain are made by Fourier series or Fourier transforms supported by Dirac sampling functions. Also, transformations are made by forming autocorrelation functions and then employing Fourier transforms to create power spectral density functions. Some of the ways in which periodic functions are described are

- In the **Time domain**:

    - as a repetitive function [equation (2.1)]
    $$s_p(t) = s_p(t - mT_1)$$
    - as an exponential Fourier series [equation (2.8)]
    $$s_p(t) = \sum_n S_n \exp(j2\pi n f_1 t)$$
    - as the outcome of replicating a generator function with a Dirac train [equation (3.3)]
    $$s_p(t) = s_g(t) \otimes \delta_{T_1}(t) = \sum_n s_g(t - nT_1)$$
    - as an autocorrelation function [equation (5.25)]
    $$\phi_p(\tau) = \sum_n |S_n|^2 \exp(j2\pi n f_1 \tau)$$

- In the **Frequency domain**:

    - as complex Fourier series coefficients [equation (2.11)]
    $$S_n = \left\langle s_p(t) \exp(-j2\pi n f_1 t) \right\rangle$$

– as the outcome of sampling a generator function with a Dirac comb [equation (3.34)]

$$S_p(f) = S_g(f) f_1 \delta_{f_1}(f)$$

– as a power spectral density function [equation (5.26)]

$$\Phi_p(f) = \sum_n |S_n|^2 \delta(f - nf_1)$$

and [equation (5.29)]

$$\Phi(f) = |S_g(f)|^2 / T$$

(2) Transient Deterministic Functions

For transient deterministic functions, transformations from time domain to frequency domain are made by Fourier transforms. Also, transformations are made by forming autocorrelation functions and then employing Fourier transforms to create energy spectral density functions. Some of the ways in which transient deterministic functions are described are

- In the **Time domain**:

  – as a limited time function [equation (3.1)]

  $$s_t(t) = s(t) \Pi \left[t - \tfrac{1}{2}(t_1 + t_2) / (t_2 - t_1) \right]$$

  – as an inverse Fourier transform [equation (3.5)]

  $$s_t(t) = \int_{-\infty}^{\infty} S(f) \exp(j 2\pi f t) df$$

  – as an autocorrelation function [equation (5.12)]

  $$\psi(\tau) = \int_{-\infty}^{\infty} s_t(t) s_t^*(t - \tau) d\tau$$

- In the **Frequency domain**:

  – as a Fourier transform [equation (3.6)]

  $$S(f) = \int_{-\infty}^{\infty} s_t(t) \exp(-j 2\pi f t) dt$$

– as an energy spectral density function [equation (5.14)]

$$\Psi(f) = \int_{-\infty}^{\infty} \psi(\tau)\exp(-j2\pi f\tau)d\tau$$

and [equation (5.16)]

$$\Psi(f) = |S(f)|^2$$

(3) Probabilistic Eternal Functions

For probabilistic eternal functions, transformations from time domain to frequency domain are made by forming autocorrelation functions and then employing Fourier transforms to create power spectral density functions. Some of the ways in which they are described are

- In the **Time domain**:
  – as an autocorrelation function [equation (8.5)]

$$\phi_X(\tau) = E[X(t+\tau)X(t)]$$

  and [equation (8.6)]

$$\phi_X(\tau) = \int_{-\infty}^{\infty} \Phi_X(f)\exp(j2\pi f\tau)df$$

- In the **Frequency domain**:
  – as a power spectral density function [equation (8.6)]

$$\Phi_X(f) = \int_{-\infty}^{\infty} \phi_X(\tau)\exp(-j2\pi f\tau)d\tau$$

Figure 1.2 diagrams the interrelation of these options to show how they may be employed to achieve the goal of determining the nature of the frequency spectrums of signals. All of the transformations are reversible across the time-frequency boundary so that time domain information can be converted to frequency domain information, and vice-versa. However, because autocorrelation functions and energy and power spectral density functions ignore phase information, re-creating the original function, or its frequency spectrum, from them, is impossible.

REVIEW QUESTIONS

1.1   With respect to functions employed to mimic signals, define the terms *analog*, *digital*, and *binary*.

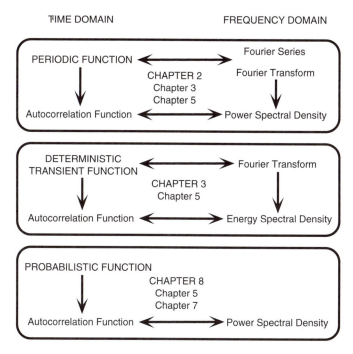

TIME DOMAIN   FREQUENCY DOMAIN

PERIODIC FUNCTION ⟷ Fourier Series
CHAPTER 2
Chapter 3
Chapter 5
Fourier Transform
Autocorrelation Function ⟷ Power Spectral Density

DETERMINISTIC
TRANSIENT FUNCTION ⟷ Fourier Transform
CHAPTER 3
Chapter 5
Autocorrelation Function ⟷ Energy Spectral Density

PROBABILISTIC FUNCTION
CHAPTER 8
Chapter 5
Chapter 7
Autocorrelation Function ⟷ Power Spectral Density

Figure 1.2 Paths to the Frequency Domain for Periodic, Deterministic
Transient, and Probabilistic Functions
*One of the purposes of signal analysis is to determine the nature of the
frequency spectrums of functions that mimic the behavior of signals.
Fourier series and autocorrelation techniques are used with periodic functions. Fourier transform and autocorrelation techniques are used with
deterministic transient functions. Autocorrelation techniques are used with
probabilistic functions.*

1.2   What parameters can carry information in an analog signal?

1.3   How is information carried in a digital signal?

1.4   With respect to functions employed to mimic signals, define the terms
      deterministic, probabilistic, and *random.*

1.5   With respect to functions employed to mimic signals, define the terms
      transient, eternal, and *periodic.*

1.6   Distinguish between *energy* and *power.*

1.7   What is the significance of expressing the magnitude of a signal in signal
      units?

1.8   Define an *energy-type function* in terms of its total energy and power averaged over all time.

1.9   Define a *power-type function* in terms of its total energy and power averaged over all time.

1.10   What class of functions do we analyze with Fourier series?

1.11   What class of functions do we analyze with Fourier transforms?

1.12   What class of functions do we analyze with random variables and processes?

2

Periodic Functions

- **Periodic function**: a function $s_p(t)$ that repeats a unique stream of values over and over again, taking the same amount of time each time so that

$$s_p(t) = s_p(t - mT_1) \qquad (2.1)$$

where m is any integer, positive, negative, or zero, and T_1 is the repetition interval or period of the function.

2.1 FOURIER SERIES

Periodic functions are decomposed into sinusoidal functions by the *trigonometric* Fourier series and the Fourier *cosine* series. Named after their discoverer, Jean Baptiste Fourier, a late 18th- and early 19th-century French mathematician, they are converted into exponential functions by the *exponential* Fourier series.

2.2 TRIGONOMETRIC FOURIER SERIES

In a trigonometric Fourier series, a periodic function is decomposed into sinusoidal functions whose frequencies are harmonically related. Thus,

$$s_p(t) = A_0 + \sum_{n=1}^{\infty} A_n \cos(2\pi nt/T_1) + \sum_{n=1}^{\infty} B_n \sin(2\pi nt/T_1) \qquad (2.2)$$

where n is a positive integer (i.e., $1 \leq n < \infty$), and the values of the coefficients A_0, A_n, and B_n are

$$A_0 = \langle s_p(t) \rangle$$
$$A_n = 2\langle s_p(t)\cos(2\pi nt/T_1) \rangle \qquad (2.3)$$
$$B_n = 2\langle s_p(t)\sin(2\pi nt/T_1) \rangle$$

A_0 is the zero frequency coefficient of the series. It is the time-average or mean of $s_p(t)$ [i.e., the area under $s_p(t)$ for a single period (in signal unit seconds) divided by the period T_1 (in seconds)]. Often, it is referred to as the *dc–term*.

If $s_p(t)$ is a *zero-mean* function, $A_0 = 0$. In many applications, A_0 needs to be zero. For one thing, the signal power is minimized [see § 2.7(4)]; for another, modulation and demodulation operations are easier [see § 18.5(6)]. The other components of the series occur at the fundamental frequency (i.e., $1/T_1$ when $n=1$) and at harmonics (i.e., n/T_1 when $n = 2, 3, \ldots$ etc.) of the fundamental frequency. A_n and B_n are twice the time-average of the product of $s_p(t)$ and sinusoidal terms.

The periodic function shown in **Figure 2.1** can be described by the *trigonometric Fourier series*

$$s_p(t)=1+\cos(2\pi t/T_1)-0.75\sin(2\pi t/T_1)+0.5\cos(6\pi t/T_1)-0.25\sin(6\pi t/T_1) \quad (2.4)$$

It consists of

- A zero-frequency coefficient $A_0 = 1$, that is the mean value of the function
- Discrete components with coefficients $A_1 = 1$ and $B_1 = -0.75$ at the fundamental frequency

$$s_p(t)=1+\cos\left(\tfrac{2\pi t}{T_1}\right)-0.75\sin\left(\tfrac{2\pi t}{T_1}\right)+0.5\cos\left(\tfrac{6\pi t}{T_1}\right)-0.25\sin\left(\tfrac{6\pi}{T_1}\right)$$

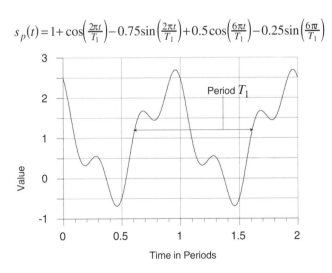

Figure 2.1 Periodic Function
The function is described by a trigonometric Fourier series.

- Discrete components with coefficients $A_3 = 0.5$ and $B_3 = -0.25$ at the third harmonic frequency

Spectrums of periodic functions consist of discrete values at frequencies that are multiples of the fundamental frequency. (Analysts are wont to say periodic functions are *monotonously repetitive* in the time domain, and *always discrete* in the frequency domain.)

2.3 FOURIER COSINE SERIES

In a Fourier cosine series

$$s_p(t) = C_0 + \sum_{n=1}^{\infty} C_n \cos\left(\frac{2\pi nt}{T_1} - \phi_n\right) \tag{2.5}$$

where

$$C_0 = A_0$$

$$C_n = \sqrt{A_n^2 + B_n^2} \tag{2.6}$$

$$\phi_n = \tan^{-1}\left[B_n / A_n\right]$$

The periodic function $s_p(t)$ consists of a dc–term (C_0), a component at the fundamental frequency, and components at harmonic frequencies that are positive, integer multiples of the fundamental frequency. A plot of the coefficients C_n versus frequency is known as the *amplitude spectrum* of $s_p(t)$. A plot of ϕ_n versus frequency is the *phase spectrum* of $s_p(t)$.

EXAMPLE 2.1 Complex Periodic Function

Given

$$s_p(t) = 1 + \cos(2\pi t/T_1) - 0.75\sin(2\pi t/T_1) + 0.5\cos(6\pi t/T_1) - 0.25\sin(6\pi t/T_1)$$

Problem Express the periodic function in a Fourier cosine series.

Solution

$$C_0 = A_0 = 1$$
$$C_1 = \sqrt{1^2 + 0.75^2} = 1.25$$
$$C_3 = \sqrt{0.5^2 + 0.25^2} = 0.56$$

$$\phi_1 = \tan^{-1}\tfrac{-0.75}{1} = \tan^{-1}-0.75 = -36.9° = -0.644 \text{ radians}$$

$$\phi_2 = \tan^{-1}\tfrac{-0.25}{0.5} = \tan^{-1}-0.5 = -26.6° = -0.464 \text{ radians}$$

so that

$$s_p(t) = 1 + 1.25\cos\left(\tfrac{2\pi t}{T_1} + 0.644\right) + 0.56\cos\left(\tfrac{6\pi t}{T_1} + 0.464\right) \tag{2.7}$$

Comment This representation produces the one-sided amplitude and phase spectrums shown in **Figure 2.2**.

2.4 COMPLEX QUANTITIES

Mathematical functions may be complex, i.e., they can have real and imaginary parts, and significant information can be obtained from the study of both of them. If $s(t)$ is complex, it can be expressed in terms of magnitude and phase, i.e.,

$$s(t) = |s(t)|\exp[j\angle s(t)]$$

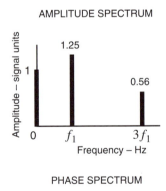

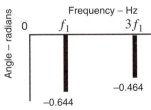

Figure 2.2 One-Sided Amplitude and Phase Spectrums of Periodic Function of Figure 2.1

and in terms of its real and imaginary parts, i.e.,

$$s(t) = a(t) + jb(t)$$
$$s*(t) = a(t) - jb(t)$$

where

$s*(t)$ is the complex conjugate of $s(t)$

$j = \sqrt{-1}$.

Manipulating these relationships gives

$$|s(t)|^2 = s(t)s*(t) = a^2(t) + b^2(t)$$

i.e., the magnitude squared of $s(t)$ is equal to the product of $s(t)$ and its complex conjugate—and to the sum of the squares of its real and imaginary parts. Furthermore,

$$\angle s(t) = \tan^{-1}[b(t)/a(t)]$$

i.e., the phase of $s(t)$ is equal to the angle whose tangent is the ratio of the imaginary and real parts of $s(t)$.

2.5 EXPONENTIAL FOURIER SERIES

The trigonometric forms of the Fourier series can be converted to more convenient representations by Euler's identities (after Leonhard Euler, an 18th-century Swiss mathematician). They are

$$\cos\theta = \tfrac{1}{2}[\exp(j\theta) + \exp(-j\theta)]; \quad \sin\theta = \tfrac{1}{2j}[\exp(j\theta) - \exp(-j\theta)] \qquad (2.8)$$

If we substitute these expressions in the trigonometric Fourier series [equation (2.2)] and simplify the result, we obtain the exponential Fourier series

$$s_p(t) = \sum_n S_n \exp(jn2\pi f_1 t) \quad \text{Synthesis Equation} \qquad (2.9)$$

where $-\infty < n < +\infty$ and f_1 is the fundamental frequency. S_n is called the *complex Fourier coefficient*. Because equation (2.9) sums a series of harmonically related components to produce the periodic function, it is known as the *synthesis* equation.

The values of S_n are found by multiplying both sides of equation (2.9) by $\exp(-jk2\pi f t)$ and time-averaging. These actions produce the equation

$$\frac{1}{T_1} \int_{\alpha}^{\alpha+T_1} s_p(t)\exp(-jk2\pi f_1 t)dt = \sum_n S_n \frac{1}{T_1} \int_{\alpha}^{\alpha+T_1} \exp[j(n-k)2\pi f_1 t]dt \qquad (2.10)$$

When $n = k$, the right-hand side (RHS) of equation (2.10) is equal to S_n. (The exponent is zero, so that the term is equal to 1. Integration produces T_1, which cancels with T_1 in the denominator). When $n \neq k$, the RHS is zero (the time-average of a set of zero-mean functions is zero). Thus, the time-averaged terms on the RHS exhibit the properties of Kronecker's *delta* function, δ_{nk} (after Leopold Kronecker, a 19th-century German mathematician), i.e.,

$$\frac{1}{T_1} \int_{\alpha}^{\alpha+T_1} s_p(t) \exp(-jk2\pi f_1 t) dt = \sum_n S_n \delta_{nk} \tag{2.11}$$

where

$$\delta_{nk} = \begin{cases} 1, & n=k \\ 0, & n \neq k \end{cases}$$

Putting $k=n$ so that n is the surviving integer, equation (2.11) becomes

$$S_n = \langle s_p(t) \exp(-jn2\pi f_1 t) \rangle \text{ Analysis Equation} \tag{2.12}$$

Because it permits the value of each complex Fourier coefficient to be determined, equation (2.12) is known as the *analysis* equation.

EXAMPLE 2.2   Cosine Function

Given

$$s(t) = A\cos(2\pi f_1 t + \theta)$$

Problem Find the complex Fourier coefficients.

Solution Expand $s(t)$ in exponential form [equation (2.8)].

$$s(t) = \frac{A}{2}\{\exp(j\theta)\exp(j2\pi f_1 t) + \exp(-j\theta)\exp(-j2\pi f_1 t)\}$$

When $n = \pm1$, the terms of the exponential Fourier series [equation (2.9)] are

$$s(t) = \sum_n S_n \exp(jn2\pi f_1 t) = S_1 \exp(j2\pi f_1 t) + S_{-1}\exp(-j2\pi f_1 t)$$

so that, by inspection

$$S_1 = \frac{A}{2}\exp(j\theta); \ S_{-1} = \frac{A}{2}\exp(-j\theta)$$

Comment The magnitudes of the complex Fourier coefficients of the cosine function are the same $(A/2)$, and the phase angles are equal in value but of opposite sign.

EXAMPLE 2.3 Sine Function

Given

$$s(t) = A\sin(2\pi f_1 t + \theta)$$

Problem Find the complex Fourier coefficients.

Solution Expand $s(t)$ in exponential form and compare with the exponential Fourier series. By inspection

$$S_1 = \frac{A}{2j}\exp(j\theta); \quad S_{-1} = -\frac{A}{2j}\exp(-j\theta)$$

or

$$S_1 = -\frac{jA}{2}\exp(j\theta); \quad S_{-1} = \frac{jA}{2}\exp(-j\theta)$$

Discussion The magnitudes of the complex Fourier coefficients of the sine function are the same $(A/2)$. Because multiplying by j is the same as increasing the phase by $\pi/2$ radians, we can restate the complex Fourier coefficients for the sine function as

$$S_1 = -\frac{A}{2}\exp[j(\theta + \pi/2)]; \quad S_{-1} = \frac{A}{2}\exp[-j(\theta + \pi/2)]$$

Thus, the values of the phase angles of the complex Fourier coefficients of the sine function differ by π (and $\pm\pi/2$ from the cosine function coefficients).

Comment **Figure 2.3** employs a three-dimensional Argand diagram to give a view of the complex coefficients of the cosine and sine functions. As the pairs of phasors of magnitude $A/2$ rotate in opposite directions (+ve frequency anti-clockwise, −ve frequency clockwise), they produce sinusoidal functions on the real plane.

2.6 ONE-SIDED AND TWO-SIDED SPECTRUMS

The use of the complex Fourier coefficient S_n introduces profound changes in the nature of the mathematical model describing a periodic function. The value of n is no longer limited to positive integers; it may take on any integer value. Since each sinusoidal component involves the frequency f_n $(= nf_1)$, in effect the analysis technique has created negative frequencies, and the frequency spectrums have become two-sided.

(1) Two-Sided Magnitude and Phase Spectrums

Because S_n is complex, it is expressed as a magnitude and a phase angle. Plotting the magnitude of S_n versus frequency produces a two-sided magni-

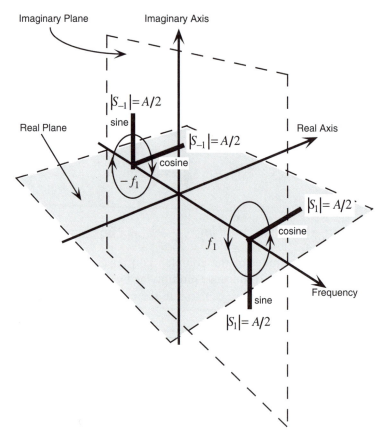

Figure 2.3 Three-Dimensional Argand Diagram of the Complex Fourier Coefficients of Sine and Cosine Functions
In the positive frequency plane, the phasors rotate counterclockwise; in the negative frequency plane, the phasors rotate clockwise. The projections on the real plane are identical.

tude spectrum, and plotting the phase of S_n versus frequency produces a two-sided phase spectrum. Spectrums of this kind for the periodic function in Figure 2.1 are shown in **Figure 2.4**.

Substituting $\cos\theta - j\sin\theta$ for $\exp(-j\theta)$ in equation (2.12) gives

$$S_n = \left\langle s(t)\left[\cos(n2\pi f_1 t) - j\sin(n2\pi f_1 t)\right]\right\rangle \tag{2.13}$$

Using this relationship, we can show that, if $s(t)$ is *real*, S_0 is *real*, $S_{-n} = S_n^*$, and $|S_{-n}| = |S_n|$, i.e., the two-sided magnitude spectrum is *even;* also, it shows that $\angle S_{-n} = -\angle S$, i.e., the two-sided phase spectrum is *odd*.

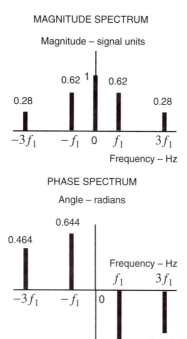

Figure 2.4 Two-Sided Magnitude and Phase Spectrums of Periodic Function of Figure 2.1
Derived from an exponential Fourier series, they are magnitude and phase components of complex Fourier coefficients.

(2) Two-Sided Power Spectrum

From equation (1.3), the signal power of a periodic, complex function is

$$\overline{P}[T_1] = \left\langle s_p^2(t) \right\rangle = \left\langle s_p(t) s_p^*(t) \right\rangle \tag{2.14}$$

Expressing $s_p^*(t)$ as an exponential Fourier series

$$s_p^*(t) = \left[\sum_{n=-\infty}^{\infty} S_n \exp(j2\pi n f_1 t) \right]^* = \sum_{n=-\infty}^{\infty} S_n^* \exp(-j2\pi n f_1 t)$$

so that

$$\overline{P}[T_1] = \left\langle s_p(t) \left[\sum_{n=-\infty}^{\infty} S_n^* \exp(-j2\pi n f_1 t) \right] \right\rangle$$

Interchanging the order of averaging and summation gives

$$\overline{P}[T_1] = \sum_{n=-\infty}^{\infty} S_n^* \left\langle s(t)\exp\left(-j2\pi n f_1 t\right)\right\rangle$$

The quantity inside the $\langle\ \rangle$ brackets is S_n, so that

$$\overline{P}[T_1] = \sum_{n=-\infty}^{\infty} S_n S_n^* = \sum_{n=-\infty}^{\infty} |S_n|^2 \tag{2.15}$$

Thus, the signal power of a periodic function $s_p(t)$ is equal to the sum of the squares of the magnitudes of its complex Fourier coefficients. This result is known as *Parseval's Power Theorem*. The two-sided power spectrum of the periodic function in Figure 2.1 is shown in **Figure 2.5**.

(3) One-Sided Power Spectrum

From equation (2.5), a real, periodic function can be represented as

$$s_p(t) = C_0 + \sum_{n=1}^{\infty} C_n \cos\left(2\pi f_n t - \phi_n\right) \tag{2.16}$$

The signal power is

$$\overline{P}[T_1] = \left\langle s_p^2(t)\right\rangle \tag{2.17}$$

POWER SPECTRUM

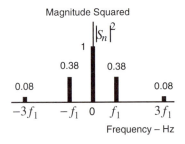

$$P[s(t)] = 1 + 2(0.38) + 2(0.08) = 1.92 \text{ signal watts}$$

Figure 2.5 Two-Sided Power Spectrum of Periodic Function of Figure 2.1
The power (averaged over all time) is the sum of the squares of the com-
ponents of the two-sided magnitude spectrum (Figure 2.4).

Squaring and time-averaging term-by-term gives

$$\overline{P}[T_1] = C_0^2 + \tfrac{1}{2}\sum_{n=1}^{\infty} C_n^2 \tag{2.18}$$

so that the signal power of a real, periodic function is the sum of the square of the dc-term and one-half of the sum of the squares of the other coefficients of the Fourier cosine series. The one-sided power spectrum of the periodic function in Figure 2.1 is shown in **Figure 2.6**. Note that the expression for signal power [equation (2.18)] does not contain the phase angle ϕ, so that $\overline{P}[T_1]$ is independent of phase.

EXAMPLE 2.4   Periodic Signal

Given A periodic signal is described as follows

$$s_p(t) = 1 + \cos(2\pi t/T_1) - \sin(2\pi t/T_1) + 0.5\cos(4\pi t/T_1) - 0.5\sin(4\pi t/T_1)$$

Problem

a) List the coefficients of the trigonometric series.
b) Express $s_p(t)$ as a Fourier cosine series.
c) Find the complex Fourier coefficients.
d) Sketch the two-sided magnitude and phase spectrums of $s_p(t)$.
e) Sketch the two-sided power spectrum of $s_p(t)$.

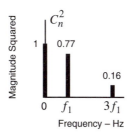

$$P[s_p(t)] = 1 + 0.77 + 0.16 = 1.93 \text{ signal watts}$$

Figure 2.6 One-Sided Power Spectrum of Periodic Function of Figure 2.1
The power (averaged over all time) is the sum of the square of the dc-term and one-half of the sum of the squares of the components of the one-sided amplitude spectrum (Figure 2.2). The value of 1.93 signal watts is slightly larger than the value in Figure 2.5, due to rounding.

f) Determine the power of $s_p(t)$ from the two-sided spectrum.

g) Sketch the one-sided power spectrum of $s_p(t)$.

h) Determine the power of $s_p(t)$ from the one-sided spectrum.

Solution

a) A zero-frequency coefficient $A_0 = 1$, that is the mean value of the function. Components with coefficients $A_1 = 1$ and $B_1 = -1$ at the fundamental frequency. Components with coefficients $A_2 = 0.5$ and $B_2 = -0.5$ at the second harmonic frequency.

b) $C_0 = A_0$, so that $C_0 = 1$; $C_n = \sqrt{A_n^2 + B_n^2}$, so that $C_1 = \sqrt{2}$ and $C_2 = \sqrt{0.5}$; $\phi_n = \tan^{-1}[B_n/A_n]$, so that $\phi_1 = \tan^{-1}[-1] = -45°$ and $\phi_2 = \tan^{-1}[-1] = -45°$. Thus,

$$s_p(t) = 1 + \sqrt{2}\cos\left(\frac{2\pi nt}{T_1} + 45°\right) + \sqrt{0.5}\cos\left(\frac{4\pi nt}{T_1} + 45°\right)$$

c) $S_n = \langle s_p(t)\exp(-jn2\pi f_1 t)\rangle$. From Example 2.2, when $s(t) = A\cos(2\pi f_1 t + \theta)$, $S_1 = \frac{A}{2}\exp(j\theta)$; $S_{-1} = \frac{A}{2}\exp(-j\theta)$ so that

$$|S_0| = 1,$$
$$S_{\pm 1} = \sqrt{2}/2\exp(\pm j45°) = 0.70711\exp(\pm j45°)$$
$$S_{\pm 2} = \sqrt{0.5}/2\exp(\pm j45°) = 0.35355\exp(\pm j45°).$$

d) From c), the magnitudes of S_n are $|S_0| = 1$, $|S_{\pm 1}| = 0.70711$, and $|S_{\pm 2}| = 0.35355$. The phases are $\phi_1 = \phi_2 = -45°$ and $\phi_{-1} = \phi_{-2} = 45°$. Hence the two-sided magnitude and phase spectrums of $s_p(t)$ are as drawn in **Figure 2.7**.

e) From d), the magnitudes squared are $|S_0|^2 = 1$, $|S_{\pm 1}|^2 = 0.5$, and $|S_{\pm 2}|^2 = 0.125$. The two-sided power spectrum of $s_p(t)$ is sketched in Figure 2.7.

f) The power of $s_p(t)$ is equal to the sum of the magnitude squared components. Thus $P[s_p(t)] = 1 + 2(0.5) + 2(0.125) = 2.25$ signal watts.

g) The power of $s_p(t)$ is equal to the sum of the square of the dc-term plus one-half the sum of the magnitudes squared of all other components. From b), we have $C_0 = 1$, $C_1 = \sqrt{2}$, and $C_2 = \sqrt{0.5}$. Thus $C_0^2 = 1$, $0.5C_1^2 = 1$, and $0.5C_2^2 = 0.25$. The one-sided power spectrum of $s_p(t)$ is sketched in Figure 2.7.

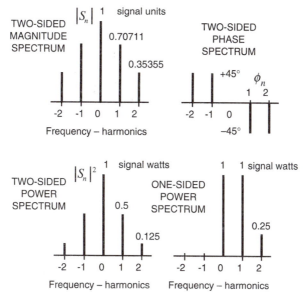

Figure 2.7 Diagrams for Example 2.4

h) The power of $s_p(t)$ is equal to the sum of the square of the dc-term plus one-half the sum of the magnitudes squared of all other components. Using g), we obtain $P[s_p(t)] = 1 + 1 + 0.25 = 2.25$ signal watts.

2.7 RECTANGLE TRAINS AND SQUARE WAVES

(1) Unit Sinc Function

Definitions of this function vary. Some authors define $\text{Sinc}(x)$ as

$$\text{Sinc}(x) = [\sin(\pi x)]/\pi x \qquad (2.19)$$

Others define $\text{sinc}(x)$ as

$$\text{sinc}(x) = [\sin(x)]/x \qquad (2.20)$$

Yet others use the term *sampling function* and write

$$\text{Sa}(x) = [\sin(x)]/x \qquad (2.21)$$

All of these functions have the value 1 when $x=0$. However, $\text{Sinc}(x)=0$ when $x = \pm 1, \pm 2, ...,$ etc., but $\text{sinc}(x)$ and $\text{Sa}(x)=0$ when $x = \pm \pi, \pm 2\pi, ...,$ etc. We use the first definition [i.e., equation (2.19)]. Two other useful properties of the Sinc function are

$$\text{Sinc}(x) = 1 - \tfrac{1}{3!}(\pi x)^2 + \tfrac{1}{5!}(\pi x)^4 - ...$$

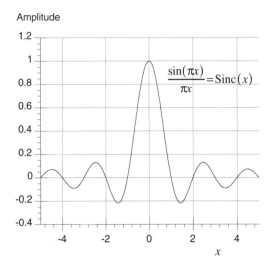

Amplitude

Figure 2.8 Unit Sinc Function
The left- and right-hand tails consist of oscillations with unit period and slowly decreasing amplitudes.

and

$$\int_{-\infty}^{\infty} \operatorname{Sinc}(x)dx = \int_{-\infty}^{\infty} \operatorname{Sinc}^2(x)dx = 1.$$

Figure 2.8 shows the central portion of a unit Sinc function. The left- and right-hand tails consist of oscillations with unit period and gradually decreasing amplitudes that continue to $\pm\infty$.

(2) Unit Rectangle Train

A unit rectangle train is a periodic function $s_p(t)$ that consists of rectangle functions of amplitude 1 signal unit and duration 1 time unit that are repeated every T_1 time units $(T_1 > 1)$. A time-domain representation of a unit rectangle function and a unit rectangle train are shown in **Figure 2.9**. They are compared with a symmetrical square wave [discussed in § 2.7(4)]. The unit rectangle train is written as

$$s_p(t) = \sum_{n} \Pi(t - nT_1) \tag{2.22}$$

where the notation $\Pi(t)$ denotes a unit rectangle function, i.e.,

$$\Pi(t) = \begin{cases} 1, & |t| < 1/2 \\ 0, & |t| > 1/2 \end{cases}$$

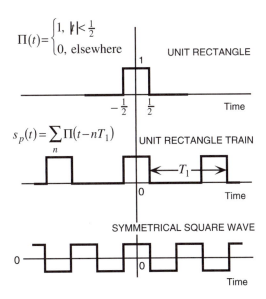

Figure 2.9 Unit Rectangle, Unit Rectangle Train, and Symmetrical Square Wave Functions

For a rectangle function of amplitude A and width b, the notation is $A\Pi(t/b)$. The notation $\sum \Pi(t - nT_1)$ denotes a sequence of unit rectangles spaced T_1 time units apart.

EXAMPLE 2.5   Complex Fourier Coefficients of Unit Rectangle Train

Given A unit rectangle train as described above.

Problem Find the complex Fourier coefficients of this function and plot two-sided amplitude, magnitude and phase, and power spectrums for the case when $T = 3$ time units.

Solution Use the analysis equation [equation (2.12)] to find S_n.

$$S_n = \left\langle \sum_n \Pi(t - nT_1) \exp(-jn2\pi f_1 t) \right\rangle$$

Time-averaging a periodic function means integrating over a period and dividing by the period. We may choose the position about which to integrate, but not the range over which we integrate. Every period will have one rectangle function; consider the rectangle that occurs around zero in Figure 2.9. Then

$$S_n = \frac{1}{T_1} \int_{-T_1/2}^{T_1/2} \Pi(t) \exp(-jn2\pi f_1 t)\, dt$$

Now, $\Pi(t)$ has a value of 1 when $|t| < 1/2$, and it is zero throughout the remainder of the period. This further simplifies the integral.

$$S_n = \frac{1}{T_1} \int_{-1/2}^{1/2} \exp(-jn2\pi f_1 t)\, dt = -\frac{\exp(-jn\pi f_1) - \exp(jn\pi f_1)}{jn2\pi f_1 T_1}$$

Using Euler's relationships [equation (2.8)], we obtain

$$S_n = \frac{1}{T_1} \frac{\sin \pi n f_1}{\pi n f_1} = f_1 \operatorname{Sinc}(n f_1)$$

Thus, the complex Fourier coefficients of the unit rectangle train exist at integer values of n. The value of each depends on the fundamental frequency $(f_1 = 1/T_1)$ and the value of the Sinc function. In this example, $T_1 = 3$ time units, so that

$$S_n = 0.333 \operatorname{Sinc}(0.333n)$$

Thus, when

$n =$	$S_n = 0.333\operatorname{Sinc}(n) =$	$|S_n|^2 =$
0	0.33333	0.11111
± 1	0.27567	0.07599
± 2	0.13783	0.01900
± 3	0.00000	0.00000
± 4	− 0.06892	0.00475
± 5	− 0.05513	0.00304
± 6	0.00000	0.00000
± 7	0.03938	0.00155
± 8	0.03446	0.00119
± 9	0.00000	0.00000
•	•	•

The values of S_n are plotted in **Figure 2.10**. They constitute the amplitude spectrum of a unit rectangle train in which the rectangle repetition period is three times the width of the rectangle. In **Figure 2.11**, you can see the plot of the magnitude and phase of the function. In forming the magnitude, negative lobes are changed to positive values. This change accounts for the phase change of π radians. Because the phase spectrum is odd, it is drawn as negative in the positive frequency half-plane and positive in the negative half-plane. In **Figure 2.12**, you can see the plot of both the two-sided power spectrum based on equation (2.15), and the one-sided power spectrum based on equation (2.18). $(|S_0| = |C_0|$ and $2|S_n| = |C_n|)$.

---

(3) Symmetrical Rectangle Train

Suppose the period of the unit rectangle train is adjusted to 2. Then we have the symmetrical train for which $S_n = 0.5\operatorname{Sinc}(0.5n)$. Even though the symmet-

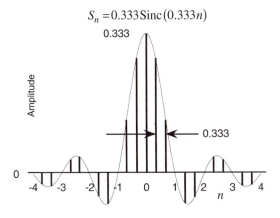

$$S_n = 0.333\,\mathrm{Sinc}(0.333n)$$

Figure 2.10 Complex Fourier Coefficients of Unit Rectangle Train in Which the Period Is Three Times the Width of a Single Rectangle Function

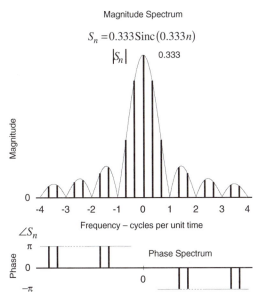

Figure 2.11 Magnitude and Phase Spectrums of Unit Rectangle Train in Which the Period Is Three Times the Width of the Unit Rectangle Function

They are the magnitude and phase components of the complex Fourier coefficients in Figure 2.10.

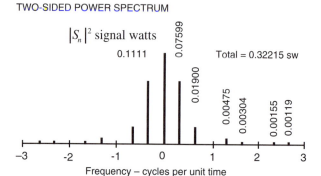

Figure 2.12 One-Sided and Two-Sided Power Spectrums of Unit Rectangle Train in Which the Period Is Three Times the Width of the Unit Rectangle Function

rical rectangle train has a greater rectangle density than the unit rectangle train described in Example 2.5, the magnitude spectrum is less dense. The frequency spacing between the lines and the magnitudes of the lines increase as the period decreases. To illustrate this effect, **Figure 2.13** compares the amplitude spectrums of rectangle trains in which the periods are two and three times the width of the rectangle function.

(4) Square Wave

From the synthesis equation (equation 2.9), you have seen that a periodic function has an amplitude spectrum that consists of a constant term (the dc-term, or the mean value) and other terms that are the amplitudes of harmonically related sinusoidal functions. If we subtract the dc-term from the amplitude spectrum of the symmetrical rectangle train, we are left with sinusoidal components whose time-average is zero, i.e., the resulting wave has as much area above the time-axis as it has below the axis. What we have done is create a

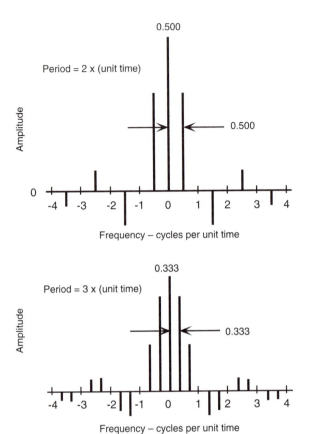

Figure 2.13 Comparison of Amplitude Spectrums of Unit Rectangle Trains with Periods of Two and Three Times the Width of the Unit Rectangle Function
As the period increases, the spectral lines move closer together and their amplitude diminishes.

symmetrical square wave of amplitude ± 0.5 signal units, rectangle width 1 time unit, and period 2 time units (it is shown in Figure 2.4). The components are

$$S_n = 0.5 \text{Sinc}(0.5n), \text{ where } 1 \leq |n| < \infty.$$

Figure 2.14 compares the amplitude spectrums of the symmetrical rectangle train (from Figure 2.13) and the symmetrical square wave created from it. The power in the symmetrical rectangle train is 0.5 signal watts, and the power in the symmetrical square wave is 0.25 signal watts. The dc-component contributes one-half the power required by the rectangle train. Small wonder then that a square wave (or other zero-mean signal) is used whenever possible.

SYMMETRICAL RECTANGLE TRAIN
Amplitude = 1 signal unit; pulse width = 1 time unit; period = 2 time units

$$S_n = 0.5 + 0.5 \sum_{n=\pm 1}^{n=\pm\infty} \mathrm{Sinc}(0.5n)$$

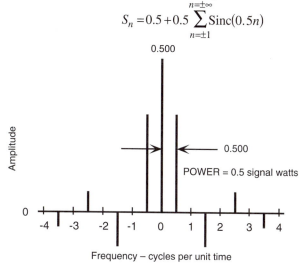

POWER = 0.5 signal watts

SYMMETRICAL SQUARE WAVE
Amplitude = ± 0.5 signal unit; pulse width = 1 time unit; period = 2 time units

$$S_n = 0.5 \sum_{n=\pm 1}^{n=\pm\infty} \mathrm{Sinc}(0.5n)$$

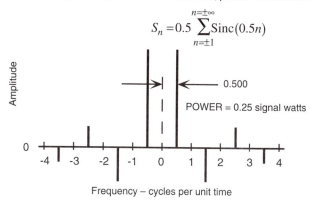

POWER = 0.25 signal watts

Figure 2.14 Comparison of Amplitude Spectrums of Symmetrical Rectangle Train and Symmetrical Square Wave Function
The symmetrical square wave is zero-mean. It does not have a dc-term. As a consequence, it consumes one-half the power of the symmetrical rectangle train.

EXAMPLE 2.6   Rectangle Train

Given An eternal rectangle train consists of rectangles of width 1 second and amplitude 2*A* signal units that are repeated every 5 seconds.

Problem

a) Write an equation [in the manner of equation (2.22)] that describes the train.

b) Calculate the power in the train.

c) Calculate the complex Fourier coefficients for $n = 0, 1, \ldots, 10$.

d) Use these coefficients to calculate an approximate value of the power in the train.

e) What percentage of the total power is carried between the first zero crossings?

f) What is the power in a symmetrical rectangle wave that consists of rectangles of width 1 second and amplitude $\pm A$ signal units that are repeated every 5 seconds? Explain your answer.

Solution

a) $s_p(t) = 2A \sum_{n=\pm\infty} \Pi(t - 5n)$

b) The energy in a rectangle is $4A^2$ signal joules. This is the energy expended every period. Thus, the power in the train is $4A^2/5 = 0.8A^2$ signal watts.

c) From Example 2.5, $S_n = \frac{2A}{T_1} \text{Sinc}(nf_1) = 0.4A\,\text{Sinc}(n/5)$. Hence, the first ten complex Fourier coefficients are

$n = 0$	$S_n = 0.40000\,A$
1	$0.37420\,A$
2	$0.30273\,A$
3	$0.20182\,A$
4	$0.09355\,A$
5	0.00000
6	$-0.06237\,A$
7	$-0.08649\,A$
8	$-0.07568\,A$
9	$-0.04158\,A$
10	0.00000

d) The power in the train is given by $\overline{P[\infty]} = |S_0|^2 + \sum_{\substack{n=-\infty \\ n\neq 0}}^{n=\infty} |S_n|^2 = |S_0|^2 + 2\sum_{n=1}^{n=\infty} |S_n|^2$.

Hence, on the basis of the values of S_n for $-10 \le n \le 10$

$$\overline{P'[\infty]} = 0.16000A^2 + 2\times0.14002A^2 + 2\times0.091646A^2$$

$$+ 2\times0.0473A^2 + 2\times0.00875A^2 + 2\times0.00389A^2$$

$$+ 2\times0.00742A^2 + 2\times0.00573A^2 + 2\times0.00173A^2$$

$$= 0.75995A^2 \text{ signal watts.}$$

e) The power between the first zero crossings is carried in the complex Fourier coefficients between $-5 \leq n \leq 5$. Hence the power is 0.72229 signal watts. This value is 90.3% of the total power in the train.

f) The symmetrical rectangle wave is equal to the rectangle train less the dc-term. Hence, the power is $0.8A^2 - 0.16A^2 = 0.64A^2$ signal watts.

REVIEW QUESTIONS

2.1 Write an expression that defines a periodic function.

2.2 Distinguish among a *trigonometric Fourier series, a Fourier cosine series,* and *an exponential Fourier series.*

2.3 State the Analysis equation, and explain why it is so called.

2.4 State the Synthesis equation, and explain why it is so called.

2.5 What fundamental change does the use of the complex Fourier coefficient S_n introduce?

2.6 Explain the difference between a *one-sided power spectrum* and a *two-sided power spectrum.* How are they related?

2.7 Define $\text{Sinc}(x)$, $\text{sinc}(x)$, and $\text{Sa}(x)$. When do each of these functions have zero values?

2.8 What notation do we use for a rectangle function of amplitude A signal units and width b seconds?

2.9 Describe the differences in amplitude spectrum you would expect to find when examining the spectrums of a rectangle train of amplitude A signal units, rectangle width b seconds ($b < T$), and period T seconds, and a rectangle train of amplitude $2A$, rectangle width b, and period $T/2$.

2.10 Describe the difference between the spectrums of a zero mean periodic function, and a periodic function with a mean value of A signal units.

3

Transient Functions

- **Transient function**: one that begins and ends. It is described by the relationship

$$s_t(t) = s(t)\, \Pi\!\left(\frac{t - \frac{1}{2}(t_2 + t_1)}{t_2 - t_1}\right) \tag{3.1}$$

or

$$s_t(t) = \begin{cases} s(t), & t_1 \le t \le t_2 \\ 0, & \text{elsewhere} \end{cases}$$

where t_1 is the time at which $s_t(t)$ begins, and t_2 is the time at which $s_t(t)$ ends.

Although $s_t(t)$ can be deterministic or probabilistic, Fourier transform techniques require integration into the future; accordingly, they can be applied only to deterministic functions.

3.1 FOURIER TRANSFORM

What happens when the *period* of the unit rectangle train shown in Figure 2.9 is increased without limit? As the period T_1 becomes very large, the rectangle train becomes a single rectangle function of unit width; the multiplier f_1 of the complex Fourier coefficients decreases to zero; and the separation between the coefficients (spectral lines) in the frequency-plane decreases to zero. These effects are illustrated in **Figure 3.1**. It shows magnitude spectrums for three unit rectangle trains: the first is the unit rectangle train of period $T_1 = 2$ shown in Figure 2.14, the second is a unit rectangle train of period $T_1 = 5$, and the third has a period of $T_1 = 25$. As $T_1 \to \infty$, the magnitudes become zero. In the limit, we destroy a periodic function and create a tran-

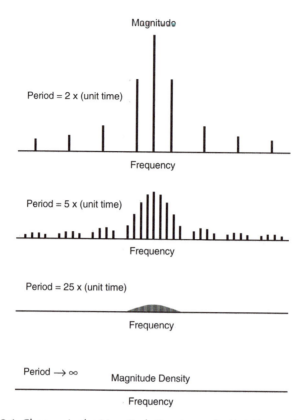

Figure 3.1 Changes in the Magnitude Spectrum of a Unit Rectangle Pulse Train as the Period Becomes Longer

At the top, the train has a period of 2x unit time and a frequency spectrum that includes a regular sequence of Fourier coefficients. As the period increases, the number of coefficients in a given frequency interval increases, and their magnitude decreases. In the limit, the spectral lines coalesce and the magnitudes become zero; they produce a magnitude density distribution—a continuous function that is the real part of the frequency spectrum of a single unit rectangle pulse.

sient function. We move from a periodic, power-type function to a deterministic, transient, energy-type function.

Despite the disappearance of the complex Fourier coefficients, it should not be assumed that a single rectangle function is without a spectrum. Along the frequency axis remains a continuous *density* function that was created by the collapse of the magnitudes and the coalescing of the coefficients.

From this exercise, you can conclude that a transient function can be modeled as a periodic function of infinite period, i.e.,

$$s_t(t) = \lim_{T_1 \to \infty} s_p(t) \tag{3.2}$$

This representation allows us to use some of the results of § 2.5. Thus, in the synthesis equation [equation (2.9)], if we substitute Δf for f_1 (and for $1/T_1$), f_n for nf_1, and $S(f_n)$ for $T_1 S_n$, we obtain

$$s_p(t) = \sum_n S(f_n) \exp(j2\pi f_n t) \Delta f \tag{3.3}$$

Exchanging Δf for f_1 recognizes that the separation between spectral lines (equal to f_1) decreases as the period increases. Under the circumstances described in equation (3.3), the analysis equation [equation (2.12)] becomes

$$S_n = \frac{S(f_n)}{T_1} = \frac{1}{T_1} \int_{-T_1/2}^{T_1/2} s_p(t) \exp(-j2\pi f_n t)\, dt$$

i.e.,

$$S(f_n) = \int_{-T_1/2}^{T_1/2} s_p(t) \exp(-j2\pi f_n t)\, dt \tag{3.4}$$

Now, let $T_1 \to \infty$. Then $s_p(t) \to s_t(t)$, $\Delta f \to 0$, (i.e., Δf becomes df), and $S(f_n)$ and $2\pi f_n$ become continuous frequency functions (i.e., $S(f_n) \to S(f)$ and $2\pi f_n \to 2\pi f$), so that equation (3.3) becomes

$$s_t(t) = \int_{-\infty}^{\infty} S(f) \exp(j2\pi ft)\, df \quad \text{Inverse Fourier Transform} \tag{3.5}$$

and equation (3.4) becomes

$$S(f) = \int_{-\infty}^{\infty} s_t(t) \exp(-j2\pi ft)\, dt \quad \text{Fourier Transform} \tag{3.6}$$

Equations (3.5) and (3.6) describe two-sided functions.

Moving from a periodic function to a transient function, we have exchanged a set of discrete coefficients for a continuous function. We express the relationship between $s_t(t)$ and $S(f)$ in the following ways:

- $S(f)$ is the *Fourier transform* of $s_t(t)$

$$S(f) = \mathbf{F}[s_t(t)]$$

where $\mathbf{F}[\ ]$ denotes the Fourier transform operator [equation (3.6)].

- $s_t(t)$ is the *inverse Fourier transform* of $S(f)$

$$s_t(t) = \mathbf{F}^{-1}[S(f)]$$

where $\mathbf{F}^{-1}[\ ]$ denotes the inverse Fourier transform operator [equation (3.5)].

- $s_t(t)$ and $S(f)$ constitute a *Fourier transform pair*

$$s_t(t) \Leftrightarrow S(f)$$

where $\Leftrightarrow$ is the symbol for the Fourier transform pair.

In general, $S(f)$ is a complex function of frequency; accordingly

$$S(f) = |S(f)| \exp(j\theta)$$

Plotting $|S(f)|$ *versus* frequency gives the continuous magnitude spectrum of $s_t(t)$; plotting θ *versus* frequency gives the continuous phase spectrum of $s_t(t)$. For *real* $s_t(t)$, the spectral density is *even* and the phase spectrum is *odd*. In the Appendix, Table A.1 lists 12 properties of the Fourier transform; some of them are used in the examples that follow.

Are limitations associated with equations (3.5) and (3.6)?[1] For sure, the integrals must converge; for convergence to occur, the function must satisfy Dirichlet's conditions

- $s(t)$ must be single-valued
- in any finite time-interval, $s(t)$ has a finite number of maximums and minimums and a finite number of discontinuities
- the infinite time-integral of $s(t)$ is finite, i.e.,

$$\int_{-\infty}^{\infty} s(t)\,dt < \infty$$

In addition, Plancherel showed that, if the value of the infinite time-integral of $|s(t)|^2$ is defined and finite, the Fourier transform of $s(t)$ exists. This condition is met by all energy-type signals. Through the use of impulses (Dirac's improper functions), Fourier transform techniques have been extended to periodic signals (see § 3.2).

EXAMPLE 3.1   Unit Rectangle and Unit Sinc Functions

Given

$$s(t) = \Pi(t).$$

---

[1]For a comprehensive discussion, see Chapter 4 of R. V. Churchill, *Fourier Series and Boundary Value Problems* (New York: McGraw-Hill, 1963).

Problem Find $\mathbf{F}\left[\Pi(t)\right]$.

Solution

$$\mathbf{F}\left[\Pi(t)\right] = \int_{-\infty}^{\infty}\Pi(t)\exp(-j2\pi f\,t)dt = \int_{-1/2}^{1/2}\exp(-j2\pi f\,t)dt$$

from which we obtain

$$\mathbf{F}\left[\Pi(t)\right] = \frac{1}{j2\pi f}\exp(j2\pi ft)\Big|_{-1/2}^{1/2} = \frac{\sin(\pi f)}{\pi f} = \mathrm{Sinc}(f)$$

Expressing the result as a transform pair

$$\Pi(t) \Leftrightarrow \mathrm{Sinc}(f) \tag{3.7}$$

Extension Using the duality principle (#2 in Table A.1) and observing that $\Pi(t)$ is an even function, i.e., $\Pi(-t) = \Pi(t)$, we obtain the dual relationship

$$\mathrm{Sinc}(t) \Leftrightarrow \Pi(f) \tag{3.8}$$

Transform pairs (3.7) and (3.8) are drawn in **Figure 3.2**.

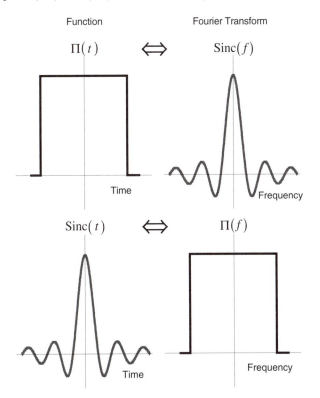

Figure 3.2 Fourier Transforms of Unit Rectangle Pulse and Unit Sinc Functions

---

EXAMPLE 3.2   Use of Time-Shifting and Time-Scaling

Given

$$\Pi(t) \Leftrightarrow \text{Sinc}(f).$$

Problem Find $\mathbf{F}\big[\Pi(2t-3)\big]$.

Solution Step 1) In $\Pi(t) \Leftrightarrow \text{Sinc}(f)$, replace t by $(t-3)$ and use the time-shifting property of the Fourier transform (#5 in Table A.1), so that

$$\Pi(t) \text{ becomes } \Pi(t-3)$$

and

$$\text{Sinc}(f) \text{ becomes } \exp(-j2\pi f 3)\text{Sinc}(f).$$

Using single primes to identify the new transform pair, we write

$$s'(t) = \Pi(t-3) \Leftrightarrow \exp(-j2\pi f 3)\text{Sinc}(f) = S'(f).$$

Step 2) In $\Pi(t-3)$, replace t by $2t$, and use the time-scaling property of the Fourier transform (#4 in Table A.1), so that

$$\Pi(t-3) \text{ becomes } \Pi(2t-3)$$

and

$$\exp(-j2\pi f 3)\text{Sinc}(f) \text{ becomes } \tfrac{1}{2}\exp(-j\pi f 3)\text{Sinc}(f/2).$$

Using double primes to identify the new transform pair, we write

$$s''(t) = \Pi(2t-3) \Leftrightarrow \tfrac{1}{2}\exp(-j\pi f 3)\text{Sinc}(f/2) = S''(f)$$

i.e.,

$$\mathbf{F}\big[\Pi(2t-3)\big] = \tfrac{1}{2}\exp(-j3\pi f)\text{Sinc}(f/2).$$

Step 3) Because $S''(f)$ is complex, it is expressed as $|S''(f)|$ and $\angle S''(f)$. Since the magnitude of $\exp(-j3\pi f)$ is 1, $|S''(f)| = |\text{Sinc}(f/2)|$. The phase angle $\angle S''(f)$ is equal to the sum of the phase shifts (π radians) due to the reversal of the even lobes of the Sinc function and $\angle -3\pi f$. Plots of $|S''(f)|$ and $\angle S''(f)$ are shown in **Figure 3.3**.

Comment $\Pi(2t-3)$ can be expressed as

$$\Pi(2t-3) = \Pi\big(2[t-3/2]\big) = \Pi\Big(\tfrac{t-3/2}{1/2}\Big).$$

It is a rectangle of amplitude 1 signal unit and width 1/2 time unit that is centered on $t=3/2$ time units.

$$\mathbf{F}\left[\Pi(2t-3)\right]=\tfrac{1}{2}\exp(-3\pi f)\mathrm{Sinc}\left(f/2\right)$$

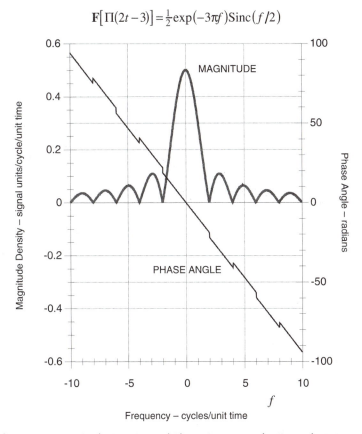

Figure 3.3 Magnitude Density and Phase Spectrums for Example 3.2

EXAMPLE 3.3   Unit Triangle Function

Given The unit triangle function is defined as

$$\Lambda(x)=\begin{cases}1-|x|, \text{ when } |x|<1\\ 0, \text{ when } |x|>1\end{cases}$$

Problem Find $\mathbf{F}\left[\Lambda(t)\right]$.

Solution The unit triangle function is created by the convolution of two unit rectangle functions, i.e., $\Lambda(t)=\Pi(t)\otimes\Pi(t)$. Using the convolution property of the Fourier transform (# 12 in Table A.1)

$$\Lambda(t)=\Pi(t)\otimes\Pi(t)\Leftrightarrow\mathrm{Sinc}(f)\mathrm{Sinc}(f)=\mathrm{Sinc}^{2}(f) \tag{3.9}$$

Comment Applying the duality property of the Fourier transform (# 2 in Table A.1)

$$\text{Sinc}^2(t) \Longleftrightarrow \Lambda(-f) = \Lambda(f) \tag{3.10}$$

Figure 3.4 shows the Fourier pairs stated above.

EXAMPLE 3.4   Gaussian (or Normal) Function

Given

$$s(t) = \exp(-\pi t^2).$$

Problem Find $\mathbf{F}\left[\exp(-\pi t^2)\right].$

Solution

$$\mathbf{F}\left[\exp(-\pi t^2)\right] = \int_{-\infty}^{\infty} \exp(-\pi t^2)\exp(-j2\pi ft)\,dt = \int_{-\infty}^{\infty} \exp\left[-\pi\left(t^2 + j2ft\right)\right]dt$$

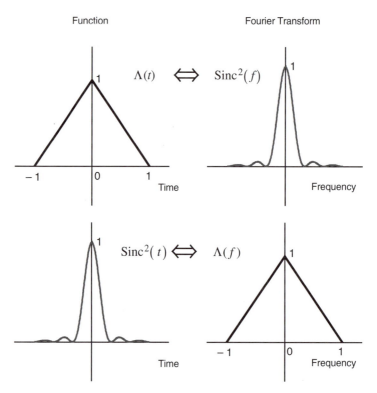

Function   Fourier Transform

$\Lambda(t) \Longleftrightarrow \text{Sinc}^2(f)$

$\text{Sinc}^2(t) \Longleftrightarrow \Lambda(f)$

Figure 3.4 Fourier Transforms of Unit Triangle and Sinc² Functions

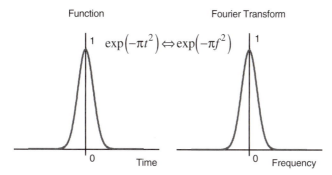

Figure 3.5 showing Function (Time domain) and Fourier Transform (Frequency domain) with $\exp(-\pi t^2) \Leftrightarrow \exp(-\pi f^2)$

Figure 3.5 Fourier Transform of Gaussian or Normal Pulse
The Gaussian function is preserved under Fourier transformation.

Completing the square under the integral and moving $\exp(-\pi f^2)$ to the left gives

$$\mathbf{F}\left[\exp(-\pi t^2)\right] = \exp(-\pi f^2) \int_{-\infty}^{\infty} \exp\left[-\pi(t + jf)^2\right] dt$$

Putting $\lambda = t + jf$ so that $d\lambda = dt$

$$\mathbf{F}\left[\exp(-\pi t^2)\right] = \exp(-\pi f^2) \int_{-\infty}^{\infty} \exp(-\pi\lambda^2) d\lambda$$

From Table A.3 (Appendix), the integral is 1; hence

$$\mathbf{F}\left[\exp(-\pi t^2)\right] = \exp(-\pi f^2)$$

or

$$\exp(-\pi t^2) \Leftrightarrow \exp(-\pi f^2) \tag{3.11}$$

Comment The form of a *Gaussian* or *normal* function is preserved under the Fourier transformation. **Figure 3.5** shows the time and frequency plane diagrams for equation (3.11).

3.2 DIRAC'S IMPROPER FUNCTION

To apply Fourier transform analysis to periodic functions requires the introduction of *delta* or impulse functions.

- **Impulse**: a very brief, very strong signal
- **Unit impulse function**: a function that has zero amplitude everywhere except at $t = 0$, and an area of 1; i.e.,

$$\delta(t)=0, \ t \neq 0 \ ; \quad \int_{-\infty}^{\infty} \delta(t)dt = 1 \tag{3.12}$$

Also known as Dirac's improper function, in the 1920s $\delta(t)$ was introduced to quantum mechanics by Paul Dirac, an English physicist. Strictly speaking, it is not a function but a symbol with special properties.[2]

(1) Fourier Transform Pair

To find the transform of a rectangle function of width b and amplitude A, we can use the scaling property of the Fourier transform (#4 in Table A.1) in equation (3.7).

$$\Pi(t) \text{ becomes } A\Pi(t/b)$$

where

and

$$A\Pi(t/b)=\begin{cases} A, \ |t/b|<1/2 \\ 0, \ |t/b|>1/2 \end{cases}$$

$$\text{Sinc}(f) \text{ becomes } Ab\text{Sinc}(bf).$$

Thus, we can write

$$A\Pi(t/b) \Leftrightarrow Ab\text{Sinc}(bf). \tag{3.13}$$

If we hold $Ab=1$ and let $b \to 0$, the rectangle function retains the same area (1 signal unit second) but becomes progressively narrower and taller, and

$$\underset{b \to 0}{\text{limit}} \ A\Pi(t/b)=\delta(t). \tag{3.14}$$

In the right-hand side of equation (3.13)

$$\underset{b \to 0}{\text{limit}} \ Ab\text{Sinc} \ (bf)=\text{Sinc}(0)=1 \tag{3.15}$$

Substituting values from equations (3.14) and (3.15) in equation (3.13)

$$\delta(t) \Leftrightarrow 1 \tag{3.16}$$

Applying the duality principle (#2 in Table A.1) to equation (3.16) and observing that $\delta(t)$ is an even function [i.e., $\delta(f)=\delta(-f)$] gives the Fourier transform pair

$$1 \Leftrightarrow \delta(f) \tag{3.17}$$

Figure 3.6 shows the transform pairs of equations (3.16) and (3.17). A function of constant unit value in the time domain is represented by an impulse

---

[2]For a full discussion of these points, see Ronald N. Bracewell, *The Fourier Transform and Its Applications*, 2nd Edition, Revised (New York: McGraw-Hill, 1986), pp. 69-97.

Function   Fourier Transform

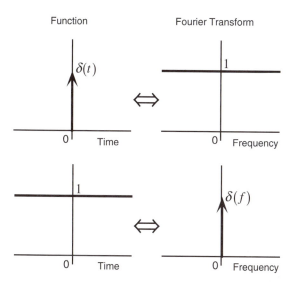

Figure 3.6 Fourier Transforms of Dirac Delta and Unit Constant Functions

function of unit area and zero frequency in the frequency domain. Also, the occurrence of an impulse function of unit area in the time domain is represented by a function of constant unit value in the frequency domain. The former describes a dc level; the latter describes the effect of releasing a finite amount of energy in a very short time—it produces energy across the entire frequency spectrum. (A practical example is provided by a lightning strike; it is heard throughout the broadcasting spectrum.)

Applying the frequency shifting principle (#6 in Table A.1) to equation (3.17) gives the Fourier transform pair

$$\exp(j2\pi f_0 t) \Leftrightarrow \delta(f - f_0) \tag{3.18}$$

An exponential time-function whose imaginary exponent contains a fixed frequency is represented by an impulse function located at that frequency on the positive frequency line.

(2) An Anomaly?

Invoking the definition of the Fourier transform [equation (3.6)], we can restate equation (3.17) as

$$\delta(f) = \int_{-\infty}^{\infty} \exp(-j2\pi ft)\, dt \tag{3.19}$$

Further, because the impulse function is an even function, we can substitute $-f$ for f and write

$$\delta(f) = \int\limits_{-\infty}^{\infty} \exp(-j2\pi ft)\, dt = \int\limits_{-\infty}^{\infty} \exp(j2\pi ft)\, dt \qquad (3.20)$$

On first exposure, equation (3.20) may appear to be a contradiction. However, study of the Argand diagram in Figure 2.3 should convince you that, irrespective of the spin direction (i.e., f or $-f$), on the real plane, the outcomes are the same.

(3) *Important Properties*

Dirac's improper function has two important properties.

- **Sifting**: integrating the product of $\delta(t-t_0)$, an impulse function that occurs at t_0, and a function $s(t)$, produces the value of $s(t)$ at time t_0, i.e.,

$$\int\limits_{-\infty}^{\infty} s(t)\, \delta(t-t_0)\, dt = s(t_0) \qquad (3.21)$$

This operation is said to *sift* out a single value of $s(t)$, namely $s(t_0)$.

When the impulse occurs at the origin, i.e., $t_0 = 0$, the operation produces $s(0)$, the value of $s(t)$ at the origin. Among other things, sifting is important in understanding the operation of *sampling* in the conversion of analog functions to digital functions.

- **Replication**: the convolution of $\delta(t-t_0)$, an impulse function located at t_0, with a function $s(t)$, delays $s(t)$ by t_0, i.e.,

$$s(t) \otimes \delta(t-t_0) = s(t-t_0) \qquad (3.22)$$

Among other things, replication is important in the construction of the Fourier transform of a periodic function and in the understanding of *aliasing* in the conversion of analog functions to digital functions. **Figure 3.7** shows the effects of sifting and replication on an arbitrary function.

EXAMPLE 3.5 Sinusoidal Functions

Given

$$\cos(2\pi f_1 t) = \tfrac{1}{2}\left[\exp(j2\pi f_1 t) + \exp(-j2\pi f_1 t)\right].$$

Problem Find $\mathbf{F}\left[\cos(2\pi f_1 t)\right]$.

Solution Employing equation (3.18) directly

$$\cos(2\pi f_1 t) = \tfrac{1}{2}\left[\exp(j2\pi f_1 t) + \exp(-j2\pi f_1 t)\right] \Leftrightarrow \tfrac{1}{2}\left[\delta(f - f_1) + \delta(f + f_1)\right] \quad (3.23)$$

Comments The magnitude density spectrum of a unit cosine function consists of two impulse functions of area $1/2$ at $\pm f_1$. Using similar reasoning, the Fourier transform of a sine function is

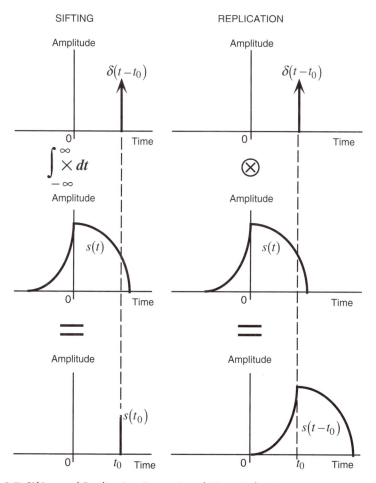

Figure 3.7 Sifting and Replication Properties of Dirac Delta Functions
By integrating the product of a delta function and a given function, the sifting property samples the given function at the point of occurrence of the delta function. The convolution of a delta function and a given function replicates the given function at the point of occurrence of the delta function.

$$\sin(2\pi f_1 t) \Leftrightarrow \tfrac{1}{2j}\left[\delta(f-f_1)-\delta(f+f_1)\right]=\tfrac{j}{2}\left[\delta(f+f_1)-\delta(f-f_1)\right]$$

These relationships are shown in **Figure 3.8**.

EXAMPLE 3.6 Truncated Periodic Function

Given A truncated periodic function is created by multiplying the periodic function $\cos(2\pi f_1 t)$ by a unit rectangle function of width mT_1, i.e.,
$s_t(t)=\Pi(t/mT_1)\cos(2\pi f_1 t)$.

Problem Find $\mathbf{F}\left[\Pi(t/mT_1)\cos(2\pi f_1 t)\right]$.

Solution Scaling equation (3.7)

$$\Pi(t/mT_1)\Leftrightarrow mT_1\,\mathrm{Sinc}(mT_1 f)$$

and, from Example 3.5

$$\cos(2\pi f_1 t)\Leftrightarrow \tfrac{1}{2}\left[\delta(f+f_1)+\delta(f-f_1)\right].$$

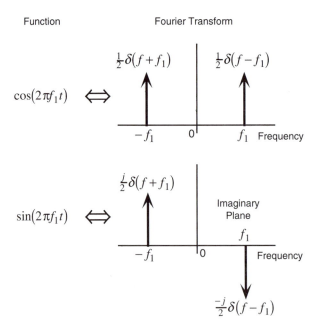

Figure 3.8 Fourier Transforms of Cosine and Sine Functions
The transform of the sine function is in the imaginary plane.

Using the multiplication in the time domain property of the Fourier transform (#11 in Table A.1) gives

$$\Pi(t/mT_1)\cos(2\pi f_1 t) \Leftrightarrow mT_1 \operatorname{Sinc}(mT_1 f) \otimes \tfrac{1}{2}[\delta(f+f_1)+\delta(f-f_1)]$$

Using the replication property of Dirac's improper function [equation (3.22)] gives

$$\Pi(t/mT_1)\cos(2\pi f_1 t) \Leftrightarrow \tfrac{mT_1}{2}\big[\operatorname{Sinc}\{mT_1(f+f_1)\}+\operatorname{Sinc}\{mT_1(f-f_1)\}\big] \quad (3.24)$$

Comment 1 The impulses of the cosine function have been replaced by Sinc functions. Changing an eternal, periodic function into a transient, quasi-periodic function causes the lines of the magnitude density spectrum to spread out. This is shown in **Figure 3.9**. See also, Example 4.3, Comment 2, Gibbs' Phenomenon.

Comment 2 When $m \to \infty$, equation (3.24) becomes

$$s_t(t)=\cos(2\pi f_1 t) \Leftrightarrow \tfrac{1}{2}[\delta(f+f_1)+\delta(f-f_1)]$$

which is equation (3.23).

EXAMPLE 3.7 Raised Cosine Pulse

Given A raised cosine pulse is created by adding 1 to a unit cosine function of period 1 time unit and multiplying by a unit rectangle function, thus:
$s_t(t)=\Pi(t)\{1+\cos(2\pi t)\}.$

Problem Find $\mathbf{F}[\Pi(t)\{1+\cos(2\pi t)\}]$.

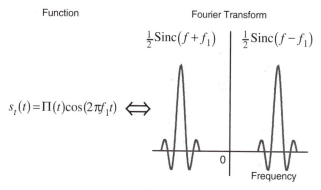

Function   Fourier Transform

$\tfrac{1}{2}\operatorname{Sinc}(f+f_1)$ $\tfrac{1}{2}\operatorname{Sinc}(f-f_1)$

$s_t(t)=\Pi(t)\cos(2\pi f_1 t) \Longleftrightarrow$

Frequency

Figure 3.9 Fourier Transform of Truncated Periodic Function
Turning a cosine function on and then off changes the Fourier transform from delta functions to Sinc functions.

Solution The Fourier transform is

$$s_t(t)=\Pi(t)\{1+\cos(2\pi t)\}\Leftrightarrow \text{Sinc}(f)\otimes\left\{\delta(f)+\tfrac{1}{2}\delta(f+1)+\tfrac{1}{2}\delta(f-1)\right\}=S(f)$$

i.e.,

$$S(f)=\text{Sinc}(f)+\tfrac{1}{2}\text{Sinc}(f+1)+\tfrac{1}{2}\text{Sinc}(f-1)$$

so that

$$S(f)=\frac{\sin(\pi f)}{\pi f}+\frac{\sin(\pi f+\pi)}{2\pi(f+1)}+\frac{\sin(\pi f-\pi)}{2\pi(f-1)}=\frac{\sin(\pi f)}{\pi f}\left\{1-\frac{f}{2(f+1)}-\frac{f}{2(f-1)}\right\}$$

Hence

$$S(f)=\text{Sinc}\ (f)\big/\left(1-f^2\right)$$

i.e.,

$$\Pi(t)\{1+\cos(2\pi t)\}\Leftrightarrow \text{Sinc}\ (f)\big/\left(1-f^2\right)\qquad(3.25)$$

Figure 3.10 shows the raised cosine function and its Fourier transform.

Comment For a cosine function of period T_1 and amplitude A,

$$A\Pi(t/T_1)\{1+\cos(2\pi f_1 t)\}\Leftrightarrow AT_1\text{Sinc}(fT_1)\big/\left(1-f^2T_1^2\right)$$

EXAMPLE 3.8 Unit Step Function

Given The unit step function is defined as

$$U(t)=\begin{cases}0,\ t<0\\ 1,\ t>0\end{cases}$$

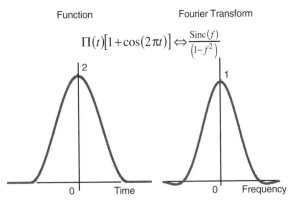

Function   Fourier Transform

$$\Pi(t)\left[1+\cos(2\pi t)\right]\Leftrightarrow \frac{\text{Sinc}(f)}{\left(1-f^2\right)}$$

Figure 3.10 Fourier Transform of Raised Cosine Pulse

Problem Find $\mathbf{F}[U(t)]$.

Solution From the definition of the unit impulse function [equation (3.12)], it follows

$$U(t) = \int_{-\infty}^{t} \delta(\tau)\,d\tau$$

Using the integration property of the Fourier transform (# 10 in Table A.1)

$$U(t) = \int_{-\infty}^{t} \delta(\tau)\,d\tau \Leftrightarrow 1/j2\pi f + \tfrac{1}{2}\delta(f) \tag{3.26}$$

Figure 3.11 shows the unit step function and its Fourier transform.

EXAMPLE 3.9 Exponential Pulse

Given A decaying exponential pulse $\exp(-t)U(t)$, i.e.,

$$s(t) = \begin{cases} \exp(-t) \text{ , when } t>0 \\ 1/2 \text{ , when } t=0 \\ 0 \text{ , when } t<0 \end{cases}$$

Problem Find $\mathbf{F}[\exp(-t)U(t)]$.

Solution The presence of the step function [$U(t)$] limits $\exp(-t)U(t)$ to the positive half-plane so that

$$\mathbf{F}[\exp(-t)U(t)] = \int_{0}^{\infty} \exp(t)\exp(-j2\pi ft)\,dt$$

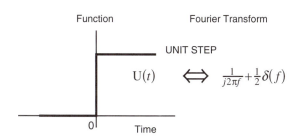

Function   Fourier Transform

UNIT STEP

$U(t)$ $\Longleftrightarrow$ $\frac{1}{j2\pi f} + \tfrac{1}{2}\delta(f)$

0   Time

Figure 3.11 Fourier Transform of Unit Step

Collecting terms

$$\int_0^\infty \exp(t)\exp(-j2\pi ft)dt = \int_0^\infty \exp\left[-t(1+j2\pi ft)\right]dt = 1/(1+j2\pi f)$$

Thus

$$\exp(-t)U(t) \Leftrightarrow 1/(1+j2\pi f) \qquad (3.27)$$

Extension For a rising exponential pulse $\exp(t)U(-t)$, i.e.,

$$s(t) = \begin{cases} 0 \text{ , when } t > 0 \\ 1/2 \text{ , when } t = 0 \\ \exp(t) \text{ , when } t < 0 \end{cases}$$

the same method shows

$$\exp(t)U(-t) \Leftrightarrow 1/(1-j2\pi f) \qquad (3.28)$$

Figure 3.12 shows the exponential pulses and their Fourier transforms.

EXAMPLE 3.10 Signum Function

Given The signum (sign) function is defined as

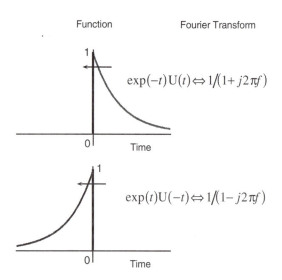

Figure 3.12 Fourier Transforms of Exponential Pulses

$$\text{sgn}(t) = \begin{cases} 1, & t > 0 \\ -1, & t < 0 \end{cases}$$

Problem Find $\mathbf{F}[\text{sgn}(t)]$.

Solution From the definition

$$\text{sgn}(t) = 2\,\text{U}(t) - 1$$

Using the linearity property of the Fourier transform (# 1 in Table A.1) and the transform of the unit step from Example 3.8

$$\text{sgn}(t) \Leftrightarrow -j/\pi f \tag{3.29}$$

Comment Applying the duality property of the Fourier transform (# 2 in Table A.1)

$$1/\pi t \Leftrightarrow -j\,\text{sgn}(f).$$

Figure 3.13 shows the Fourier pairs stated above.

3.3 FOURIER TRANSFORM OF A PERIODIC FUNCTION

A periodic function is produced from the convolution of an ideal sampling function and a generating function. Using this strategy, the Fourier transforms of many periodic functions can be derived.

(1) Ideal Sampling Function

Denoted by $\delta_{T_1}(t)$, an *ideal sampling function* is a periodic function that consists of a train of unit impulses spaced T_1 apart, i.e.,

$$\delta_{T_1}(t) = \sum_k \delta(t - kT_1) \tag{3.30}$$

Sometimes $\delta_{T_1}(t)$ is known as a Dirac *train*. Because it is a periodic function, we may express it in terms of the synthesis equation [equation (2.9)] as

$$\delta_{T_1}(t) = \sum_k S_k \exp(jk2\pi f_1 t)$$

From the analysis equation [equation (2.12)], the value of S_k is

$$S_k = \frac{1}{T_1} \int_{-T_1/2}^{T_1/2} \delta_{T_1}(t) \exp(-jk2\pi f_1 t)\,dt$$

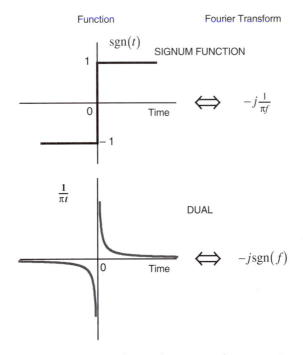

Figure 3.13 Fourier Transforms of Signum and $1/\pi t$ Functions

In the period $-T_1/2 \to T_1/2$, $\delta_{T_1}(t) = \delta(t)$. Making this substitution, then expanding the limits of integration to $\pm\infty$, and using the sampling property of $\delta(t)$

$$S_k = \frac{1}{T_1}\int\limits_{}^{\infty} \delta(t)\exp(-jk2\pi f_1 t)\,dt = \frac{1}{T_1}\exp(0) = 1/T_1$$

Hence

$$\delta_{T_1}(t) = \frac{1}{T_1}\sum_k \exp(jk2\pi f_1 t) \tag{3.31}$$

Taking the Fourier transform of both sides of equation (3.31) gives

$$\mathbf{F}\left[\delta_{T_1}(t)\right] = \frac{1}{T_1}\sum_k \int\limits_{-\infty}^{\infty}\exp\left[-j(2\pi f - k2\pi f_1)t\right]dt = \frac{1}{T_1}\sum_k \delta(f - kf_1)$$

Following the convention for $\delta_{T_1}(t)$, this may be written

$$\mathbf{F}\left[\delta_{T_1}(t)\right] = f_1\delta_{f_1}(f)$$

so that

$$\delta_{T_1}(t) \Leftrightarrow f_1 \delta_{f_1}(f) = \tfrac{1}{T_1} \delta_{1/T_1}(f) \qquad (3.32)$$

Expanding the shorthand, equation (3.32) states

$$\sum_{n=-\infty}^{\infty} \delta(t - nT_1) \Leftrightarrow f_1 \sum_{n=-\infty}^{\infty} \delta(f - nf_1) = \tfrac{1}{T_1} \sum_{n=-\infty}^{\infty} \delta\left(f - \tfrac{n}{T_1}\right)$$

Thus, an ideal sampling function of period T_1 in the time domain becomes a set of impulse functions of area f_1 ($=1/T_1$) separated by f_1 in the frequency domain. Often, the set is called a Dirac *comb*.

(2) Generating Function of a Periodic Function

A generating function $s_g(t)$ represents one period of a periodic function $s_p(t)$

$$s_g(t) = \begin{cases} s_p(t), & -T_1/2 \le t \le T_1/2 \\ 0, & \text{elsewhere} \end{cases}$$

Recalling the replication property of an impulse function [equation (3.22)]

$$s_p(t) = \sum_{n} s_g(t - nT_1) = s_g(t) \otimes \delta_{T_1}(t) \qquad (3.33)$$

Remembering that convolution in the time domain is equal to multiplication in the frequency domain (#12 in Table A.1)

$$s_p(t) = s_g(t) \otimes \delta_{T_1}(t) \Leftrightarrow S_g(f) f_1 \delta_{f_1}(f) \qquad (3.34)$$

Expanding the shorthand, equation (3.34) states

$$s_p(t) = \sum_{n=-\infty}^{\infty} s_g(t - nT_1) \Leftrightarrow f_1 \sum_{n=-\infty}^{\infty} S_g(f) \delta(f - nf_1) = \tfrac{1}{T_1} \sum_{n=-\infty}^{\infty} S_g(f) \delta\left(f - \tfrac{n}{T_1}\right)$$

Thus,

- in the *time domain*, a periodic function can be represented as an infinite sequence of generating functions spaced T_1 apart.
- in the *frequency domain*, a periodic function can be represented as an infinite sequence of impulse functions whose values are $f_1 \times$ samples of the Fourier transform of the generating function taken at frequencies that are integer multiples of f_1. The envelope of this Dirac comb follows the shape of the Fourier transform of the generating function.

Figure 3.14 illustrates the creation of a unit rectangle train from a generating function that is a unit pulse. The steps are consistent in the time domain and the frequency domain. In the time domain, by the convolution of a unit rectangle function and a train of impulse functions, we produce a unit rectangle train, and, in the frequency domain, by the multiplication of a unit Sinc function and a Dirac comb, we produce the magnitude spectrum of the unit rectangle train—a set of delta functions that imitates the shape of a Sinc function.

EXAMPLE 3.11 Symmetrical Square Wave

Given A symmetrical square wave with period T_1 that alternates between $\pm A$.

Problem Find the Fourier transform.

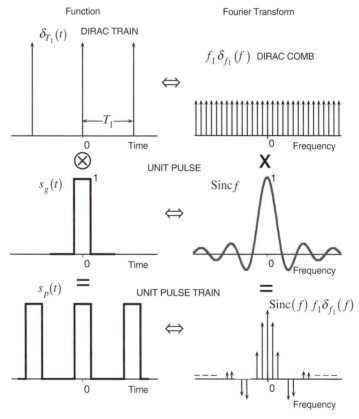

Figure 3.14 Creation of Unit Rectangle Pulse Train and Its Frequency Spectrum
In the time domain, the unit rectangle pulse train is created by the convolution of a Dirac train and a unit rectangle pulse. In the frequency domain, it is created by multiplying a Dirac comb by a unit Sinc function. The Sinc function is the Fourier transform of the unit rectangle pulse.

Solution The square wave is a periodic function. We can represent it as a generating function convolved with a Dirac train. The direct application of this principle leaves us with the generating function $s_g(t)$ shown in **Figure 3.15**. Except for being zero-mean, it has the parameters of a rectangle function. We will define an interim function $s'_p(t) = s_p(t) + A$; it is a symmetrical rectangle train for which the generating function is a rectangle of amplitude $2A$ and width $T_1/2$. Accordingly, we write

$$s'_g(t) = 2A\Pi(2t/T_1)$$

so that, using the time-scaling property

$$s'_p(t) = 2A\Pi(2t/T_1) \otimes \delta_{T_1}(t) \Leftrightarrow A\mathrm{Sinc}(T_1 f/2)\,\delta_{f_1}(f) = S'(f).$$

To find the Fourier transform of the symmetrical square wave, we use the relationship

$$s_p(t) = s'_p(t) - A$$

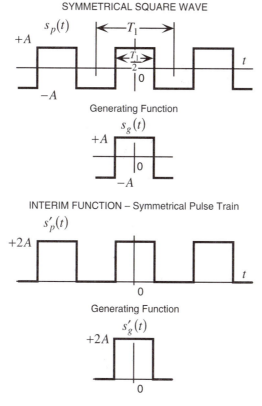

Figure 3.15 Diagrams for Example 3.11

i.e.,

$$s_p(t) = s'_p(t) - A \Leftrightarrow S'(f) - A\delta(f) = S(f).$$

Hence

$$S(f) = A\operatorname{Sinc}(T_1 f/2)\,\delta_{f_1}(f) - A\delta(f)$$

and we have

$$2A\Pi(2t/T_1) \otimes \delta_{T_1}(t) - A \Leftrightarrow A\operatorname{Sinc}(T_1 f/2)\,\delta_{f_1}(f) - A\delta(f) \qquad (3.35)$$

Thus, the Fourier transform of the symmetrical square wave (i.e., $S(f)$) is

equal to the Fourier transform of the symmetrical rectangle train (i.e., $S'(f)$) minus the impulse at $f = 0$ (i.e., the dc-term).

Comment This result supports the discussion in § 2.7(4). If we put $2A = 1$, $T_1 = 2$ time units, and replace $\delta_{f_1}(f)$ by the sum of individual impulse functions, we obtain the spectrum in Figure 2.14.

3.4 ADDITIONAL EXAMPLES

The following additional examples provide an opportunity to emphasize some properties of the Fourier transform that will be of use later in the book.

EXAMPLE 3.12 Selected Transforms

Given

$$s_t(t) = \mathbf{F}^{-1}[S(f)] \text{ and } S(f) = \mathbf{F}[s_t(t)].$$

Problem Find

a) $\mathbf{F}[1 - \cos(2\pi f_0 t)]$

b) $\mathbf{F}[\exp(-|t|)]$

c) $\mathbf{F}^{-1}[A\Pi(2f/w)]$

d) $\mathbf{F}\left[\Pi\left(\frac{t - T/2}{T}\right)\right]$

e) $\mathbf{F}^{-1}[A\operatorname{Sinc}(f - f_0)]$

Solution

a) Using Property #1 of Table A.1, i.e.,

$$As_1(t) + Bs_2(t) \Leftrightarrow AS_1(f) + BS_2(f),$$

and the transform pairs

$$1 \Leftrightarrow \delta(f)$$

$$\cos(2\pi f_0 t) \Leftrightarrow \tfrac{1}{2}\left[\delta(f + f_0) + \delta(f - f_0)\right],$$

we see

$$1 - \cos(2\pi f_0 t) \Leftrightarrow \delta(f) - \tfrac{1}{2}\left[\delta(f + f_0) + \delta(f - f_0)\right].$$

b) The two-sided exponential pulse is the sum of $\exp(-t)U(t)$ and $\exp(t)U(-t)$. Using Property #1 of Table A.1 and the transforms from Example 3.9, i.e.,

$$\exp(-t)U(t) \Leftrightarrow 1/(1 + j2\pi f)$$

and

$$\exp(t)U(-t) \Leftrightarrow 1/(1 - j2\pi f),$$

we can write

$$\exp(-|t|) \Leftrightarrow 1/(1 + j2\pi f) + 1/(1 - j2\pi f) = 2\Big/\left\{1 + (2\pi f)^2\right\}.$$

c) Using Property #4 of Table A.1, i.e.,

$$s(at) \Leftrightarrow \tfrac{1}{|a|}S(f/a),$$

and

$$\text{Sinc}(t) \Leftrightarrow \Pi(f),$$

we obtain

$$\text{Sinc}(at) \Leftrightarrow \tfrac{1}{|a|}\Pi(f/a)$$

Multiplying both sides by $A|a|$ gives

$$A|a|\text{Sinc}(at) \Leftrightarrow A\Pi(f/a).$$

Comparing $A\Pi(2f/w)$ with $A\Pi(f/a)$ we see $a = w/2$ so that

$$A\tfrac{w}{2}\text{Sinc}(wt/2) \Leftrightarrow A\Pi(2f/w).$$

d) The expression $\Pi\!\left(\frac{t - T/2}{T}\right)$ describes a rectangular pulse of width T that is centered at $T/2$. To find the Fourier transform, we use the time-scaling property (#4) to give

$$\Pi(t/T) \Leftrightarrow T\text{Sinc}(fT),$$

and the time shifting property (#5), i.e.,

$$s(t-t_0) \Leftrightarrow S(f)\exp(-j2\pi f t_0)$$

to give

$$\Pi\left(\tfrac{t-T/2}{T}\right) \Leftrightarrow T\text{Sinc}(fT)\exp(-j\pi fT).$$

e) Using Property #6 of Table A.1, i.e.,

$$s(t)\exp(j2\pi f_0 t) \Leftrightarrow S(f-f_0),$$

and

$$\Pi(t) \Leftrightarrow \text{Sinc}(f)$$

gives

$$\Pi(t)\exp(j2\pi f_0 t) \Leftrightarrow \text{Sinc}(f-f_0).$$

Multiplying both sides by A, we see that

$$A\Pi(t)\exp(j2\pi f_0 t) \Leftrightarrow A\text{Sinc}(f-f_0).$$

EXAMPLE 3.13 Dirac Trains

Given

$$\delta_{T_1}(t) \Leftrightarrow f_1\delta_{f_1}(f), \text{ where } f_1 = 1/T_1.$$

Problem

a) Explain the difference between a Dirac comb and a Dirac train.
b) Find the Fourier transforms of $\delta_{2T_1}(t)$ and $\delta_{T_1/2}(t)$.

Solution

a) A Dirac comb consists of equal value impulses (of energy) situated at frequencies spaced at regular intervals such that $f = nf_1$ where n is the set of integers $-\infty < n < \infty$. f_1 is the fundamental frequency of the Dirac train producing the comb. A Dirac train is a sequence of impulses that occur at regular intervals such that $t = nT_1$ where n is the set of integers $-\infty < n < \infty$, and T_1 is the difference in the time of occurrence of contiguous impulses. A Dirac train is a dynamic event in which impulses succeed each other at regular intervals; a Dirac comb is a static array of impulses that form a frequency spectrum.

b) Using property #4

$$\delta_{aT_1}(t) \Leftrightarrow \frac{f_1}{|a|}\delta_{f_1/a}(f)$$

Hence

$$\delta_{2T_1}(t) \Leftrightarrow \tfrac{f_1}{2}\delta_{f_1/2}(f)$$

$$\delta_{T_1/2}(t) \Leftrightarrow 2f_1\delta_{2f_1}(f)$$

The trains and their transforms are drawn in **Figure 3.16**.

EXAMPLE 3.14 Rectangle Pulses and Sinc Functions

Given

$$\Pi(t) \Leftrightarrow \text{Sinc}(f).$$

Problem Find the Fourier transforms of $\Pi(t/2)$ and $\Pi(2t)$.

Solution Using the time-scaling property, we find

$$\Pi(at) \Leftrightarrow \frac{1}{|a|}\text{Sinc}(f/a)$$

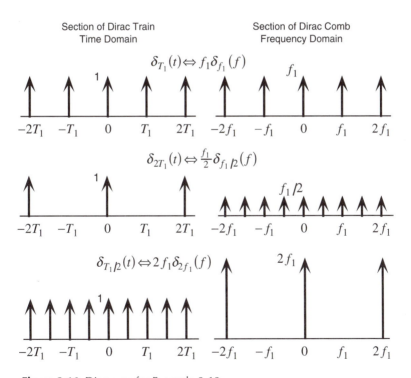

Figure 3.16 Diagrams for Example 3.13

so that

$$\Pi(t/2) \Leftrightarrow 2\operatorname{Sinc}(2f)$$

$$\Pi(2t) \Leftrightarrow \tfrac{1}{2}\operatorname{Sinc}(f/2)$$

The pulses and their transforms are drawn in **Figure 3.17**.

EXAMPLE 3.15 Cosine Functions

Given

$$\cos(2\pi f_1 t) \Leftrightarrow \tfrac{1}{2}\left[\delta(f+f_1) + \delta(f-f_1)\right].$$

Problem Find the Fourier transforms of $\cos(\pi f_1 t)$ and $\cos(4\pi f_1 t)$.

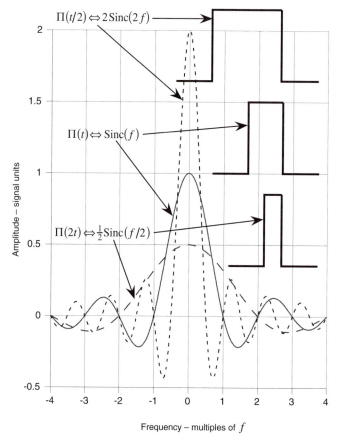

Figure 3.17 Diagrams for Example 3.14

Solution We are scaling frequency, not time. For $\cos(\pi f_1 t)$, $f_1' = f_1/2$ so that

$$\cos(\pi f_1 t) \Leftrightarrow \tfrac{1}{2}\left[\delta(f + f_1/2) + \delta(f - f_1/2)\right]$$

and for $\cos(4\pi f_1 t)$, $f_1' = 2f_1$ so that

$$\cos(4\pi f_1 t) \Leftrightarrow \tfrac{1}{2}\left[\delta(f + 2f_1) + \delta(f - 2f_1)\right].$$

The cosine functions and their transforms are shown in **Figure 3.18**.

EXAMPLE 3.16 Pairs of Impulse Functions

Given

$$\sin(2\pi f_0 t) \Leftrightarrow \tfrac{j}{2}\left[\delta(f + f_0) - \delta(f - f_0)\right].$$

Problem Find the Fourier transforms of $\delta(t + T_1) - \delta(t - T_1)$, $\delta(t + T_1/2) - \delta(t - T_1/2)$, and $\delta(t + 2T_1) - \delta(t - 2T_1)$.

Solution Equation (3.18) states

$$\exp(2\pi f_0 t) \Leftrightarrow \delta(f - f_0)$$

Applying Property #2 in Table A.1, i.e.,

$$S(t) \Leftrightarrow s(-f)$$

gives

$$\delta(t - t_0) \Leftrightarrow \exp(-j2\pi f_0 t).$$

Therefore,

$$\delta(t + t_0) - \delta(t - t_0) \Leftrightarrow \exp(j2\pi ft_0) - \exp(-j2\pi ft_0) = 2j\sin(2\pi ft_0).$$

Substituting $t_0 = T_1$, $T_1/2$ and $2T_1$ gives

$$\delta(t + T_1) - \delta(t - T_1) \Leftrightarrow 2j\sin(2\pi fT_1)$$
$$\delta(t + T_1/2) - \delta(t - T_1/2) \Leftrightarrow 2j\sin(\pi fT_1)$$
$$\delta(t + 2T_1) - \delta(t - 2T_1) \Leftrightarrow 2j\sin(4\pi fT_1)$$

The impulse functions and the magnitudes of their transforms are shown in **Figure 3.19**. In the Appendix, Table A.2 lists some common transform pairs for future reference.

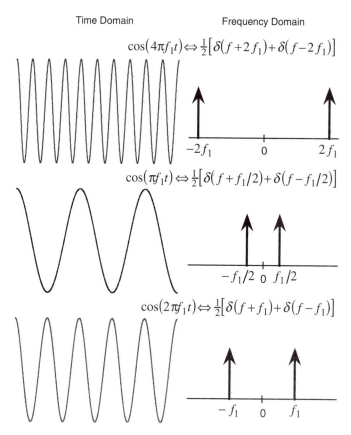

Time Domain   Frequency Domain

$$\cos(4\pi f_1 t) \Leftrightarrow \tfrac{1}{2}\big[\delta(f+2f_1)+\delta(f-2f_1)\big]$$

$$-2f_1 \qquad 0 \qquad 2f_1$$

$$\cos(\pi f_1 t) \Leftrightarrow \tfrac{1}{2}\big[\delta(f+f_1/2)+\delta(f-f_1/2)\big]$$

$$-f_1/2 \;\; 0 \;\; f_1/2$$

$$\cos(2\pi f_1 t) \Leftrightarrow \tfrac{1}{2}\big[\delta(f+f_1)+\delta(f-f_1)\big]$$

$$-f_1 \quad 0 \quad f_1$$

Figure 3.18 Diagrams for Example 3.15

REVIEW QUESTIONS

3.1 Explain why, to use Fourier transform analysis, $s_t(t)$ must be deterministic.

3.2 Given $s_t(t) \Leftrightarrow S(f)$, state the inverse Fourier transform relationship.

3.3 Given $s_t(t) \Leftrightarrow S(f)$, state the Fourier transform relationship.

3.4 For real $s_t(t)$ what can be said about the magnitude spectrum and the phase spectrum of $s_t(t)$?

3.5 In Example 3.3, what is the critical piece of information that makes $F[\Lambda(t)]$ computable?

3.6 What function is preserved under Fourier transformation?

3.7 Define Dirac's improper function.

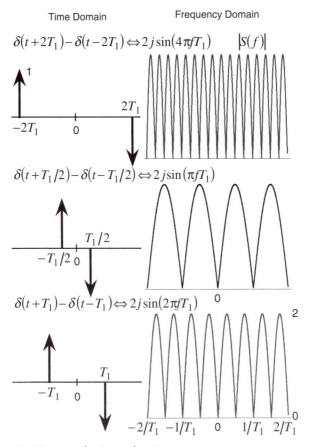

Figure 3.19 Diagrams for Example 3.16

3.8　Given $\delta(t-t_0)$ and $s(t)$, write down the expression for the sifting operation. What is its result?

3.9　Given $\delta(t-t_0)$ and $s(t)$, write down the expression for the replication operation. What is its result?

3.10　Given $S(f)=\frac{1}{2}\big[\delta(f+f_1)+\delta(f-f_1)\big]$, what is $s(t)$?

3.11　Given $s(t)=\sin(2\pi f_1 t)$, what is $S(f)$?

3.12　A cosine function is truncated; what effect does this action have on the magnitude density spectrum?

3.13　Define an ideal sampling function.

3.14　Define a generating function.

3.15 Define a Dirac comb.

3.16 The Fourier transform of a symmetrical rectangle train of period T_1 in which the rectangle pulses are of amplitude $2A$ and width $T_1/2$ is $S(f) = A\,\text{Sinc}(fT_1/2)\delta_{f_1}(f)$. What is the Fourier transform of a symmetrical square wave with period T_1 that alternates between $\pm A$?

4

LTI Systems

- **System**: an identifiable device that transforms signals in such a way that the output is related to the input. From an analytical point of view, a system can be described by the rule $f[\ ]$ that relates the output $y(t)$ to the input $x(t)$, i.e.,

$$y(t) = f[x(t)]$$

The rule can be an algebraic or logical operation, a differential or integral equation, or other function. An important class of systems is *linear time-invariant (LTI) systems*.

- **LTI System**: an identifiable device that transforms signals so that
  - *linearity*: the principle of superposition applies. Adding a second input signal to an existing input signal results in an output signal that is proportional to the sum of the two signals; multiplying one, or more, input signals by constant(s) produces an output signal whose components are multiplied by corresponding quantities, i.e.,

$$a_1 y_1(t) + a_2 y_2(t) = f[a_1 x_1(t) + a_2 x_2(t)]$$

  - *time invariance*: a time shift in the input results in a similar shift in the output.

$$y(t - \tau) = f[x(t - \tau)]$$

LTI systems are

- **Causal** (or realizable): if the input signal starts at zero time and exists only in the interval $0 < t < \infty$, the output signal exists only in the same interval

$$y(t) = f[x(t)], \quad 0 < t < \infty$$

A realizable system cannot produce an output until an input is applied.

- **Memoryless**: the output depends only on the corresponding input

$$y_n = f[x_n]$$

The output signal depends only on a single value (the immediate value) of the input signal, not on any sequence of input values.

4.1 LAPLACE AND FOURIER TRANSFORMS

The analysis of LTI systems is a subject to which it is usual to apply Laplace transforms. The Laplace transform of $x(t)$ is defined as

$$x(t) \leftrightarrow X(s) = \mathbf{L}[x(t)] = \int_{-\infty}^{\infty} x(t)\exp(-st)dt \qquad (4.1)$$

where $\leftrightarrow$ stands for the Laplace transform pair, and s is a complex variable of the form $\sigma + j\omega$. σ and ω are real quantities. When $\sigma = 0$ and $s = j2\pi f$, equation (4.1) becomes

$$x(t) \leftrightarrow X(f) = \mathbf{F}[x(t)] = \int_{-\infty}^{\infty} x(t)\exp(-j2\pi ft)$$

which is the Fourier transform relationship. Since we are interested principally in frequency spectrums, we will continue to employ that relationship.

4.2 IMPULSE RESPONSE

- **Impulse response**: if, at time $t = 0$, the input to an LTI system is a unit impulse function $\delta(t)$, the impulse response $h(t)$ of the system is

$$h(t) = f[\delta(t)] \qquad (4.2)$$

The output signal is equal to the convolution of the input signal and the impulse response of the system, i.e.,

$$y(t) = x(t) \otimes h(t) \qquad (4.3)$$

The proof of equation (4.3) is as follows. Because of the replication property of an impulse function [see § 3.2(3)], a continuous input signal $x(t)$ can be expressed as the convolution of a unit impulse function and itself so that

$$y(t) = f[x(t) \otimes \delta(t)] = f\left[\int_{-\infty}^{\infty} x(\tau)\delta(t-\tau)d\tau\right] \qquad (4.4)$$

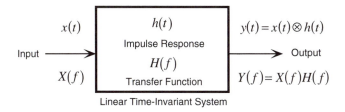

Figure 4.1 LTI System Parameters
A summary of the input-output relationships of an LTI system.

Because the system is linear, we may interchange operations in equation (4.4), so that

$$y(t) = \int_{-\infty}^{\infty} x(\tau)f[\delta(t-\tau)]d\tau$$

Because the system is time-invariant, equation (4.2) gives

$$f[\delta(t-\tau)] = h(t-\tau)$$

so that

$$y(t) = x(t) \otimes h(t)$$

which is equation (4.3).

4.3 TRANSFER FUNCTION

- **Transfer function**: the ratio of the Fourier transform of the output signal and the Fourier transform of the input signal, i.e.,

$$H(f) = Y(f)/X(f) \text{ or } Y(f) = X(f)H(f) \tag{4.5}$$

In **Figure 4.1**, I summarize the input-output relationships of an LTI system.

EXAMPLE 4.1 Impulse Response

Given The LTI system shown in **Figure 4.2**.

Problem Find the impulse response.

Solution When $x(t) = \delta(t)$, $x'(t) = \delta(t) - \delta(t-t_0)$, and

$$h(t) = \int_{-\infty}^{t} x'(\lambda)d\lambda = \int_{-\infty}^{t} \delta(\lambda) - \delta(\lambda - t_0)d\lambda = U(t) - U(t-t_0)$$

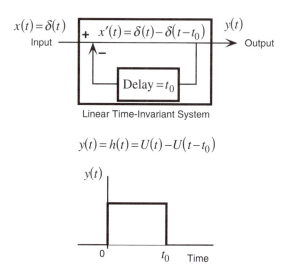

$$y(t) = h(t) = U(t) - U(t - t_0)$$

Figure 4.2 LTI System for Example 4.1

The result is drawn in Figure 4.2.

EXAMPLE 4.2 LTI System Output Signal

Given An LTI system for which $h(t) = U(t)$.

Problem If $x(t) = \exp(-\alpha t) U(t)$, find $y(t)$.

Solution

$$y(t) = x(t) \otimes h(t) = \int_{-\infty}^{t} x(\lambda) h(t - \lambda) d\lambda = \int_{0}^{t} \exp(-\alpha t) dt = \tfrac{1}{\alpha}[1 - \exp(-\alpha t)]$$

The signals for $\alpha = 2$ are drawn in **Figure 4.3**.

4.4 BANDWIDTH AND PASSBAND

When the passband is less than the bandwidth of the input, the output signal $[y(t)]$ is a distorted version of the input signal $[x(t)]$.

- **Bandwidth**: the frequency interval that just contains all of the energy present in the signal
- **Passband**: the range of frequencies that is transmitted by an LTI system.

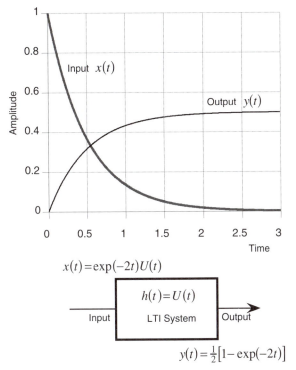

Figure 4.3 LTI System Input and Output Signals for Example 4.2

LTI systems whose sole purpose is to limit the frequencies contained in the output signal are known as *filters*. Basically, they are of two kinds:

- **Low-pass filter**: the range of frequencies that may be transmitted extends from 0 to f_{co} Hz. f_{co} is the cut-off frequency. The transfer function of a low-pass filter is

$$H(f) = \Pi\left[\frac{f \pm \frac{1}{2}f_{co}}{f_{co} + \Delta f}\right] \qquad (4.6)$$

Δf is a small quantity added to the width of the rectangle function to ensure that it is wide enough to pass the discrete components of the frequency spectrum that coincide with its nominal width.

- **Bandpass filter**: the range of frequencies that may be transmitted extends from f_{min} to f_{max} Hz. The transfer function of a bandpass filter is

$$H(f) = \Pi\left[\frac{f \pm \frac{1}{2}(f_{max} + f_{min})}{f_{max} - f_{min} + \Delta f}\right] \qquad (4.7)$$

In theory, another filter is possible—a *high*-pass filter—one in which the range of possible frequencies in the output functions extends from a frequency greater than zero to infinity. In practice, a limit always exists to the maximum frequency that can be supported by a system. High-pass filters, then, are a special case of bandpass filters.

EXAMPLE 4.3 Effect of Passband on a Symmetrical Square Wave

Given A symmetrical square wave for which $S_n = 0.5\,\text{Sinc}(0.5n)$, when $1 \le |n| < \infty$.

Problem Determine what percentage of the total signal power is lost when the wave is applied to an LTI system that passes

1) only the fundamental
2) only the fundamental and third harmonic component
3) only the fundamental and third and fifth harmonic components
4) only the fundamental and third, fifth, and seventh harmonic components

Solution Determine the values of S_n and $|S_n|^2$ up to the seventh harmonic (see Example 2.5).

when $n = \pm 1$, $S_{\pm 1} = \frac{1}{\pi}\sin\left(\frac{\pi}{2}\right) = 0.31831$ su and $|S_{\pm 1}|^2 = 0.10132$ sw

when $n = \pm 3$, $S_{\pm 3} = \frac{1}{3\pi}\sin\left(\frac{3\pi}{2}\right) = -0.10610$ su and $|S_{\pm 3}|^2 = 0.01126$ sw

when $n = \pm 5$, $S_{\pm 5} = \frac{1}{5\pi}\sin\left(\frac{5\pi}{2}\right) = 0.06366$ su and $|S_{\pm 5}|^2 = 0.00405$ sw

when $n = \pm 7$, $S_{\pm 7} = \frac{1}{7\pi}\sin\left(\frac{7\pi}{2}\right) = -0.04547$ su and $|S_{\pm 7}|^2 = 0.00207$ sw

The signal power is the sum of the squares of the complex coefficients [equation (2.15)], i.e.,

for problem 1) $P = |S_{-1}|^2 + |S_1|^2 = 2|S_1|^2 = 0.20264$ sw

for problem 2) $P = 2|S_1|^2 + 2|S_3|^2 = 0.22516$ sw

for problem 3) $P = 2|S_1|^2 + 2|S_3|^2 + 2|S_5|^2 = 0.23326$ sw

for problem 4) $P = 2|S_1|^2 + 2|S_3|^2 + 2|S_5|^2 + 2|S_7|^2 = 0.23740$ sw

For a symmetrical square wave of amplitude 0.5 signal units, the signal power is $(0.5)^2 = 0.25$ signal watts. Hence, the percentage of power lost is for problem 1) $P = \left[1 - \frac{0.20264}{0.25}\right] \times 100 = 18.9\%$

for problem 2) $P = \left[1 - \frac{0.22516}{0.25}\right] \times 100 = 9.9\%$

for problem 3) $P = \left[1 - \frac{0.23326}{0.25}\right] \times 100 = 6.7\%$

for problem 4) $P = \left[1 - \frac{0.23740}{0.25}\right] \times 100 = 5.0\%$

Comment 1 A symmetrical square wave has an infinite bandwidth. This example shows the effect of attempting to transmit such a signal over a practical communication system that has a limited passband. Even when the passband extends to the seventh harmonic, a significant amount of signal power is lost. Manifest as a loss of *squareness* in the time domain; the effect is shown in **Figure 4.4**. The passband expands from one that transmits the fundamental component only (in the top left quadrant) to one that transmits the fundamental component up to the 7th harmonic (in the bottom right quadrant).

Comment 2 Functions such as a square wave that exhibit sharp discontinuities are subject to *Gibbs' phenomenon* at the points of discontinuity. When represented by a Fourier series of n terms, the sum converges to the midpoint of the discontinuity. However, on each side is an oscillatory overshoot of period $T_1/2n$ (T_1 is the period of the wave) and peak value of 18% of the step height. As $n \to \infty$, the oscillations collapse into spikes above and below the discontinuity. The phenomenon is attributable to the truncation of the representation. Effectively, by limiting the series to n terms, we are subjecting the function to a low-pass filter—or multiplying it by a rectangle function $[\Pi(t/b)]$. As demonstrated in Example 3.6, in the frequency domain, this results in the occurrence of a Sinc function at the points of discontinuity.

EXAMPLE 4.4 Complex Periodic Function

Given The complex periodic function defined by equation (2.3) and shown in Figure 2.1, i.e.,

$$s_p(t) = 1 + \cos(2\pi t/T_1) - 0.75\sin(2\pi t/T_1) + 0.5\cos(6\pi t/T_1) - 0.25\sin(6\pi t/T_1).$$

Problem

a) What is the bandwidth of the complex sinusoidal function shown in Figure 2.1?

b) What is the signal power?

c) Write the transfer function of a filter that just passes the complex sinusoidal function.

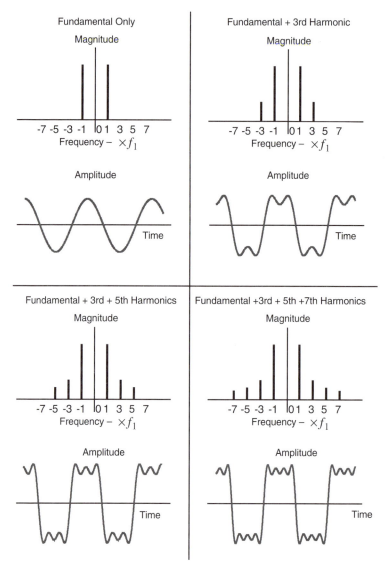

Figure 4.4 Effect of Limiting Passband on Transmission of Symmetrical Square Wave Examined in Example 4.3

d) Write the transfer function of a filter that just passes the dc term and fundamental frequency component of the complex sinusoidal function.

e) What is the signal power of the output signal coming from the filter in d)?

f) What percentage of the input power is lost in the filter in d)?

g) Write the transfer function of a filter that just passes the third harmonic component of the complex sinusoidal function.

h) What is the signal power of the output signal coming from the filter in g)?

i) What percentage of the input power is lost in the filter in g)?

Solution

a) For the complex sinusoidal function shown in Figure 2.1, energy exists at dc, f_1 Hz, and $3f_1$ Hz. Therefore, the signal bandwidth is $3f_1$ Hz.

b) From Figures 2.5 and 2.6, the power in the output function is 1.93 signal watts.

c) The low-pass filter must pass frequencies from dc to $3f_1$ Hz. The two-sided transfer function is

$$H(f) = \Pi\left[\frac{f \pm 1.5 f_1}{3 f_1 + \Delta f}\right]$$

This is shorthand for an LTI device that has a system transfer function consisting of rectangular functions situated at $f = \pm 1.5 f_1$ Hz, with gains of 1 and widths of $3 f_1 + \Delta f$ Hz. The entire signal will appear at the output of the low-pass filter, i.e.,

$$y(t) = 1 + \cos(2\pi f_1 t) - 0.75\sin(2\pi f_1 t) + 0.5\cos(6\pi f_1 t) - 0.25(6\pi f_1 t)$$

d) To just pass the dc term and fundamental frequency components of the complex sinusoidal function shown in Figure 2.1, the widths of the rectangle functions are reduced to $f_1 + \Delta f$ Hz, and the center frequency of each passband is moved to $\pm 0.5 f_1$, i.e.,

$$H(f) = \Pi\left[\frac{f \pm 0.5 f_1}{f_1 + \Delta f}\right]$$

e) The signal at the output of the low-pass filter will be

$$y(t) = 1 + \cos(2\pi f_1 t) - 0.75\sin(2\pi f_1 t)$$

From Figures 2.5 and 2.6, the power in the output function will be 1.77 signal watts

f) 0.16 signal watts is lost; this is 8.3% of the power at the input.

g) To just pass the third harmonic components, the widths of the rectangle functions are reduced to $2\Delta f$ Hz, and they are moved to $\pm 3 f_1$ Hz, i.e.,

$$H(f) = \Pi\left[\frac{f \pm 3 f_1}{2\Delta f}\right]$$

h) Only the third harmonic will appear at the output

$$y(t) = 0.5\cos(6\pi f_1 t) - 0.25(6\pi f_1 t)$$

From Figures 2.5 and 2.6, the power in the output function is 0.16 signal watts

i) 1.77 signal watts are lost; this is 91.7% of the power at the input.

REVIEW QUESTIONS

4.1   Define an *LTI System*.

4.2   What is the meaning of *causal* when applied to an LTI system?

4.3   What is the meaning of *memoryless* when applied to an LTI system?

4.4   In an LTI system, express the output signal in terms of the impulse response.

4.5   In an LTI system, express the Fourier transform of the output signal in terms of the transfer function.

4.6   Write expressions for the transfer functions of low-pass and bandpass filters.

5

Autocorrelation and Spectral Density

Correlation describes a relationship between two (or more) things. It is used to obtain expressions for the frequency distribution of energy- and power-type functions.

5.1 CORRELATION FUNCTIONS

Correlation functions measure specific relationships between two (or more) functions. They are the foundation for autocorrelation—a measure applied to a function and its time-displaced twin.

(1) Cross-Correlation Functions

$\psi_{ij}(\tau)$, the energy-type (or psi-type) *cross-correlation function* (CCF) of two deterministic energy-type functions $s_i(t)$ and $s_j(t)$ is defined as

$$\psi_{ij}(\tau) = \int_{-\infty}^{\infty} s_i(t)s_j^*(t-\tau)dt \quad \text{or} \quad \int_{-\infty}^{\infty} s_i(t+\tau)s_j^*(t)dt \tag{5.1}$$

The quantity τ is a *scanning* or *searching* parameter; it shifts (delays) one function with respect to the other and establishes a time difference between them. Time is a dummy variable that disappears on performing the integration. Putting $\tau=0$ in equation (5.1), we obtain

$$\psi_{ij}(0) = \int_{-\infty}^{\infty} s_i(t)s_j^*(t)dt = E_{ij} \tag{5.2}$$

The right-hand side of equation (5.2) is known as the cross-energy E_{ij} of $s_i(t)$ and $s_j(t)$.

In a similar manner, $\phi_{ij}(\tau)$, the *power*-type (or *phi*-type) cross-correlation function of two deterministic power-type functions is

$$\phi_{ij}(\tau)=\left\langle s_i(t)s_j^*(t-\tau)\right\rangle \quad \text{or} \quad \left\langle s_i(t+\tau)s_j^*(t)\right\rangle \tag{5.3}$$

and

$$\phi_{ij}(0)=\left\langle s_i(t)s_j^*(t)\right\rangle = P_{ij} \tag{5.4}$$

The right-hand side of equation (5.4) is known as the cross-power P_{ij} of $s_i(t)$ and $s_j(t)$.

(2) Parseval's Theorem

* **Parseval's theorem**: the inner product of two energy-type functions, $s_i(t)$ and $s_j(t)$, is equal to the inner product of their Fourier transforms, i.e.,

$$\int_{-\infty}^{\infty}s_i(t)\,s_j^*(t)\,dt = \int_{-\infty}^{\infty}S_i(f)\,S_j^*(f)\,df \tag{5.5}$$

The proof of this statement is as follows. The inverse transform of $s_j^*(t)$ is

$$s_j^*(t)=\left[\int_{-\infty}^{\infty}S_j(f)\exp(j2\pi ft)df\right]^* = \int_{-\infty}^{\infty}S_j^*(f)\exp(-j2\pi ft)df \tag{5.6}$$

Substituting in equation (5.2)

$$E_{ij} = \int_{-\infty}^{\infty}s_i(t)\left[\int S_j^*(f)\exp(-j2\pi ft)df\right]dt \tag{5.7}$$

Interchanging the order of integration with respect to time and frequency

$$E_{ij} = \int_{-\infty}^{\infty}\left[s_i(t)\exp(-j2\pi ft)dt\right]S_j^*(f)df \tag{5.8}$$

The quantity in [ ] brackets is $S_i(f)$ so that

$$E_{ij} = \int_{-\infty}^{\infty}S_i(f)S_j^*(f)df \tag{5.9}$$

Hence

$$\int_{-\infty}^{\infty}s_i(t)s_j^*(t)dt= \int_{-\infty}^{\infty}S_i(f)S_j^*(f)df$$

This result is known as *Parseval's theorem* (after August von Parseval, a 19th-century German mathematician).

Putting $i = j$ in equation (5.5) gives

$$\int_{-\infty}^{\infty} s(t)s^*(t)dt = \int_{-\infty}^{\infty} S(f)S^*(f)df = E[\infty] \tag{5.10}$$

where $E[\infty]$ is the total signal energy [see equation (1.5)], or

$$\int_{-\infty}^{\infty} |s(t)|^2 dt = \int_{-\infty}^{\infty} |S(f)|^2 df = E[\infty] \tag{5.11}$$

The total signal energy $E[\infty]$ of a deterministic transient function $s(t) \Leftrightarrow S(f)$ may be determined by integrating the product of the function and its complex conjugate, or integrating the modulus squared, in either time or frequency domains. In one or the other of these forms, it is known as Parseval's theorem or Rayleigh's theorem (after John William Strutt, 3rd Baron Rayleigh, a 19th-century English physicist). For periodic functions, a related result, known as Parseval's power theorem, is given in § 2.6(2).

5.2 ENERGY-TYPE FUNCTIONS

(1) Energy-Type Autocorrelation Function

In § 1.4, I explained the difference between energy- and power-type functions and showed how to evaluate their infinite integrals. In this section I deal with energy-type functions.

In equation (5.1), putting $i = j$, we obtain

$$\psi(\tau) = \int_{-\infty}^{\infty} s(t)s^*(t-\tau)dt \quad \text{or} \quad \int_{-\infty}^{\infty} s(t+\tau)s^*(t)dt \tag{5.12}$$

The function $\psi(\tau)$ is known as the energy-type (or *psi*-type) *autocorrelation function* (ACF) of the deterministic energy-type function $s(t)$. For *real* $s(t)$, $|\psi(\tau)|$ is even, $\angle\psi(\tau)$ is odd, and the maximum value occurs at the origin. Also, when $\tau = 0$,

$$\psi(0) = \int_{-\infty}^{\infty} s(t)s^*(t)dt = E[\infty] \tag{5.13}$$

The value of the energy-type ACF at $\tau = 0$ is equal to the total energy of the function $s(t)$.

(2) Energy Spectral Density Function

If $\psi(\tau) \Leftrightarrow \Psi(f)$, i.e.,

$$\Psi(f) = \int_{-\infty}^{\infty} \psi(\tau)\exp(-j2\pi\tau)d\tau \tag{5.14}$$

The area under $\Psi(f)$ is $\psi(0)$ (see property #8, Table A.1), so that, from equation (5.13)

$$\psi(0) = \int_{-\infty}^{\infty} \Psi(f)df = E[\infty] \tag{5.15}$$

$\Psi(f)$ is called the *energy spectral density* (ESD) of $s(t)$. Measured in signal joules per Hertz, it is the energy spectrum of $s(t)$. For real functions, the energy spectral density is always real, never negative, and even.

- **Autocorrelation theorem**: if $s(t) \Leftrightarrow S(f)$, the Fourier transform of the autocorrelation function and the energy spectral density of $s(t)$ are equal to the square of the modulus of $S(f)$, i.e.,

$$\psi(\tau) \Leftrightarrow |S(f)|^2 = \Psi(f) \tag{5.16}$$

The proof of this statement is as follows. By definition [equation (5.12)],

$$\psi(\tau) = \int_{-\infty}^{\infty} s(t+\tau)s*(t)dt$$

The RHS is the convolution of $s(t)$ and $s*(-t)$; the Fourier transform of $s(t) \otimes s^*(-t)$ is $S(f)S*(f)$. Hence

$$\int_{-\infty}^{\infty} s(t+\tau)s*(t)dt = s(t) \otimes s*(-t) \Leftrightarrow S(f)S*(f)$$

so that, because $S(f)S*(f)=|S(f)|^2$ and $\psi(\tau) \Leftrightarrow \Psi(f)$,

$$\psi(\tau) \Leftrightarrow |S(f)|^2 = \Psi(f)$$

EXAMPLE 5.1 Unit Rectangle Function

Given

$$s(t) = \Pi(t).$$

Problem Find the ESD and ACF of $s(t)$.

Solution

$$s(t) = \Pi(t) \Leftrightarrow \text{Sinc}(f) = S(f)$$

so that the ESD is

$$\Psi(f) = |S(f)|^2 = \text{Sinc}^2(f) \tag{5.17}$$

and the ACF is

$$\psi(\tau) = F^{-1}[\Psi(f)] = F^{-1}[\text{Sinc}^2(f)] = [\Pi(\tau) \otimes \Pi(\tau)] = \Lambda(\tau) \tag{5.18}$$

The function $\Lambda(\tau)$ stands for a unit triangle function of magnitude 1 unit and base ± 1 units. These relationships are shown in **Figure 5.1**.

EXAMPLE 5.2   Unit Sinc Function

Given

$$s(t) = \text{Sinc}(t).$$

Problem Find the ESD and ACF of $s(t)$.

Solution

$$s(t) = \text{Sinc}(t) \Leftrightarrow \Pi(f) = S(f)$$

so that the ESD is

$$\Psi(f) = |\Pi(f)|^2 = \Pi(f) \tag{5.19}$$

and the ACF is

$$\psi(\tau) = F^{-1}[\Psi(f)] = \text{Sinc}(\tau) \tag{5.20}$$

These relationships are shown in **Figure 5.2**.

EXAMPLE 5.3   Exponential Function

Given

$$s(t) = A\exp(-\alpha t)\, U(t) \text{ where } \alpha \text{ is real.}$$

Problem Find the ACF and ESD of $s(t)$.

Solution

1) Find the ACF.

By definition

$$\psi(\tau) = \int_{-\infty}^{\infty} s(t)\, s^*(t-\tau)\, dt$$

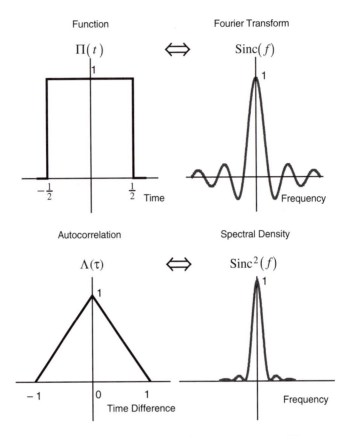

Function   Fourier Transform

$\Pi(t)$ $\Longleftrightarrow$ $\mathrm{Sinc}(f)$

Time   Frequency

Autocorrelation   Spectral Density

$\Lambda(\tau)$ $\Longleftrightarrow$ $\mathrm{Sinc}^2(f)$

Time Difference   Frequency

Figure 5.1 Fourier Transform, Autocorrelation Function, and Energy Spectral Density of Unit Rectangle Function (Example 5.1)

Since $s(t)$ is real, we may ignore the conjugate symbol, so that

$$\psi(\tau) = \int_{-\infty}^{\infty} s(t)\, s(t-\tau)\, dt = \int_{-\infty}^{\infty} A^2 \exp(-\alpha t) \exp\{-\alpha(t-\tau)\}\, U(t-\tau)dt$$

The presence of $U(t-\tau)$ means that the composite function only exists in the positive half plane, in the range $\tau \to \infty$. These relationships are shown in **Figure 5.3**. Thus

$$\psi(\tau) = A^2 \exp(\alpha\tau) \int_{\tau}^{\infty} \exp(-2\alpha t)\, dt = \frac{A^2 \exp(\alpha\tau)}{-2\alpha} [\exp(-2\alpha t)]_{\tau}^{\infty}$$

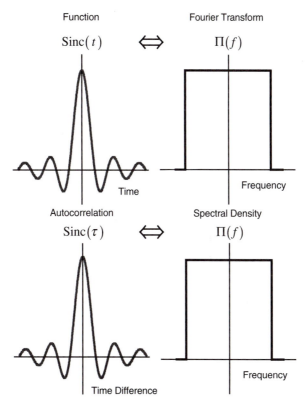

Figure 5.2 Fourier Transform, Autocorrelation Function, and Energy Spectral Density of Unit Sinc Function (Example 5.2)

Hence

$$\psi(\tau) = \frac{A^2 \exp(-\alpha\tau)}{2\alpha}$$

Since the function is real, the ACF is even, so that

$$\psi(\tau) = \frac{A^2}{2\alpha} \exp(-\alpha|\tau|) \tag{5.21}$$

2) Find the ESD.

$$\Psi(f) = \mathbf{F}\big[\psi(\tau)\big] = \frac{A^2}{2\alpha} \int\limits_{-\infty}^{\infty} \exp(-\alpha|\tau|) \exp(-j2\pi f\tau)\, d\tau$$

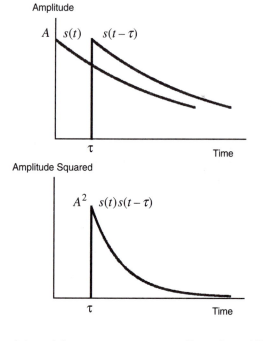

Figure 5.3 Diagrams for Example 5.3

Since $\Psi(f)$ is even for real signals

$$\Psi(f)=\tfrac{A^2}{2\alpha}2\int\limits_{0}^{\infty}\exp\!\left(-\left[\alpha+j2\pi f\right]\tau\right)d\tau=\tfrac{A^2}{2\alpha}\left(\frac{\alpha}{\alpha^2+4\pi^2 f^2}-\frac{j2\pi f}{\alpha^2+4\pi^2 f^2}\right)$$

Because $\Psi(f)$ is always real, the imaginary part must be zero, so that

$$\Psi(f)=A^2\big/\!\left(\alpha^2+4\pi^2 f^2\right) \qquad\qquad (5.22)$$

5.3 POWER-TYPE FUNCTIONS

Power-type functions must be averaged over all time to produce finite integrals. This leads to different expressions for autocorrelation and spectral density.

(1) Power-Type Autocorrelation Function

Putting $i=j$ in equation (5.3) produces the power-type autocorrelation function

$$\phi(\tau)=\left\langle s(t)s^*(t-\tau)\right\rangle \ \text{ or }\ \left\langle s(t+\tau)s^*(t)\right\rangle \tag{5.23}$$

For *real* $s(t)$, $\phi(\tau)$ is even, $\angle\phi(\tau)$ is odd, and the maximum value occurs at the origin. Also, putting $\tau=0$

$$\phi(0)=\left\langle s(t)s^*(t)\right\rangle=\left\langle |s(t)|^2\right\rangle=\overline{P}[\infty] \tag{5.24}$$

i.e., the value of the power-type ACF at $\tau=0$ is equal to the power of the function $s(t)$ averaged over all time. If the function is periodic, the autocorrelation function is periodic with the same period.

(2) Power Spectral Density Function

If $\phi(\tau)\Leftrightarrow\Phi(f)$, then the area under $\Phi(f)$ is $\phi(0)$, so that

$$\phi(0)=\int_{-\infty}^{\infty}\Phi(f)df=\overline{P}[\infty] \tag{5.25}$$

$\Phi(f)$ is called the *power spectral density* (PSD) of $s(t)$. Measured in signal watts per Hertz, it is the power spectrum of $s(t)$. Always real and never negative, it is even for real functions.

For a periodic function, we can substitute equation (2.9) in equation (5.23) to give

$$\phi_p(\tau)=\left\langle \sum_n S_n\exp[j2\pi nf_1(t+\tau)]\sum_n S_n^*\exp(-j2\pi nf_1t)\right\rangle$$

Evaluating the RHS for integer values of n, one at a time

$$\phi_p(\tau)=\frac{1}{T_1}\int_{\alpha}^{\alpha+T_1}\sum_n |S_n|^2\exp(j2\pi nf_1\tau)dt$$

Because the integrand is a constant with respect to time, we obtain

$$\phi_p(\tau)=\sum_n |S_n|^2\exp(j2\pi nf_1\tau) \tag{5.26}$$

From equation (3.18)

$$\exp(j2\pi n f_1 \tau) \Leftrightarrow \delta(f - nf_1)$$

so that taking the Fourier transform of both sides of equation (5.26) leads to

$$\Phi_p(f) = \sum_n |S_n|^2 \delta(f - nf_1) \tag{5.27}$$

i.e., the PSD of a periodic function consists of impulses of areas equal to the squares of the magnitudes of its complex Fourier coefficients. In effect, equation (5.27) is equivalent to equation (2.15).

The operation defined on the RHS of equation (5.27) is known as *impulse magnitude squaring* and is sometimes denoted by $|\ |^{2i}$ so that

$$\Phi_p(f) = \sum_n |S_n|^{2i}$$

EXAMPLE 5.4   Cosine Function

Given

$$s(t) = A\cos(2\pi f_1 t + \theta) \text{ and } A, \ f_1, \text{ and } \theta \text{ are constants.}$$

Problem Find the ACF and PSD of $s(t)$.

Solution The function is periodic; therefore we use power-type functions. The function is real; therefore we ignore the imaginary parts and the conjugate symbol disappears. As a result,

$$\phi(\tau) = \langle s(t)s^*(t+\tau)\rangle = A^2\langle\cos(2\pi f_1 t + \theta)\cos\{2\pi f_1(t+\tau)+\theta\}\rangle$$

Expanding the cosine product produces two terms. One is $\cos(2\pi f_1 \tau)$. It is constant (with respect to t) and survives time-averaging. The other is $\cos(4\pi f_1 t + 2\theta - 2\pi f_1 \tau)$; it is a periodic function and does not survive time-averaging. Hence

$$\phi(\tau) = \tfrac{A^2}{2}\cos(2\pi f_1 \tau) \tag{5.28}$$

and

$$\Phi(f) = \mathbf{F}\left[\tfrac{A^2}{2}\cos(2\pi f_1 \tau)\right] = \tfrac{A^2}{4}\left[\delta(f+f_1)+\delta(f-f_1)\right] \tag{5.29}$$

Comment Integrating equation (5.29) gives the power of a cosine wave, i.e.,

$$\int_{-\infty}^{\infty}\Phi(f)df = A^2/2 \text{ signal watts}$$

This is a well-known result.

EXAMPLE 5.5   Symmetrical Square Wave

Given From Example 3.11, a symmetrical square wave with period T_1 that alternates between $\pm A$, for which

$$S(f) = A\operatorname{Sinc}(T_1 f/2)\, \delta_{f_1}(f) - A\delta(f)$$

Problem Find the PSD and ACF.

Solution

1) PSD

Impulse magnitude squaring

$$\Phi(f) = A^2 \operatorname{Sinc}^2(T_1 f/2)\, \delta_{f_1}(f) - A^2 \delta(f) \qquad (5.30)$$

2) ACF

Taking the inverse Fourier transform

$$\phi(\tau) = 2A^2 \Lambda(2\tau/T_1) \otimes \delta_{T_1}(\tau) - A^2$$

This is a symmetrical triangular wave oscillating between $\pm A^2$ with period T_1. It is shown in **Figure 5.4**.

Comment The power in the wave is found by integrating equation (5.30); the result can be written in the form

$$\overline{P}[\infty] = \int_{-\infty}^{\infty} \Phi(f)\,df = \sum_{\substack{n=-\infty \\ n\neq 0}}^{n=\infty} A^2 \operatorname{Sinc}^2(n/2) \text{ signal watts}$$

This is similar to results obtained in § 2.7.

---

(3) Generating Functions

In § 5.2(2), I showed that

$$\Psi(f) = |S(f)|^2$$

i.e., the energy spectral density of $s(t)$ is equal to the square of the modulus of its Fourier transform. For a periodic function represented by a generating function repeated periodically, i.e., $s_g(t)$ exists over the interval T_1 and is zero elsewhere, we can write

$$\Psi_g(f) = |S_g(f)|^2 \qquad (5.31)$$

Dividing both sides by T_1 we obtain

$$\Psi_g(f)/T_1 = |S_g(f)|^2 /T_1 \qquad (5.32)$$

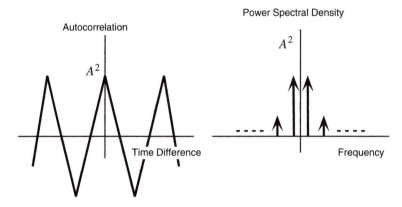

$$\phi(\tau) = 2A^2 \Lambda\!\left(\tfrac{2\tau}{T_1}\right) \otimes \delta_{T_1}(t) - A^2 \qquad \Phi(f) = A^2 \operatorname{Sinc}\!\left(\tfrac{T_1}{2}f\right)\delta_{f_1}(f) - A^2\delta(f)$$

Figure 5.4 Diagrams for Example 5.4

This is the power spectral density (averaged over the interval T_1) of $s_g(t)$; it is also the power spectral density (averaged over all time) of a wave composed of an eternal, periodic sequence of $s_g(t)$ so that

$$\Phi(f) = |S_g(f)|^2 / T_1 \qquad (5.33)$$

We make use of this result in § 13.4(3).

REVIEW QUESTIONS

5.1   State Parseval's theorem for two energy-type functions.

5.2   State Parseval's theorem for a single energy-type function.

5.3   Define the *energy-type autocorrelation function.*

5.4   Under what circumstance is the autocorrelation function of an energy-type function equal to its total energy?

5.5   Define the *energy spectral density* of an energy-type function.

5.6   Under what circumstance is the energy spectral density of an energy-type function equal to its total energy?

5.7   State the autocorrelation theorem.

5.8   Define the *power-type autocorrelation function.*

5.9   Under what circumstance is the autocorrelation function of a power-type function equal to its power averaged over all time?

5.10   Define the *power spectral density* of a power-type function.

5.11   Under what circumstance is the power spectral density of a power-type function equal to its power averaged over all time?

5.12   For a periodic function, state the power spectral density in terms of its complex Fourier coefficients.

6

Probability

The theory of probability is key to describing the diversity of outcomes from activities in which resources are employed in a probabilistic fashion. First applied to games of chance, contemporary applications include signal processing, message handling, and information theory.

Fundamental to the development of useful results is the concept of randomness. In common use, the term *random* describes something that proceeds without definite aim, purpose, or reason so that no order is apparent. A synonym is *haphazard*. As a mathematical term, it is precisely defined:

- **Random**: if, in a long series of trials, all possible outcomes occur the same number of times, the outcomes are said to be randomly-distributed.

Prime examples of random behavior are tossing a *fair* coin, and rolling a *true* die. In tossing a coin, after a great many tosses, the number of heads is equal to the number of tails. In rolling a die, after a great many rolls, the number of times a number (1 through 6) has been uppermost is the same for all numbers (i.e., $n_1 = n_2 = \dots n_6$).

To achieve these results, over-emphasizing the need for a large number of trials is impossible. **Figure 6.1** shows the results of simulating tossing a coin an increasing number of times. (Instead of performing toss, after toss, after toss, we use the entries in a table of random numbers to represent the outcome of successive tosses.) In Run #1, heads dominate the early tosses; in Run #2, tails dominate the early tosses. For small numbers of tosses, the ratio of the number of heads to the number of all tosses behaves erratically. For more than 100 tosses, the ratio begins to settle close to the theoretical value of 0.5. Thus, when an experiment has two possible outcomes, more than 100 trials may be required to achieve values close to the theoretical average. For experiments that produce more than two outcomes, the number of trials needed to approach agreement with theory increases substantially.

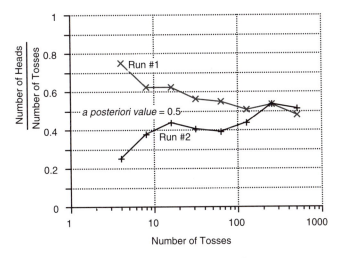

Figure 6.1 Results of Simulated Coin-Tossing Experiments
*In Run #1, heads predominated in the early trials. In Run #2, tails pre-
dominated in the early trials. In both cases, over 100 trials were required
before the* a priori *value of 0.5 was approached.*

6.1 SETS AND VENN DIAGRAMS

To understand probability we must formalize the concept of a collection of
elements and the representation of logical relationships by two-dimensional
figures.

(1) Sets

- **Set**: a collection of particular things. Called *elements*, they are likely to
  be numbers (arithmetic or literal)—thus, the set of: integer numbers
  between 6 and 16; points on a segment of a line; or points in a given
  area. A set M consisting of elements a, b, c, ... k, is written
  $M\{a, b, c, ... k\}$. Sets may be classified as:
  - *countable*: The set of elements corresponds in some way with inte-
    ger numbers, e.g., $M\{1, 2, 3, ... k\}$
  - *uncountable*: The elements of the set form a continuum, e.g.,
    $M\{0 < n \leq k\}$
  - *null*: Identified by the symbol $\emptyset$, the set has no elements, e.g.,
    $M\{\,\} = \emptyset$
  - *finite*: The set contains a limited number of elements, e.g.,
    $M\{a, b, c, ... k\}$

  – *infinite*: The set contains an unlimited number of elements, e.g., $M\{n \to \infty\}$
  – *countably infinite*: The set contains an unlimited number of countable elements, e.g., $M\{1, 2, 3, ... \infty\}$
  – *universal*: Frequently denoted by S, the set of all elements associated with a given situation

(2) *Venn Diagrams*

Sets can be represented by closed figures that contain their elements. Known as *Venn* diagrams (invented by John Venn, a 19th-century English mathematician), in **Figure 6.2** we show a universal set S and four subsets of S identified as A, B, C, and D. Three of the subsets (A, B, and C) overlap each other. The fourth (D) is described as disjoint from A, B, and C; its elements are unrelated to them except insofar as they are all subsets of S. We use the diagram to identify and illustrate some of the properties of sets.

- **Equality**: two sets are equal if all the elements in one are present in the other, and vice-versa. In symbols, if $Q \subseteq R$ [the sign $\subseteq$ stands for all elements (of Q) are contained in (R)] and $R \subseteq Q$, then $Q = R$.

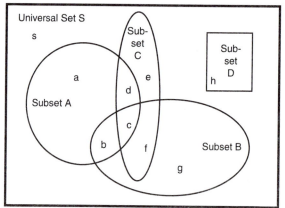

a, b, c, ... s, identify component areas within the rectangle
Area representing Universal Set S = (s + a + b + c + d + e + f + g + h)
Area representing Subset A = (a + b + c + d)
Area representing Subset B = (b + c + f + g)
Area representing Subset C = (c + d + e + f)
Area representing Subset D = (h)

Figure 6.2 Venn Diagram Representation of Subsets and the Universal Set
The diagram shows a universal set (S) and four subsets (A, B, C, and D). Three of the subsets (A, B, and C) overlap each other. The fourth set (D) is described as disjoint from A, B, and C; its elements are unrelated to them except insofar as they are all subsets of S.

- **Difference**: written Q−R, the difference between Q and R is the set containing all of the elements of Q that are not present in R. In Figure 6.2, A−B is represented by the area (a+d). In similar fashion, B−A is represented by the area (f+g).
- **Union**: written Q∪R, the union of Q and R is the set of all elements contained in either Q or R, or both. In Figure 6.2, B∪C is represented by area (b+c+f+g+d+e).
- **Intersection**: written Q∩R, the intersection of Q and R is the set whose elements are common to both. In Figure 6.2, A∩C is represented by the area (c+d). Sets that have no intersection are said to be *mutually exclusive.* Thus, in Figure 6.2, subsets B and D are mutually exclusive.
- **Complement**: the set consisting of all elements that are not contained in Q is called the complement or negation of Q. Written $\overline{Q}$, the complement is given by S−Q, where S is the universal set. It follows from the definition of S that $\overline{S}=\varnothing$. In Figure 6.2, $\overline{B}$ is represented by the area (s+a+d+e+h).

6.2 PROBABILITY

Engineers have an intuitive feel for probability and can apply it readily in practical situations. Mathematicians are not so lucky. A formal description of probability requires much more work.

(1) Definition

A formal description of probability depends on the following definitions:

- **Experiment**: physical activity devoted to the observation of some phenomenon subject to probabilistic behavior. It consists of a primitive activity that is repeated many times. Thus, tossing a coin a large number of times is an experiment that consists of repetitions of a coin toss (primitive activity).
- **Trial**: single execution of the primitive activity, e.g., a single toss of the coin. When only two values are possible for the outcome of each trial (e.g., head or tail, or 0 or 1) and their probabilities are constant over the number of trials that constitute the experiment, the trials are said to be *independent* and are called *Bernoulli* trials.
- **Outcome**: the result of a trial, e.g., the state of the tossed coin (head or tail).
- **Sample point** (s_i): value of the outcome, e.g., head or tail.

- **Sample space** (S): the universal set of all sample points generated by an experiment. In the coin-tossing experiment, sample space consists of two subsets: one contains all sample points that are heads, and the other contains all sample points that are tails. Sample space is described as discrete or continuous:
  - *discrete sample space*: Contains a finite set (or countably infinite set) of sample points
  - *continuous sample space*: Contains a continuum of sample points (infinite set)
- **Event**: a subset of sample space that contains one, or more, sample points distinguished from the rest by a specific property. Thus, the event *heads* corresponds to one of the domains identified above; the event *tails* (or *not* heads) corresponds to the other. An event may be described as
  - *a simple event*: A single sample point
  - *a compound event*: The aggregate of several related sample points, e.g., in Figure 6.2, subset A, B, C, or D could represent compound events

Using these constructs, we define probability as follows:

- **Probability**: in the context of a specific experiment, the number assigned to an event to describe the likelihood of its occurrence

Probability is calculated in three ways:

- *a posteriori* **probability**: if after N trials, the event A is observed to occur n_A times, the probability of event A [denoted by $P(A)$] is

$$P(A) = n_A/N \qquad (6.1)$$

For example, if, in tossing a coin 100 times, the simple event *heads* is observed to occur 40 times, the *a posteriori* probability of the event heads is $40/100 = 0.4$. Because N trials occur, and only one outcome (sample point) results from each trial

$$\sum_{i=1}^{N} s_i = S, \text{ and } \sum_{i=1}^{N} P(s_i) = 1 \qquad (6.2)$$

i.e., the sum of all sample points is sample space (S), and the sum of the probabilities of all outcomes is 1. Sometimes, in recognition of the experimental nature of *a posteriori* probability, it is called *empirical* probability.

- *a priori* **probability**: if all outcomes of an experiment are equally likely, and an event A can occur in n_A different ways out of the total of n possible ways, the probability of the event A is

$$P(A) = n_A/N \qquad (6.3)$$

For example, in a coin-tossing experiment: *heads* and *tails* are equally likely, the simple event *heads* occurs in only one way, and a total of two possible events exists (heads or tails). Hence, the *a priori* probability of heads is $1/2$. Because $0 \leq n_A \leq n$, $P(A)$ is valued between 0 and 1. When $n_A = 0$, $P(A) = 0$ and A never occurs (it is said to be *impossible*). When $n_A = n$, $P(A) = 1$ and A occurs all the time (it is said to be *certain*). As N becomes very large, the *a posteriori* value given by equation (6.1) approaches the *a priori* value of equation (6.3). Sometimes, in recognition of the analytical nature of *a priori* probability, it is called *classical* probability.

- **Axiomatic probability**: using the nomenclature of *sets*, with each event A in a given class of events, we associate a number $P(A)$ such that

$$P(A) \geq 0 \tag{6.4}$$

$$P(S) = 1 \text{ ; where S is sample space, and } A \subset S \tag{6.5}$$

$$P(A_1 \cup A_2 \cup ...) = P(A_1) + P(A_2) + ... \tag{6.6}$$

where A_1, A_2, ... etc., are mutually exclusive events.

Under these conditions, P is called a *probability* function, and $P(A)$ is the probability of the event A. From these axioms, it follows that

$$0 \leq P(A) \leq 1 \tag{6.7}$$

and

$$P(A) = \sum_A P(s_i) \tag{6.8}$$

i.e., the probability of event A is the sum of the probabilities of the sample points contained within A.

EXAMPLE 6.1   *A Priori* and *A Posteriori* Probabilities

Given Six-bit words are formed from a stream of 0s and 1s that are distributed in a random fashion.

Problem Determine the *a priori* probabilities of words containing 1, 2, 3, 4, 5, and 6 1s.

Solution By listing the set of all possible 6-bit words (64), we identify all of the sample points of the experiment. Because they are generated by a random stream of symbols, all 64 possible words will be equally likely. In **Figure 6.3**, we show sample space divided into seven events that contain sample points related by the number of 1s they contain. Each is a subset of sample space and represents Event (n,1). They range from sets with one member to a set with 20 mem-

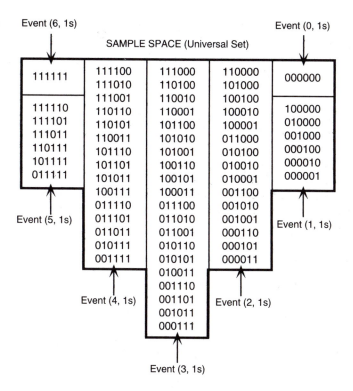

Figure 6.3 Derivation of *a priori* Probability of Words with a Given Number of 1s

In the sample space containing all possible 6-bit words, 64 sample points exist. They can be divided into six events that contain sample points related by the number of 1s they contain. Each event is a subset of sample space. They are finite, countable, and mutually exclusive.

bers. In all cases they are finite, countable, and mutually exclusive. When all outcomes are equally likely and all events are mutually exclusive, the probability of Event ($n,1$) is the number of ways in which the event ($n,1$) can happen divided

by the sum of the ways in which it can *and* cannot happen, i.e., the number of sample points in subset (*n*,1) divided by the total number of sample points in sample space. We calculate P(*word with given number of 1s*) in Figure 6.3. For instance, the number of words in which three 1s occur is 20. This is 20/64 = 31.25% of the total. Thus, the *a priori* probability of event (3, 1) occurring is 0.3125. The number of words in which two 1s occur is 15, so that P(*word with two 1s*) = 15/64 = 0.234375, and so on.

Comment Another way of achieving our goal is to generate a string of 6-bit words at random and assign them to sample space in Figure 6.3. To do this, we used a table of random numbers to create 256, 6-bit words and calculated *a posteriori* P(*word with given number of 1s*) by dividing the number of samples in each event by the total number of trials performed. In **Figure 6.4** the *a posteriori* probabilities of the six events are compared with the *a priori* probabilities of Figure 6.3. Not surprisingly, differences exist. For instance: *a poste-*

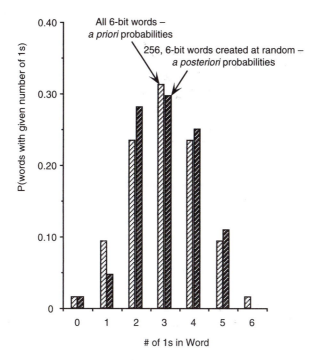

Figure 6.4 Comparison of *a priori* and *a posteriori* Probabilities of Words with Number of 1s in 6-Bit Words
The a posteriori *probabilities are calculated from a run of 256, 6-bit words. Because 256 trials is a small number for an experiment with 6 events and 64 outcomes,* a priori *and* a posteriori *values do not agree.*

riori P(*word with one 1*) is less than the *a priori* value; *a posteriori* P(*word with two 1s*) is greater than the *a priori* value; and no words with six 1s exist. (The *a priori* probability of such a word appearing is 0.015625; thus, in a random sequence of 256 words, 4 should occur.) Such discrepancies are to be expected; 256 is a relatively small number of trials for an experiment with 6 events and 64 possible outcomes.

EXAMPLE 6.2   4-Bit Words

Given 4-bit words are formed in a continuous fashion from a stream of binary symbols that contains 0s and 1s distributed in a random fashion.

Problem Calculate the *a priori* probabilities of 0, 1, 2, 3, and 4, 1s in the 4-bit words.

Solution

Word	# of 1s	# of Words with	*a priori* P(n, 1s)
0000	0	0, 1s = 1	0.0625
0001	1	1, 1s = 4	0.25
0010	1	2, 1s = 6	0.375
0011	2	3, 1s = 4	0.25
0100	1	4, 1s = 1	0.0625
0101	2		
0110	2	# of All Words = 16	
0111	3		
1000	1		
1001	2		
1010	2		
1011	3		
1100	2		
1101	3		
1110	3		
1111	4		

---

(2) Joint Probabilities

In some situations, the sample space can be divided into mutually exclusive events (as in Figure 6.3); in other situations, some elements are common to two, or more, events (as in Figure 6.2). In Figure 6.2

$$(b+c)=(a+b+c+d)+(b+c+f+g)-(a+b+c+d+f+g)$$

Each of the areas is part of event A or event B. Converting to subsets we have

$$A \cap B = A + B - (A \cup B)$$

Dividing both sides by the universal set gives

$$P(A \cap B) = P(A) + P(B) - P(A \cup B) \tag{6.9}$$

where $P(A \cap B)$ is the *joint* probability of events A *and* B. In a similar fashion, we can develop the expression

$$P(A \cup B) = P(A) + P(B) - P(A \cap B) \tag{6.10}$$

where $P(A \cup B)$ is the *joint* probability of events A *or* B. If A and B are mutually exclusive, $A \cap B = \varnothing$. Substituting in equation (6.10) yields

$$P(A \cup B) = P(A) + P(B) \tag{6.11}$$

i.e., *the probability of mutually exclusive events is the sum of their individual probabilities.*

(3) Conditional Probabilities

In other situations, the probability of event A may depend on the occurrence of event B. We define the *conditional* probability of an event A given event B has occurred as

$$P(A|B) = \frac{P(A \cap B)}{P(B)} \tag{6.12}$$

where $P(A|B)$ is used to represent the statement: *the probability of event A occurring when event B has occurred.* If A and B are mutually exclusive, $(A \cap B) = \varnothing$, and $P(A|B) = 0$. If A and B are independent events, A does not depend on B, so that $P(A|B) = P(A)$, and

$$P(A \cap B) = P(A)P(B) \tag{6.13}$$

i.e., *the probability of independent events occurring is the product of their individual probabilities.*

Using equation (6.12), we can write

$$P(A \cap B) = P(A|B)P(B) \text{ and } P(B \cap A) = P(B|A)P(A) \tag{6.14}$$

and, since $P(A \cap B) = P(B \cap A)$

$$P(B|A)P(A) = P(A|B)P(B)$$

or

$$P(B|A) = \frac{P(A|B)P(B)}{P(A)} \tag{6.15}$$

Equation (6.15) is known as *Bayes Rule* or *Bayes Theorem* (after Thomas Bayes, an 18th-century English philosopher). It relates the *a priori* probabilities of A or B occuring and the *a posteriori* probabilities of event B occurring, given event A has occurred, and of event A occurring, given event B has occurred. Example 6.3 illustrates its use.

EXAMPLE 6.3   Binary Symmetric Channel

Given A binary communication system consists of a transmitter that sends 0s or 1s to a receiver over a communication channel. Occasionally, errors occur so that when a 1 is sent, a 0 is received, and vice-versa. **Figure 6.5** shows the physical model and the key probabilities that govern the communication process. In the diagram, p_0 and p_1 are the probabilities of sending 0s or 1s, and p is the probability of receiving 0 when 1 is sent, or receiving 1 when 0 is sent.

Problem

    a) Develop expressions for $P(R_1)$, the probability of receiving 1s, and $P(R_0)$, the probability of receiving 0s.

    b) Find expressions for $P(S_1|R_1)$ and $P(S_0|R_0)$, i.e., given 1 is received, what is the probability that 1 was sent, and given 0 is received, what is the probability that 0 was sent?

    c) Find an expression for the probability of error of the system.

Solution

    a) The event R_1, 1s received, has two components: $S_1|R_1$, 1s are received when 1s were sent, and $S_0|R_1$, 1s are received when 0s were sent. Together, they represent all possible ways in which R_1 can occur; hence, their probabilities add to 1, i.e.,

$$P(S_1|R_1)+P(S_0|R_1)=1$$

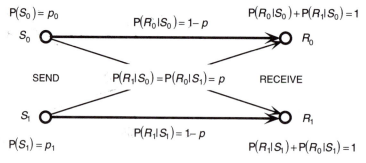

Figure 6.5 Binary Symmetric Channel (Example 6.3)
A binary communication channel consists of a transmitter sending 0s and 1s to a receiver; it can be modeled as shown. Performance depends on the conditional probabilities of receiving 1 or 0 given 1 or 0 was sent, and receiving 0 or 1 given 1 or 0 was sent.

Expanding the conditional probabilities with Bayes Rule (equation 6.15) and rearranging with respect to $P(R_1)$, we obtain the probability of receiving 1s.

$$P(R_1) = P(R_1|S_1)P(S_1) + P(R_1|S_0)P(S_0) = (1-p)p_1 + pp_0$$

In a similar manner, we can find the expression for the probability of receiving 0s.

$$P(R_0) = P(R_0|S_0)P(S_0) + P(R_0|S_1)P(S_1) = (1-p)p_0 + pp_1$$

b) Using Bayes Rule,

$$P(S_1|R_1) = \frac{P(R_1|S_1)P(S_1)}{P(R_1)} = \frac{(1-p)p_1}{(1-p)p_1 + pp_0}$$

$$P(S_0|R_0) = \frac{P(R_0|S_0)P(S_0)}{P(R_0)} = \frac{(1-p)p_0}{(1-p)p_0 + pp_1}$$

c) The probability of error is $P(R_1|S_0) \cup P(R_0|S_1)$. From equation (6.10)

$$P(R_1|S_0) \cup P(R_0|S_1) = P(R_1|S_0) + P(R_0|S_1) - P(R_1|S_0) \cap P(R_0|S_1)$$

But $P(R_1|S_0)$ and $P(R_0|S_1)$ are mutually exclusive so that $P(R_1|S_0) \cap P(R_0|S_1) = 0$; hence

$$P(R_1|S_0) \cup P(R_0|S_1) = P(R_1|S_0) + P(R_0|S_1) = 2p$$

Extension When $p_0 = p_1$, i.e., 0s and 1s are equally likely

$$P(S_1|R_1) = P(S_0|R_0) = (1-p)$$

6.3 PROBABILISTIC FUNCTIONS

Variables whose values are expressed in terms of probability of occurrence give rise to probabilistic functions.

(1) Random Variable

- **Random variable**: also known as a *stochastic* variable and usually denoted by a capital letter $(X, Y, \text{etc.})$, a function, $X(s_i)$, that assigns a real number to every sample point

The *rule* employed in the transformation from sample points (s_1, s_2, ... s_n) in sample space to numbers (x_1, x_2, ... x_k) on the number line defines the random variable. Sample points may share the same value (i.e., x_j may be assigned to several sample points); however, no sample point may have more than one value. The action of a random variable in a specific situation (Example 6.3) is conceptualized in **Figure 6.6**.

EXPERIMENT: Find average number of calls in the hour, and determine CDF and PDF of calling activity.
TRIAL: Measure number of simultaneous calls at two-minute intervals. The results are shown below, left.

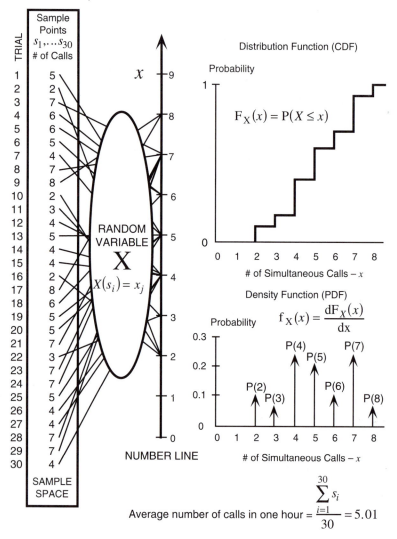

$$\text{Average number of calls in one hour} = \frac{\sum_{i=1}^{30} s_i}{30} = 5.01$$

Figure 6.6 Application of Probabilistic Functions to Analysis of Telephone Calling Activity
The random variable makes the transformation from sample space to the number line so that the PDF and CDF of the activity can be determined.

Random variables are divided into *discrete* random variables—they form a finite countable set of values on the real number line, and *continuous* random variables—they form a finite uncountable set of values on the real number line.

(2) Distribution Function

- **Distribution function**: also known as *cumulative probability* distribution function (CDF), a real valued function related to the probability of occurrence of x_1, x_2, ... etc., on the real number line. Usually denoted by $F_X(x)$, it is the probability of the event $(X \leq x)$, i.e.,

$$F_X(x) = P(X \leq x). \tag{6.16}$$

The distribution function has the following properties:

$$0 \leq F_X(x) \leq 1$$
$$F_X(-\infty) = 0; F_X(\infty) = 1$$
$$F_X(x_1) \leq F_X(x_2) \text{ when } x_1 \leq x_2 \tag{6.17}$$
$$P(x_1 < X \leq x_2) = F_X(x_2) - F_X(x_1)$$
$$\lim_{\varepsilon \to 0} F_X(x + \varepsilon) = F_X(x)$$

Thus, $F_X(x)$ is a function that has values between 0 and 1, increases from left to right, and is continuous.

If the underlying random variable is *discrete,* the distribution function will have a step-like distribution function

$$F_X(x) = \sum_{j=1}^{N} P(x_j) U(x - x_j) \tag{6.18}$$

where $U(\ )$ is the unit step function and $P(x_j)$ is the discrete change in the distribution function at the point x_j. An example of a discrete distribution function is shown in Figure 6.6. If the underlying random variable is *continuous,* the distribution function will be continuous. If the random variable is *mixed,* i.e., part discrete and part continuous, the random variable will have a distribution function that exhibits steps joined by continuous segments.

(3) Density Function

- **Density function**: also known as *probability* density function (PDF), a real valued function related to the probability of occurrence of x_1, x_2, ... etc., on the real number line. Usually denoted by $f_X(x)$, it is the derivative of the distribution function, i.e.,

$$f_X(x) = \frac{dF_X(x)}{dx} \qquad (6.19)$$

The density function has the following properties

$$0 \le f_X(x); \; 0 \le x < \infty$$

$$\int_{-\infty}^{\infty} f_X(x)dx = 1 \qquad (6.20)$$

$$P(x_1 < X \le x_2) = \int_{x_1}^{x_2} f_X(x)dx$$

Thus, $f_X(x)$ is a function that is always positive, and of unit area.

If the underlying random variable is *discrete*

$$f_X(x) = \frac{dF_X(x)}{dx} = \sum_{j=1}^{k} P(x_j)\delta(x - x_j) \qquad (6.21)$$

so that the density function will consist of impulse functions of area $P(x_j)$ at $x_1, x_2, \dots x_k$. An example of a discrete density function is shown in Figure 6.6. If the underlying random variable is *continuous* or *mixed,* the density function will exist where the derivatives of the distribution function exist.

EXAMPLE 6.4   Telephone Calling Activity

Given   In an office, telephones are in use sporadically throughout the day. During the busiest hour, observations are made every two minutes. They show the number of calls in progress to be 5, 2, 7, 6, 6, 5, 4, 7, 8, 2, 3, 4, 5, 4, 4, 2, 8, 6, 5, 5, 7, 3, 7, 7, 5, 4, 4, 7, 7, and 4.

Problem   Determine the distribution and density functions of the calling activity in the busy hour.

Solution   Figure 6.6 shows the steps employed. They are

1) *Define Sample Space.* Because we measure activity every two minutes for one hour, our experiment consists of 30 trials. The outcome of each trial is a sample point that represents the number of calls in progress at the time of the trial. Because 30 trials occur, 30 sample points exist (s_1, $s_2, \dots s_{30}$).

2) *Form the Number Line.* Using the notion of random variable, each sample point is mapped onto the number line so that all points are arrayed in order (ascending magnitude).

3) *Form the Distribution Function.* The distribution function is the probability of the event $X < x$. On the number line, the thirty values (x_1, x_2, ... x_{30}) are distributed on the integers 2 through 8. Starting from the bottom:
- No sample points are mapped below 2; hence $P(X < 2) = 0$
- Three sample points (s_2, s_{10} and s_{16}) are mapped at 2; hence
$$P(X < 3) = \frac{3}{30} = 0.1$$

- Two sample points (s_{11} and s_{22}) are mapped at 3; hence
$$P(X < 4) = \frac{3}{30} + \frac{2}{30} = 0.167$$

- Seven sample points (s_7, s_{12}, s_{14}, s_{15}, s_{26}, s_{27}, and s_{30}) are mapped at 4; hence
$$P(X < 5) = \frac{3}{30} + \frac{2}{30} + \frac{7}{30} = 0.4$$

and so on.

The distribution function is a staircase in which the steps represent the sample points added at each integer. The maximum probability is 1 when $x \geq 8$.

4) *Form the Density Function.* The density function is the derivative of the distribution function with respect to x. Differentiating the distribution function yields the impulse functions shown. Their areas represent the probabilities of x simultaneous calls.

---

(4) Expectation

- **Expectation**: denoted by $E[X]$, it is the mean or expected value of the random variable $X(s_i)$. When the random variable assigns numbers x_1, x_2, ... x_n on the number line with probabilities $P(x_1)$, $P(x_2)$, ... $P(x_n)$
$$E[X] = \sum_n x_k P(x_k) \tag{6.22}$$

As a matter of convenience, the mean value of a random variable is often denoted by μ or $\bar{X}$, i.e., $\mu = \bar{X} = E[X]$.

It follows from the definition of the density function that
$$f_X(x_k) = P(X = x_k)$$
so that, when X is discrete
$$E[X] = \sum_n x_k f_X(x_k) \tag{6.23}$$

and, when X is continuous

$$E[X] = \int_{-\infty}^{\infty} x f_X(x) dx \tag{6.24}$$

In equation (6.22), if x_1, x_2,... x_n are randomly distributed so that they are equiprobable, i.e., $P(x_k) = 1/n$, we obtain the *arithmetic* mean of the values

$$E[X] = (x_1 + x_2 + ... + x_n)/n \tag{6.25}$$

Commonly, this is called the *average* value.

When a is a constant, X_1, X_2, ... X_n are random variables whose expectations exist, and X and Y are mutually independent random variables, the expectation function has the following properties:

$$E[aX] = a E[X] \tag{6.26}$$

i.e., the expectation of a constant times a random variable is equal to the constant times the expectation of the random variable,

$$E[X_1 + X_2 + ... + X_n] = E[X_1] + E[X_2] + ... + E[X_n] \tag{6.27}$$

i.e., the expectation of the sum of a number of random variables is the sum of their individual expectations, and

$$E[XY] = E[X] E[Y] \tag{6.28}$$

i.e., the expectation of the product of two independent random variables is the product of their individual expectations.

EXAMPLE 6.4   Telephone Calling Activity (Continued)

Further Problem Find the average number of simultaneous calls in the busy hour.

Solution Form the average using equation (6.25). Over the 30 trials, the value of all observations was 153. Hence, the average observation was $153/30 = 5.1$ simultaneous calls.

---

(5) Variance, Standard Deviation, and Standardized Random Variable

- **Variance**: a measure of the *dispersion* or *scatter* of the values of the random variable about the mean

$$Var(X) = E\left[(X - \mu)^2\right] \tag{6.29}$$

$\text{Var}(X)$ is a non-negative number, i.e., $0 \le \text{Var}(X)$. When X is discrete

$$\text{Var}(X) = \sum_n (x_k - \mu)^2 f_X(x_k) \tag{6.30}$$

and, when X is continuous

$$\text{Var}(X) = \int_{-\infty}^{\infty} (x_k - \mu)^2 f_X(x) dx \tag{6.31}$$

Figure 6.7 illustrates the difference between continuous density functions with large and small variances. In equation (6.30), if the probabilities are all equal, we obtain the variance of the set of numbers $x_1, x_2, \dots x_n$, i.e.,

$$\text{Var}(n) = \frac{1}{n}\left[(x_1 - \mu)^2 + (x_2 - \mu)^2 + \dots + (x_n - \mu)^2\right] \tag{6.32}$$

- **Standard deviation**: the positive square root of $\text{Var}(X)$

$$\sigma_X^2 = \text{Var}(X) \text{ or } \sigma_X = \sqrt{\text{Var}(X)} \tag{6.33}$$

- **Standardized random variable**: if X is a random variable with mean μ and standard deviation σ, we create the standardized random variable $X*$ by subtracting the mean and dividing by the standard deviation, i.e.,

$$X* = (X - \mu)/\sigma . \tag{6.34}$$

The standardized random variable is a dimensionless quantity situated about the origin (zero mean) with $X*$ measured in units of standard deviation. **Figure 6.8** shows an example of an *actual* distribution function with mean 4 and standard deviation 2 and its standardized counterpart. In tables of numerical values, the standard function is tabulated.

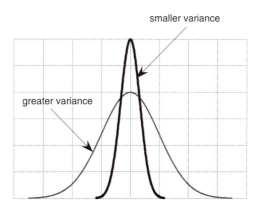

Figure 6.7 Meaning of Variance with Respect to Statistical Distributions

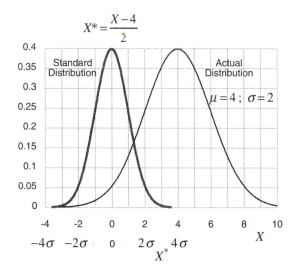

Figure 6.8 Comparison of Actual and Standardized Distributions of a Random Variable

A standardized random variable is a zero mean dimensionless quantity. In tables of numerical values, tabulating the standardized function is usual.

EXAMPLE 6.4   Telephone Calling Activity (Continued)

Further Problem Find the variance and standard deviation of the calls in the busy hour.

Solution

1) Using equation (6.30) and $\mu = 5$, find and sum the individual components of the variance function.

x_j	$f_X(x_j)$	$(x_j - \mu)^2$	$(x_j - \mu)^2 f_X(x_j)$
2	$\frac{3}{30} = 0.100$	9	0.9
3	$\frac{2}{30} = 0.067$	4	0.268
4	$\frac{7}{30} = 0.233$	1	0.233
5	$\frac{6}{30} = 0.200$	0	0
6	$\frac{3}{30} = 0.100$	1	0.100
7	$\frac{7}{30} = 0.233$	4	0.932
8	$\frac{2}{30} = 0.067$	9	0.603

$$\sum_{2}^{8} = 3.063$$

i.e., the variance of the busy hour calls is 3.063 erlangs.

2) Using equation (6.33), the standard deviation is given by

$$\sigma = \sqrt{\text{Var}(X)} = 1.750$$

i.e., the standard deviation of the busy hour calls is 1.750 erlangs.

EXAMPLE 6.5　More Telephone Calling Activity

Given The experiment of Example 6.4 is repeated. The 30 new sample points are: 6, 7, 7, 5, 3, 3, 6, 6, 4, 4, 5, 5, 7, 5, 3, 6, 7, 8, 2, 7, 4, 3, 6, 8, 2, 5, 3, 5, 4, and 3.

Problem

a) Find the average number of simultaneous calls in the hour

b) Find the CDF and PDF of the calling activity

c) To illustrate the function performed by the random variable, construct a diagram similar to Figure 6.6.

Solution

a) Total number of calls observed in 30 samples = 149. Thus, the average number of simultaneous calls over the entire sampling time (i.e., 1 hour) = 149/30 = 4.97 calls.

b) Sort the sample points to determine the number of trials that result in a given number of simultaneous calls. Multiply the number of trials by the number of calls for each event and divide by 149 to give the PDF. Sum the PDF values to give the CDF.

Event (*n* calls)	# of Trials	PDF	CDF
0	0	0	0
1	0	0	0
2	2	0.0268	0.0268
3	6	0.1208	0.1476
4	4	0.1074	0.2550
5	6	0.2013	0.4563
6	5	0.2013	0.6576
7	5	0.2350	0.8926
8	2	0.1074	1.0000

c) The PDF and CDF of the calling activity and the function performed by the random variable appear in **Figure 6.9**.

Comment With so few trials, the result that the density distribution functions in Figures 6.6 and 6.9 are significantly different is no surprise.

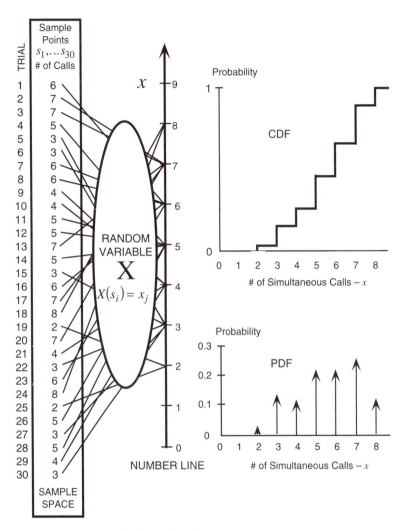

Figure 6.9 Diagram for Example 6.5

(6) Moments

The expression $E\left[(X-\mu)^2\right]$ on the right-hand side of equation (6.29) is described as the *2nd moment* of the random variable X about the mean. Also, it is called the 2nd *central* moment. In general

- **rth central moment**: the rth moment of a random variable X about the mean

$$\mu_r = E\left[(X - \mu)^r\right] \tag{6.35}$$

When X is discrete

$$\mu_r = \sum_n (x_j - \mu)^r f_X(x_j) \tag{6.36}$$

and, when X is continuous

$$\mu_r = \int_{-\infty}^{\infty} (x - \mu)^r f_X(x) dx \tag{6.37}$$

In a fashion analogous to (6.35), the *r*th moment of the random variable X *about the origin* is

$$\mu'_r = E\left[X^r\right]$$

The first five central moments have special significance:

- The zero central moment is 1, i.e.,

$$\mu_0 = E[1] = 1.$$

- The first central moment is the mean, i.e.,

$$\mu_1 = E[X] = \mu$$

- The second central moment is the variance, i.e.,

$$\mu_2 = E\left[(X - \mu)^2\right] = \sigma^2 \text{ or } \text{Var}(X)$$

- The third central moment gives a measure of *skewness*—the symmetry (or lack of symmetry) about the mean. Thus,

$$\mu_3 = E\left[(X - \mu)^3\right]$$

α_3, the coefficient of skewness $= \mu_3 / \mu_2^{3/2} = \mu_3 / \sigma^3$. In a normal distribution [see § 7.3(1)], $\alpha_3 = 0$. When $\alpha_3 < 0$, the greater values are concentrated on the left of the mean, and when $\alpha_3 > 0$, the greater values are concentrated on the right of the mean.

- The fourth central moment gives a measure of *kurtosis*—a term that describes the concentration of values around the mean. Thus,

$$\mu_4 = E\left[(X - \mu)^4\right]$$

α_4, the coefficient of kurtosis $= \mu_4/\mu_2^2 = \mu_4/\sigma^4$. In a normal distribution, $\alpha_4 = 3$. When $\alpha_4 < 3$, the values are less heavily concentrated about the mean, and when $\alpha_4 > 3$, the values are more heavily concentrated about the mean.

(7) Moment Generating Function of a Random Variable

A way to find the rth moment of a random variable is to use the moment generating function.

- **Moment generating function of a random variable**: for a random variable with density function $f_X(x)$, the moment generating function $M_X(t)$ is the expected value of $\exp(tx)$, where t is a real number $(-\infty < t < \infty)$, i.e., when X is continuous

$$M_X(t) = E[\exp(tx)] = \int_{-\infty}^{\infty} \exp(tx) f_X(x)\, dx \tag{6.38}$$

and, when X is discrete

$$M_X(t) = \sum_{j=1}^{n} \exp(tx_j) f_X(x_j) \tag{6.39}$$

The values of the moments of random variable X can be determined by expanding the exponential function in a Taylor series (after Brook Taylor, a late-17th- and early-18th-century English mathematician)

$$M_X(t) = 1 + tE[X] + \tfrac{t^2}{2!}E[X^2] + \tfrac{t^3}{3!}E[X^3] + \dots$$

so that

$$M_X(t) = 1 + \mu't + \mu_2' t^2/2! + \mu_3' t^3/3! + \dots$$

To find μ_r', we differentiate the above expression r times and put $t = 0$ to give

$$\mu_r' = \frac{d^r}{dt^r} M_X(t)\Big|_{t=0} \tag{6.40}$$

(8) Characteristic Function of a Random Variable

Another way to find the rth moment of a random variable is to use the characteristic function.

- **Characteristic function of a random variable**: for a random variable with density function $f_X(x)$, the characteristic function $\phi_X(t)$ is the expected value of $\exp(jtx)$, where t is a real number $(-\infty < t < \infty)$.

$$\phi_X(t) = E\left[\exp(jtx)\right] = \int_{-\infty}^{\infty} \exp(jtx) f_X(x) dx \qquad (6.41)$$

The right-hand side of equation (6.41) shows that $\phi_X(t)$ is the Fourier transform of $f_X(x)$ and that the characteristic function can be found by taking the inverse Fourier transform of the density distribution, i.e.,

$$f_X(x) = \int_{-\infty}^{\infty} \phi_X(t) \exp(-jtx) dx \qquad (6.42)$$

In a manner similar to that used to derive equation (6.40), we can use the characteristic function to find μ_r'; we expand the exponential function, differentiate the expression r times, and put $t = 0$, i.e.,

$$\mu_r' = (-j)^r \frac{d^r}{dt^r} \phi_X(t)\Big|_{t=0} \qquad (6.43)$$

The use of $M_X(t)$ or $\phi_X(t)$ to find the moments of a random variable will depend on a number of factors, not the least of which is $\phi_X(t)$ always exists, whereas $M_X(t)$ may not.

REVIEW QUESTIONS

6.1 What is a *set*?

6.2 What is a *Venn* diagram?

6.3 Define the union of sets Q and R.

6.4 Define the intersection of two sets Q and R.

6.5 Define *probability*.

6.6 Distinguish between *a priori* and *a posteriori* probability.

6.7 What is the probability of mutually exclusive events occuring?

6.8 What is the probability of independent events occuring together?

6.9 Define *conditional probability*.

6.10 State Bayes rule.

6.11 Define a *random variable.*

6.12 Define a *cumulative probability distribution function.*

6.13 Define a *probability density function.*

6.14   Define the *expectation* of a random variable.

6.15 Define the *variance* of a random variable.

6.16 Define the *standard variation* of a random variable.

6.17 Define a *standardized random variable.*

6.18 Describe the *r*th central moment of a random variable.

6.19 Distinguish between the moment generating function of a random variable and its characteristic function.

7

Random Variables

The distribution of the values of a random variable can be represented by many analytic functions. In this chapter, I discuss three general purpose distributions—Binomial, Poisson, and Normal—and three specialized distributions—Uniform, Exponential and Erlang.

7.1 BINOMIAL-DISTRIBUTED RANDOM VARIABLE

When a random variable can assume only two values, and every outcome is independent of every other outcome, the distribution of values can be represented by the binomial function.

(1) Bernoulli Trials and Binomial Coefficients

A Bernoulli trial (after Jacob Bernoulli, a 17th-century Swiss mathematician) is one in which the outcome is two-valued and the probability of each remains constant throughout the experiment (series of trials). Usually, the states are designated *success* (S) and *failure* (F) and

$$P(S) = p; \ P(F) = q = (1 - p)$$

For an experiment of n independent trials, the probability of success of any sample point is

$$P(s_i) = p^k (1 - p)^{n-k} \tag{7.1}$$

where k is the number of successes in n trials. If our interest is in the number of successes in n trials and not in the position of each success in the sequence of trials, the probability of k successes will be the sum of the number of ways in which they can be achieved multiplied by the probability of a single outcome. The former is the binomial coefficient (i.e., the number of ways in which k thing can be taken from n things); the latter is given by equation (7.1), so that

$$b(k;n,p) = P(S=k) = C_k^n\, p^k (1-p)^{n-k} \tag{7.2}$$

where C_k^n is the binomial coefficient. Sometimes written $\binom{n}{k}$, it is equal to

$$C_k^n = \frac{n!}{k!(n-k)!}$$

The notation $b(k:n,p)$ is used to mean the probability that n Bernoulli trials with $P(S)=p$ and $P(F)=q=(1-p)$ result in k successes. For $k=0$, substituting in equation (7.2) gives

$$b(0;n,p) = (1-p)^n \tag{7.3}$$

and for k and $k-1$, we can write the recursive relationship

$$\frac{b(k;n,p)}{b(k-1;n,p)} = \frac{\frac{n!}{k!(n-k)!}p^k q^{n-k}}{\frac{n!}{(k-1)!(n-k+1)!}p^{k-1} q^{n-k+1}} = \frac{(n-k+1)p}{kq} \tag{7.4}$$

(2) Binomial-Distributed Random Variable

A discrete random variable is called *binomial* when the probability density function is

$$f_X(x) = \sum_{k=0}^{n} b(k;n,p)\,\delta(x-k) \tag{7.5}$$

and the distribution function is

$$F_X(x) = \int f_X(x)\,dx = \sum_{k=0}^{n} b(k;n,p)\,U(x-k) \tag{7.6}$$

Binomial-distributed random variables can be shown to have

Mean $= \mu = np$

Variance $= \sigma^2 = npq$

Characteristic function $= \phi_X(t) = [q + p\exp(jt)]^n$

If p and q are not close to zero, and $n\to\infty$, in the limit, equation (7.5) becomes *continuous* and

$$\lim_{n\to\infty} P(a \le X^* \le b) = \frac{1}{\sqrt{2\pi}} \int_a^b \exp\!\left(-u^2/2\right) du \tag{7.7}$$

EXAMPLE 7.1 Binomial PDF

Given A random stream of 0s and 1s is divided into segments six bits long.

Problem Find the probabilities of segments that contain 0, 1, 2, 3, 4, 5, and 6 1s, i.e., find the analytical description of the PDF for the situation described in Example 6.1.

Solution Because the experiment is concerned with some property of a *random* stream of 0s and 1s, $p = (1-p) = 1/2$, and

$$f_X(x) = \sum_{k=0}^{n} C_k^n 2^{-n} \delta(x-k)$$

Evaluating $C_k^6 2^{-6}$ when $k = 0, 1, 2, ..., 6$.

k	$C_k^6 2^{-6}$
0	0.015625
1	0.09375
2	0.234375
3	0.3125
4	0234375
5	0.09375
6	0.015625

Substituting values in equation (7.5)

$$f_X(x) = 0.016\delta(x) + 0.094\delta(x-1) + 0.234\delta(x-2) + 0.312\delta(x-3)$$
$$+0.234\delta(x-4) + 0.094\delta(x-5) + 0.0168\delta(x-6)$$

In **Figure 7.1**, I plot both PDF and CDF. The probabilities $P(x, 1s)$ are the coefficients of the terms in the PDF.

7.2 POISSON-DISTRIBUTED RANDOM VARIABLE

For large numbers of trials, evaluating the binomial function becomes overwhelming. Fortunately, under certain circumstances, it can be approximated by a Poisson distribution.

(1) Poisson Approximation to Binomial Distribution

When the number of trials (n) is large, and the probability of success (p) is small, to seek approximate solutions is convenient. Putting $\lambda = np$, and substituting in equation (7.2).

$$f_X(x) = 0.016\,\delta(x) + 0.094\,\delta(x-1) + 0.234\,\delta(x-2) + 0.312\,\delta(x-3)$$
$$+0.234\,\delta(x-4) + 0.094\,\delta(x-5) + 0.0168\delta(x-6)$$
$$F_X(x) = 0.016\,U(x) + 0.094\,U(x-1) + 0.234\,U(x-2) + 0.312\,U(x-3)$$
$$+0.234\,U(x-4) + 0.094\,U(x-5) + 0.0168\,U(x-6)$$

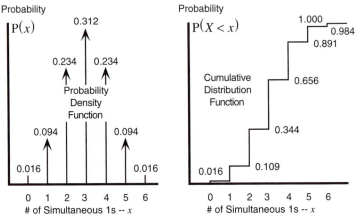

Figure 7.1 Probability Density Function and Cumulative Distribution Function for Example 7.1

$$b(0;n,p) = (1-\lambda/n)^n \qquad (7.8)$$

Taking logarithms of both sides and expanding the right-hand side as a Taylor Series

$$\log b(0;n,p) = \log(1-\lambda/n)^n = -\lambda - \lambda^2/2n - \lambda^3/3n - \ldots \qquad (7.9)$$

So that for large n

$$b(0;n,p) \cong \exp(-\lambda) \qquad (7.10)$$

In equation (7.2), putting $\lambda = np$, with $n \rightarrow \infty$ and $p \rightarrow 0$, so that λ is finite

$$\frac{b(k;n,p)}{b(k-1;n,p)} = \frac{\lambda - (k-1)p}{kq} \cong \frac{\lambda}{k} \qquad (7.11)$$

Hence

$$b(0;n,p) = \exp(-\lambda)$$
$$b(1;n,p) = \lambda \exp(-\lambda)$$
$$b(2;n,p) = \tfrac{1}{2}\lambda^2 \exp(-\lambda)$$

and, generally

$$b(k;n,p) = \frac{\lambda^k}{k!}\exp(-\lambda)$$

This is the Poisson approximation to the binomial distribution (after Siméon Poisson, an early-19th-century French mathematician). In shorthand

$$p(k;\lambda) = \frac{\lambda^k}{k!}\exp(-\lambda) \tag{7.12}$$

It is used in the case of *rare* events, i.e., $n \geq 50$ and $\lambda \leq 5$. **Figure 7.2** shows values for the Poisson probability density function for $\lambda = 1, 2,$ and 5.

(2) Poisson-Distributed Random Variable

A discrete random variable is called *Poisson* when the probability density function is

$$f_X(x) = \sum_{k=0}^{\infty} p(k;\lambda)\,\delta(x-k) \tag{7.13}$$

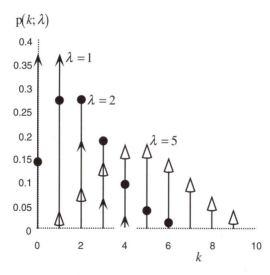

Figure 7.2 Poisson Probability Density Function
The Poisson distribution is an approximation to the binomial distribution for the case of rare events. The probability density function [p(k;λ)] consists of discrete values at k = 0, 1, ... n that vary with the value of λ.

and the distribution function is

$$F_X(x) = \int f_X(x)\,dx = \sum_{k=0}^{\infty} p(k;\lambda)\,U(x-k). \tag{7.14}$$

Poisson-distributed random variables can be shown to have

Mean $= \mu = \lambda$

Variance $= \sigma^2 = \lambda$

Characteristic function $= \phi_X(t) = \exp\left[\lambda\{\exp(jt)-1\}\right]$

If $p \rightarrow 0$, and $\lambda \rightarrow \infty$, the limit equation (7.13) becomes *continuous* and

$$\lim_{\lambda \to \infty} P(a \leq X^* \leq b) = \frac{1}{\sqrt{2\pi}} \int_a^b \exp\left(-u^2/2\right) du \tag{7.15}$$

It is no coincidence that the right-hand side of equation (7.15) is the same as the right-hand side of equation (7.7). I explain in § 7.3(4).

(3) Poisson Process

Suppose we wish to find the probability of exactly k events occurring within a fixed time interval. Consider a sequence of random events. As shown in **Figure 7.3**, they can be represented by points on the *time* line. Further, suppose the time interval $(t \rightarrow t + \tau)$ is divided into n subintervals and that each subinterval has an identical probability (p_n) of being occupied. Because the subintervals are *occupied* or *empty,* they can be considered to be the result of Bernoulli trials, and the probability of exactly k occupied subintervals is $b(k;np_n)$. If $n \rightarrow \infty$ and $p_k \rightarrow 0$ with $np_n = \lambda\tau$, we can employ the Poisson approximation so that the probability of exactly k occupied subintervals is $p(k;\lambda\tau)$. As n increases, the number of occupied cells will be the same as the number of events in the interval (we assume no simultaneous events). Let N be the number of intervals examined, and n_k be the number of times exactly k events are observed; then

$$N = n_1 + n_2 + \ldots + n_k + \ldots \tag{7.16}$$

and the total number of events observed (T) is

$$T = n_1 + 2n_2 + \ldots + kn_k + \ldots \tag{7.17}$$

For large n, $n_k = np(k;\lambda\tau)$, so that, substituting in equation (7.17)

$$T = N\left[p(1;\lambda\tau) + p(2;\lambda\tau) + \ldots + p(k;\lambda\tau) + \ldots\right]$$

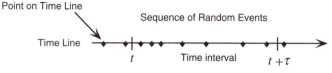

Point on Time Line

Sequence of Random Events

Time Line

t Time interval $t + \tau$

Divide time into n subintervals that have
equal probability of being occupied

Time Line

$p_n = $ probability of being occupied
$b(k;np_n) = $ probability of k subintervals being occupied

If $n \to \infty$, $p_n \to 0$, and $np_n = \lambda\tau$,
probability of k intervals being occupied $= p(k; \lambda\tau)$

Time Line

Number of occupied cells same as number of events

$$p(k; \lambda\tau) = \frac{(\lambda\tau)^k}{k!} \exp(-\lambda\tau)$$

where $\lambda\tau$ is the average number of events
observed in interval τ

Figure 7.3 Conceptual Development of Poisson Process
*Because the Poisson distribution concerns rare events, an important
assumption is that two events do not occur simultaneously. As a consequence, the subintervals are either occupied or empty and can be considered to be the outcomes of Bernouilli trials.*

Expanding the Poisson shorthand

$$T = N\lambda\tau\exp(-\lambda\tau)\left[1 + \lambda\tau + (\lambda\tau)^2/2! + ... + (\lambda\tau)^k/k! + ...\right]$$

so that, because the quantity in [ ] brackets is $\exp(\lambda\tau)$

$$T = N\lambda\tau$$

i.e.,

$$\lambda\tau = T/N \qquad\qquad (7.18)$$

T/N is the average number of events observed in each interval. Hence, the
probability of exactly k events occurring in the interval $(t, t+\tau)$ is

$$p(k;\lambda\tau) = \frac{(\lambda\tau)^k}{k!}\exp(-\lambda\tau) \qquad\qquad (7.19)$$

where $\lambda\tau$ is the average number of events observed in all intervals.

EXAMPLE 7.2   Density of Telephone Calls

Given In Example 6.4, telephone activity in an office was observed over a one-hour period by sampling the number of calls in progress at two-minute intervals. The total call activity was 153, i.e., the sum of the number of calls observed in progress in each two-minute interval.

Problem If the calls are Poisson distributed, construct a density diagram showing the probability of finding exactly k calls in progress in any two-minute interval.

Solution Because $T = 153$ and $N = 30$, $\lambda\tau \cong 5$. Evaluating $p(k;5)$ gives

k	$p(k;5)$
0	0.0067
1	0.0337
2	0.0842
3	0.1404
4	0.1755
5	0.1755
6	0.1462
7	0.1044
8	0.0653
9	0.0363
10	0.0181

These values are plotted in the left half of **Figure 7.4**.

Comments In the right half of Figure 7.4, we plot the values of $p(k;5)$ and compare them with the values computed for the activity observed in Example 6.4. The degree of agreement between the sets of values is not high. The theoretical distribution (Poisson) is more symmetrical and less peaked than the real distribution. Nevertheless, a Poisson distribution may be an acceptable representation of the distribution of calls during a busy hour.

Further Comment In using the Poisson distribution, we have changed the model from the physical one of counting simultaneous calls of unknown duration at specific times, to a mathematical model that is based on the probability of occurrence of independent instantaneous events during the inter-measurement period. In effect, we observe small segments of calls, assume they exist independently, and assume that they occur at times in the inter-measurement period so that simultaneous arrivals do not occur. Because the number of events per interval in the Poisson process is the same

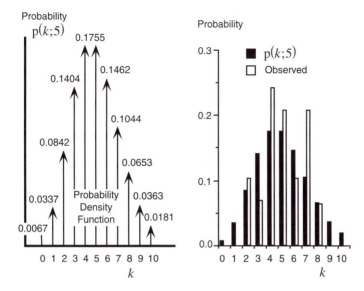

Figure 7.4 Diagrams for Example 7.2

as the number of simultaneous calls sampled in the physical model, the calculated results resemble the observed results.

---

(4) Interarrival Process

Sometimes knowing the distribution of the times between events is important; it is determined by the probability of no arrivals in a period τ and an arrival $\Delta\tau$ later, i.e.,

$$P(\tau) = P(0,\tau)P(1,\Delta\tau)$$

so that, substituting for the probabilities

$$P(\tau) = \exp(-\lambda\tau)[\lambda\Delta\tau] = \lambda\exp(-\lambda\tau)\Delta\tau \tag{7.20}$$

If we denote $P(\tau)/\Delta\tau$ by $p(\tau)$, equation (7.20) becomes

$$p(\tau) = \lambda\exp(-\lambda\tau) \tag{7.21}$$

$p(\tau)$ is the probability density of the time during which no event occurs; it is the interarrival time. The average value is

$$E(\tau) = \bar{\tau} = 1/\lambda \tag{7.22}$$

7.3 NORMALLY-DISTRIBUTED RANDOM VARIABLE

The normal distribution enjoys an important position in probability theory. The best-known of all distributions of a random variable, it can be used to represent binomial, Poisson, and other distributions, when the number of trials is very large.

(1) Normal Distribution

The *normal* distribution (also called the *Gaussian* distribution, after Johann Friedrich Carl Gauss, an early-19th-century German mathematician) is a continuous probability function that approximates the binomial distribution when the number of trials (n) is large and p and q (the probabilities of success and failure of a single trial) display finite values not approaching zero. In addition, as $n \to \infty$, it becomes the limiting distribution of both the binomial and Poisson distributions, and of a large class of other distributions that have finite mean and variance.

(2) Normally-Distributed Random Variable

A continuous random variable is called *normal* when the probability density function is

$$f_X(x) = \frac{1}{\sigma\sqrt{2\pi}} \exp\left(-\frac{1}{2\sigma^2}(x-\mu)^2\right) \tag{7.23}$$

and the distribution function is

$$F_X(x) = \frac{1}{\sigma\sqrt{2\pi}} \int_{-\infty}^{x} \exp\left(-\frac{1}{2\sigma^2}(u-\mu)^2\right) du \tag{7.24}$$

where $-\infty < x < \infty$. Normally-distributed random variables have

Mean $= \mu$

Variance $= \sigma^2$

Characteristic function $= \phi_X(t) = \exp\left(j\mu t - \frac{1}{2}\sigma^2\omega^2\right)$

(3) Standardized Normally-Distributed Random Variable

If we replace the random variable X with the standardized random variable X^*, the mean is 0 and the variance is 1. Under this condition

$$f_{X^*}(x) = \frac{1}{\sqrt{2\pi}} \exp\left(-x^2/2\right) \tag{7.25}$$

and

$$F_{X*}(x) = \frac{1}{\sqrt{2\pi}} \int_{-\infty}^{x} \exp\left(-u^2/2\right) du \tag{7.26}$$

Equation (7.25) defines the *standard* normal density function, and equation (7.26) defines the *standard* normal distribution function.

The graph of the standard normal density function is a symmetrical, bell-shaped curve of unit area. It has a maximum value of $1/\sqrt{2\pi} = 0.399$. The graph of the standard normal distribution function is an *S*–shaped curve with values between 0 and 1 that crosses the $x=0$ axis at a value of 0.5. They are shown in **Figure 7.5**. Because standardized normal functions are used in many applications, a great deal of attention is paid to their numerical values. Thus

- The area under the density curve is equal to 1.
- The area under the density curve between $x=\pm 0.67$ is equal to 0.5, i.e., $P(-0.67 \leq X* \leq 0.67) = 0.5$. Expressed in another way, 50% of events fall within 0.67σ of the mean.
- The area under the density curve between $x=\pm 1$ is equal to 0.6827, i.e., $P(-1 \leq X* \leq 1) = 0.6827$. Expressed in another way, approximately 68% of events fall within 1σ of the mean.
- The area under the density curve between $x=\pm 2$ is equal to 0.9545, i.e., $P(-2 \leq X* \leq 2) = 0.9545$. Expressed in another way, approximately 95% of events fall within 2σ of the mean.
- The area under the density curve between $x=\pm 3$ is equal to 0.9973, i.e., $P(-3 \leq X* \leq 3) = 0.9973$. Expressed in another way, almost all events fall within 3σ of the mean.

Equation (7.25) allows us to express the probability of $a < X* < b$ as

$$P(a < X* < b) = \frac{1}{\sqrt{2\pi}} \int_{a}^{b} \exp\left(-u^2/2\right) du \tag{7.27}$$

Using the fact that

$$\frac{1}{\sqrt{2\pi}} \int_{0}^{\infty} \exp\left(-u^2/2\right) du = 1/2$$

equation (7.26) is sometimes written

$$F_{X*}(x) = 1/2 + \frac{1}{\sqrt{2\pi}} \int_{0}^{x} \exp\left(-u^2/2\right) du$$

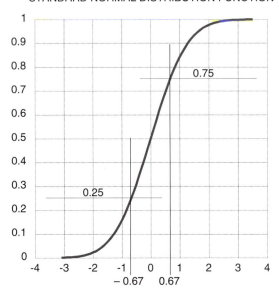

STANDARD NORMAL DISTRIBUTION FUNCTION

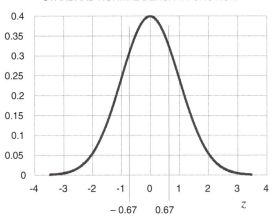

STANDARD NORMAL DENSITY FUNCTION

Figure 7.5 Distribution and Density Functions of a Standard Normal-Distributed Random Variable

The normal (or Gaussian) distribution is a continuous probability function that approximates the binomial distribution when the number of trials is large and the probabilities of success or failure of a single trial have values not approaching zero. As n → ∞, the binomial and Poisson distributions become normal.

(4) Central Limit Theorem

The sums of independent normally-distributed random variables are themselves normally-distributed. Moreover, even if individually they are not normally-distributed, their sum still tends to a normal distribution.

- **Central limit theorem**: if $X_1, X_2, ...X_n$ are independent random variables with finite means $\mu_1, \mu_2, ...\mu_n$ and variances $\sigma_1^2, \sigma_2^2, ...\sigma_n^2$, their sum is

$$S_n^* = \left\{ \sum_{i=1}^{n} X_i - \sum_{i=1}^{n} \mu_i \right\} \bigg/ \sqrt{\sum_{i=1}^{n} \sigma_i^2}$$

Under certain conditions, S_n^* approaches a normal distribution with zero mean and unit variance as $n \to \infty$, i.e.,

$$\lim_{n \to \infty} P\left(a \le S_n^* \le b\right) = \frac{1}{\sqrt{2\pi}} \int_a^b \exp\left(-t^2/2\right) dt \qquad (7.28)$$

The right-hand side of equation (7.28) is identical to the right-hand sides of equations (7.7) and (7.15). Thus, for a very large number of trials (limit $n \to \infty$), the standardized normal distribution is equal to the standardized binomial and standardized Poisson distributions. This result is a manifestation of the more general relationship stated above; it is used to justify the application of the normal distribution to a wide range of experimental situations. From a practical point of view, the condition $n \ge 10$ may be sufficient to do this.

EXAMPLE 7.3   Simple Case of Central Limit Theorem

Given $S_n = X_1 + X_2 + ... + X_n$, where $X_1, X_2, ... X_n$ are independent, identically distributed random variables with finite mean μ and variance σ^2.

Problem   Show that, in limit $n \to \infty$, S_n is a normally-distributed, random variable.

Solution   The expectation of the sum of a number of independent, identically distributed random variables is the sum of their individual expectations, i.e.,

$$E\left[S_n\right] = E\left[X_1\right] + E\left[X_2\right] + ... + E\left[X_n\right] = n\mu$$

Similarly,

$$\text{Var}(S_n) = \text{Var}(X_1) + \text{Var}(X_2) + ... + \text{Var}(X_n) = n\sigma^2$$

Under these circumstances, the standardized random variable corresponding to S_n is

$$S_n^* = \frac{S_n - n\mu}{\sigma\sqrt{n}}$$

Creating the moment generating function for S_n^* and expanding it into component random variables that are identically-distributed and independent

$$E\left[\exp\left(S_n^* t\right)\right] = E\left[\exp\left(\frac{S_n - n\mu}{\sigma\sqrt{n}}t\right)\right]$$

Expanding S_n

$$E\left[\exp\left(S_n^* t\right)\right] = E\left[\exp\left(\frac{X_1 - n\mu}{\sigma\sqrt{n}}t\right)\exp\left(\frac{X_2 - n\mu}{\sigma\sqrt{n}}t\right)\cdots\exp\left(\frac{X_n - n\mu}{\sigma\sqrt{n}}t\right)\right]$$

so that

$$E\left[\exp\left(S_n^* t\right)\right] = E\left[\exp\left(\frac{X_1 - n\mu}{\sigma\sqrt{n}}t\right)^n\right] = E\left[\exp\left(S_1^* t\right)^n\right] \qquad (7.29)$$

Expressing the exponential function as a power series

$$E\left[\exp\left(\frac{X_1 - \mu}{\sigma\sqrt{n}}t\right)\right] = E\left[1 + \frac{X_1 - \mu}{\sigma\sqrt{n}}t + \frac{(X_1 - \mu)^2}{2\sigma^2 n}t^2 + \frac{(X_1 - \mu)^3}{6\sigma^3 n^{3/2}}t^3 \cdots\right]$$

Substituting individual expected values

$$E\left[\exp\left(\frac{X_1 - \mu}{\sigma\sqrt{n}}t\right)\right] = E[1] + \frac{1}{\sigma\sqrt{n}}E[X_1 - \mu]t + \frac{1}{2\sigma^2 n}E\left[(X_1 - \mu)^2 t^2\right]$$

$$+ \frac{1}{6\sigma^3 n^{3/2}}E\left[(X_1 - \mu)^3\right]t^3 \cdots$$

Now, $E[1] = 1$, $E[X_1 - \mu] = 0$, $E\left[(X_1 - \mu)^2\right] = \sigma^2$, $E\left[(X_1 - \mu)^3\right] = 0$, etc., so that

$$E\left[\exp\left(\frac{X_1 - \mu}{\sigma\sqrt{n}}t\right)\right] = 1 + t^2/2!n + t^4/4!n^2 + t^6/6!n^3 \cdots$$

Substituting these values in equation (7.29)

$$E\left[\exp\left(S_n^* t\right)\right] = \left(1 + t^2/2!n + t^4/4!n^2 + t^6/6!n^3 \cdots\right)^n$$

Multiplying out

$$E\left[\exp\left(S_n^* t\right)\right] = \left(1 + t^2/2 + t^4/8 + t^6/48 \ldots + \text{terms containing } 1/n\right)$$

In limit $n \to \infty$

$$E\left[\exp\left(S_n^* t\right)\right] = \left(1 + t^2/2 + t^4/8 + t^6/48 \ldots\right) = \exp\left(t^2/2\right)$$

Now, the moment generating function of the standardized normal distribution is $\exp(t^2/2)$. Hence, S_n^* and S_n are normally-distributed, random variables.

Comment Proving the *central limit theorem* for more general cases is complicated and tedious; nevertheless, it has been done in sufficient volume to demonstrate that the concept is probably correct, and the idea is useful.

7.4 UNIFORMLY-DISTRIBUTED RANDOM VARIABLE

A continuous random variable is called *uniform* when the probability density function is

$$f_X(x)=\begin{cases}1/(b-a)\text{ , when }a\le x\le b\\0\text{ , otherwise}\end{cases}\qquad(7.30)$$

and the distribution function is

$$F_X(x)=\int f_X(x)dx=\begin{cases}0\text{ , when }x<a\\(x-a)/(b-a)\text{ , when }a\le x<b.\\1\text{ , when }x\ge b\end{cases}\qquad(7.31)$$

Uniformly-distributed random variables can be shown to have

$$\text{Mean} = \mu = \tfrac{1}{2}(a+b)$$

$$\text{Variance} = \sigma^2 = \tfrac{1}{12}(b-a)^2$$

$$\text{Characteristic function} = \phi_X(t)=\frac{\exp(jtb)-\exp(jta)}{jt(b-a)}$$

Figure 7.6 shows the nature of this distribution.

7.5 EXPONENTIALLY-DISTRIBUTED RANDOM VARIABLE

A continuous random variable is called *exponential* when the probability density function is

$$f_X(x)=\begin{cases}\alpha\exp(-\alpha x)\text{ , when }x>0\\0\text{ , when }x\le 0\end{cases}\qquad(7.32)$$

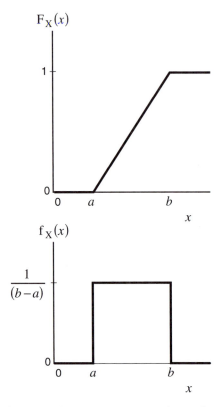

Figure 7.6 Distribution and Density Functions of a Uniform-Distributed Random Variable

and the distribution function is

$$F_X(x) = \int f_X(x)dx = 1 - \exp(-\alpha x). \qquad (7.33)$$

Exponentially-distributed random variables can be shown to have

Mean $= \mu = 1/\alpha$

Variance $= \sigma^2 = \mu^2 = 1/\alpha^2$

Characteristic function $= \phi_X(t) = \alpha/(\alpha - jt)$

Figure 7.7 shows the nature of this distribution.

7.6 ERLANG-DISTRIBUTED RANDOM VARIABLE

A continuous random variable is called *Erlang* when the probability density function is

$$f_X(x) = \frac{(\mu k)^k}{(k-1)!} x^{k-1} \exp(-\mu kx) \tag{7.34}$$

where k is a positive, non-zero integer, $x \geq 0$, and μ is a positive, real number (*Note*: this is not the mean). Because the distribution depends on the integer value of k, equation (7.34) is frequently referred to as the *Erlang-k* distribution. When $k = 1$

$$f_X(x) = \mu \exp(-\mu x)$$

i.e., the Erlang-1 distribution is an exponential distribution [compare equation (7.32)]. When k is very large ($k \to \infty$), Erlang$-\infty$ becomes a uniform (constant) probability distribution. **Figure 7.8** shows the Erlang probability density function for $k = 1, 2, 3,$ and 4, and $\mu = 1$ and 3.

The Erlang distribution function is

$$F_X(x) = \left[1 - \exp(-\mu kx) \sum_{n=0}^{n=k-1} (\mu kx)^n / n! \right] \tag{7.35}$$

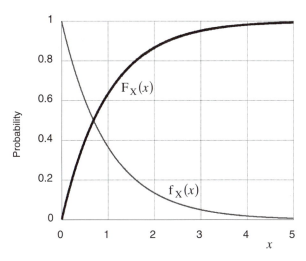

Figure 7.7 Distribution and Density Functions of an Exponential-Distributed Random Variable

Figure 7.9 shows the Erlang probability distribution function for $\mu=1$, $k=1$; $\mu=3$, $k=3$; and $\mu=5$, $k=5$. As noted above, when $k=1$, the distribution function is exponential. Other properties are

$$\text{Mean } = \overline{X} = 1/\mu$$

$$\text{Variance } = \sigma^2 = 1/\mu^2 k$$

$$\text{Characteristic function } = \phi_X(t) = \left(\frac{\mu k}{\mu k - jt}\right)^k$$

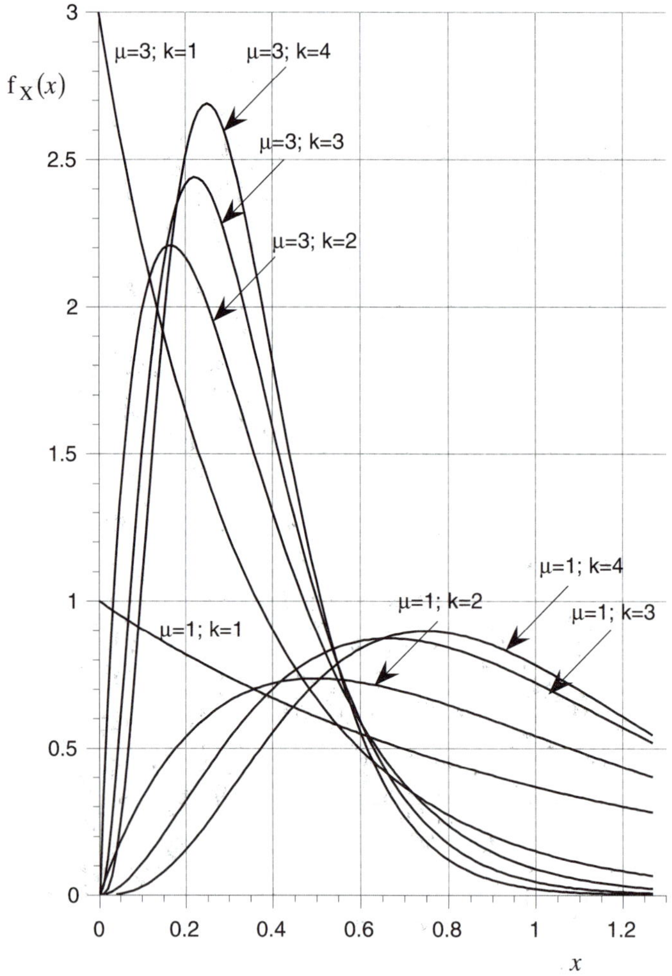

Figure 7.8 Density Functions of an Erlang-k Distributed Random Variable

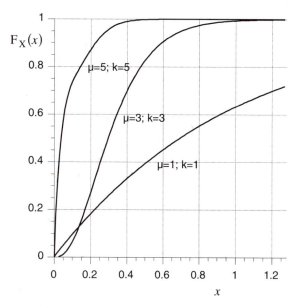

Figure 7.9 Distribution Functions of an Erlang-k Distributed Random

Early in the 20th century, this family of probability distributions was developed by A. K. Erlang, a Danish engineer. By adjusting the values of μ and k, the Erlang distribution can be made to fit real-world data for which neither the exponential nor the uniform probability distributions are adequate approximations. In § 9.2(4) I show examples of the way in which the Erlang distribution bridges between the uniform distribution and the exponential distribution. In § 9.3 I describe Erlang's formulas for blocked or delayed telephone traffic.

REVIEW QUESTIONS

7.1 What is a *Bernouilli trial?*

7.2 What is the probability of success of any sample k in a series of k independent Bernouilli trials?

7.3 The binomial coefficient is the number of ways k things can be taken from n things. State its value.

7.4 Given a binomial-distributed random variable. Is it discrete or continuous? State the probability density function and the distribution function.

7.5   When is the Poisson-distributed random variable a good approximation to the binomial-distributed random variable?

7.6   Given a Poisson-distributed random variable, is it discrete or continuous? State the probability density function and the distribution function.

7.7   The Poisson-distributed random variable can be used to find the probability of exactly k events occurring within a given time interval. State its value.

7.8   Given a normally-distributed random variable, is it discrete or continuous? State the probability density function and the distribution function.

7.9   State the probability density function and the distribution function of a standardized normally-distributed random variable.

7.10   State the central limit theorem. What is its significance?

7.11   Given a uniformly-distributed random variable, is it discrete or continuous? State the probability density function and the distribution function.

7.12   Given an exponentially-distributed random variable, is it discrete or continuous? State the probability density function and the distribution function.

7.13   Given an Erlang-distributed random variable, is it discrete or continuous? State the probability density function and the distribution function.

8

Random Processes

To characterize real-world data, voice, and video signals, I must account for their random values and the fact they are functions of time. The two-dimensional sample space they create defines a random process.

- **Random process**: denoted by $X(t,s)$, a family of time-dependent, statistical functions (sample functions), each member of which is one of the outcomes of an experiment directed to determining the behavior of a random signal. At time t_k, the values of the sample functions, i.e.,

  $x_1(t_k,s_1)$, $x_2(t_k,s_2)$, ... $x_n(t_k,s_n)$, form the random variable $X(t_k)$.

- **Sample function**: a time-dependent function whose future values are described in terms of probabilities estimated from statistics associated with past values and the assumption that future behavior is related somehow to the past.

8.1 RANDOM OR STOCHASTIC PROCESSES

In general, a random process has many properties. I limit discussion to three points that are important for those that represent random signals.

(1) Ensemble of Sample Functions

To model real-world, non-deterministic signals, we associate points in sample space with time-dependent, statistical functions. Known as *sample functions,* they represent the outcomes of trials associated with the behavior of a random signal. As a consequence, sample space becomes an *ensemble* of time-dependent functions. An example of such an ensemble is shown in **Figure 8.1**. It may be interpreted as the sample space resulting from an experiment in which n trials are conducted simultaneously with n identical sources. (This is the equivalent of simultaneously tossing n coins rather than tossing a coin n times in succession, as postulated for Figure 6.1.)

135

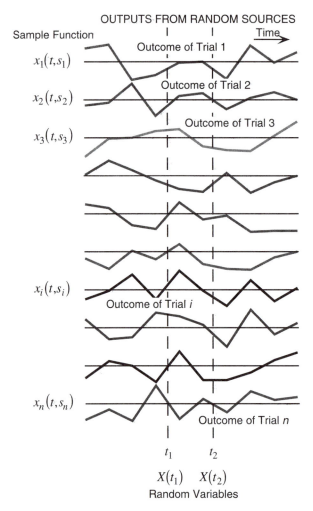

Figure 8.1 An Ensemble of Sample Functions—A Random Process
*Sample functions are the outcomes of trials associated with the behavior
of a random signal. The diagram shows the sample space resulting from
experiments conducted simultaneously on identical sources.*

When they are *random,* and the number of trials is very large, the ensemble is called a *random* or *stochastic* process. Denoted by $X(t,s)$, a random process is a family of time-dependent functions (sample functions), each of which is one of the outcomes of an experiment. At time t_1, the values of the sample functions, i.e., $x_1(t_1,s_1)$, $x_2(t_1,s_2)$, ...$x_n(t_1,s_n)$, form the random variable $X(t_1)$. Similarly, at time t_2, the values of the sample functions, i.e., $x_1(t_2,s_1)$, $x_2(t_2,s_2)$, ...$x_n(t_2,s_n)$, form the random variable $X(t_2)$. Thus, a random process may be represented as a time-ordered set of random variables. The expected values of $X(t_1)$ and $X(t_2)$ are known as *ensemble* averages of the random process at t_1 and t_2.

(2) Stationary Processes

* **Strict-sense stationary random process**: a random process is said to be *strict-sense stationary* when each of the central moments remains the same irrespective of any change in time.

For most purposes, considering all possible moments is unnecessary, and a useful limited condition is described as *wide-sense stationary*.

* **Wide-sense stationary random process**: a random process is said to be *wide-sense stationary* when the expectation (mean) of the random process is a constant and the autocorrelation of the random process is not dependent on time, only time differences, i.e.,

$$E[X(t,s)] = \overline{X} = \text{constant} \tag{8.1}$$

and

$$E[X(t,s)X(t+\tau,s)] = \phi_X(\tau) \tag{8.2}$$

(3) Ergodic Processes

* **Ergodic random process**: a random process is said to be *ergodic* when the time-average of every measurable sample function equals the corresponding ensemble average. In this circumstance any sample function will define the random process uniquely. An ergodic process is necessarily a stationary process; a stationary process is not necessarily ergodic.

Equating the time and ensemble averages of an ergodic random process that represents a random signal, we obtain some significant relationships. Thus

* The time average (or dc-level) of the sample function $x_k(t)$ is equal to the expectation of the random variable $X(t_k)$, i.e.,

$$\lim_{T\to\infty} \frac{1}{T} \int_{-T/2}^{T/2} x(t)dt = E[X] = \overline{X} = \int_{-\infty}^{\infty} xp(x)dx \tag{8.3}$$

In turn, because the process is ergodic, we can apply these statements to the whole, i.e., the dc-level of the random process is equal to the mean of the process.

- The mean-square value $E[X^2]$ of the random variable $X(t)$ is equal to the signal power of the sample function $x(t)$, i.e.,

$$E[X^2] = \int_{-\infty}^{\infty} x^2 p(x) dx = \lim_{T \to \infty} \frac{1}{T} \int_{-T/2}^{T/2} x^2(t) dt \qquad (8.4)$$

In turn, because the process is ergodic, we can apply these statements to the whole, i.e., the mean-square value of the random process is equal to the signal power of the process. $\sqrt{E[X^2]}$ is the root-mean-square value of the process.

- If the ergodic random process is zero-mean so that $E[X]=0$, the variance $\sigma_X^2 = E[X^2]$. Thus, the variance of a zero-mean, ergodic random process is equal to the average (over all time, $\overline{P}[\infty]$) power of the process, and the standard deviation σ_X of a zero-mean, ergodic random process is the root-mean-square value of the process.

In modeling random signals, the use of an ergodic random process provides a significant simplification—all of its statistical characteristics can be determined from a single sample function. For this reason, whenever possible, wide-sense stationary, ergodic processes are employed. The stationary nature of such a process implies that it is eternal and requires the use of power-type functions.

8.2   AUTOCORRELATION AND POWER DENSITY OF RANDOM PROCESSES

Random processes are power-type functions. In this section, I show the relationship between the autocorrelation and power spectral density functions of a random process. It allows the frequency distribution of power in the process to be determined.

(1) Autocorrelation Function

The autocorrelation function $[\phi_X(\tau)]$ provides a means of describing the interdependence of two random variables obtained by observing the random process $X(t)$ at times t_1 and t_2. When $X(t)$ is stationary, the ACF depends only on the time difference (τ); thus,

$$\phi_X(\tau) = E[X(t+\tau)X(t)] \qquad (8.5)$$

When $\tau=0$, this relationship becomes

$$\phi_X(0) = E[X^2(t)] = \overline{P}[\infty]$$

i.e., the value of the ACF at $\tau=0$ is the power in the process. In addition

$$\phi_X(\tau) = \phi_X(-\tau)$$

i.e., the ACF exhibits even symmetry, and

$$|\phi_X(\tau)| \le \phi_X(0)$$

i.e., the maximum magnitude of the ACF occurs when $\tau=0$.

(2) Wiener-Khinchine Relation

In the 1930s, Alexsandr Khinchine, a German mathematician, and Norbert Wiener, an American mathematician, developed the theoretical basis for the statement:

- **Wiener-Khinchine relation**: the power spectral density and the autocorrelation function of a wide-sense stationary process form a Fourier pair, i.e.,

$$\phi_X(\tau) = \int_{-\infty}^{\infty} \Phi_X(f)\exp(j2\pi f\tau)df \iff \int_{-\infty}^{\infty} \phi_X(\tau)\exp(-j2\pi f\tau)d\tau = \Phi_X(f) \quad (8.6)$$

Putting $f=0$ in the RHS of equation (8.6) gives

$$\Phi_X(0) = \int_{-\infty}^{\infty} \phi_X(\tau)d\tau$$

i.e., the zero-frequency value of the PSD is equal to the area under the graph of the ACF. Also, substituting $-f$ for f, in the RHS of equation (8.6), gives

$$\Phi_X(f) = \Phi_X(-f)$$

i.e., the PSD exhibits even symmetry. Putting $\tau=0$ in the LHS of equation (8.6) and using equation (8.5) gives

$$\phi_X(0) = \int_{-\infty}^{\infty} \Phi_X(f)df = E[X^2(t)] = \overline{P}[\infty]$$

i.e., the mean-square value (power) of the zero-mean random process is equal to the area under the graph of the PSD. Finally, for all f,

$$\Phi_X(f) \ge 0$$

i.e., the PSD is never negative. The Wiener-Khinchine relation provides the link between the ACF and PSD of a random process in the same way that § 5.3(2) links deterministic functions.

EXAMPLE 8.1 Random Binary Wave

Given A binary, wide-sense stationary process in which: 1s and 0s are equally likely and statistically independent; 1s are represented by $+A$, and 0s are represented by $-A$; the pulse width is T seconds; the pulses are non-return-to-zero; and the starting point of $x(t)$ is uniformly distributed between 0 and T seconds. A sample function is shown in **Figure 8.2**.

Problem Find the ACF and PSD of the process.

Solution Because the process is wide-sense stationary, the ACF and PSD of a sample function are the ACF and PSD of the process.

1) Find the ACF. In Figure 8.2, the starting time of the first pulse is equally likely to lie between 0 and T seconds. Hence, the density function is

$$f_{t_0}(t_0) = \begin{cases} 1/T, & \text{when } 0 \le t_0 \le T \\ 0, & \text{elsewhere} \end{cases}$$

1s and 0s are equally likely and independent. Hence the process is zero-mean and

$$E[X(t)] = 0$$

Also, the power in the process is A^2 so that

$$E[X^2(t)] = A^2$$

To find the ACF, we must consider two cases, $\tau > T$ and $\tau < T$. When

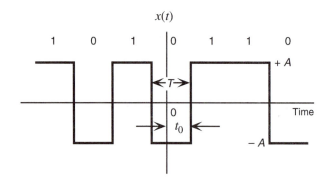

Figure 8.2 Sample Function of Random Binary Wave for Example 8.1

$\tau > T$, $X(t)$ and $X(t + \tau)$ occur in different pulse intervals—so they are independent, and

$$E[X(t)X(t + \tau)] = E[X(t)]E[X(t + \tau)] = 0$$

When $\tau < T$, $X(t)$, and $X(t + \tau)$ occur in the same pulse interval provided $t_0 < T - |\tau|$. Averaging over all possible values of t_0

$$E[X(t)X(t + \tau)] = \int_0^{T-|\tau|} A^2 f_{t_0}(t_0)dt = A^2(1 - \tau/T), \quad \tau < T$$

Hence, introducing modulus signs to permit negative arguments, the ACF is

$$\phi_X(\tau) = \begin{cases} A^2(1 - |\tau|/T), & \text{when } |\tau| < T \\ 0, & \text{when } |\tau| > T \end{cases} \tag{8.7}$$

This is the triangle function shown in **Figure 8.3**.

2) Find the PSD. Use the Wiener-Khinchine relationship, i.e.,

$$\Phi_X(f) = \int_{-T}^{T} A^2(1 - \tau/T)\exp(-j2\pi f\tau)d\tau = A^2 T\text{Sinc}^2(fT) \tag{8.8}$$

This is the Sinc^2 function shown in Figure 8.3.

Comment In Example 3.11 [equation (3.35)], I showed the Fourier transform of a symmetrical square wave of amplitude $\pm A$ and period T_1 to be

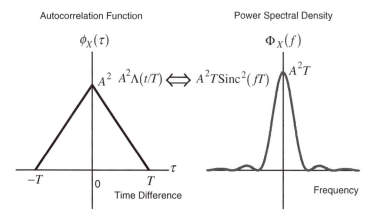

Figure 8.3 Autocorrelation Function and Power Spectral Density for Example 8.1
The figures are the ACF and PSD of a Random Binary Wave.

$$S(f) = A\operatorname{Sinc}(fT_1/2)\delta_{f_1}(f) - A\delta(f)$$

Impulse magnitude squaring [see equation (5.27)] gives

$$\Phi(f) = A^2\operatorname{Sinc}^2(fT_1/2)\delta_{f_1}(f) - A^2\delta(f) \tag{8.9}$$

In this example, I have shown the PSD of a random binary wave (1s and 0s equally likely and amplitude $\pm A$) to be

$$\Phi_X(f) = A^2 T \operatorname{Sinc}^2(fT)$$

In this expression, T is the timeslot width; it is equal to $T_1/2$. Hence

$$\Phi_X(f) = \tfrac{1}{2}A^2 T_1 \operatorname{Sinc}^2(fT_1/2) \tag{8.10}$$

Equations (8.9) and (8.10) show the difference between a periodic symmetrical square wave and a random (probabilistic) square wave.

- The PSD of a *periodic* wave consists of discrete components separated by T_1 that exhibit a Sinc^2 envelope, except at $f = 0$.
- The PSD of a *random* wave is a continuous distribution whose envelope is a Sinc^2 function and whose magnitude depends on the width of the timeslot.

REVIEW QUESTIONS

8.1 What is a sample function?

8.2 Describe a *random process.*

8.3 Distinguish between a *strict-sense stationary process* and a *wide-sense stationary process.*

8.4 What is an *ergodic random process?*

8.5 To what is the mean value of an ergodic random process equal?

8.6 To what is the mean-square value of an ergodic random process equal?

8.7 To what is the variance of a zero mean ergodic random process equal?

8.8 To what is the standard deviation of a zero mean ergodic random process equal?

8.9 Is a random process an energy-type or power-type process?

8.10 State the Wiener-Khinchine relation.

9

Queues

In systems that carry messages, resources must be shared among contending customers. To do so, some messages may be delayed until resources are available and others may be destroyed. Queueing theory is directed toward determining the characteristics of the delay process.

9.1 QUEUEING THEORY

Queueing theory provides a large number of alternative models that describe a waiting line situation. My discussion is limited to those that can be applied to simple communication systems.

- **Queueing theory**: mathematical discipline that provides techniques for determining the characteristics of waiting lines
- **Waiting line**: phenomenon that occurs whenever the demand for a service exceeds the capacity available to provide the service

(1) Simple Queueing Model

The basic structure and the parameters associated with a simple queueing model are shown in **Figure 9.1**. Over time, the *input source* generates units that *call* for a service. They enter a *queue* to wait their turn to receive it. At times determined by the *service facility* and according to some rule known as the *service discipline,* a member of the queue is selected for service. When the service has been performed, the unit is released to the outside world and another calling unit is selected from the queue. The number of units in the queueing system depends on the rate at which calling units are generated, the capacity of the queue, the number of servers in the service facility, and the service rate achieved by the *service mechanism*. To analyze a queueing system, we make certain assumptions:

Line length = number of calling units in queueing system
Queue length = number of calling units waiting for service

E_n = state in which there are n calling units in system
$P_n(t)$ = probability that exactly n calling units are in system at time t
P_n = probability that exactly n calling units are in system
s = number of parallel service channels (servers) in system
λ_n = mean arrival rate of new calling units when n units in system
μ_n = mean service rate when n units are in system
L = expected line length (in calling units)
L_q = expected queue length (in calling units)
W = expected waiting time in system
W_q = expected waiting time in queue

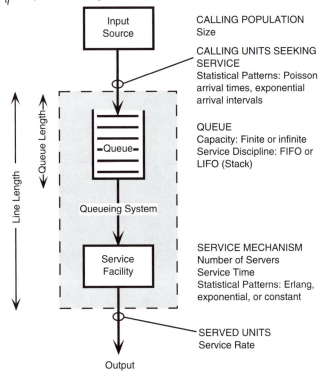

Figure 9.1 Basic Structure and Parameters of Queueing Model

- **Calling unit arrival pattern**: the input source contains the units that will call for service from time to time. If the number of units is very large, we assume an infinite population. In this way, the characteristics of the source population are not altered by the movement of calling units to the queue. If the statistics of unit arrival at the queue are Poisson, the probability of having n units arrive one at a time in t seconds is

$$P_n(t) = \frac{(\lambda t)^n}{n!} \exp(-\lambda t)$$

and the times between the arrivals of consecutive calling units are distributed exponentially.

- **Service discipline**: the queue stores arriving units until they can receive service. We assume a single queue; however, several queues can be operated in parallel. If the capacity of the queue is limited, calling units that arrive when it is full will be discarded. Units are selected for service (leave the queue) according to the service discipline employed. Common rules are first in, first out (FIFO) and last in, first out (LIFO). Usually, we assume that the units leave the queue in the order in which they were received (i.e., FIFO).

- **Service mechanism**: a service facility consists of one or more servers that provide the required service (implement the service mechanism). We assume a single server; however, servers can be linked in series, so that each unit is serviced in several ways consecutively, or used in parallel, so that several units are served by single servers concurrently. Service times (also known as holding times) are assumed to be distributed in Erlangian, exponential, or constant fashion.

An important quantity is called the *traffic intensity ratio* (v). It is the ratio of the average rate (i.e., units per unit time) at which units arrive at the queue (λ) and the average rate at which units are serviced and leave the queueing system (μ), i.e.,

$$v = \lambda/\mu \qquad (9.1)$$

Measured in Erlangs, when $\lambda > \mu$, i.e., $v > 1$, units are arriving at a faster average rate than the average rate at which a single server can serve. The system fills up and units are lost, or we must add servers. When $\lambda < \mu$, i.e., $v < 1$, units are arriving at a slower average rate than the average rate at which a single server can serve and the system is able to cope. When the mean arrival rate of calling units at the queue is *constant* and less than or equal to the mean service rate, the queueing process is said to be *steady-state*. Under this circumstance

$$L = \lambda W \qquad (9.2)$$

where

L = expected line length (in calling units)
W = expected waiting time in the queueing system

Also

$$L_q = \lambda W_q \tag{9.3}$$

where

L_q = expected queue length (in calling units)
W_q = expected waiting time in the queue

Further, if the mean service time $(1/\mu)$ is constant

$$W = W_q + 1/\mu \tag{9.4}$$

(2) Kendall Classification System

In 1951, D. G. Kendall, a British mathematician, published a classification system that describes the essential components of a queueing system. Consisting of three symbols (A/B/c), it permits easy identification of the input process (A), the service time distribution process (B), and the number of servers (c). The symbols are identified in **Table 9.1**. Examples of common systems described in Kendall's shorthand are:

```
+--------------------------------------+
|        Format: A/B/c                 |
|                                      |
|  A = input process                   |
|  B = service time distribution       |
|  c = number of servers               |
+--------------------------------------+
```

Symbol   Description

M = Poisson arrival process *OR* exponential interarrival time
     OR exponential service time
E_k= Erlangian arrival process *OR* Erlangian service time
G = General service time
GI = General independent service time
D = Deterministic or constant interarrival time *OR* deterministic
     or constant service time

Table 9.1 Kendall Notation to Describe Queueing Systems
D. G. Kendall devised a notation to describe queueing systems. It consists of three symbols that identify the input process, the service time distribution process, and the number of servers.

- **M/M/1**: a queueing system with a Poisson arrival process, an exponential service time distribution, and a single server
- **M/D/1**: a queueing system with a Poisson arrival process, a constant service time distribution, and a single server
- **M/E$_k$/1**: a queueing system with a Poisson arrival process, an Erlang-k service time distribution, and a single server.

An extension of Kendall's notation includes three other parameters: maximum queue length (K), number of potential calling units in the source population (m), and the queue discipline (Z). Thus

- **M/M/1/8/∞/FIFO**: a queueing system with a Poisson arrival process, an exponential service time distribution, a single server, an 8-unit queue, an infinite number of calling units in the source population, and a FIFO queue discipline.

For most applications, Kendall's three-symbol notation is adequate.

(3) Birth–Death Process

A birth–death process (a Markov process) can be used to describe the state of a queueing system. In it, *birth* refers to the arrival of a calling unit at the queueing system, and *death* refers to the departure of a serviced unit. Births and deaths are assumed to occur randomly, and their mean occurrence rate depends only on the present state of the system [number of calling units (n) in the queueing system]. **Figure 9.2** shows that state transitions occur only to

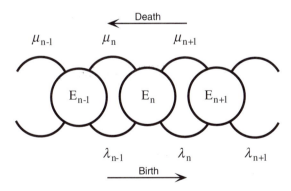

Figure 9.2 State Transition Rate Diagram of Birth–Death Process
A birth-death process (a Markov process) is used to describe the states and transitions in a queueing process. Births and deaths are assumed to occur at random, and the mean occurrence rate depends only on the current state of the system.

nearest neighbors, i.e., to the next state $(n+1)$ in the event of a birth, or to the previous state $(n-1)$ in the event of a death.

During the time interval $t \to t+\Delta t$, four events can occur: no births or deaths, one birth, one death, and the number of events exceeds one. We will restrict the value of Δt so that only one event, or no event, occurs in the interval. With this condition, the individual probabilities are:

- If the system is in state E_{n-1}, the probability that between t and $t+\Delta t$ the transition $E_{n-1} \to E_n$ occurs is $[\lambda_{n-1}\Delta t]P_{n-1}(t)$.
- If the system is in state E_{n+1}, the probability that between t and $t+\Delta t$ the transition $E_n \leftarrow E_{n+1}$ occurs is $[\mu_{n+1}\Delta t]P_{n+1}(t)$.
- If the system is in state E_n, the probability that between t and $t+\Delta t$ neither a birth nor a death occurs is $[1-\lambda_n\Delta t - \mu_n\Delta t]P_n(t)$.

All of these events lead to state E_n at $t+\Delta t$ so that

$$P_n(t+\Delta t)=[\lambda_{n-1}\Delta t]P_{n-1}(t)+[\mu_{n+1}\Delta t]P_{n+1}(t)+[1-\lambda_n\Delta t-\mu_n\Delta t]P_n(t) \quad (9.5)$$

Combining terms, subtracting $P_n(t)$ from both sides and dividing by Δt, equation (9.5) becomes

$$\frac{P_n(t+\Delta t)-P_n(t)}{\Delta t}=\lambda_{n-1}P_{n-1}(t)+\mu_{n+1}P_{n+1}(t)-(\lambda_n+\mu_n)P_n(t) \quad (9.6)$$

In the limit $\Delta t \to 0$ equation (9.6) becomes

$$\frac{dP_n(t)}{dt}=\lambda_{n-1}P_{n-1}(t)+\mu_{n+1}P_{n+1}(t)-(\lambda_n+\mu_n)P_n(t) \quad (9.7)$$

When $n=0$, $\lambda_{-1}=0$ and $\mu_0=0$ so that the initial equation in the series represented by equation (9.7) is

$$\frac{dP_0(t)}{dt}=\lambda_1P_1(t)-\lambda_0P_0(t) \quad (9.8)$$

Equations (9.7) and (9.8) are a set ($n=0, 1, 2, \ldots$) of differential equations that provides the value of $P_n(t)$, i.e., the probability that exactly n calling units are in the queueing system at time t. Unfortunately, a general solution is not possible. What is possible is a solution to the case of a system that has reached steady-state.

When the queueing system has reached a steady-state, the state probabilities are constant so that $P_n(t)$ becomes P_n and $dP_n(t)/dt = 0$. Under these conditions, equations (9.7) and (9.8) become

$$\lambda_{n-1}P_{n-1}+\mu_{n+1}P_{n+1}-(\lambda_n+\mu_n)P_n=0, \quad \text{when } n>0 \quad (9.9)$$

and

$$\mu_1P_1-\lambda_0P_0=0, \quad \text{when } n=0$$

i.e.,

$$P_1 = (\lambda_0/\mu_1)P_0 \tag{9.10}$$

When $n>0$, equation (9.9) yields

$$P_{n+1} = (\lambda_n/\mu_{n+1})P_n + \frac{\mu_n P_n - \lambda_{n-1}P_{n-1}}{\mu_{n+1}} \tag{9.11}$$

Manipulating equation (9.11), and employing equation (9.10), you can show that

$$\mu_n P_n - \lambda_{n-1}P_{n-1} = \mu_1 P_1 - \lambda_0 P_0 = 0$$

so that, by progressively reducing the value of n, we obtain

$$P_n = \frac{\lambda_{n-1}}{\mu_n}P_{n-1} = \frac{\lambda_{n-1}\lambda_{n-2}}{\mu_n\mu_{n-1}}P_{n-2} = \ldots = \frac{\lambda_{n-1}\lambda_{n-2}\ldots\lambda_0}{\mu_n\mu_{n-1}\ldots\mu_1}P_0$$

which can be written

$$P_n = \left[\prod_{i=0}^{n-1}\lambda_i \bigg/ \prod_{i=1}^{n}\mu_i\right]P_0 \tag{9.12}$$

The value of P_0 is obtained by remembering that the sum of all probabilities is one, i.e.,

$$\sum_{n=0}^{\infty}P_n = 1$$

so that

$$P_0 + \sum_{n=1}^{\infty}P_n = 1$$

and substituting the value of P_n from equation (9.12)

$$P_0 = \frac{1}{1 + \sum\limits_{n=1}^{\infty}\left[\prod\limits_{i=0}^{n-1}\lambda_i \bigg/ \prod\limits_{i=1}^{n}\mu_i\right]} \tag{9.13}$$

Hence, from equations (9.12) and (9.13)

$$P_n = \left[\prod_{i=0}^{n-1}\lambda_i \bigg/ \prod_{i=1}^{n}\mu_i\right] \bigg/ 1 + \sum_{n=1}^{\infty}\left[\prod_{i=0}^{n-1}\lambda_i \bigg/ \prod_{i=1}^{n}\mu_i\right] \tag{9.14}$$

Now, the expected value of the number of calling units in the queueing system (line length, L) is given by

$$\overline{L} = \sum_{n=0}^{\infty} n P_n \tag{9.15}$$

and, if s servers exist, the expected number of calling units in the queue (queue length, L_q) is given by

$$\overline{L_q} = \sum_{n=s}^{\infty} (n-s) P_n \tag{9.16}$$

For some special cases, these summations yield analytical solutions; when they do not, they can be evaluated by numerical means.

9.2 QUEUEING MODELS

To reduce analytical complexity, further discussion is limited to single-server systems that have achieved steady-state conditions. The three models described cover a range of arrival and service strategies.

(1) *M/M/1 Queueing System*

An M/M/1 queueing system is a single-server system with Poisson arrival times and exponentially-distributed service times. The arrival rate $\lambda_n = \lambda$, a constant, and the service rate $\mu_n = \mu$, a constant; also, in the steady state, $\lambda < \mu$. Under these conditions,

$$P_0 = \frac{1}{\displaystyle\sum_{n=0}^{\infty} (\lambda/\mu)^n} = 1 - \lambda/\mu \tag{9.17}$$

and

$$P_n = P_0 (\lambda/\mu), \quad n > 0 \tag{9.18}$$

Substituting in equation (9.15), we obtain the expected line length, i.e.,

- average number of calling units in the queueing system

$$\overline{L}_{M/M/1} = \sum_{n=0}^{\infty} n(1 - \lambda/\mu)(\lambda/\mu)^n = \lambda/(\mu - \lambda) \tag{9.19}$$

Substituting in equation (9.16), we obtain the expected queue length, i.e.,

- average number of calling units in the queue

$$\overline{L_q}_{M/M/1} = \lambda^2 / [\mu(\mu - \lambda)] \qquad (9.20)$$

Substituting in equation (9.4), we obtain

- average waiting time in the queue

$$\overline{W_q}_{M/M/1} = \overline{L_q}_{M/M/1} / \lambda = \lambda / [\mu(\mu - \lambda)] \qquad (9.21)$$

And substituting in equation (9.3), we obtain

- average waiting time in the queueing system

$$\overline{W}_{M/M/1} = \overline{W_q}_{M/M/1} + 1/\mu = 1/(\mu - \lambda) \qquad (9.22)$$

EXAMPLE 9.1   M/M/1 Queueing System

Given A single-server, M/M/1 queueing system receives an average of 10 requests per second for service. The requesting population is infinite, and the requests can be served up to a rate of 12 per second.

Problem Find

a) the average delay in service

b) the average number of calling units in the system

Solution

a) To find the average delay in the queueing system, substitute in equation (9.21).

$$\overline{W}_{M/M/1} = 1/(\mu - \lambda) = 1/(12 - 10) = 0.5 \text{ seconds}$$

b) To find the average number of calling units in the system, substitute in equation (9.19).

$$\overline{L}_{M/M/1} = \lambda / (\mu - \lambda) = 10/(12 - 10) = 5 \text{ units}$$

---

(2) M/D/1 Queueing System

An M/D/1 queueing system is a single-server system with Poisson arrival times and constant service times. Without analysis, we state the results. The values are around one-half of the M/M/1 model above.

- average number of calling units in the queueing system is

$$\overline{L}_{M/D/1} = \frac{\lambda^2}{2\mu(\mu - \lambda)} + \lambda/\mu \qquad (9.23)$$

- average number of calling units in the queue is

$$\overline{L_q}_{M/D/1} = \frac{\lambda^2}{2\mu(\mu - \lambda)} \tag{9.24}$$

- average waiting time in the queueing system is

$$\overline{W}_{M/D/1} = \frac{\lambda}{2\mu(\mu - \lambda)} + 1/\mu \tag{9.25}$$

- average waiting time in the queue is

$$\overline{W_q}_{M/D/1} = \frac{\lambda}{2\mu(\mu - \lambda)} \tag{9.26}$$

EXAMPLE 9.2   M/D/1 Queueing System

Given   A single-server, M/D/1 queueing system receives an average of 10 requests per second for service. The requesting population is infinite, and the requests can be served up to a rate of 12 per second.

Problem Find

   a) the average delay in service

   b) the average number of calling units in the system

Solution

   a) To find the average delay in the queueing system, substitute in equation (9.25).

$$\overline{W}_{M/D/1} = \frac{\lambda}{2\mu(\mu - \lambda)} + 1/\mu = 10/48 + 1/12 = 14/48 = 0.29167 \text{ seconds}$$

   b) To find the average number of calling units in the system, substitute in equation (9.23).

$$\overline{L}_{M/D/1} = \frac{\lambda^2}{2\mu(\mu - \lambda)} + \lambda/\mu = 100/48 + 10/12 = 140/48 = 2.91667 \text{ units}$$

Comment These results are approximately 60% of the results for M/M/1.

---

(3) M/E_k/1 Queueing System

An $M/E_k/1$ queueing system is a single-server system with Poisson arrival times and Erlangian service times. Without analysis, we state the results.

- average number of calling units in the queueing system is

$$\overline{L}_{M/E_k/1} = \frac{1+k}{2k}\frac{\lambda^2}{\mu(\mu-\lambda)} + \lambda/\mu \tag{9.27}$$

- average number of calling units in the queue is

$$\overline{L}_{q\,M/E_k/1} = \frac{1+k}{2k}\frac{\lambda^2}{\mu(\mu-\lambda)} \tag{9.28}$$

- average waiting time in the queueing system is

$$\overline{W}_{M/E_k/1} = \frac{1+k}{2k}\frac{\lambda}{\mu(\mu-\lambda)} + 1/\mu \tag{9.29}$$

- average waiting time in the queue is

$$\overline{W}_{q\,M/E_k/1} = \frac{1+k}{2k}\frac{\lambda}{\mu(\mu-\lambda)} \tag{9.30}$$

When $k=1$, the Erlang distribution becomes an exponential distribution. Putting $k=1$ in these equations gives the corresponding equations for an M/M/1 system. When $k\rightarrow\infty$, the Erlang distribution becomes a uniform (constant) distribution and the equations are equal to the M/D/1 system.

EXAMPLE 9.3 M/E$_k$/1 Queueing System

Given A single-server, M/E$_k$/1 queueing system receives an average of 10 requests per second for service. The requesting population is infinite, and the requests can be served up to a rate of 12 per second. Take the value of $k=3$.

Problem Find

a) the average delay in service

b) the average number of calling units in the system

Solution

a) To find the average delay in the queueing system, substitute in equation (9.29).

$$\overline{W}_{M/E_k/1} = 0.66667 \times 0.41667 + 0.08333 = 0.36111 \text{ seconds}$$

b) To find the average number of calling units in the system, substitute in equation (9.27).

$$\overline{L}_{M/E_k/1} = 0.66667 \times 4.16667 + 0.83333 = 3.61112 \text{ units}$$

Comment 1 These values fall between those for M/M/1 and M/D/1.

Comment 2 When $k=1$, the results become

$$\overline{W}_{M/E_k/1} = 0.41667 + 0.08333 = 0.5 \text{ seconds}$$

$$\overline{L}_{M/E_k/1} = 4.16667 + 0.83333 = 5 \text{ units}$$

They are equal to those for M/M/1, and confirm that one extreme of the Erlang distribution is an exponential distribution.

When $k \to \infty$, the results become

$$\overline{W}_{M/E_k/1} = 0.5 \times 0.41667 + 0.08333 = 0.29166 \text{ seconds}$$

$$\overline{L}_{M/E_k/1} = 0.5 \times 4.16667 + 0.83333 = 2.91666 \text{ units}$$

They are equal to those for M/D/1, and confirm that the other extreme of the Erlang distribution is a uniform function.

---

(4) Comparisons

The average time a calling unit spends in the queue is a measure of the delay it encounters in the queueing system. **Figure 9.3** compares the relative waiting

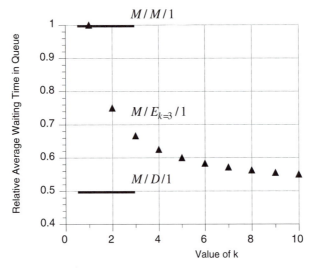

Figure 9.3 Comparison of Average Time in Queue for Three Queueing Models

The average time a calling unit spends in the queue is a measure of the delay it encounters in the queueing system. The waiting time performance of an M/E$_k$/1 (Erlang-k service time distribution) system bridges the values of M/M/1 (exponential service time distribution) and M/D/1 (constant service time distribution) systems.

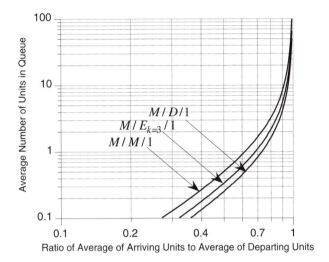

Figure 9.4 Average Number of Calling Units in Queue as a Function of the Ratio of Average Number of Arriving Units to Average Number of Departing Units

times for $M/M/1$, $M/D/1$, and $M/E_k/1$ systems. If the average time spent in the queue is one time unit for $M/M/1$, it is one-half for $M/D/1$ and between these limits for $M/E_k/1$.

The average number of calling units in the queue is a measure of queueing system complexity. **Figure 9.4** compares the average number of calling units in the queue as a function of the ratio of the average number of units arriving to the average number of units departing. As arrival rate begins to equal departure rate, the queue grows rapidly. The queue is longest for $M/M/1$ and shortest for $M/D/1$; the values for $M/E_{k=3}/1$ are between them.

9.3 ERLANG'S FORMULAS

Several formulas that predict the performance of multi-channel communication systems under varying input loads (offered traffic) were developed by A. K. Erlang. In developing these results, he assumed

- Call requests are Poisson-distributed; users may request a channel at any time.
- Service times are exponentially-distributed.
- The number of channels is limited; all idle channels are available to every call request.

They share common ground with the queueing models described in § 9.2. The Erlang B and Erlang C formulas[1] are applied to many trunking situations.

- **Erlang B Formula**: determines the probability of call blocking in a communication (trunking) system. Because the system provides no queueing capability, the blocked call is *lost* and must be *re-tried* at a later time. This strategy is known as *blocked call cleared* (BCC) or *lost call cleared* (LCC). In a steady state,
  - the probability of the next arriving call being blocked is

$$P[\text{blocking}] = \frac{A^C/C!}{\sum_{k=0}^{C} A^k/k!} \tag{9.31}$$

  where A is the mean value of the offered traffic (in Erlangs) and C is the number of service channels in the trunking system.
  - the probability of finding x calls in progress simultaneously is

$$P[x \text{ calls in progress}] = \frac{A^C/x!}{\sum_{k=0}^{C} A^k/k!} = P[\text{blocking}]\, C!/x! \tag{9.32}$$

Figure 9.5 plots the probability of the next arriving call request being blocked [equation (9.31)] and the probabilities of specific numbers of calls in progress [equation (9.32)] as functions of the ratio of offered traffic to the maximum traffic served by the communication system. Under a BCC strategy, a certain number of calls will always be carried (C erlangs). Arriving calls that are blocked are immediately discarded and do not remain in the system. When the offered traffic is equal to the system capacity the probability of the next call being blocked is approximately 28%. When the offered traffic is twice the system capacity it is approximately 58%.

- **Erlang C Formula**: determines the probability of call delay in a communication system that includes a queue to hold all call requests that cannot be assigned immediately to an idle channel. The strategy is known as *blocked call delayed* (BCD) or *lost call delayed* (LCD). In a steady state,
  - the probability of the next arriving call being delayed is

$$P[\text{call delayed}] = \frac{A^C}{A^C + C!(1 - A/C)\sum_{k=0}^{C-1} A^k/k!} \tag{9.33}$$

[1]For a modern derivation, see Theodore S. Rappaport, *Wireless Communications* (Upper Saddle River, NJ: Prentice Hall, 1996), pp. 555-64.

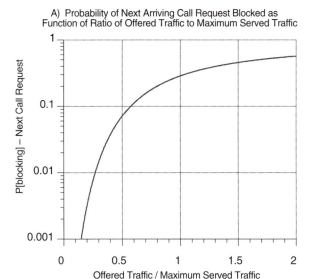

A) Probability of Next Arriving Call Request Blocked as Function of Ratio of Offered Traffic to Maximum Served Traffic

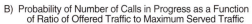

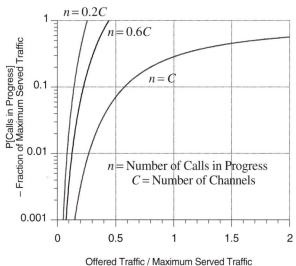

B) Probability of Number of Calls in Progress as a Function of Ratio of Offered Traffic to Maximum Served Traffic

$n = 0.2C$

$n = 0.6C$

$n = C$

n = Number of Calls in Progress
C = Number of Channels

Offered Traffic / Maximum Served Traffic

Figure 9.5 Probability of Blocking and Probability of Number of Simultaneous Calls Using Erlang B Formulas
Erlang B describes a system in which blocked calls are lost, but can be retried later. The service strategy is known as blocked call cleared (BCC) or lost call cleared (LCC).

– the probability that any call is delayed (in the queue) for more than t seconds is

$$P[\text{wait} > t] = P[\text{call delayed}]\exp\{-(C-A)t/H\} \qquad (9.34)$$

where H is the average duration of a call (holding time).
– the average delay for all calls in the system is

$$D = P[\text{call delayed}]\{H/(C-A)\} \qquad (9.35)$$

Figure 9.6 plots the probability of the next arriving call request being delayed [equation (9.33)] and the probabilities of the average call delay exceeding specific multiples of the average holding time [equation (9.34)], as functions of the ratio of offered traffic to the maximum traffic served by the communication system. Under steady-state conditions, when the offered traffic is equal to the system capacity, the average delay for all calls becomes very large. With an 80% load: the probability of the average wait exceeding 0.3 holding times is 41%, the probability of the average wait exceeding one holding time is 20%, and the probability of the average wait exceeding three holding times is 3%. (The probability of the average wait exceeding ten holding times is 0.003%.)

EXAMPLE 9.4   Probability of Receiving a Busy Signal

Given   Your telephone is connected to a system that can handle five simultaneous calls. When the offered traffic exceeds this capability, a trunks-busy signal is returned, i.e., a BCC policy is employed.

Problem   What is the probability of receiving a busy signal when the average loads on the system are 3, 5, and 7 erlangs?

Solution   The probability of the next arriving call being blocked is given by equation (9.31). Substituting $C=5$ gives the following

$$P[\text{blocking}] = \frac{A^5/5!}{\left(1 + A + A^2/2! + A^3/3! + A^4/4! + A^5/5!\right)}$$

so that, when $A=3, 5$, or 7 erlangs

Mean Value Offered Traffic	Probability of Busy Signal
3 E	0.11005
5 E	0.28486
7 E	0.42470

EXAMPLE 9.5   Probability of Call Delay

Given   Your telephone is connected to a system that can handle five simultaneous calls. When the offered traffic exceeds this capability, the incoming calls

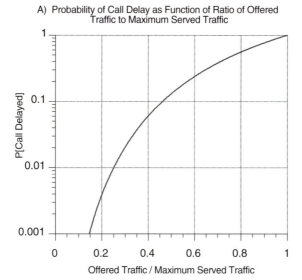

A) Probability of Call Delay as Function of Ratio of Offered Traffic to Maximum Served Traffic

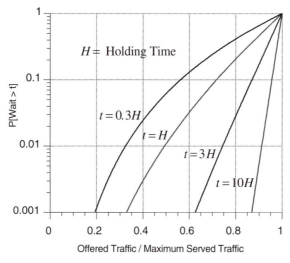

B) Probability of Call Delay Exceeding Specific Value as Function of Ratio of Offered Traffic to Maximum Served Traffic

Figure 9.6 Probability of Call Delay and Probability of Call Delay More Than Specific Time Using Erlang C Formulas
Erlang C describes a system in which calls are held until they can be serviced. The service strategy is known as blocked call delayed (BCD) or lost call delayed (LCD).

are delayed until they can be carried, i.e., a BCD policy is employed.

Problem In the steady state, what is the probability of call delay when the average loads on the system are 1, 3, and 5 erlangs?

Solution The probability of the next arriving call being delayed is given by equation (9.33). Substituting $C=5$ gives the following

$$P[\text{call delayed}] = \frac{A^5}{A^5 + 5!(1-A/5)(1+A+A^2/2!+A^3/3!+A^4/4!)}$$

so that, when $A=1, 3$, or 5 erlangs

Mean Value Offered Traffic	Probability of Call Delayed
1 E	0.0038314
3 E	0.23615
5 E	1.0

EXAMPLE 9.6   Average Call Delay

Given Your telephone is connected to a system that can handle five simultaneous calls. When the offered traffic exceeds this capability, the incoming calls are delayed until they can be carried, i.e., a BCD policy is employed.

Problem In Example 9.5, we calculated the probability of call delay when the average loads on the system are 1, 3, and 5 erlangs. What is the average delay for all calls in the system if the average holding time is 3 minutes?

Solution The average delay for all calls in the system is given by equation (9.35). Substituting $C=5$ and $H=3$ gives the following:

$$D = P[\text{call delayed}]\{3/(5-A)\}$$

so that, when $A=1, 3$, or 5 erlangs, the average call delay in seconds is

Mean Value Offered Traffic	Average Call Delay
1 E	0.0028736×3×60=0.51725 s
3 E	0.35423×3×60=63.7614 s
5 E	very large

REVIEW QUESTIONS

9.1 List the basic components of a queueing model.

9.2 Define the *traffic intensity ratio.*

9.3 Explain the Kendall 3-symbol classification system.

9.4 Describe a birth-death process.

9.5 What are the major characteristics of an M/M/1 queueing system?

9.6 What are the major characteristics of an M/D/1 queueing system?

9.7 How are M/M/1 and M/D/1 related?

9.8 What are the major characteristics of an $M/E_k/1$ queueing system?

9.9 How is $M/E_k/1$ related to M/M/1 and M/D/1?

9.10 How are calls managed with a BCC strategy? Which Erlang formula applies?

9.11 How are calls managed with a BCD strategy? Which Erlang formula applies?

10

Noise

In the real world of telecommunication, messages are accompanied by noise so that, at the output of the receiver, the signal is a distorted replica of the signal that the sender intended to send.

- **Noise**: the difference between the (normalized) corrupted signal present at the output of the receiver and the (normalized) signal the sender presented to the transmitter; it is the sum of all unwanted signals that are added to the message signal in the generation, transmission, reception, and detection processes employed in achieving telecommunication.

10.1 SOURCES OF NOISE

Noise consists of:

- Signals picked up at the receiver such as
  - another modulated wave that interferes with the wanted signal
  - signals generated by natural, impulsive phenomena such as lightning, or by man-made impulsive sources, such as automobile ignition systems
  - galactic radiation
- Signals generated in the processes and equipment used to consummate telecommunication such as
  - the random motion of electrons, ions, or holes in the materials that compose the (receiving) equipment
- Signals that represent processing errors or approximations such as
  - those made in processing the signal for transmission that result in sending a signal that differs from the signal that the sender intended to send
  - those made in detecting and reconstructing the signal at the receiver

10.2 SIGNAL-TO-NOISE RATIO

- **Signal-to-Noise Ratio** (SNR): the ratio of the power in the message signal to the power in the noise signal, i.e.,

$$\text{SNR} = \frac{\text{Signal power }(S)}{\text{Noise power }(N)}$$

Because of the wide range of signal-to-noise ratios, employing the logarithm (to the base 10) of the power ratio is usual. Expressed in this way, the unit is the *bel*. For most work, a more convenient unit is the *decibel* or $bel \times 10^{-1}$, so that

$$(\text{SNR})_{\text{db}} = 10\log_{10}(\text{S/N}) \text{ decibels} \tag{10.1}$$

The idea of the signal-to-noise ratio can be extended to ratios expressed in signal units. For a signal of $s(t)$ signal units and noise of $n(t)$ signal units,

$$(\text{SNR})_{\text{db}} = 10\log_{10}\frac{|s(t)|^2}{|n(t)|^2} = 20\log_{10}\frac{|s(t)|}{|n(t)|}$$

Power ratios expressed in decibels are listed in the Appendix (Table A.4). Good-quality communication systems achieve SNRs of 35 to 45 db, or better. Many persons have difficulty recognizing words when the SNR in a telephone system drops below 15 db.

10.3 WHITE NOISE

When we cannot take account of the sources of noise individually, we can assume that they produce a single, random signal in which power is distributed uniformly at all frequencies. By analogy to white light (that contains all visible frequencies), this signal is called *white* noise.

- **White noise**: a sample function $n(t)$ of a wide-sense stationary random process $N(t)$ whose power spectral density $[\Phi_N(f)]$ is $N_0/2$ signal watts/Hz, i.e.,

$$\Phi_N(f) = N_0/2 \tag{10.2}$$

and the autocorrelation function $[\phi_N(\tau)]$ is

$$\phi_N(\tau) = \tfrac{1}{2}N_0\delta(\tau) \tag{10.3}$$

Diagrams of these quantities are shown in the upper half of **Figure 10.1**.

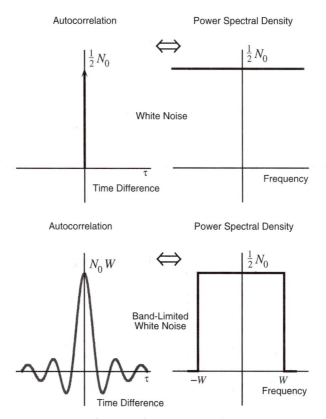

Figure 10.1 Autocorrelation and Power Spectral Density Functions of White Noise and Band-Limited White Noise

White noise has the property that different samples are uncorrelated, and that, if the probability density function of the amplitude spectrum is Gaussian (normal), they are statistically independent.

Because the spectral density is constant at *all* frequencies, the signal consumes infinite power, i.e.,

$$P_N = \int_{-\infty}^{\infty} \tfrac{1}{2} N_0 f \, df = \infty \tag{10.4}$$

Thus, white noise is an unrealizable signal. Fortunately, all practical systems are low-pass or bandpass, and we need to consider only noise within the frequency ranges over which they operate. A more useful concept is band-limited white noise:

- **Band-limited white noise**: noise that has a constant power spectral density over a finite range of frequencies, i.e.,

$$\Phi_N(f) = \begin{cases} N_0/2 \, , & \text{when } -W < f < W \\ 0 \, , & \text{elsewhere} \end{cases} \tag{10.5}$$

Under this circumstance,

$$\phi_N(\tau) = N_0 W \text{Sinc}(2W\tau) \tag{10.6}$$

i.e., the autocorrelation function is a Sinc function. Diagrams of these quantities are shown in the lower half of Figure 10.1.

10.4 NOISE AND BINARY DECISIONS

Noise corrupts the wanted signal and can produce errors in decisions made on the basis of the amplitude of the received signal. Suppose we have a stream of unipolar digital pulses [denoted by $f(t)$] in which the levels are 0 and A. The observed waveform $y(t)$ will consist of $f(t)$ and a noise signal $n(t)$, i.e.,

$$y(t) = f(t) + n(t) \tag{10.7}$$

At time t_1, the possible signals are

$$y(t_1) = \begin{cases} A + n(t_1), & \text{signal one present} \\ n(t_1), & \text{signal zero present} \end{cases}$$

Because $n(t_1)$ is a sample function of a random process, it may add to or subtract from the signals and destroy our certainty as to which level is present. To determine the probability of error, we define the decision threshold, μ, as

$$\text{when } y(t_1) \text{ is } \begin{cases} > \mu, & \text{signal one present} \\ < \mu, & \text{signal zero present} \\ = \mu, & \text{unknown} \end{cases}$$

The probability of $y(t_1) = \mu$ is very small and we will ignore it. If $n(t_1)$ is gaussian-distributed with zero mean and mean-square value of

$$\overline{n^2}(t_1) = \sigma^2$$

the probability density function of $y(t)$ when signal zero is present is

$$f_y(y = 0) = \frac{1}{\sqrt{2\pi}\sigma} \exp\left\{ -y^2/2\sigma^2 \right\} \tag{10.8}$$

and the probability density function of $y(t)$ when signal one is present is

$$f_y(y = 1) = \frac{1}{\sqrt{2\pi}\sigma} \exp\left\{ -(y - A)^2/2\sigma^2 \right\} \tag{10.9}$$

Figure 10.2 shows the relative positions of these functions. By inspection, if ones and zeros are equally likely, the decision threshold should be positioned at $y = A/2$ so that the two possible false readings are equally likely. The shaded areas in the figure represent cases of false identification; the probabilities are

$$P_{\varepsilon_0} = \int_{A/2}^{\infty} \frac{1}{\sqrt{2\pi}\sigma} \exp\{-y^2/2\sigma^2\} dy \qquad (10.10)$$

and

$$P_{\varepsilon_1} = \int_{-\infty}^{A/2} \frac{1}{\sqrt{2\pi}\sigma} \exp\{-(y-A)^2/2\sigma^2\} dy \qquad (10.11)$$

If P_0 and P_1 are the probabilities of zero and one signals being sent, then

$$P_0 = P_1 = \tfrac{1}{2}$$

and the total probabity of error P_ε is

$$P_\varepsilon = P_0 P_{\varepsilon_0} + P_1 P_{\varepsilon_1} = \tfrac{1}{2}\left(P_{\varepsilon_0} + P_{\varepsilon_1}\right) \qquad (10.12)$$

With $\mu = A/2$ and $P_0 = P_1 = \tfrac{1}{2}$, $P_{\varepsilon_0} = P_{\varepsilon_1}$. Therefore, we can write

$$P_\varepsilon = \int_{A/2}^{\infty} \frac{1}{\sqrt{2\pi}\sigma} \exp\{-y^2/2\sigma^2\} dy \qquad (10.13)$$

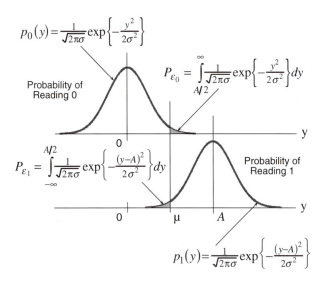

Figure 10.2 Probability Density Functions for 0 (level 0) and 1 (level A) with a Decision Threshold at μ

The shaded areas represent cases of false identification.

This expression has the form of the complementary error function, so that

$$P_\varepsilon = \text{Erfc}(A/2\sigma) \qquad (10.14)$$

The complementary error function is defined as

$$\text{Erfc}(x) = 1 - \text{Erf}(x)$$

where Erf is the error function. Its value is

$$\text{Erf}(x) = \frac{2}{\sqrt{\pi}} \int_0^x \exp(-u^2)\,du$$

where u is a variable of integration. The complimentary error function cannot be evaluated analytically; tables of numerical values exist, and approximations have been found. As an example, when x is large and positive, the bounds of $\text{Erfc}(x)$ can be shown to be[1]

$$\frac{1}{\sqrt{\pi}x}\left(1 - \frac{1}{2x^2}\right)\exp(-x^2) < \text{Erfc}(x) < \frac{1}{\sqrt{\pi}x}\exp(-x^2) \qquad (10.15)$$

For a unipolar non-return to zero (NRZ) binary signal (see § 13.2) of amplitude A, the signal power S is $A^2/2$. For zero-mean, gaussian-distributed noise, the noise power N is σ^2. Substituting these values in equation (10.14) gives the probability of error in terms of the signal-to-noise ratio, i.e.,

$$P_{\varepsilon_{\text{unipolarNRZ}}} = \text{Erfc}\sqrt{S/2N} \qquad (10.16)$$

For a bipolar NRZ binary signal of amplitude $A/2$, i.e., $S = A^2/4$,

$$P_{\varepsilon_{\text{bipolarNRZ}}} = \text{Erfc}\sqrt{S/N} \qquad (10.17)$$

Figure 10.3 shows approximate values of probability of error and SNR derived from equations (10.16) and (10.17). The probabilities decrease rapidly as the signal-to-noise power ratio exceeds 10 db.

EXAMPLE 10.1   SNR and Probability of Decision Error

Given Signal power is measured at the output of an ideal bandpass filter of bandwidth $2W$ Hz. When no message signal is present, a power of 1×10^{-6} watts is measured. When a NRZ unipolar message signal is present, a power of 1.1×10^{-5} watts is measured. The noise signal is characterized as white noise.

---

[1]N. M. Blachman, *Noise and Its Effect on Communications* (New York: McGraw-Hill, 1966), pp. 5–6.

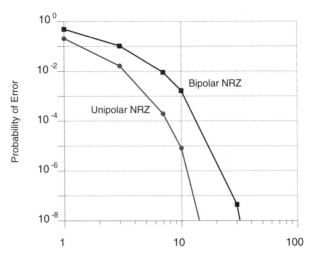

Signal-to-Noise Power Ratio – decibels

Figure 10.3 Probability of Error Due to Noise for Unipolar NRZ and Bipolar NRZ Pulse Trains
The probabilities of error decrease rapidly as the signal-to-noise power ratio exceeds 10 db.

Problem

   a) Express the signal-to-noise ratio in db.

   b) What is the probability that the receiver will read a pulse wrong?

Solution

   a) Let the message signal power be S microwatts, and the noise power be N microwatts; then $N=1$ and $S+N=11$ so that

$$(S+N)/N=11$$

   Simplifying

$$S/N=10=10 \text{ db.}$$

   b) From equation (10.16), the probability of error for a NRZ unipolar pulse is[2]

$$P_\varepsilon = \text{Erfc}\sqrt{S/2N} = 1 - \text{Erf}\sqrt{5} = 1 - 0.99843 = 0.00157$$

---

[2]Tables of the Error Integral can be found in Eugene Jahnke, Fritz Emde, *Tables of Functions*, 4th Edition (Dover: New York, 1945) p. 24.

i.e.,

$$P_\varepsilon = 1.57 \times 10^{-3}$$

This is in agreement with Figure 10.3.

EXAMPLE 10.2 SNR and Probability of Decision Error

Given With the same signal and noise power levels as in Example 10.1, signal power is measured at the output of an ideal bandpass filter of bandwidth $4W$ Hz. The noise signal is characterized as white noise.

Problem

  a) What are the power readings when the signal is absent and when the signal is present?

  b) What is the signal-to-noise ratio?

  c) What is the probability that the receiver will read a pulse wrong?

Solution

  a) Increasing the bandwidth of the filter by two increases the noise power by two, so that $N=2$ signal microwatts and $S+N=12$ signal microwatts.

  b) Under these circumstances,

$$(S+N)/N = 6$$

and

$$S/N = 5 = 6.98970 \text{ db.}$$

  c) From equation (10.16), the probability of error for a NRZ unipolar pulse is

$$P_\varepsilon = \text{Erfc}\sqrt{S/2N} = 1 - \text{Erf}\sqrt{2.5} = 1 - 0.9646 = 0.0354$$

so that the probability that the receiver will read a pulse wrong is 3.54×10^{-2}.

EXAMPLE 10.3 Filter Passband

Given With the same signal and noise power levels as in Example 10.1, the message signal is a pure tone centered at the middle of the passband.

Problem What is the passband of the filter for an SNR of 30 db?

Solution Given $S/N = 30 \text{ db} = 1000$, and the noise power is 1×10^{-6} watts in $2W$ Hz, i.e., the noise power density is $\frac{1}{2W} \times 10^{-6}$ watts/Hz. If the passband of the filter is W', the noise power is $\dfrac{W'}{2W} \times 10^{-6}$. From Example 10.1, the signal power is 1×10^{-5} watts; hence

$$\frac{S}{N} = 1000 = \frac{1 \times 10^{-5}}{\frac{W'}{2W} \times 10^{-6}} = \frac{2W}{W'} \times 10$$

so that

$$W' = 2W \times 10^{-2}$$

EXAMPLE 10.4   Decision Errors with Unipolar and Bipolar Signals

Given   A very long sequence of binary pulses that represent a random distribution of 1s and 0s.

Problem   Calculate the probability of error in assigning a value to

a) unipolar NRZ pulses when SNR = 10 db

b) bipolar NRZ pulses when SNR = 5 db

Solution

a) An SNR of 10 db is a signal ratio of 10. Substituting in equation (10.16)

$$P_{\varepsilon_{unipolarNRZ}} = \text{Erfc}\sqrt{5} = \text{Erfc}(2.23607)$$

From equation (10.15)

$$\frac{1}{\sqrt{5\pi}}\left(1 - \frac{1}{2 \times 5}\right)\exp(-5) < \text{Erfc}\left(\sqrt{5}\right) < \frac{1}{\sqrt{5\pi}}\exp(-5)$$

so that

$$0.24771 \times \exp(-5) < \text{Erfc}\left(\sqrt{5}\right) < 0.27524 \times \exp(-5)$$

i.e.,

$$0.0015300 < \text{Erfc}\left(\sqrt{5}\right) < 0.0017000$$

The probability of error lies between 1.53×10^{-3} and 1.70×10^{-3}.

b) An SNR of 5 db is a signal ratio of 3.162. Substituting in equation (10.17)

$$P_{\varepsilon_{bipolarNRZ}} = \text{Erfc}\sqrt{3.162} = \text{Erfc}(1.77820)$$

From equation (10.15)

$$\frac{1}{\sqrt{3.162\pi}}\left(1 - \frac{1}{6.324}\right)\exp(-3.162) < \text{Erfc}(1.77820) < \frac{1}{\sqrt{3.162\pi}}\exp(-3.162)$$

so that

$$0.011310 < \text{Erfc}(1.77820) < 0.013434$$

i.e., the probability of error lies between 1.13×10^{-2} and 1.34×10^{-2}.

REVIEW QUESTIONS

10.1 Define *noise*.

10.2 Define *signal-to-noise ratio*.

10.3 Explain why decibels are used, and state the relationship between SNR and $(\text{SNR})_{db}$.

10.4 Define *white noise* and *band-limited white noise*.

10.5 Explain how noise can produce errors in decisions made on the basis of the amplitude of the received signal.

11

Information Theory

Stimulated by the need to assess the information-carrying capacity of a communication system, information theory provides the analytical tools to evaluate the amount of information contained in message signals, and to compare the performance of actual systems.

- **Information**: a commodity contained in the symbols transferred from source to receiver over a communication channel.

The quantity of *information* carried by a symbol was defined by Claude Shannon[1] (a 20th-century American mathematician) in terms of the likelihood that it will be sent.

11.1 SELF INFORMATION

- **Self information**: when a source emits a sequence of symbols x_1, x_2, ... and symbol x_j occurs with probability P_j, x_j conveys a quantity of information (self information) I_j given by

$$I_j = -\log P_j = \log 1/P_j \qquad (11.1)$$

According to equation (11.1), if the same symbol is received all the time, the probability of being sent is 1 and the self information is 0, i.e.,

$$P_j = 1 \text{ and } I_j = 0$$

---

[1]Claude E. Shannon, "A mathematical theory of communications," *Bell System Technical Journal,* 27 (1948), 379–423 and 623–56.

If a symbol is sent infrequently, the probability of being sent is close to zero and the self information is high, i.e.,

$$P_j \to 0 \text{ and } I_j \to \infty$$

These relations mimic common experience. A message that is repeated constantly is ignored; it provides very little new information. On the other hand, attention is paid to a message that is received infrequently; it provides new information.

Because they assume only two states, for sequences of binary symbols, we use the logarithm to the base 2 so that equation (11.1) is written

$$I_j = -\log_2 P_j = \log_2 1/P_j \qquad (11.2)$$

and the unit of self information is 1 *bit*. Unfortunately, this term is used also for a binary digit. When confusion is possible, custom reserves the term *bit* for self information and employs the term *binit* for binary digit.

For a binary source that emits equiprobable symbols, i.e.,

$$P(x_j) = P(x_k) = 1/2$$

equation (11.2) gives

$$I_j = I_k = \log_2 2 = 1 \text{ bit.} \qquad (11.3)$$

Thus, the information conveyed by every binary symbol when 1s and 0s are equally likely is 1 bit.

If a source emits two successive binary symbols with joint probability $P(x_j, x_k)$

$$I_{jk} = \log_2 1/P(x_j, x_k) \qquad (11.4)$$

and, if x_j and x_k are independent

$$I_{jk} = \log_2 1/P_j P_k = \log_2 1/P_j + \log_2 1/P_k = I_j + I_k \qquad (11.5)$$

Thus, the total information due to two successive, independent binary symbols is equal to the sum of their self information. This result can be generalized to n symbols emitted from a common source.

11.2 ENTROPY AND INFORMATION RATE

A term borrowed from statistical thermodynamics, entropy measures the average information contained in the stream of symbols emitted by a memoryless source.

- **Discrete memoryless source**: an information source that emits symbols at a rate of r symbols/s. Each symbol x_j occurs with probability P_j and conveys self information I_j. The symbol probabilities remain constant over time and are mutually independent. No symbol depends on the occurrence of any other symbol.

If a discrete memoryless source emits a sequence of symbols $x_1, x_2, ... x_M$, the amount of information produced by the source during any symbol interval is a discrete random variable (X) having possible values $I_1, I_2, ... I_M$.

- **Source entropy**: the average value (expected value) of information per symbol (in bits) emitted by a source. Denoted by $H(X)$, for a discrete memoryless source that emits a sequence of symbols $x_1, x_2, ... x_M$, the source entropy is

$$H(X) = \sum_{j=1}^{M} P_j I_j = \sum_{j=1}^{M} P_j \log_2 1/P_j \qquad (11.6)$$

When the source always emits the same symbol, i.e., $P_j = 1$ and $P_k = 0$ for all $k \neq j$,

$$H(X) = 0.$$

When the source emits all symbols with the same probability, i.e., $P_j = 1/M$ for all j, and $0 \leq j \leq M$,

$$H(X) = \log_2 M.$$

These conditions represent absolute certainty, and total uncertainty. They are the bounds of $H(X)$, so that

$$0 \leq H(X) \leq \log_2 M \qquad (11.7)$$

- **Information rate**: when a source emits a sequence containing a very large number of symbols ($n \gg 1$), the total information transferred is $nH(X)$ bits. If the symbols are produced at a rate r per second, the source information rate R, in bits/s is

$$R = rH(X) \qquad (11.8)$$

EXAMPLE 11.1   Entropy of a Binary Source

Given A memoryless binary source in which 1s occur with probability p_1, 0s occur with probability p_0, and $p_1 = 1 - p_0$.

Problem Find the entropy of the source.

Solution Since the source is memoryless, the symbols are statistically independent, so that

$$H(S) = -p_1 \log_2 p_1 - p_0 \log_2 p_0$$

Rewriting in terms of p_1

$$H(S) = -p_1 \log_2 p_1 - (1 - p_1) \log_2 (1 - p_1)$$

Comment **Figure 11.1** shows how the entropy of a binary memoryless source varies with the probability of symbol 1 occurring. The entropy attains a maximum value of 1 bit when $p_1 = p_0 = \frac{1}{2}$, and is zero when $p_1 = 0$ (i.e., $p_0 = 1$). This agrees with equation (11.3) and the discussion of self information in § 11.1.

11.3 ENTROPY CODING

For a discrete memoryless source that emits M equilikely symbols, all the symbols have the same value of self information and can be represented by codewords of equal length (N binits) under the restriction $M \leq 2^N$, or $N \geq \log_2 M$. In this case, the average length $(\overline{N})$ is N.

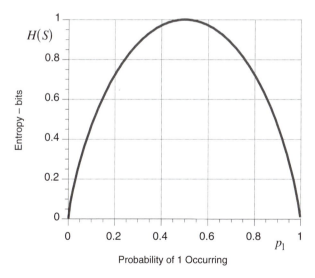

Figure 11.1 Entropy of a Binary Source (from Example 11.1)
When the probability of 1 occurring is the same as the probability of 0 occurring, the entropy of the source is one bit per symbol. As the balance degrades, the entropy drops off until, when all 1s or all 0s are sent, the entropy of the source is zero.

For a discrete memoryless source that emits M symbols that have different values of self information,

$$\overline{N} = \sum_{j=1}^{M} P_j N_j \tag{11.9}$$

where N_j is the length of the codeword for the jth symbol.

- **Coding efficiency**: for a binary source, the ratio of the average information (in bits) per symbol and the average length of the symbols (in binits), i.e.,

$$\eta = H(X)/\overline{N} \tag{11.10}$$

Thus, for a source of constant entropy, coding efficiency increases as the average length of the codewords decreases. Optimum efficiency occurs when $\overline{N} = H(X)$.

For coding to be a success, the code must be decipherable. For codewords of equal length, this requirement is easy to fulfill—start at the beginning and divide the symbol stream into equal length words $\left(\overline{N} = N\right)$. For symbol streams with variable length codewords, the task is more difficult. One approach is to require that no member of the code set be the prefix of any other member.

An efficient code in which the codeword lengths increase as the symbol self information decreases, and no member of the code is the prefix of any other member, is known as *Shannon-Fano* coding.[2] **Figure 11.2** shows its application to the coding of 8 symbols (S1, S2, ... S8) that have probabilities of being sent ranging from 0.35 to 0.02, and an entropy of 2.57 bits. To construct a Shannon-Fano code, the symbols are arranged in descending order of probability of being sent. The column is divided so as to split the sum of the probabilities in half—or as close to half as possible. To the symbols above the split, 0s are assigned, and to the symbols below the split, 1s are assigned. Each subdivision is then divided again so as to have approximately half the sum of the probabilities above the dividing line, and half below. 0s and 1s are assigned as before. The process continues until all symbols have been coded. In the example, the entropy

$$H(X) = \sum_{1}^{8} P_k \log_2 1/P_k = 2.57105 \text{ bits}$$

and the average number of binits per codeword is

$$\overline{N} = \sum_{1}^{8} P_k N = 2.63$$

---

[2]R. M. Fano, "A Heuristic Discussion of Probabilistic Coding," *IEEE Transactions on Information Theory*, IT-9 (1963), 64–74.

Symbol	P_k	$P_k \log_2 \frac{1}{P_k}$	SHANNON-FANO CODING TABLE					Code	$P_k N$
S1	0.35	0.53011	0	0				00	0.70
S2	0.20	0.46440	0	1				01	0.40
S3	0.15	0.41055	1	0	0			100	0.45
S4	0.10	0.33220	1	0	1			101	0.30
S5	0.08	0.29151	1	1	0			110	0.24
S6	0.06	0.24354	1	1	1	0		1110	0.24
S7	0.04	0.18576	1	1	1	1	0	11110	0.20
S8	0.02	0.11288	1	1	1	1	1	11111	0.10

Entropy = 2.57105 bits
Average Number of Binits per Codeword = 2.63
Coding Efficiency = 97.8%

Figure 11.2 Construction of a Shannon-Fano Code
The diagram illustrates the development of a code in which the length of a codeword is inversely related to the probability of occurrence of the symbol it represents.

so that the coding efficiency is 97.8%. If we chose to code the symbols with an equal number of binits (3), the efficiency is 85.7%. Thus, for this particular set of symbols and statistical mix, Shannon-Fano coding improves coding efficiency by 14.1% over equal length, 3 binit codewords.

EXAMPLE 11.2   Shannon-Fano Coding and Decoding

Given A set of eight symbols occurs with probabilities S1, 0.03; S2, 0.17; S3, 0.08; S4, 0.12; S5, 0.15; S6, 0.1; S7, 0.11; and S8, 0.24.

Problem

a) Construct a Shannon-Fano code for the set.

b) Determine the entropy of the code and the average number of binits per codeword.

c) What is the efficiency of the code?

d) Decipher the data stream

⇐ 11011000110100111011010110001111000101110010011101...

Solution

a), b), and c) In **Figure 11.3**, we construct a diagram similar to Figure 11.2 to answer the questions.

d) We must match the bits in sequence against the eight members of the code set, i.e., S8 = 00, S2 = 010, S5 = 011, S4 = 100, S7 = 101, S6 = 110, S3 = 1110, and S1 = 1111. The stream begins with 1; this eliminates S8, S2, and S5. The second bit is 1; this eliminates S4 and S7. The third bit is 0; this eliminates S3 and S1, and matches S6. Thus, the first symbol is S6. Bits 4, 5, and 6 repeat 110, so the second symbol is S6.

Bit 7 is 0, eliminating all but S8, S2, and S5. Bit eight is 0 so that the third symbol is S8.

Bit 9 is 1, eliminating S8, S2, and S5. Bit 10 is 1, eliminating S4 and S7. Bit 11 is 0, making the fourth symbol S6, etc.

The first four symbols in the message are S6 S6 S8 S6. We leave it to the reader to confirm that the entire message is

S6 S6 S8 S6 S4 S3 S6 S7 S4 S5 S6 S8 S7 S3 S2 S5 S7

Comment That we can decipher the data stream into a unique sequence of symbols confirms that the Shannon-Fano technique produces unique codewords.

Symbol	P_k	$P_k \log_2 \frac{1}{P_k}$	SHANNON-FANO CODING TABLE				Code	$P_k N$
S8	0.24	0.49413	0	0			00	0.48
S2	0.17	0.43459	0	1	0		010	0.51
S5	0.15	0.41055	0	1	1		011	0.45
S4	0.12	0.36707	1	0	0		100	0.36
S7	0.11	0.35029	1	0	1		101	0.33
S6	0.10	0.33219	1	1	0		110	0.30
S3	0.08	0.29151	1	1	1	0	1110	0.32
S1	0.03	0.15177	1	1	1	1	1111	0.12

Entropy = 2.8321 bits
Average Number of Binits per Codeword = 2.87
Coding Efficiency = 98.6%

Figure 11.3 Diagram for Example 11.2

11.4   CHANNEL CAPACITY

- **Channel capacity**: the maximum rate of reliable information transfer over a communication channel (in bits/s)
- **Discrete channel**: a communication channel whose input consists of a symbol from an alphabet of K symbols, and whose output consists of a symbol belonging to the same alphabet

(1) Shannon's Capacity Theorem

- **Capacity theorem**: if C is the capacity of a discrete memoryless channel, $H(X)$ is the entropy of a discrete information source emitting r symbols per second, and $rH(X) \leq C$, then a coding scheme exists such that the output of the source can be transmitted over the channel with an arbitrarily small probability of error. Conversely, if $rH(X) > C$, the channel cannot transmit messages without error.

Under the condition $rH(X) \leq C$, the capacity theorem asserts that you can encode the source symbols so as to achieve virtually error-free transmission over a noisy channel. Regarded as a remarkable result at the time of its development, it has spurred a half-century of work directed to discovering key coding algorithms.

(2) Maximum Entropy of Continuous Random Variable

- **Entropy of Continuous Random Variable**: in a fashion analogous to equation (11.6), the entropy (in bits) of a continuous random variable X with probability density function $f_X(x)$ is

$$H(X) = \int_{-\infty}^{\infty} f_X(x) \log_2 [f_X(x)] dx \qquad (11.11)$$

Provided the mean-square value of X is constant, $H(X)$ achieves a maximum when

$$f_X(x) = \frac{1}{\sigma\sqrt{2\pi}} \exp\left(-x^2/2\sigma^2\right)$$

i.e., the continuous random variable is zero-mean and Gaussian-distributed. Under these conditions

$$H(X)_{max} = \log_2\left(\sigma\sqrt{2\pi e}\right) = 2 + \log_2 \sigma \qquad (11.12)$$

(3) Hartley-Shannon Law

- **Continuous channel**: a communication channel whose input and output signals are continuous time functions.

- **Hartley-Shannon law**: for a continuous channel with finite average power and additive, band-limited, white Gaussian noise, the information capacity (in bits/s) is

$$C = B_T \log_2(1 + S/N) \tag{11.13}$$

where B_T is the transmission bandwidth of the channel, S is the average power of the transmitted signal, and N is the average power of the noise signal.

If $x(t)$ and $y(t)$ are the continuous signals at the input and output of the channel, we can define them as sample functions of a zero-mean, white Gaussian noise process and represent them by samples taken at the Nyquist rate, i.e., $\sum x_\delta(t)$ and $\sum y_\delta(t)$. If Y is the random variable composed of $\sum y_\delta(t)$, substituting in equation (11.12) gives the maximum entropy (in bits/sample) of the output signal

$$H(Y)_{max} = \log_2\left(\sqrt{2\pi e(S+N)}\right) \tag{11.14}$$

- **Equivocation** $[H(Y|X)]$: conditional entropy or measure of the average uncertainty of received sample Y given that sample X was transmitted, i.e.,

$$H(X) = H(Y) - H(Y|X) \tag{11.15}$$

where

$$H(Y \mid X) = -\int\limits_{-\infty}^{\infty}\int f_{X,Y}(x,y)\log_2\left[f_{Y|X}(y\mid x)\right]dxdy$$

$f_{X,Y}(x,y)$ is the joint probability density function of X and Y, and $f_{Y|X}(y\mid x)$ is the conditional probability density function of Y given X.

In this case, equivocation is the entropy of the added noise, so that

$$H(Y|X) = \log_2\left(\sqrt{2\pi eN}\right) \tag{11.16}$$

and the average transmitted information (in bits/sample) is

$$H(Y) - H(Y|X) = \log_2\left(\sqrt{2\pi e(S+N)}\right) - \log_2\left(\sqrt{2\pi eN}\right)$$

which becomes

$$H(Y) - H(Y|X) = \tfrac{1}{2}\log_2(1 + S/N) \tag{11.17}$$

Now, the samples are taken at $2B_T$ samples/s, so that the information capacity of the channel (in bits/s) is

$$C = B_T \log_2(1 + S/N)$$

which is the Hartley-Shannon Law stated as equation (11.13). It sets an upper limit to the rate of transmission over a channel characterized by additive white Gaussian noise, and, for a specified channel capacity, defines the way in which B_T may be traded for signal-to-noise ratio, and *vice-versa*.

Figure 11.4 shows the variation of B_T/C versus signal-to-noise ratio. For a fixed transmission bandwidth, the channel capacity increases with S/N. Thus

$$C_{40db} \cong 2C_{20db}$$

and

$$C_{30db} \cong 3C_{10db}$$

EXAMPLE 11.3   Channel Capacity

Given   A 4 kHz channel with additive white Gaussian noise.

Problem

a) If the signal-to-noise ratio is 20 db, calculate the channel capacity in bits/s.

b) If the signal-to-noise ratio is 40 db, calculate the channel capacity.

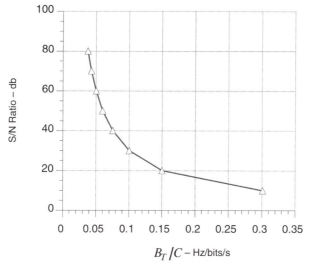

Figure 11.4 Variation of B_T/C versus signal-to-noise ratio

For a fixed transmission bandwidth, the channel capacity increases with S/N.

Solution

    a) Substituting in equation (11.13)

$$C = 4 \times 10^3 \log_2(101) = 26.578 \text{ kbits/s}$$

    b) Substituting in equation (11.13)

$$C = 4 \times 10^3 \log_2(10001) = 53.156 \text{ kbits/s}$$

Comment A 4 kHz channel is a telephone channel. According to the Hartley-Shannon law, under the constraint of white Gaussian noise, the maximum information capacity is 26.6 kbits/s at an SNR of 20 db, and 53.2 kbits/s at an SNR of 40 db. If we assume a binary datastream with 1s and 0s equally likely, so that each binit carries 1 bit of information, the maximum binit rates are 26.6 and 53.2 kbinits/s. Approaching these speeds requires very sophisticated coding techniques.

---

(4) Shannon Bound

- **Shannon bound**: the capacity (in bits/s) of a channel with additive white noise, with given average transmitted power, and unlimited transmission bandwidth, i.e., $B_T \rightarrow \infty$.

$$\lim_{B_T \to \infty} C = \frac{S}{N_0} \log_2 e \tag{11.18}$$

The average power of white noise with power spectral density $N_0/2$ signal watts/Hz is $N = B_T N_0$ signal watts. Substituting in equation (11.13)

$$C = B_T \log_2(1 + S/N_0 B_T) \tag{11.19}$$

To simplify taking the limit $B_T \rightarrow \infty$, a convenient aproach is to rewrite equation (11.18) as

$$C = \frac{S}{N_0} \log_2\left[(1 + S/N_0 B_T)^{N_0 B_T / S} \right]$$

In taking the limit, we use the relationship

$$\lim_{x \to 0}(1 + x)^{1/x} = e$$

so that

$$\lim_{B_T \to \infty} C = \frac{S}{N_0} \log_2 e = 1.443 \frac{S}{N_0}$$

This is the expression quoted above in equation (11.18). It defines the maximum rate of information transmission for an additive white noise limited communication channel with infinite bandwidth and finite average power. The nature of this relationship is shown in **Figure 11.5**.

(5) Bandwidth Efficiency

- **Bandwidth efficiency**: a measure of the number of bits per second transmitted per Hertz of channel bandwidth.

Given by C/B_T, it can be expressed in terms of signal energy per bit (E_b) and bit duration (T_b)

$$C/B_T = \log_2\left(1 + A\frac{E_b/N_0}{T_b B_T}\right) \tag{11.20}$$

where A is a constant and equal to 1 for a Gaussian channel, and to $12/k^2$ for a PCM channel [see § 12.5(3)].

EXAMPLE 11.4   Bandwidth Efficiency

Given   Two 4 kHz channels, one characterized by additive white Gaussian noise, the other a PCM channel for which $A = 12/k^2 = 0.1$ [see equation

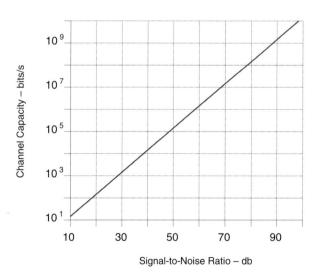

Figure 11.5 Shannon Bound
The capacity (in bits/s) of a channel with additive white Gaussian noise, given average transmitted signal power and unlimited bandwidth.

(11.20)]. The channels transport binary datastreams in which 1s and 0s are equally likely, so that each binit carries 1 bit of information.

Problem

   a) If the signal-to-noise ratio is 20 db, calculate the channel capacity in binits/s of the PCM channel.

   b) If the signal-to-noise ratio is 40 db, calculate the channel capacity in binits/s of the PCM channel.

   c) Compare the results to those from Example 11.3.

   d) What is the bandwidth efficiency of these situations?

Solution

   a) Substituting in equation (11.13)

$$C = 4 \times 10^3 \log_2(10.1) = 13.289 \text{ kbits/s} = 13.289 \text{ kbinits/s}$$

   b) Substituting in equation (11.13)

$$C = 4 \times 10^3 \log_2(1001) = 39.867 \text{ kbits/s} = 39.867 \text{ kbinits/s}$$

   c) Comparing binits/s for the same SNR

$$C_{PCM}/C_{Gauss} = 50\% \text{ at SNR} = 20 \text{ db}$$

$$C_{PCM}/C_{Gauss} = 75\% \text{ at SNR} = 40 \text{ db}$$

   d) Bandwidth efficiency in kbinits/Hz is

		PCM	Gaussian
At 20 db	$C/B_T =$	3.322	6.644
At 40 db		9.967	13.289

Comments The advantage of the Gaussian channel decreases with improvement in SNR. For SNR of 60 db,

$$C_{PCM}/C_{Gauss} = 66.445/79.734 = 0.833$$

However, most telephone channels have SNRs around 30 db.

REVIEW QUESTIONS

11.1   Define *information* and *self information*.

11.2   Distinguish between a *bit* and a *binit*.

11.3   For a binary stream in which 1s and 0s are equally likely, what is the self information of one binit?

11.4   What is a *discrete memoryless source?*

11.5   Define *source entropy*.

11.6   What are the bounds of source entropy?

11.7   Define *information rate.*

11.8   What is the entropy of a memoryless binary source in which 1s and 0s occur with equal probability?

11.9   Define *coding efficiency.*

11.10   State Shannon's capacity theorem.

11.11   State the Hartley-Shannon law.

11.12   State Shannon's bound on channel capacity.

12

Digital Voice Signals

Voice signals consist of bursts of electrical energy. In a conversation, the participants share the task of talking. On average, each one talks about 40% of the time. Occasionally, the talking activities will overlap as one interrupts the other, but most of the remaining 20% represents the time neither party is talking.

Speech consists of elementary sounds called *phonemes* that are uttered at rates up to 10 per second. They are produced by bursts of air that originate in the voice box, acquire pitch by passing through the vocal chords, and excite resonances in the vocal tract. The sequence of sound waves is detected by the ear and interpreted in the brain as speech spoken in a voice unique to the speaker.

By means of a microphone, the sound pressure waves are transformed into electrical signals that exhibit *frequency, phase,* and *amplitude.* Roughly, they correspond to the pitch, intelligibility, and intensity of the original sound. Measurements on speech signals produced by a large population of talkers indicate that they consist of energy between approximately 100 Hz and 7000 Hz, with a peak around 250 Hz to 500 Hz. The dynamic range is some 35 or 40 db. The amplitude distribution favors low signal levels and can be represented by a Laplace probability density function [see § 12.3(1)]. For transmission over standard telephone facilities, the signal energy is restricted to a range from 200 Hz to 3400 Hz. On toll facilities, the signal-to-noise ratio is generally expected to be 30 db, or better.

Naturally analog, voice signals are converted to digital signals for transport over today's digital communication systems, and returned to analog form at their destination. Within limits, the reconstructed analog signal mimics the original analog signal.

12.1 SAMPLING AN ANALOG SIGNAL

To digitize an analog signal, it is reduced to a sequence of values that can be expressed as digital codes. The values are obtained from signal samples taken for this purpose.

(1) Band-Limited Signal

Telecommunication signals are limited to specific bandwidths by the equipment that transports and processes them. Accordingly, we consider only band-limited signals, i.e., signals that contain energy in the range $0 < f \le W$ Hz. Elsewhere the signal value is zero.

(2) Sampling

- **Sampling**: the process of determining the value of the amplitude of an analog signal at an instant in time

Usually, sampling is done in a repetitive manner at equal time intervals. An *ideal* sampling function consists of a periodic train of Dirac delta functions [see § 3.3(1)]. Denoted by $\delta_{T_s}(t)$, it consists of delta functions spaced apart by the sample time T_s. Thus

$$\delta_{T_s}(t) = \sum_{k=-\infty}^{\infty} \delta(t - kT_s) \tag{12.1}$$

The Fourier transform of this function is [equation (3.32)]

$$\delta_{T_s}(t) \Leftrightarrow f_s \delta_{f_s}(f) \tag{12.2}$$

where $f_s = 1/T_s$. This relation tells us that, in the frequency domain, the ideal sampling function becomes a comb of delta functions of area f_s, which are separated from each other by f_s. Applying the sampling function to the wave $s(t)$, we can obtain expressions for the sampled wave in time and frequency domains

$$s_\delta(t) = s(t)\delta_{T_s}(t) = \sum_{k=-\infty}^{\infty} s(kT_s)\delta(t - kT_s) \tag{12.3}$$

$$S_\delta(f) = S(f) \otimes \delta_{f_s}(f) = f_s \sum_{k=-\infty}^{\infty} S(f - kf_s) \tag{12.4}$$

Equation (12.3) represents a train of delta functions (of period T_s) that assume the value of the signal wave at the sampling instants. Equation (12.4) represents the spectrum replicated about the zero frequency point at intervals of f_s. **Figure 12.1** shows an example of these signals for the case of a

Original Signal

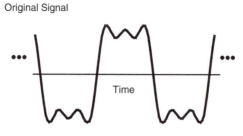

Sampled Signal

$$s_\delta(t) = s(t)\delta_{T_s}(t) = \sum_{k=-\infty}^{\infty} s(kT_s)\delta(t - kT_s)$$

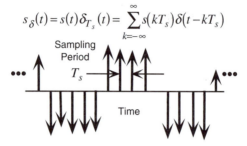

Spectrum of Sampled Signal

$$S_\delta(f) = S(f) \otimes \delta_{f_s}(f) = f_s \sum_{k=-\infty}^{\infty} S(f - kf_s)$$

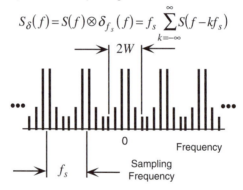

Figure 12.1 The Sampling of a Periodic Waveform
The original signal is a square wave that is limited to the fundamental, third, and fifth harmonics. It is sampled by a Dirac train at 110% of the Nyquist rate. The spectrum of the sampled signal duplicates the spectrum of the original signal at intervals of 2W + 10% Hz.

square wave that is limited to the fundamental, third, and fifth harmonics (taken from Figure 4.4). In drawing the spectrum, we oversample by 10% [see § 12.1(3)].

- **Nyquist rate**: a sampling rate (in samples/s, f_s) that is exactly twice the highest frequency (in Hertz, $2W$) in a band-limited signal $(0 < f \leq W)$.

When $f_s \geq 2W$, sampling yields a set of delta functions that contains the information necessary to reconstruct the original, band-limited signal. To demonstrate this result for the case of $f_s = 2W$, imagine passing $S_\delta(f)$ through a low-pass filter that limits the frequencies to $\pm W$. Reference to Figure 12.1 shows that this range just encompasses the frequency components necessary to reconstruct the original signal. Sampling at $f_s > 2W$ increases the interval between the repeated blocks of harmonic components in the bottom diagram in Figure 12.1; it does not change the internal relationship between the components.

(3) Aliasing

If $f_s < 2W$, i.e., the sampling rate is less than the Nyquist rate, the reconstructed signal will be a distorted version of the original with the energy that falls in the frequency range $f_s/2 < f \leq W$ Hz masquerading at lower frequencies. This effect is known as

- **Aliasing**: condition that exists when signal energy that is present over one range of frequencies in the original signal appears in another range of frequencies after sampling.

Figure 12.2 shows the effect on the spectrum of sampling the signal of Figure 12.1 at a rate that is less than the Nyquist rate ($f_s = 1.8W$). Energy from the spectrums on either side of the baseband appears in the band $\pm W$. The result is the distorted signal shown in the lower diagram of Figure 12.2. To prevent aliasing, to *oversample* the band-limited signal is usual (perhaps 110% to 120% of the Nyquist rate), and a low-pass filter of bandwidth W Hz may be included in the signal chain before the sampling operation.

EXAMPLE 12.1   Sampling Above and Below the Nyquist Rate

Given A cosine function $A\cos(2\pi 1000t)$ is sampled by a Dirac train $\delta_{T_1}(t)$ and passed through an ideal low-pass filter of bandwidth 1050 Hz (i.e., $-1050 \leq f \leq 1050$ Hz) and gain $1/f_1$, i.e., T_1, the sampling period.

Problem Derive expressions for the output signals when sampling occurs at

a) 2100 samples/s

b) 1900 samples/s

c) Show that sampling at 1900 samples/s produces a function that is limited to the band $0 > f \geq 950$ Hz.

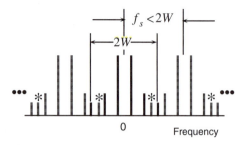

* Aliasing Energy from Adjacent Spectrums

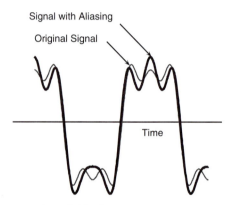

Figure 12.2 Demonstration of Aliasing
*Undersampling the signal causes the spectrums in the sampled wave to
overlap. The result is that energy that is present over one range of frequencies in the original signal appears in another range of frequencies after
sampling. This aliasing energy causes the reconstructed signal to be a distorted version of the original signal.*

Solution
   a) From Example 3.4,

$$A\cos(2\pi 1000t) \Leftrightarrow \tfrac{A}{2}\big[\delta(f-1000)+\delta(f+1000)\big]$$

and, from § 3.3(1)

$$\delta_{T_1}(t) \Leftrightarrow f_1 \delta_{f_1}(f)$$

When sampling occurs at 2100 samples/s, $T_1 = 1/2100$ seconds and
$f_1 = 2100$ Hz. In the frequency domain, sampling produces

$$S(f) = \tfrac{A}{2}\big[\delta(f-1000)+\delta(f+1000)\big] \otimes 2100\delta_{2100}(f)$$

i.e.,

$$S(f)=1050A\left[...+\delta(f+2100\pm1000)+\delta(f\pm1000)+\delta(f-2100\pm1000)+...\right]$$

or

$$S(f)=1050A\left[...+\delta(f+1100)+\delta(f+1000)+\delta(f-1000)+\delta(f-1100)+...\right]$$

When applied to the low-pass filter with gain $1/2100$, only the center terms survive, i.e., $\frac{1}{2}\delta(f\pm1000)$ and we have $A\cos(2\pi1000t)$.

b) When sampling occurs at 1900 samples/s,

$$T_1=1/1900 \text{ seconds and } f_1=1900 \text{ Hz}$$

In the frequency domain, sampling produces

$$S(f)=950A\left[...+\delta(f+900)+\delta(f+1000)+\delta(f-1000)+\delta(f-900)+...\right]$$

When applied to the low-pass filter with gain $1/1900$, the four center terms survive, and we have $A\cos(2\pi1000t)+A\cos(2\pi900t)$.

c) In the form given above, the fact is not obvious that b) is band-limited to one-half the Nyquist rate. However,

$$A\cos(2\pi1000t)+A\cos(2\pi900t)=2\sin(2\pi950t)\sin(2\pi50t)$$

so that the highest frequency is 950, one-half the sampling rate in Hertz, and

$$0>f\geq950 \text{ Hz.}$$

EXAMPLE 12.2 Aliasing

Given The function

$$s(t)=A\cos(2\pi1000t)+\tfrac{A}{2}\cos(2\pi2000t)+\tfrac{A}{4}\cos(2\pi3000t)$$

is sampled at 4000 samples/s, and passed through a low-pass reconstruction filter (gain $1/f_1$) whose cutoff frequency is 3100 Hz.

Problem Determine the percentage of aliasing power in the reconstructed function.

Solution The Fourier transform of the function is

$$S(f)=\tfrac{A}{2}\left[\delta(f\pm1000)\right]+\tfrac{A}{4}\left[\delta(f\pm2000)\right]+\tfrac{A}{8}\left[\delta(f\pm3000)\right]$$

When sampled at 4000 samples/s, the frequency domain includes the following impulse functions

$$\frac{S_{\delta_{4000}}(f)}{4000} = \dots \frac{A}{2}\left[\delta(f+4000\pm1000)\right]+\frac{A}{4}\left[\delta(f+4000\pm2000)\right]+\frac{A}{8}\left[\delta(f+4000\pm3000)\right]$$

$$+\frac{A}{2}\left[\delta(f\pm1000)\right]+\frac{A}{4}\left[\delta(f\pm2000)\right]+\frac{A}{8}\left[\delta(f\pm3000)\right]$$

$$+\frac{A}{2}\left[\delta(f-4000\pm1000)\right]+\frac{A}{4}\left[\delta(f-4000\pm2000)\right]+\frac{A}{8}\left[\delta(f-4000\pm3000)\right]+\dots$$

When applied to a low-pass filter with gain $1/4000$, the following survive

$$S_{LPF}(f)=\frac{A}{2}\,\delta(f+3000)+\frac{A}{4}\delta(f+2000)+\frac{A}{8}\delta(f+1000)$$

$$+\frac{A}{2}\delta(f\pm1000)+\frac{A}{4}\delta(f\pm2000)+\frac{A}{8}\delta(f\pm3000)$$

$$+\frac{A}{2}\delta(f-3000)+\frac{A}{4}\delta(f-2000)+\frac{A}{8}\delta(f-1000)$$

The reconstructed function consists of cosine functions with frequencies of $1000, 2000,$ and 3000 Hz,

$$s'(t)=1.25A\cos(2\pi1000t)+A\cos(2\pi2000t)+1.25A\cos(2\pi3000t)$$

The power of the reconstructed signal is equal to the sum of the squares of the individual components, i.e.,

$$\overline{P}'[\infty]=A^2\left[\frac{1}{2}(1.25)^2+1+\frac{1}{2}(1.25)^2\right]=2.5625A^2 \text{ signal watts}$$

and the power of the original signal is

$$\overline{P}[\infty]=A^2\left[1+1/4+1/16\right]=1.3125A^2 \text{ signal watts}$$

Thus, the aliasing power is

$$\overline{P}''[\infty]=\overline{P}'[\infty]-\overline{P}[\infty]=1.25A^2 \text{ signal watts}$$

In **Figure 12.3**, compare the original signal with the signal after sampling and filtering.

Comment This is an extreme example of undersampling (67% of the Nyquist rate). It shows clearly that aliasing can severely distort the sampled signal.

12.2 QUANTIZATION

Using a process known as quantization, sample values are expressed in octets to produce digital signals. An octet is able to represent $2^8 = 256$ states. If we use them to describe the values of the train of impulses that is the sampled signal, we cannot describe each impulse value precisely, but must round it to the nearest binary value; the octets will only approximate the values of the real signal. This approach is known as:

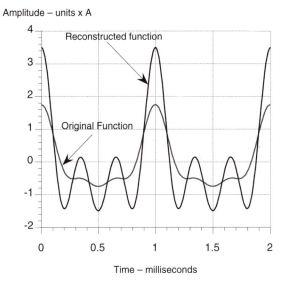

Figure 12.3 Diagram for Example 12.2

- **Quantization**: the process that segregates sample values into ranges and assigns a unique, discrete identifier to each range. Whenever a sample value falls within a range, the output is the discrete identifier assigned to the range.

In addition, we normalize the process by declaring the greatest sample value to be ± 1 and making it equal to the range to which the greatest discrete value is assigned.

(1) Linear Quantization

The process of linear quantization is illustrated in **Figure 12.4**. The samples are represented by impulses whose values correspond to the amplitudes of the band-limited signals at the sample times. A set of uniformly spaced bands is shown. Each of the q bands is identified by an octet selected from those that can be made from an n bit binary code. When the levels span ±1, the distance between each level is $2/q$. As the pulses arrive, they are assigned the octet that identifies the band in which they fall. The result is a series of octets that represents bands of sample values derived from the original signal. In transmitting the octet, the most significant bit is sent first. It is the bit that defines whether the sample is positive (1) or negative (0).

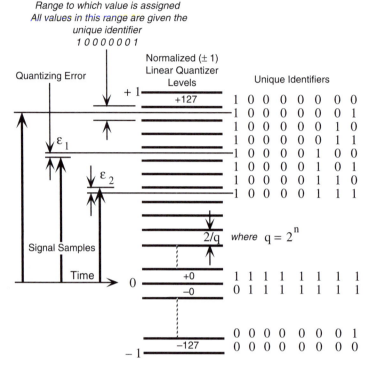

Figure 12.4 Principle of Linear Quantizing
Quantization is the process that segregates sample values into ranges and assigns a unique identifier to each range. The output is a series of octets that represent bands of sample values derived from the original signal. From Telecommunications Primer *by E. Bryan Carne, © 1995. Reprinted by permission of Prentice-Hall, Inc., Upper Saddle River, NJ.*

(2) Quantization Noise Signal

At the quantizer, the signal consists of two components, the quantized signal and a signal that represents the difference between the quantized signal and the input signal. The latter is the amount by which the quantization process has degraded the output—it is the quantization *noise* signal. Assuming the quantization error is zero mean and uniformly distributed, and the quantizer is normalized with q levels, you can calculate the magnitude of the quantization noise power. Suppose ε_k is the quantization error, and $s_q(kT_s)$ is the quantized signal; then

$$\varepsilon_k = s_q(kT_s) - s(kT_s) \tag{12.5}$$

and the mean-square quantization noise, i.e., the noise power, is

$$\overline{\varepsilon_k^2} = \sigma_q^2 = \frac{q}{2}\int_{-1/q}^{1/q}\varepsilon^2 d\varepsilon = 1/3q^2 \tag{12.6}$$

For an input signal $s(kT_s)$, the signal power is $s^2(kT_s)$. Hence, the signal-to-noise ratio is

$$S/N = \frac{s^2(kT_s)}{\sigma_q^2} = 3q^2 s^2(kT_s) \tag{12.7}$$

As a consequence of normalization, $|s(kT_s)|_{max} = 1$, so that $s^2(kT_s)_{max} = 1$; putting $q = 2^n$ we obtain

$$S/N \leq 3 \times 2^{2n} \text{ or } S/N \leq 4.8 + 6.0n \text{ db.} \tag{12.8}$$

In **Figure 12.5** you see the SNR experienced by a sinusoidal signal quantized in 256 equally spaced levels ($n = 8$). The signal power is expressed in db relative to a full-load signal (i.e., one that just sweeps through all 256 levels). From the graph, you see that the quantization SNR experienced by a full-load signal is 52.8 db (i.e., a ratio of almost 200,000 to 1); that an input signal with approximately one-hundredth (i.e., −20 db) of the power of a full load signal will experience a quantization SNR of 32.8 db (i.e., a ratio of almost

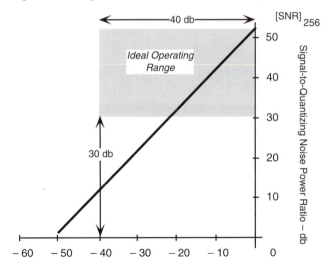

Figure 12.5 Noise Performance of Linear Quantizer
When the quantizing ranges are uniformly spaced, quantizing noise limits the ability of the quantizer to produce acceptable signal-to-noise power ratios at low levels of input signal. From Telecommunications Primer *by E. Bryan Carne,* © *1995. Reprinted by permission of Prentice-Hall, Inc., Upper Saddle River, NJ.*

2000 to 1); and that an input signal with approximately one-ten-thousandth (i.e., −40 db) of the power of a full-load signal will experience a quantization SNR of 12.8 db (i.e., a ratio of almost 20 to 1). Recalling that the dynamic range of voice signals can be as much as 40 db, and that toll-quality voice requires an SNR of 30 db, a linear quantizer with 256 levels can handle only part of the required range with better than 30 db SNR.

EXAMPLE 12.3   SNR and Number of Quantization Levels

Given A full-load sinusoidal signal is presented to linear quantizers with 2^n levels where $n = 4, 8, 12$, and 16.

Problem Find the quantizing SNR.

Solution From equation (12.8), the quantization SNR for a full-load sinusoidal signal is

$$S/N = 4.8 + 6.0n \ \text{db}.$$

Substituting $n = 4, 8, 12$ and 16 gives

n	$[S/N]_0$ db
4	28.8
8	52.8
12	76.8
16	100.8

Extension If the dynamic range of the input signal is 40 db, the quantization SNR when the signal is at a minimum (i.e., 40db below full load) is

n	$[S/N]_{-40}$ db
4	−11.2
8	12.8
12	36.8
16	60.8

Comment Equation (12.8) states that SNR is improved by 6 db for each unit increase in n, i.e., for 2^9 quantization levels, a full-load, sinusoidal tone signal has a quantization SNR of 58.8 db, and an input signal with approximately one-hundredth of the power of a full-load signal will experience a quantization SNR of 38.8 db. If we increase n to 12, an input signal with approximately one-ten-thousandth (i.e., − 40 db) of the power of a full-load signal will experience a quantization SNR of 36.8 db. This is the sort of performance required to achieve toll-quality with linear quantization. To do it, the input signal samples must be matched against 4096 levels, and each discrete identifier will be 12 bits long.

12.3 COMPANDING

To achieve acceptable quality using attainable quantizing and reconstruction techniques, the designers of early digital voice systems had to distort the signal samples by compressing and then expanding them. Known as companding, the essence of their work is still in use today.

(1) Distributing Signals Evenly

Speech signal samples show a preponderance of low amplitudes, and many samples have zero, or close to zero, values. The distribution can be approximated by a zero-mean Laplace distribution (after the late-18th-century and early-19th-century French mathematician Marquis Pierre Simon de Laplace). For $b > 0$, it is defined as follows

$$f_X(x) = \frac{b}{2} \exp(-b|x|)$$

$$F_X(x) = \begin{cases} \frac{1}{2} \exp(bx), & -\infty < x < 0 \\ 1 - \frac{1}{2} \exp(bx), & 0 \le x < \infty \end{cases}$$

$$\overline{X} = 0$$

$$\sigma_X^2 = \frac{2}{b^2}$$

and

$$\Phi_X(f) = \frac{b^2}{b^2 + 4\pi^2 f^2}$$

Normalized probability density functions for zero-mean distributions with $b = 1, 2, 4,$ and 8 appear in **Figure 12.6**. They are characterized by a peak at zero amplitude and an exponential fall-off towards ±1. Because most of the samples are of low amplitude, they will fall in levels close to zero where the effect of quantizing noise is greatest. Only a few samples will be encoded in the intervals approaching one, where the effect of quantizing noise is least. Example 12.4 illustrates this effect.

EXAMPLE 12.4   Distribution of Samples in Equal Quantizing Levels

Given Sampling a certain signal produces a set of samples whose amplitudes create a triangular probability density function (PDF) given by

$$f_X(x) = \begin{cases} 1 - |x|, & 0 \le |x| \le 1 \\ 0, & \text{elsewhere} \end{cases}$$

or $f_X(x) = \Lambda(x)$. The samples are quantized in 8 equal levels between ±1.

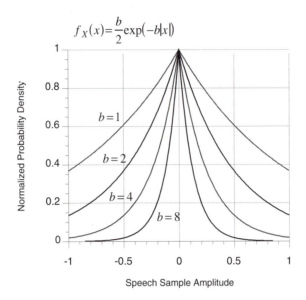

$$f_X(x) = \frac{b}{2}\exp(-b|x|)$$

Figure 12.6 Zero-Mean Laplace Distribution Density
The samples derived from speech signals show a Laplace distribution density.

Problem Find the percentage of samples that fall in each level.

Solution Taking advantage of symmetry, we will work in the positive quadrant. **Figure 12.7** shows the positive half of the PDF of the signal samples. The number of samples that fall in each quantizing level will be proportional to the four areas created by dividing the amplitudes equally in four (to simulate linear quantization). Thus 43.75% of the samples fall in the lowest level, 31.25% fall in the second level, 18.75% fall in the third level, and only 6.25% of samples fall in the highest level.

EXAMPLE 12.5   Information Rate

Given The signal from Example 12.4. It consists of a regular train of samples taken at $2W$ samples/s. They are quantized in 8 equal levels so that

$$p_1 = p_8 = 0.03125; \quad p_2 = p_7 = 0.09375; \quad p_3 = p_6 = 0.15625; \text{ and } \quad p_4 = p_5 = 0.21875.$$

Problem
   a) Find the information rate of the source, i.e., the information rate at the output of the quantizer.

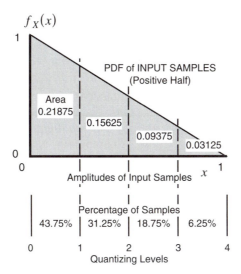

Figure 12.7 Diagram for Example 12.4

b) If the quantizer levels are changed so that $p_1 = p_2 = ... = p_8$, find the new information rate.

Solution

a) Equation (11.8) gives the information rate of the source as $R = rH(X)$. In this case, the sampling rate r is $2W$ samples/s. From equation (11.6), the entropy is

$$H(X) = \sum_{j=1}^{M} P_j \log 1/P_j$$

so that

$$H(X) = 2 \times 0.03125 \log_2(32) + 2 \times 0.09375 \log_2(10.67) + 2 \times 0.15625 \log_2(6.4)$$
$$+ 2 \times 0.21875 \log_2(4.57)$$

and $H(X) = 2.749$ bits/symbol. Thus, the information rate is

$$R = 5.498W \text{ bits/s}$$

b) The quantizing levels are changed so that the samples are distributed in an equally likely manner among the eight levels. Under this condition

$$H(X) = 8 \times 0.125 \log_2(8) = 3$$

and the information rate is $6W$ bits/s.

Comment Changing the quantizer levels so that the samples are distributed equally among them improves the entropy of the source, i.e., increases the average amount of information per sample.

As Example 12.5 shows, better performance is achieved if the quantizing levels are arranged so that the signal samples have equal probability of falling into any interval. To do this, we employ *companding* (a contraction of the two words *com*pressing and ex*panding*). A compressor is used at the transmitting end of a circuit to decrease the amplitudes of the larger signals relative to the smaller signals. At the receiving end, an expander reverses the compression and restores the samples to their original amplitude relationships. The shape of the companding curve depends on the probability density function that describes the signal samples. Example 12.6 derives the shape of a companding characteristic for the triangular PDF first stated in Example 12.4.

EXAMPLE 12.6   Simple Companding Curve

Given The signal from Example 12.4. After compression, it is applied to an eight-level, uniform quantizer.

Problem Determine the compressor characteristic that will cause the signal samples to be distributed evenly across the quantizing intervals.

Solution **Figure 12.8** shows the positive half of the PDF. To determine the four quantizing intervals between 0 and 1 in which the signal samples have equal probabilities of occurring, divide the area under $f_x(x)$ into four equal parts, as follows.

$$\text{Area under positive half of PDF} = \tfrac{1}{2}$$

Hence,

$$\frac{(1-c)^2}{2} = \frac{(1-b)^2}{2} - \frac{1}{8} = \frac{(1-a)^2}{2} - \frac{1}{4} = \frac{1}{8}$$

so that

$$a = 0.134; b = 0.293: c = 0.5$$

To find the quantizing levels, divide each of the four equal areas in two equal parts. If we designate the boundary between them as the quantizing level, we will have signals in each quantizing interval with equal likelihoods of being greater than or less than the quantizing level. Using simple mensuration

$$a' = 0.062; b' = 0.209; c' = 0.387; d' = 0.646$$

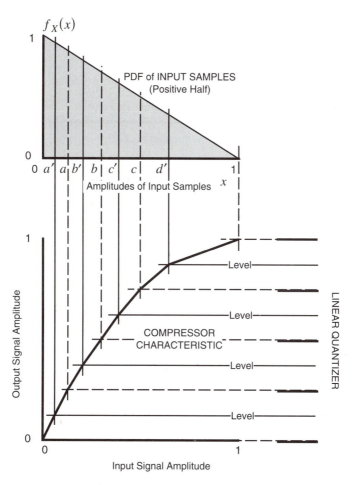

Figure 12.8 Diagram for Example 12.5

In the bottom diagram in Figure 12.8, you see the intersections of the equiprobable intervals with the uniform quantizing intervals, and the equiprobable quantizing levels with the uniform quantizing levels. These points give shape to the compressor characteristic.

---

(2) μ-Law and A-Law Companding

Among the telephone administrations of the world, two companding functions are employed.

- **μ-law companding**: used in North America and regions of the world that follow North American practices. If the amplitude of the input signal is denoted by x and the amplitude of the compressed output signal is $z(x)$, they are related by

$$z(x) = \text{sgn}(x) \frac{\log_e(1+\mu|x|)}{\log_e(1+\mu)} \tag{12.9}$$

where $\text{sgn}(x)$ denotes the signum function, and $\mu = 255$. For all values of x, the relationship is non-linear. The positive quadrant of the compressor characteristic is shown in **Figure 12.9**.

• **A-law companding**: used in Europe and regions of the world that follow European practices,

$$z(x) = \text{sgn}(x) \frac{A|x|}{1+\log_e(A)}, \quad 0 \le |x| \le \frac{1}{A}$$

$$= \text{sgn}(x) \frac{1+\log_e(A|x|)}{1+\log_e(A)}, \quad \frac{1}{A} \le |x| \le 1 \tag{12.10}$$

where $A = 87.6$. For values of $|x| \le 0.01$, the relation is linear; for larger values, it is non-linear.

In **Figure 12.10**, the signal-to-quantization noise ratio of a companded signal is plotted against the power of the input signal expressed in db relative to a full-load signal. Comparing this curve to that obtained earlier for linear quan-

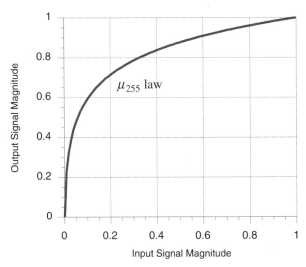

Figure 12.9 *μ*-Law Compressor Characteristic

tization (Figure 12.5), you can see that companding improves the SNR at lower signal levels and degrades it at higher levels and that it is a solution to achieving better than 30 db SNR over a 40 db dynamic range.

12.4 RECONSTRUCTING THE SIGNAL

At the expander, a train of narrow pulses is created from the string of octets representing the compressed samples. Their amplitudes are determined by the quantizer level to which they were assigned, multiplied by the inverse of the compressor characteristic. The pulse repetition rate is equal to the sampling rate. With an exact match between these two rates, the frequencies present in the original signal are duplicated. Next, this pulse train is applied to a filter (reconstruction filter) that passes frequencies $\leq W$ Hz. Provided that sampling was conducted at, or above, the Nyquist rate, the signal at the output of the filter will be a close mimic to the original, band-limited signal. (From Figure 12.1 we see that a low-pass filter of passband $\pm W$ Hz will isolate the spectrum of the input wave.)

In modern systems, the functions I have described have been combined and are implemented in digital signal processors (DSPs) or custom integrated

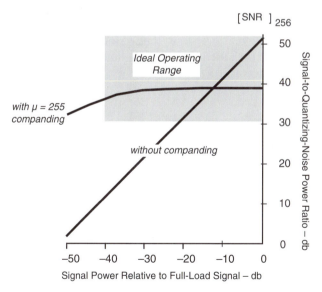

Figure 12.10 Companding Makes Operation over the Ideal Operating Range Possible

Companding improves the SNR at lower signal levels and degrades it at higher signal levels. It is a solution to obtaining 30 db SNR over a 40 db dynamic range. From Telecommunications Primer *by E. Bryan Carne, © 1995. Reprinted by permission of Prentice-Hall, Inc., Upper Saddle River, NJ.*

circuits. At the transmitter, the voice signal may be sampled and quantized in 4096 (2^{12}) levels, in a linear fashion. Then, using a table stored in ROM, each 12-bit word is replaced in a biased fashion by one of 256 octets to produce an approximately uniform distribution of signal values. Usually, the companding characteristic is approximated by linear segments that are divided into equal quantizing levels.

Because many of the speech samples are of low amplitude, the quantization levels are coded in the manner of Figure 12.5. In T-carrier systems, 0s are represented by the absence of a pulse; therefore, to avoid loss of synchronization and other difficulties due to long strings of zeros, the greater number of 1s is used for the smaller signals. The most significant bit denotes whether the level is positive (1) or negative (0). It is the leading bit when the code is transmitted.

12.5   PCM VOICE

Because telephone voice must withstand the rigors of worldwide service that may include several analog-to-digital and digital-to-analog conversions, the telephone administrations of the world adopted a robust coding scheme that employs a sampling rate of 8000 samples/s, codes each sample in 8 bits to produce a signal with a bit-rate of 64,000 bits/s, and includes μ-law or A-law companding. Called *pulse code modulation* (PCM), it provides communications-quality voice signals that are easily understood and readily attributable to particular speakers. Because it is based on sampling at 8000 times per second, the analog signal reconstructed from the PCM stream cannot contain frequencies above 4000 Hz.

(1) Decoding Noise

Signals that interfere with the digital signal stream can affect regeneration and produce erroneous values in the receiver. Inspection of Figure 12.5 shows that the change in quantization levels due to an error in bit position k is 2^{k-1} levels; i.e., an error in the kth bit shifts the decoded level by

$$\varepsilon_k = \pm\left(\tfrac{2}{q}\right)2^k \qquad (12.11)$$

so that the decoding noise power is

$$\overline{\varepsilon_k^2} = \frac{1}{n}\sum_{k=0}^{k=n-1}\left(\tfrac{2}{q}2^k\right)^2 = \frac{4}{nq^2}\sum_0^{n-1}4^k \qquad (12.12)$$

Evaluating the geometric progression, equation (12.12) becomes

$$\overline{\varepsilon_k^2} = \frac{4}{nq^2}\cdot\frac{4^n-1}{3} = \frac{4}{3n}\cdot\frac{q^2-1}{q^2} \qquad (12.13)$$

For $q >> 1$, equation (12.13) becomes

$$\overline{\varepsilon_k^2} \cong 4/3n \tag{12.14}$$

If P_ε is the probability of error in any bit position, nP_ε is the probability of error in any word, and the decoding noise power is

$$N_D = nP_\varepsilon \overline{\varepsilon_k^2} \cong 4P_\varepsilon/3 \tag{12.15}$$

i.e., if the errors are uniformly distributed over the bit positions, decoding introduces an error signal whose noise power is approximately 1.33 P_ε signal watts.

(2) Signal-to-Total Noise Power Ratio

The analog voice signal delivered by a PCM channel is corrupted by the sum of quantization noise and decoding noise so that the total noise power N_T is

$$N_T = N_Q + N_D = \frac{1 + 4q^2 P_\varepsilon}{3q^2} \tag{12.16}$$

When $P_\varepsilon << 1/4q^2$, $N_T \cong 1/3q^2$ and quantization noise is the dominant corruption mechanism; on the other hand, when $P_\varepsilon >> 1/4q^2$, $N_T \cong 4P_\varepsilon/3$ and decoding noise is the dominant corruption mechanism.

In **Figure 12.11**, you see values for the signal-to-total noise power ratio for a range of probabilities of bit errors, and power of a sinusoidal signal at the input to the quantizer. To achieve a channel with $[SNR]_T \geq 30$ db, and dynamic range ≥ 40 db, the probability of bit error must be greater than 1 bit in error in 100 million bits (1×10^{-8}). This is a value that is hard to achieve on wire lines, although it is readily achievable over optical fibers. If we reduce $[SNR]_T$ to 15 db—a value that many persons would describe as difficult to use—then a wire line with a probability of 1 bit in error in 100 thousand bits (1×10^{-5}) can accommodate a dynamic range of approximately 35 db. If we limit dynamic range to 20 db—a value that is practical in many situations—wire line error rates as low as 1×10^{-4} will produce an $[SNR]_T$ of approximately 20 db. Thus, the probability of error has a significant effect on the operating conditions that can be tolerated on a PCM channel.

(3) Capacity of a PCM Channel

A message signal $m(t)$ is bandwidth limited to W Hz and sampled at the Nyquist rate to produce $2W$ samples/s. The samples are applied to a quantizer with L levels. If the levels are chosen so that the samples are evenly distributed among them, the probability of a sample falling in any one of them is

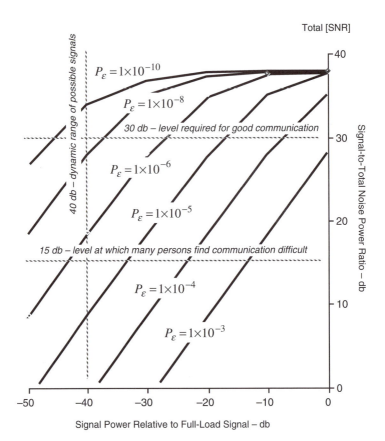

Figure 12.11 Effect of Errors on Performance of PCM Voice
The probability of error has a significant impact on the range of operating conditions that can be tolerated on a PCM channel. To achieve a channel with $[SNR]_T > 30$ db and dynamic range of 40 db, the probability of bit error must be greater than 1×10^{-8}. This value is hard to achieve on wire lines, but is readily available on optical fibers. If we can operate with $[SNR]_T > 20$ db and dynamic range of 20 db, the probability of bit error can be as low as 1×10^{-4}. This value is readily achieved on wire lines. From Telecommunications Primer *by E. Bryan Carne, © 1995. Reprinted by permission of Prentice-Hall, Inc., Upper Saddle River, NJ.*

$1/L$. Thus, the amount of information carried by each sample is $\log_2 L$ bits, and the channel capacity (in bits/s) is

$$C = 2W \log_2 L \qquad (12.17)$$

If the samples are encoded using words consisting of n code elements, each having one of M discrete amplitude levels

$$L = M^n \tag{12.18}$$

Substituting in equation (12.17) gives

$$C = 2Wn\log_2 M \tag{12.19}$$

If the separation between adjacent amplitude levels is $k\sigma$, where k is a constant, σ is the root-mean square amplitude of the noise, and the levels are arranged symmetrically about zero, the individual levels have the values $\pm k\sigma/2, \pm 3k\sigma/2, \dots \pm(M-1)k\sigma/2$. With the levels being equally likely, the average transmitted signal power is

$$S = \frac{2}{M}\left[\left(\tfrac{k\sigma}{2}\right)^2 + \left(\tfrac{3k\sigma}{2}\right)^2 + \dots + (M-1)^2\left(\tfrac{k\sigma}{2}\right)^2\right]$$

Simplifying and replacing σ^2 by N, the average noise power, gives

$$S = \frac{k^2 N}{12}\left(M^2 - 1\right) \tag{12.20}$$

Rewriting equation (12.20) in terms of M

$$M = \sqrt{1 + \frac{12S}{k^2 N}} \tag{12.21}$$

Substituting equation (12.21) in equation (12.19)

$$C = Wn\log_2\left(1 + \frac{12S}{k^2 N}\right) \tag{12.22}$$

For a message bandwidth of W Hz and a codeword length of n, the minimum transmission bandwidth B_T is Wn, so that

$$C = B_T \log_2\left(1 + \frac{12S}{k^2 N}\right) \tag{12.23}$$

Comparing equation (12.23) with the Hartley-Shannon law [equation (11.13)], you see that, to achieve the same capacity, the signal power in a PCM system must be $k^2/12$ times the signal power in a Gaussian channel. If a typical value of k is 10, the PCM channel is approximately one-eighth as efficient as the Gaussian channel.

(4) Differential PCM

If a band-limited voice signal is sampled at the Nyquist rate, the sample values will not vary greatly from one sample to the next. Put another way, a high correlation exists between contiguous samples. Taking advantage of this situation, two variations of basic PCM have been developed. Thus

- **Differential PCM** (DPCM): a technique that reduces redundancy between successive PCM samples. At the kth sampling time, we determine the difference between $s(kT_s)$ and $s[(k-1)T_s]$, the value at the $(k-1)$th sampling time. The value is encoded and transmitted. By integrating these differences, the receiver is able to construct an approximation $[s'(kT_s)]$ to the transmitted signal. To make the process reliable, the transmitter also constructs $s'(kT_s)$. Should an unacceptable difference develop between the transmitter's predicted value and the real signal at the transmitter, the transmitter updates its predicted value and sends a value based on $s(kT_s)-s'(kT_s)$ to correct the receiver's predicted value.

- **Adaptive DPCM** (ADPCM): a DPCM technique in which input signal samples are encoded in 14-bit linear PCM. The current sample is subtracted from a predicted value derived from the weighted average of several previous samples and the last two predicted values. The difference is applied to a 15-level adaptive quantizer that assigns a 4-bit code. At the receiver, an inverse quantizer forms a quantized difference signal. Together with the output of an adaptive predictor, it is used to update the value of the estimated signal.

ITU-T has developed ADPCM recommendations (*ad hoc* standards) for 4 kHz voice at bit rates of 40, 32, 24, and 16 kbits/s. With minor modifications to suit the North American network, ANSI-T1 has adopted these recommendations, placing particular emphasis on 32 and 16 kbits/s.

REVIEW QUESTIONS

12.1 Identify the frequency, phase, and amplitude of a voice signal with features of the original sound.

12.2 A signal is sampled by an ideal sampling function. Write time-domain and frequency-domain expressions for the sampled signal.

12.3 What is the Nyquist rate?

12.4 What is *aliasing?* How is it produced by sampling?

12.5 What is *quantizing?*

12.6 What is *linear quantizing?*

12.7 Relate the signal to quantizing noise power ratio to the number of quantizing levels.

12.8 What is a *full-load signal?*

12.9 Describe the amplitude distribution of samples derived from analog voice signals.

12.10 Explain the statement "Better performance is achieved if the quantizing levels are arranged so that the signal samples have equal probability of falling into any interval."

12.11 Discuss the action of companding.

12.12 Describe PCM voice.

12.13 What is *decoding noise?*

12.14 Under what condition is quantizing noise the dominant corruption mechanism of PCM voice?

12.15 Under what condition is decoding noise the dominant corruption mechanism of PCM voice?

12.16 What is the capacity of a PCM voice channel?

12.17 What is *differential PCM?* Why is it used?

12.18 What is *adaptive DPCM?* Why is it used?

13

Data Signals

Between machines, data are exchanged by the transfer of streams of binary digits. Since each bit may assume only two values—0 or 1—the bits need to be used in sequences so that a set of bits can represent a range of data. Thus, 8 bits can represent 256 states. Two codes in common use are:

- **ASCII** *(American Standard Code for Information Interchange):* employs sequences of seven bits and contains 128 unique codewords. Some are assigned to represent the alphanumeric characters and symbols found in messages, and others represent special characters required to control communication.
- **EBCDIC** *(Extended Binary Coded Decimal Interchange Code):* employs sequences of eight bits and contains 256 unique codewords. Some are assigned to represent the alphanumeric characters and symbols found in messages, others represent the special characters required to control communication, and the remainder may be used for purposes defined by the users.

A listing of alphabetic, numeric, and control words used in ASCII and EBCDIC is given in the Appendix (Table A.5).

13.1 OPERATIONS

In some fashion, the ASCII and EBCDIC codewords generated by keystrokes (and similar actions) must be delimited so that the receiver can recognize them. Demarcation is accomplished by start and stop bits, or packing several words into a frame.

210

(1) Start and Stop Bits

From the keyboards of many terminals, characters are generated one after another and transmitted singly to a local processor. Given that no two persons type at the same speed and that individual keystrokes may be separated by different time delays, the stream will be irregularly paced with spaces between each character. In other terminals, the characters are collected until a complete line of text is created, or the return key is pressed, causing the line to be sent as a burst of contiguous characters. Whether sent one-by-one as they are generated, or sent line-by-line as each line is completed, each character is framed by a start bit (0) and one or two stop bits (1 or 11); thus an eight-bit codeword will be sent as

$$\leftarrow 0_s \, xxxxxxxx 1_s 1_s$$

The start bit 0_s provides a reference from which the receiver measures time in order to sample the succeeding bits approximately in their center. The stop bit(s) 1_s provide(s) a positive indication of the end of the character. At the processor, timing begins anew with each character received. Thus, the receiver does not have to maintain precise timing over more than ten bits [8-bit character + one or two stop bit(s)].

(2) Frames, Packets, and Envelopes

In more sophisticated processing devices, characters are run together without start and stop bits to form a block of data (text). Then the data block is framed by a start sequence and a stop sequence. The start sequence is called the *header*—it contains synchronizing, address, and control information. The stop sequence is called the *trailer*—it contains error checking and terminating information. The entire data entity is called a *frame*. For one frame in common use, precise timing must be maintained over 1088 bits (5 characters in the header + 128 characters in the text + 3 characters in the trailer = 136 characters x 8 bits = 1088 bits).

An alternative nomenclature for a frame is *packet*. Invented for an early form of data network, it is the block of data (header + text + trailer) that is transmitted from a sender to a destination across a packet network. A later concept is an *envelope*. It transports *payload* between nodes in high-speed networks. The payload consists of packets and frames, and other digital sequences.

(3) Coding Efficiency

A significant difference exists in the coding efficiency of these two transmission methods:

- **Start and stop bits:** assuming an 8-bit codeword, one start bit, and one stop bit, the ratio of information bits to total bits is $8/10 = 0.8$, i.e., the coding efficiency is 80%. With two stop bits, it is 73%.

- **Header and trailer:** assuming a 5-character header, a 128-character text, and a 3-character trailer, the ratio of information bits to total bits is $1024/1088 = 0.94$, i.e., the coding efficiency is 94%.

Thus, working with start and stop bits, the data stream is made up of 80 or 73% information bits and 20 or 27% overhead bits. When working with the frame described above, the data stream is made up of 94% information bits and 6% overhead bits.

EXAMPLE 13.1   Message Deciphering and Coding Efficiency

Given   A message is typed on a keyboard connected to an instrument that sends a line of text at a time. Characters are coded in 8-bit EBCDIC, and one start bit and two stop bits are used.

Problem   A particular line reads

0101000111101010100111001001001110101000011101101000111101000000111

a) What message does it convey?
b) What is the coding efficiency?

Solution

a) The start bit is zero. Thus, the first member of the datastream is on the left-hand end. Characters begin with the least significant bit on the left. Codes for the characters are contained in Table A.5. The bit stream reads

← start **E** stop stop start **u** stop stop start **r** stop stop start **e** stop stop start **k** stop stop start **a** stop stop

b) $8 \times 6 = 48$ bits are character bits; the remainder $3 \times 6 = 18$ bits are overhead.

$$\text{Coding efficiency} = \frac{48}{64} = 75\%$$

13.2 BINARY SIGNAL FORMATS

Some common examples of binary signal formats are shown in **Figure 13.1**. As drawn, they are current or voltage pulses that carry the bit sequence ←101100111000. For specificity, we describe them as current pulses.

- **Unipolar:** a 1 is represented by a current of $2A$ signal units, and a 0 is represented by a current of 0 signal units. Two implementations are possible:

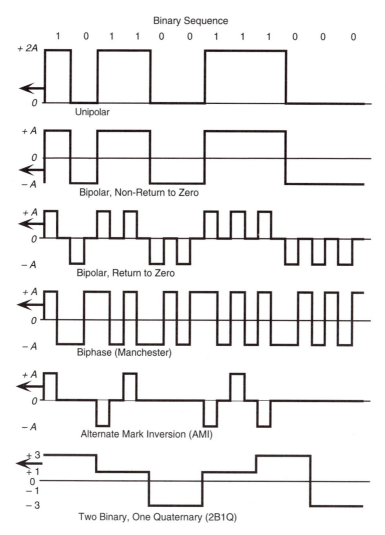

Figure 13.1 Examples of Binary Signal Formats

- *non-return to zero* (NRZ). Currents are maintained for the entire bit period (timeslot). In a long sequence in which 1s and 0s are equally likely, the power in a unipolar function is $\frac{1}{2}(2A)^2$, i.e., $2A^2$ signal watts.

- *return to zero* (RZ). Currents are maintained for a fraction of the timeslot. In a long sequence in which 1s and 0s are equally likely, and current is maintained for one-half the timeslot, the power in a return to zero unipolar function is $\frac{1}{2} \times 2A^2 = A^2$—one-half of that in NRZ unipolar.

With NRZ operation,

- long strings of 0s produce periods in which no current is generated
- long strings of 1s produce periods in which only positive current is generated
- when 1s and 0s are equally likely, the mean signal value is A signal units

All three situations present problems to electronic receivers. When a constant current flows, or no current flows, no timing information is available, so that synchronization is impossible. When the current has a mean value (not zero-mean), the receiver must be able to follow long-term level shifts, which can be difficult.

With RZ operation,

- long strings of 0s produce periods in which no current is generated
- long strings of 1s produce periods in which positive current flows for a fraction of a timeslot and returns to zero between 1s, so that the receiver can detect a change
- when 1s and 0s are equally likely, and the pulses are $T/2$ wide, the mean signal value is $A/2$ signal units

Thus, RZ eliminates one of the receiver's problems, i.e., timing, but not long-term level shifts.

- **Bipolar:** a 1 is represented by a current of A signal units, and a 0 is represented by a current of $-A$ signal units. Two implementations are possible:
  - *non-return to zero* (NRZ). Currents are maintained for the entire timeslot. In a long sequence in which 1s and 0s are equally likely, the power in an NRZ bipolar function is A^2 signal watts, i.e., one-half the unipolar NRZ value.
  - *return to zero* (RZ). Currents are maintained for a fraction of the timeslot. In a long sequence in which 1s and 0s are equally likely, and current is maintained for one-half the timeslot, the power in a return to zero bipolar function is $A^2/2$—one-half of that in NRZ bipolar—and one-fourth the unipolar NRZ value.

Long strings of 1s or 0s produce constant currents in NRZ bipolar; they are a problem for electronic circuits. The difficulties are largely eliminated with RZ bipolar because the receiver detects the current returning to zero in each pulse period. Moreover, when 1s and 0s are equally likely, the mean signal value is zero.

- **Biphase or Manchester:** a 1 is a positive current pulse of amplitude A signal units that changes to a negative current pulse of equal magnitude, and a 0 is a negative currrent pulse that changes to a positive current pulse. The changeover occurs exactly at the middle of the timeslot so that the receiver is better able to sample the signal—this reduces the chances of errors. In a long sequence in which 1s and 0s are equally likely, the power in a biphase function is A^2 signal watts. It is equal to that require by bipolar signaling; thus it is one-half the power of the unipolar representation.

Manchester signaling is employed between equipment that must operate at high speeds and maintain close synchrony; it is a popular signaling technique for local-area networks. Long strings of 1s or 0s do not result in constant currents, and the signal is zero mean under all circumstances.

- **Alternate mark inversion** (AMI): 1s are represented by return-to-zero current pulses of magnitude A that alternate between positive and negative. 0s are represented by the absence of current pulses. In a long sequence in which 1s and 0s are equally likely, the power in an AMI function is $A^2/4$—one-half of that is RZ bipolar—and one-eighth the unipolar representation.

In a long sequence in which 1s and 0s are equally likely, the average pulse repetition rate of AMI signaling is one-half the bit rate (i.e., the same as unipolar). Further, because the polarity of the pulses alternates, the power spectral density is centered at approximately one-half the bit rate (expressed in Hz) of unipolar, and virtually the entire power is contained within a bandwidth equal to the bit rate (see Figures 13.8 and 13.9). With a pulse shape that approximates a raised cosine, AMI is used extensively in T-1 carrier systems for high-speed digital connections.

- **Two binary, one quaternary** (2B1Q): four signal levels (±3 and ±1 volts) each represent a pair of bits. Of each pair, the first bit determines whether the level is positive or negative (1 = +ve, 0 = −ve), and the second bit determines the magnitude of the level (1 = |1|, 0 = |3|). For the purpose of comparison, if we designate level 3 as A, in a long sequence in which 0s and 1s are equally likely, the power of 2B1Q function is $5A^2/9$—approximately twice the power of AMI.

2B1Q signaling is employed to provide ISDN Basic Rate service at 160 kbits/s and ISDN digital subscriber loop services.

13.3   SCRAMBLING

For reasons noted in § 13.2, in many signal formats, datastreams that contain long strings of 1s or 0s are to be avoided. Some of the techniques employed are:

- Customer data are carried in seven of the eight bits in each octet, and the eighth bit is set to 1 (principally T-1 carrier and associated channels used for data services).
- When the datastream consists of multiple streams that are interleaved bit-by-bit (higher-order T-carrier channels), one of the component streams may be logically inverted before transmission and logically reverted to its original condition on reception.
- In other cases, the datastream is *scrambled.*

Figure 13.2 shows a simple hardware scheme for accomplishing scrambling.[1] Before the datastream is transmitted, it is added to a stream derived from itself by delaying it in a shift register and performing two exclusive-or opera-

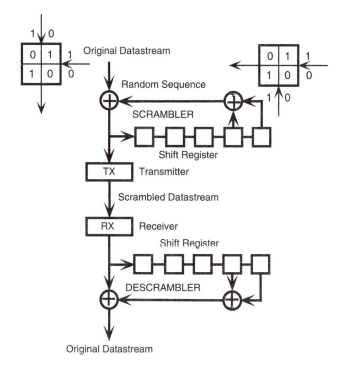

Figure 13.2 Simple Self-Synchronizing Data Scrambler
Before transmission, the datastream is added to a stream derived from itself. As a result, the transmitted datastream is more random; long sequences of 1s or 0s are broken up and repetitive patterns are changed. At the receiver, the operation is repeated to restore the original datastream.

---

[1]J. E. Savage, "Some simple self-synchronizing digital data scramblers," *Bell System Technical Journal,* 37(1958), 1501–42.

tions. The result of using the scrambler is a new signal sequence that is related to the original, but is more random. Long sequences of 1s or 0s are broken up, and repetitive patterns are changed. At the receiver, the operations are repeated to produce the original datastream.

EXAMPLE 13.2   Scrambling

Given The scrambling device shown in Figure 13.2. With no input signal, the initial condition of the shift register is 10101.

Problem Show that an all-zeros datastream is converted into a datastream that contains an approximately equal number of 1s and 0s.

Solution The scrambled datastream for the first 24 0s is shown in **Figure 13.3**. It contains 13 1s and 11 0s.

EXAMPLE 13.3   Unscrambling

Given The unscrambling device shown in Figure 13.2. With no input signal, the initial condition of the shift register is 10101.

Problem Show that the all-zeros datastream scrambled in Example 13.2 is unscrambled to become all zeros once more.

Solution The descrambled datastream is shown in **Figure 13.4**. It contains all 0s.

13.4   ESTIMATING SPECTRAL DENSITIES

Several computational techniques can be used to estimate the power spectral densities of random digital signals. In this section, two common techniques are described.

(1) Periodic and Non-Periodic Components

A random digital sequence that has a mean value (not zero) can be represented by a periodic component and a random component. In **Figure 13.5** I illustrate this statement for a random unipolar pulse train with timeslots of

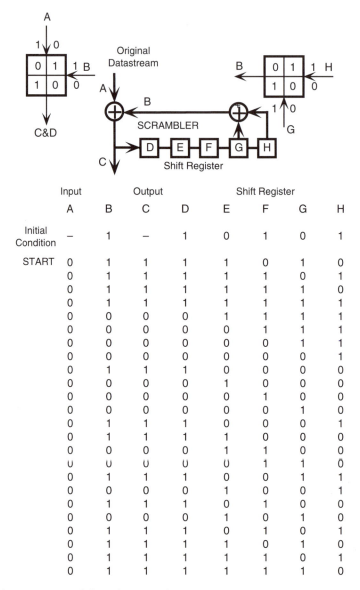

	Input	Output		Shift Register			
	A	B	C	D	E	F	G	H
Initial Condition	–	1	–	1	0	1	0	1
START	0	1	1	1	1	0	1	0
	0	1	1	1	1	1	0	1
	0	1	1	1	1	1	1	0
	0	1	1	1	1	1	1	1
	0	0	0	0	1	1	1	1
	0	0	0	0	0	1	1	1
	0	0	0	0	0	0	1	1
	0	0	0	0	0	0	0	1
	0	1	1	1	0	0	0	0
	0	0	0	0	1	0	0	0
	0	0	0	0	0	1	0	0
	0	0	0	0	0	0	1	0
	0	1	1	1	0	0	0	1
	0	1	1	1	1	0	0	0
	0	0	0	0	1	1	0	0
	0	0	0	0	0	1	1	0
	0	1	1	1	0	0	1	1
	0	0	0	0	1	0	0	1
	0	1	1	1	0	1	0	0
	0	0	0	0	1	0	1	0
	0	1	1	1	0	1	0	1
	0	1	1	1	1	0	1	0
	0	1	1	1	1	1	0	1
	0	1	1	1	1	1	1	0

Figure 13.3 Worksheet for Example 13.2

width T seconds. If a pulse is in the slot, it has amplitude $2A$ signal units, width t_w seconds ($t_w \leq T$), and the mean value of the train is At_w/T. The periodic component is a regular train of unipolar pulses of amplitude A, width t_w, repetition period T, and mean value At_w/T. The random component is a

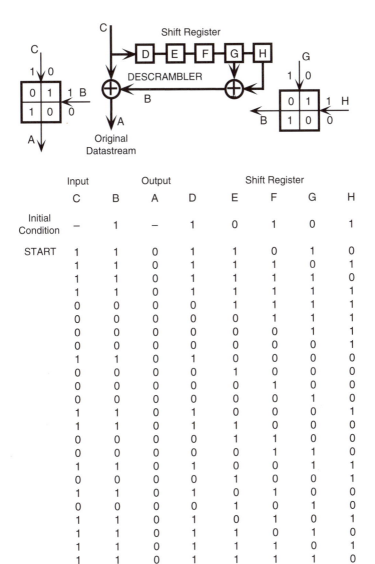

Figure 13.4 Worksheet for Example 13.3

	Input			Output	Shift Register			
	C	B	A	D	E	F	G	H
Initial Condition	–	1	–	1	0	1	0	1
START	1	1	0	1	1	0	1	0
	1	1	0	1	1	1	0	1
	1	1	0	1	1	1	1	0
	1	1	0	1	1	1	1	1
	0	0	0	0	1	1	1	1
	0	0	0	0	0	1	1	1
	0	0	0	0	0	0	1	1
	0	0	0	0	0	0	0	1
	1	1	0	1	0	0	0	0
	0	0	0	0	1	0	0	0
	0	0	0	0	0	1	0	0
	0	0	0	0	0	0	1	0
	1	1	0	1	0	0	0	1
	1	1	0	1	1	0	0	0
	0	0	0	0	1	1	0	0
	0	0	0	0	0	1	1	0
	1	1	0	1	0	0	1	1
	0	0	0	0	1	0	0	1
	1	1	0	1	0	1	0	0
	0	0	0	0	1	0	1	0
	1	1	0	1	0	1	0	1
	1	1	0	1	1	0	1	0
	1	1	0	1	1	1	0	1
	1	1	0	1	1	1	1	0

zero-mean irregular bipolar train of pulses of amplitude $\pm A$ signal units and width t_w seconds. To demonstrate this property analytically, we compute the ACF of the random train.

If each time slot has probability 0.5 of being occupied, the average power in the train is $2A^2 t_w/T$ signal watts. Hence, when $\tau = 0$, the value of

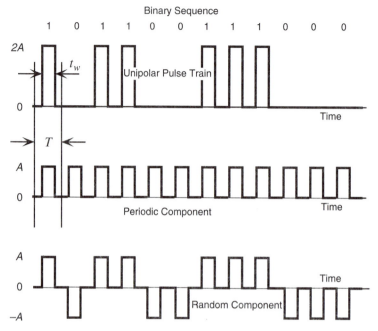

Figure 13.5 Decomposition of a Random Unipolar Pulse Train into Periodic and Random Components

A random digital sequence that has a mean value (not zero) can be represented by a periodic sequence and a random sequence.

the ACF is

$$\phi(0) = 2A^2 t_w / T \tag{13.1}$$

Remembering that the ACF is defined as

$$\phi(\tau) = \lim_{T \to \infty} \frac{1}{T} \int_{-T/2}^{T/2} f(t) f(t - \tau) dt$$

between $\tau = t_w$ and $\tau = (T - t_w)$, one or other term $[f(t) \text{ or } f(t - \tau)]$ is zero so that the ACF is zero. When $\tau = nT$, the probability of both terms being present is

$$C_1^2 p(1 - p) = \tfrac{1}{2}$$

so that

$$\phi(nT) = A^2 t_w / T \tag{13.2}$$

From Example 5.1, we know that the ACF of a rectangular pulse is a triangle function; hence, the ACF of the random sequence of rectangular unipolar pulses we described above consists of the triangular sequence shown in **Figure 13.6**. It can be divided into a random $\left[\phi_R(\tau)\right]$ and a periodic $\left[\phi_p(\tau)\right]$ component as follows:

$$\phi_R(\tau) = \frac{A^2 t_w}{T} \Lambda\left(\tau/t_w\right) \tag{13.3}$$

and

$$\phi_p(\tau) = \frac{A^2 t_w}{T} \Lambda\left(\tau/t_w\right) \otimes \delta_T(\tau) \tag{13.4}$$

Using the Wiener-Khinchine relationship, the power spectral density of the random component is

$$\Phi_R(f) = \frac{A^2 t_w^2}{T} \mathrm{Sinc}^2\left(f t_w\right) \tag{13.5}$$

and the periodic component yields a line spectrum given by

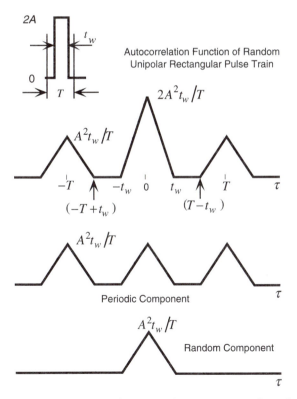

Figure 13.6 Decomposition of Autocorrelation Function of Random Unipolar Rectangular Pulse Train into Periodic and Random Components

$$\Phi_p(f) = \sum_{n=-\infty}^{\infty} \frac{A^2 t_w^2}{T^2} \text{Sinc}^2(nt_w/T) \tag{13.6}$$

The sum of these expressions is the spectral density of a random sequence of rectangular unipolar pulses.

When $t_w = T$, i.e., we have a unipolar NRZ train of amplitude $2A$, equations (13.5) and (13.6) can be written

$$\Phi_{\text{UnipolarNRZ}}(f) = A^2 T \text{Sinc}^2(fT) + A^2 \delta(f) \tag{13.7}$$

The line spectrum is reduced to a single term at the origin. All other spectral lines are zero because they occur at points where the Sinc function is zero. Recall that $A^2 T \text{Sinc}^2(fT)$ is the power spectral density of a random binary wave (see Example 8.1). Equation (13.7) adds an impulse function at the origin to convert it to a unipolar train (see Example 3.11).

When $t_w = T/2$, i.e., we have a unipolar RZ train of amplitude $2A$, equations (13.5) and (13.6) can be written

$$\Phi_{\text{UnipolarRZ}}(f) = \frac{A^2 T}{4} \text{Sinc}^2(fT/2) + \frac{A^2}{4}\left[\delta(f) + \frac{4}{\pi^2}\delta(f \pm 1/T) + \right.$$

$$\left. \frac{4}{9\pi^2}\delta(f \pm 3/T) + ...\right] \tag{13.8}$$

Figure 13.7 shows half-plane (i.e., $fT \geq 0$) values of the sum for a full-width pulse $(t_w = T)$ and a half-width pulse $(t_w = T/2)$. The spectral lines of the periodic component of the full-width pulse are zero when $n = \pm 1, \pm 2$, etc., and A^2 when $n=0$. For a half-width pulse, discrete lines have the value $A^2/4$, A^2/π^2, $A^2/9\pi^2$, etc., at $n=0, 1, 3$, etc. If operating conditions require that a timing signal be reconstructed from the received signal, the frequency information carried by these lines can be important, and a return to zero pulse is better suited to this application.

Expressing a random digital sequence as the sum of a periodic component and a non-periodic (random) component is possible when the sequence is not a zero-mean function. When it is, the decomposition in Figure 13.5 cannot take place; zero-mean random functions have no periodic components. To retain zero-mean status, 0s and 1s must be equally likely. If the balance shifts one way or the other, the function will acquire a mean value, and periodic terms. Figure 13.5 makes the two components obvious; however, the example is extreme.

(2) *Using Fourier Transforms*

In § 5.3(3), I showed that, for a periodic function represented by a generating function repeated periodically

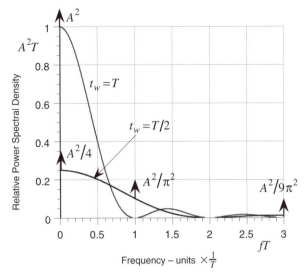

Figure 13.7 Spectral Densities of Random Rectangular Pulse Trains
The trains are full-width and half-width random rectangular pulse trains. The random components produce the continuous functions; the periodic components produce the impulses shown. They exist at the harmonic frequencies of the periodic function when the random function is not zero.

$$\Phi(f)=\left|S_g(f)\right|^2 \big/ T$$

To estimate the power spectral density of a random wave that is composed of one, or more, generating functions, we multiply $\left|S_g(f)\right|^2 \big/ T$ by the probability of occurrence of each $s_g(t)$ and sum them, i.e.,

$$\Phi(f)=\sum_n p_i \left|S_{g_i}(f)\right|^2 \big/ T \tag{13.9}$$

This technique is used in Examples 13.4 through 13.6.

EXAMPLE 13.4   Spectral Density of Bipolar Random Rectangular Pulse Train

Given A bipolar random rectangular pulse train that consists of timeslots of width T seconds in which a pulse of amplitude $\pm A$ signal units exists. In each slot, with probability 0.5, a pulse of amplitude A and duration T exists, or, with probability 0.5, a pulse of amplitude $-A$ and duration T.

Problem Find the power spectral density.

Solution The pulse train is zero-mean so that no periodic component exists. The generating function is a rectangle function

$$s_g(t) = A\Pi(t/T)$$

for which

$$S_g(f) = AT\text{Sinc}(fT)$$

Applying equation (13.9), weighting the power spectral density terms by the probabilities of their occurrence, and summing them

$$\Phi(f) = \frac{1}{T}\left\{0.5\left[AT\text{Sinc}(fT)\right]^2 + 0.5\left[-AT\text{Sinc}(fT)\right]^2\right\}$$

so that

$$\Phi_{\text{BipolarNRZ}}(f) = A^2 T\text{Sinc}^2(fT) \tag{13.10}$$

Comment This is the same result as obtained in Example 8.1.

EXAMPLE 13.5   Spectral Density of Unipolar Random Rectangular Pulse Train

Given A unipolar pulse train that consists of timeslots of width T seconds; with probability 0.5, a pulse exists in the slot with amplitude $2A$ signal units and width t_w seconds ($t_w \leq T$).

Problem Find the power spectral density.

Solution Because it is not zero-mean, the pulse train consists of a random and a periodic component. The random component has a generating function

$$s_g(t) = A\Pi(t/t_w) \Leftrightarrow At_w\text{Sinc}(ft_w) = S_g(f)$$

As in Example 13.4, we apply equation (13.9), weight the power spectral density terms by the probabilities of their occurrence, and sum them to estimate the spectral density of the random component

$$\Phi_R(f) = \frac{A^2 t_w^2}{T}\text{Sinc}^2(ft_w) \tag{13.11}$$

The periodic component is a return-to-zero rectangular wave of amplitude A. We use equation (13.6) directly. It gives

$$\Phi_p(f)=\sum_n \frac{A^2 t_w^2}{T^2}\,\text{Sinc}^2(nt_w/T) \tag{13.12}$$

so that

$$\Phi_{\text{UnipolarRZ}}(f)=\frac{A^2 t_w^2}{T^2}\left[T\,\text{Sinc}^2(ft_w)+\sum_n \text{Sinc}^2(nt_w/T)\right] \tag{13.13}$$

Comment When $t_w=T$, equation (13.13) becomes

$$\Phi_{\text{UnipolarNRZ}}(f)=A^2 T\,\text{Sinc}^2(fT)+A^2\delta(f)$$

which is equation (13.7). When $t_w=T/2$, equation (13.13) becomes

$$\Phi_{\text{UnipolarRZ}}(f)=\tfrac{A^2 T}{4}\text{Sinc}^2(fT/2)+\tfrac{A^2}{4}\left[\delta(f)+\tfrac{4}{\pi^2}\delta(f\pm1/T)+\tfrac{4}{9\pi^2}\delta(f\pm3/T)+...\right]$$

which is equation (13.8).

EXAMPLE 13.6   Spectral Density of 2B1Q Train

Given A 2B1Q non-return-to-zero train with timeslots of width $2T$ seconds; the four levels are $\pm A$ and $\pm A/3$.

Problem Find the power spectral density.

Solution The train is zero-mean so that we have only a random component. The generating functions are

$$s_g'(t)=\pm A\Pi(t/2T)\Leftrightarrow \pm2AT\,\text{Sinc}(2fT)=S_g'(f)$$

and

$$s_g''(t)=\pm\tfrac{4}{3}\Pi(t/2T)\Leftrightarrow \pm\tfrac{2}{3}AT\,\text{Sinc}(2fT)=S_g''(f)$$

Applying equation (13.9), weighting the power spectral density terms by the probabilities of their occurrence (0.25), squaring, dividing by T, and summing them

$$\Phi_{2B1Q}(f)=\tfrac{5}{9}A^2 T\,\text{Sinc}^2(2fT) \tag{13.14}$$

---

(3) Using Autocorrelation Functions

Some formats employ pulses whose polarity depends on the polarity of the previous pulse. In AMI, under normal circumstances, a positive 1 in a timeslot is always followed by a negative 1 in the next timeslot that contains a pulse.

In Manchester, each timeslot contains a half-width positive pulse followed immediately by a half-width negative pulse, or a half-width negative pulse followed immediately by a half-width positive pulse. To find their spectral densities, we create a train using the relationship

$$s(t) = s_g(t) \otimes a_n \delta_T(t)$$

where $a_n = \pm 1$ or 0 and $\delta_T(t)$ is a regular train of unit impulses spaced T seconds apart. The Fourier transform gives

$$s(t) = s_g(t) \otimes a_n \delta_T(t) \Leftrightarrow S_g(f) \mathbf{F}[a_n \delta_T(t)] = S(f) \tag{13.15}$$

and Wiener-Khinchine gives

$$\phi(\tau) = \psi_g(\tau) \otimes \phi\{a_n \delta_T(t)\} \Leftrightarrow \Psi_g(f) \mathbf{F}[\phi\{a_n \delta_T(t)\}] = \Phi(f) \tag{13.16}$$

From equation (5.16), $\Psi(f) = |S_g(f)|^2$ so that equation (13.16) becomes

$$\Phi(f) = |S_g(f)|^2 \mathbf{F}[\phi\{a_n \delta_T(t)\}] \tag{13.17}$$

The power spectral density depends on the square of the modulus of the Fourier transform of the generating function and the Fourier transform of the ACF of the train of unit impulses.

Now

$$\phi\{a_n \delta_T(t)\} = \langle a_n' \delta_T(t) a_n'' \delta_T(t+\tau) \rangle \tag{13.18}$$

As noted above, in an AMI train, the polarity of the impulse functions alternates. Under this circumstance

$$\langle a_n' \delta_T(t) a_n'' \delta_T(t+\tau) \rangle = \langle \{\delta(t+T/2) - \delta(t-T/2)\}\{\delta(t+\tau+T/2) - \delta(t+\tau-T/2)\} \rangle$$

Evaluating the right-hand side at $\tau = 0$, and $\pm T$

$$\phi\{a_n \delta_T(t)\} = \tfrac{1}{T}[\delta(t) - \tfrac{1}{2}\{\delta(t+T) + \delta(t-T)\}] \tag{13.19}$$

and

$$\mathbf{F}[\phi\{a_n \delta_T(t)\}] = \tfrac{1}{T}[1 - \cos(2\pi fT)] \tag{13.20}$$

Substituting in equation (13.17) and employing the identity [from Table A.6 (Appendix)]

$$2\sin^2(\theta) = [1 - \cos(2\theta)]$$

gives

$$\Phi(f) = |S_g(f)|^2 \tfrac{1}{T}[1 - \cos(2\pi fT)] = \tfrac{2}{T}|S_g(f)|^2 \sin^2(\pi fT) \tag{13.21}$$

Employing equation (13.21), we can estimate the power spectral density of AMI and Manchester trains.

- If the AMI train is formed from rectangle pulses of width T, i.e.,

$$A\Pi(t/T) \Leftrightarrow AT\,\text{Sinc}(fT)$$

so that

$$\left|S_g(f)\right|^2 = A^2 T^2 \,\text{Sinc}^2(fT)$$

and if each timeslot has a 50% probability of containing a pulse

$$\Phi_{AMI-T} = A^2 T\,\text{Sinc}^2(fT)\sin^2(\pi fT) \qquad (13.22)$$

- If the AMI train is formed from half-width pulses, i.e.,

$$A\Pi(2t/T) \Leftrightarrow \tfrac{1}{2}AT\,\text{Sinc}(fT/2)$$

and if each timeslot has a 50% probability of containing a pulse

$$\Phi_{AMI-T/2} = \tfrac{1}{4}A^2 T\,\text{Sinc}^2(fT/2)\sin^2(\pi fT) \qquad (13.23)$$

- If the AMI train is formed from raised cosine pulses, i.e.,

$$A\left[1+\cos(2\pi t/T)\right]\Pi(t/T) \Leftrightarrow \left\{AT\,\text{Sinc}(fT)\right\}/\left(1-f^2 T^2\right)$$

and if each timeslot has a 50% probability of containing a pulse

$$\Phi_{AMI-RCos} = \left\{A^2 T\,\text{Sinc}^2(fT)\sin^2(\pi fT)\right\}/\left(1-f^2 T^2\right)^2 \qquad (13.24)$$

- For a Manchester train with rectangular pulses of amplitude A and timeslots of width T in which a positive half-width pulse is followed by a negative half-width pulse, or a negative half-width pulse is followed by a positive half-width pulse, and each is equally likely

$$\Phi_{\text{Mnchstr}}(f) = A^2 T\,\text{Sinc}^2(fT/2)\sin^2(\pi fT/2) \qquad (13.25)$$

(4) Comparison of Spectrums

Figure 13.8 shows normalized power spectral density curves (positive frequency half-plane only) for bipolar NRZ, 2B1Q, AMI raised cosine and Manchester formats; to give more details of the sidelobe structure, **Figure 13.9** shows them on a logarithmic plot.

- **Bipolar NRZ:** the full-width, zero-mean function is used as reference.
- **2B1Q:** because of the double-width timeslots of 2B1Q, power is distributed at the lower frequencies, and the first zero occurs at one-half the frequency of the first zero of bipolar NRZ. Significant power is present in the sidelobes.
- **AMI-RCosine:** zeros occur at $fT=0$ and at the zeros of bipolar NRZ. The sidelobes fall off rapidly so that only two lobes appear in Figure

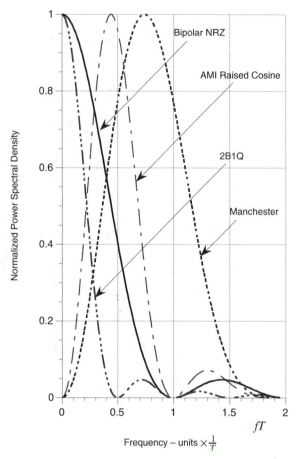

Figure 13.8 Normalized Power Spectral Density Curves for Some Random Signal Formats

In 2B1Q, pulses are double width compared to Bipolar NRZ, and the power density spectrum is concentrated at lower frequencies. In Manchester, the basic pulse is half-width compared to Bipolar NRZ, and the power is distributed at higher frequencies. AMI Raised Cosine and Bipolar NRZ are full-width pulses; their power density falls between the PSDs of 2B1Q and Manchester.

13.9. Peak power density occurs approximately midway between the first two zeros.

- **Manchester:** zeros occur at dc and at the even zeros of bipolar NRZ. Peak power density occurs approximately at the first zero of bipolar NRZ. Significant power is present in the sidelobes.

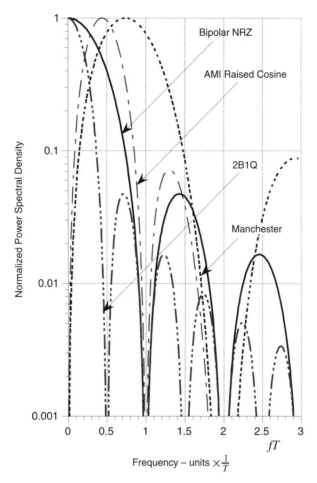

Figure 13.9 Normalized Power Spectral Density Curves for Some Random Signal Formats
To show more detail of the higher frequency lobes in Figure 13.7, the ordinate scale is made logarithmic. The first lobes of all signals are between 10 and 15 db below their peak values. AMI Raised Cosine has a second lobe more than 30 db below its peak value.

Figure 13.8 shows a frequency progression from the double-width pulses of 2B1Q to the half-width pulses of Manchester. Because of their relatively narrow, lower-frequency, power density spectrums, 2B1Q and AMI-RCosine are employed in the bandwidth-limited environment of telephone connections. Manchester is employed in local-area networks, and other applications where bandwidth is not a limiting factor and precise synchronization is important.

REVIEW QUESTIONS

13.1   Distinguish between *ASCII* and *EBCDIC.*

13.2   Under what circumstances are start and stop bits used?

13.3   What is a *frame,* a *packet,* and an *envelope?*

13.4   Why is operating with frames more efficient than operating with start and stop bits?

13.5   Distinguish between *synchronous* and *asynchronous operation.*

13.6   Distinguish between *return-to-zero* and *non-return-to-zero* binary signal formats.

13.7   When operating with NRZ pulses, what problems do long strings of 1s and 0s produce for electronic equipment?

13.8   Which of these problems does RZ operation eliminate?

13.9   When 1s and 0s are equally likely, Manchester and AMI binary signal formats are zero mean; does this change when 1s and 0s are not equally likely? Why?

13.10   What options are available to ensure that enough 1s are present in a digital stream?

13.11   Why must a certain number of 1s be present in the data stream?

13.12   Under what condition can a random digital stream be divided into a periodic and a random component?

13.13   Explain the existence of impulses on the curves in Figure 13.7.

13.14   Explain the difference in the arguments of the $\sin^2$ terms in equations (13.26) and (13.28).

14

Intersymbol Interference

- **Intersymbol interference** (ISI): the modification of one pulse (symbol) by others on a channel characterized by attenuation and dispersion.

Early telegraphers soon realized that the *dots* and *dashes* of Morse code could be sent only so fast. What telegraphers did not find out until later was that their pulses became rounded, diminished, and stretched out, and individual pulses ran into one another before they reached the receiver. To reduce the level of *intersymbol interference,* the farther the distance, the longer the sender must wait between pulses (i.e., the slower the sender can signal). Expressed another way, the maximum signaling rate a bearer can support decreases as the transmission distance increases. Indeed, a useful measure is

- **Distance × bandwidth product:** for a bearer of length L (miles or kilometers) and bandwidth B_T (Hertz), the product LB_T is approximately constant over a wide range of values. By extension, the product (distance) × (bit rate) is approximately constant also.

Figure 14.1 is a section of a transmission line bounded by a transmitter and a receiver. It lists the factors that degrade digital signals and the techniques employed to correct them. To alleviate ISI, at the transmitter we are concerned with signal formats and pulse shaping; at the receiver, where the incoming signal has been attenuated and dispersed by the channel, and perturbed by noise, we are concerned with equalization of the effect of the channel transfer function, and with the errors introduced by noise and other factors in the regeneration process.

Consider a bipolar, non-return-to-zero, data signal in which ones and zeros are equally likely. It can be written

$$s(t) = \sum_{k=-\infty}^{\infty} A_k s_g(t - kT) \qquad (14.1)$$

231

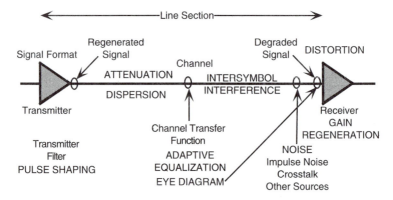

Figure 14.1 Digital Transmission Environment
Signals flow from left to right. Only one direction of transmission is shown. A complete circuit will have a similar channel in which signals flow from right to left.

where

$$A_k = \begin{cases} +1, \text{when signal one is present} \\ -1, \text{ when signal zero is present} \end{cases}$$

$s_g(t)$ is a normalized pulse

T is the timeslot duration (in seconds)

At the receiver, suppose that $s_g(t)$ is corrupted to become $s'_g(t)$; then, at $t = iT$, we can write

$$s(iT) = \sum_{k=-\infty}^{\infty} A_k s'_g\big[(i-k)T\big]$$

Separating out the ith bit

$$s(iT) = A_i s'_g(iT) + \sum_{\substack{k \neq i \\ k=-\infty}}^{\infty} A_k s'_g\big[(i-k)T\big] \tag{14.2}$$

$A_i s'_g(iT)$ is the ith received bit. The other term in the RHS of equation (14.2) represents the residue from all other bits; this is intersymbol interference. The principal cause of ISI is *delay distortion*. Because the individual frequency components propagate at different speeds over the transmission media, they become dispersed, causing changes in amplitude and phase that produce pulse distortion. Ways of combating ISI include pulse shaping, equalization, and regeneration.

14.1 PULSE SHAPING

Controlling the shape of the pulse is a way to reduce ISI. In this section, three common shapes are described. They range from theoretical to practical solutions.

(1) Sinc Function

A waveform that produces *zero* ISI is the Sinc function. In equation (14.1), we can substitute

$$s_g(t) = \text{Sinc}(t/T) \tag{14.3}$$

where T seconds is the time between pulses. $\text{Sinc}(t/T)$ has a peak value at the origin and is equal to zero at integer multiples of T. Substituting in equation (14.1) gives

$$s(t) = \sum_{k=-\infty}^{\infty} A_k \text{Sinc}\left(\frac{t-kT}{T}\right) \tag{14.4}$$

If $s(t)$ is sampled at $t = 0, \pm T, \pm 2T$, etc., all of the Sinc functions except one will be zero. For instance, at the ith sampling instant

$$s(iT) = A_i \text{Sinc}\left(\frac{t-iT}{T}\right) = A_i$$

and all of the other terms in equation (14.2) are zero, i.e.,

$$\sum_{\substack{k \neq i \\ k=-\infty}}^{\infty} A_k \text{Sinc}\left[\frac{t-(i-k)T}{T}\right] = 0$$

In the frequency domain, equation (14.3) becomes

$$S_g(f) = T \Pi(fT) \tag{14.5}$$

i.e.,

$$S_g(f) = \begin{cases} T \text{ when } |f| < 1/2T \\ 0 \text{ otherwise} \end{cases} \tag{14.6}$$

This is an ideal low-pass filter with passband amplitude of T, and passband of $1/2T$. It represents the minimum bandwidth required to support the Sinc function. If we designate the passband of this filter as the transmission bandwidth, B_T, then

$$B_T = 1/2T = r/2 \tag{14.7}$$

where r is the bit rate in bits/s.

Because of the abrupt transitions at $\pm B_T$, the low-pass filter described by equation (14.6) cannot be realized in practice. Furthermore, the lobes of Sinc functions decay relatively slowly so that many tails can be significant around any sampling time. This situation reduces the margin for error in sampling times and forces sampling to be done precisely at the zero point.

(2) Nyquist Pulse Shaping

In 1928, another solution to achieving zero ISI was described by Nyquist.[1] It employs a low-pass filter with a raised cosine spectrum. Shown in **Figure 14.2**, the frequency characteristic is defined as

$$S(f) = \begin{cases} 1/2B_T, & \text{when } |f| < f_1 \\ \frac{1}{4B_T}\left\{1+\cos\left[\frac{\pi(|f|-f_1)}{2B_T-2f_1}\right]\right\}, & \text{when } f_1 < |f| < 2B_T - f_1 \\ 0, & \text{when } |f| > 2B_T - f_1 \end{cases} \qquad (14.7)$$

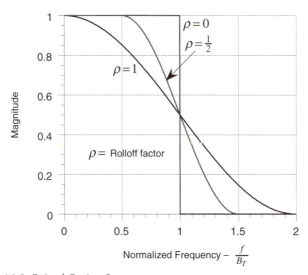

Figure 14.2 Raised Cosine Spectrum
To achieve a pulse that produces zero intersymbol interference and improves on the Sinc pulse, Nyquist devised a low-pass filter with a raised cosine spectrum. When the rolloff factor is 1, the impulse response is a pulse with very low sidelobes (see Figure 14.3).

---

[1]H. Nyquist, "Certain topics in telegraph transmission theory," *Transactions AIEE*, 47 (1928), 617–644.

The frequency f_1 is related to B_T and a quantity ρ, called the *rolloff* factor, by

$$\rho = 1 - f_1/B_T$$

If we regard the filter as an LTI system, equation (14.7) defines the system transfer function. We obtain the impulse response by taking the inverse Fourier transform. It is

$$s(t) = \mathrm{Sinc}(2B_T t)\frac{\cos(2\pi\rho B_T t)}{1 - 16\rho^2 B_T^2 t^2} \tag{14.8}$$

Figure 14.3 shows solutions for $\rho = 0$ and 1. When $\rho = 0$, $f_1 = B_T$. Under this circumstance, we have the Sinc function solution described above. When $\rho = 1$, $f_1 = 0$, and we have a Sinc-like function with zeros at integer values and dramatically diminished sidelobes that can be produced by exciting a filter with attainable characteristics.

With both the Sinc pulse and Nyquist pulse, an important *caveat* exists: in time, both extend to $\pm\infty$ and require the processing system to be *anticipatory*. In other words, neither of them is *causal* so they cannot be realized by real systems (see Chapter 4).

(3) Raised Cosine Pulse

Is there a realistic, causal approximation to the Nyquist pulse? In the time domain, **Figure 14.4** compares the central portions of a Nyquist $\rho = 1$ pulse with a normalized raised cosine pulse; they match very closely in this

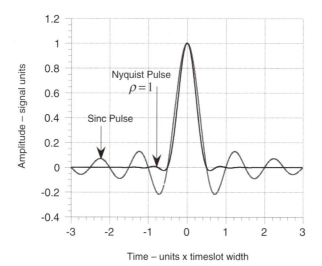

Figure 14.3 Time Domain Comparison of Sinc Pulse and Nyquist Pulse
The tails of the Nyquist pulse are practically zero at $\pm T$.

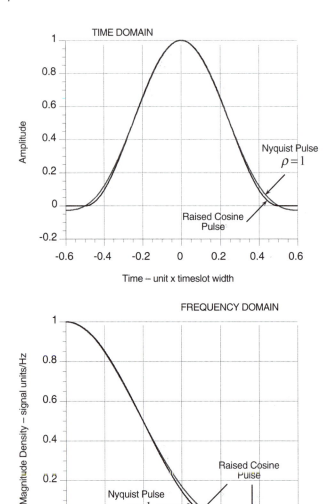

Figure 14.4 Time and Frequency Domain Comparisons of Nyquist Pulse and Raised Cosine Pulse

In the time domain, a raised cosine pulse is 0 when $|t| > T/2$. In the frequency domain, the spectrum of a Nyquist pulse is 0 when $|f| > 2B_T$.

region. In the positive half-plane of the frequency domain, Figure 14.4 compares a Nyquist $\rho = 1$ pulse with a normalized raised cosine pulse; they match closely in this region, also. The raised cosine pulse, then, can be a practical approximation to the Nyquist $\rho=1$ pulse that should reduce intersymbol interference.

In signal transport systems, the pulse shape is defined by a template that includes the range of variations (tolerances) acceptable on important features. **Figure 14.5** compares a raised cosine pulse with the template that defines the pulse shape employed in T-1 carrier systems. At best, the raised cosine pulse is a convenient approximation to the real thing.

14.2 EQUALIZATION

Another technique employed to reduce intersymbol interference uses a process that corrects the distortion produced by the transmission channel. Called *equalization,* it passes the received signals through a filter that approximates the inverse of the transfer function of the channel. **Figure 14.6** shows the principle.

(1) Linear Filter

The bandwidth and pulse shape of the transmitted signal is set by the characteristics of the transmit filter (Π_{tx}). During transmission over the channel, the frequency distribution of the energy is modified by the characteristics of the

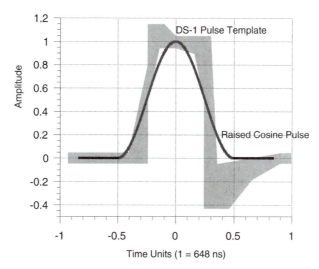

Figure 14.5 Time Domain Comparison of Raised Cosine Pulse and Template That Defines Limits of DS-1 Signal
The template includes the degradation due to transmission.

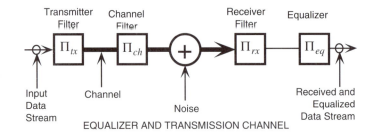

EQUALIZER AND TRANSMISSION CHANNEL

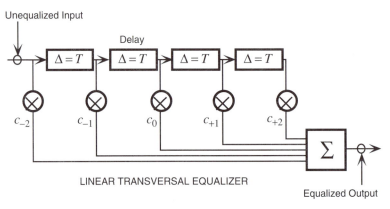

LINEAR TRANSVERSAL EQUALIZER

Equalized Output

Figure 14.6 Principle of Channel Equalization
Equalization is a process that corrects the distortion due to transmission. The received signals are passed through a filter that approximates the inverse of the transfer function of the channel.

channel [represented by the channel filter (Π_{ch})]. At the receiver, energy passes through the front-end, which is represented by a third filter (Π_{rx}). Finally, the energy is applied to a fourth filter (Π_{eq}), the equalizer that corrects for the effects of the channel filter. Π_{tx} and Π_{rx} are likely to be raised cosine filters of the kind shown in Figure 14.2. If the response of the channel filter is expressed as

$$\Pi_{ch}(f)=|\Pi_{ch}(f)|\exp\{j\theta(f)\} \tag{14.9}$$

where $|\Pi_{ch}(f)|$ is the amplitude response and $\theta(f)$ is the phase response, then, if it is to compensate for the effect of the channel, the response of the equalizer filter must be

$$\Pi_{eq}(f)=\frac{1}{|\Pi_{ch}(f)|}\exp\{-j\theta(f)\} \tag{14.10}$$

(2) Implementation

A literal interpretation of equation (14.2) implies that ISI is caused by all of the pulses that have used, or will use, the channel. In fact, ISI is usually caused by a limited number of pulses to either side of the ith pulse. Accordingly, the equalizer can be implemented by a transversal filter with a finite number of taps such as is shown in the bottom half of Figure 14.6. In this particular embodiment, a string of five unequalized pulses is captured in the tapped delay line. Applied to individual weighting elements, the coefficients are adjusted so that the sum of the pulses produces an (almost) undistorted center pulse. In low-speed operation, the filter coefficients are adjusted manually at the time of setup. In high-speed operation, adaptive equalization[2] is employed; the coefficients are periodically checked and adjusted automatically throughout the life of the connection.

Training is done by exchanging a known sequence of pulses between the transmitter and receiver. During the training sequence, the filter characteristics are adjusted so as to produce a pulse train that the receiver can decode correctly. The training process may last from a few seconds up to as long as a minute. When training is completed, the filter is able to correct (most of) the channel-induced distortion suffered by pulses sent over the channel.

(3) Eye Diagrams

To assess the performance of the equalization process, viewing the waveforms at the receiver is customary. On an oscilloscope whose time sweep is synchronized with the signaling rate, you can produce a display that superimposes all of the waveform combinations over adjacent signaling intervals. Called an *eye* diagram, **Figure 14.7** shows the sort of display produced by a random raised cosine pulse AMI train. The clear areas in the middle are the *eyes*. They show the signal differences between 1s and 0s for the particular installation being measured. Poor equalization is shown by smaller eyes; good equalization corresponds to larger eyes. Eye diagrams are a convenient way to assess the performance of a digital transmission channel, and to evaluate the efficacy of equalization.

14.3 REGENERATION

A different strategy that overcomes the effects of ISI and achieves higher signaling rates is to reduce the distance over which the pulse stream is transmitted before processing it. This is done by introducing regenerators (also known as repeaters) at regular intervals along the transmission path. Their purpose is

---

[2]R. W. Lucky, "Automatic equalization for digital communication," *Bell System Technical Journal*, 44 (1965), 547–88.

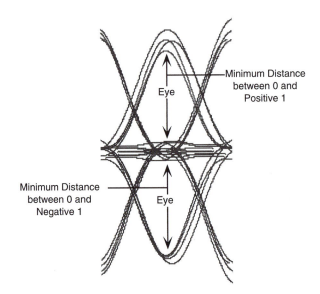

Figure 14.7 Eye Diagram for Raised Cosine Pulse AMI Train
To assess the success of equalization, a display that superimposes all of the waveform combinations over adjacent timeslots is used. Poor equalization is shown by smaller eyes; good equalization is shown by larger eyes.

to read the pulse stream before it degrades substantially, generate a new stream, and pass it on to the next unit.

(1) Functional Model

Figure 14.8 shows a functional model of a regenerator. It consists of sections that amplify, extract timing, sample and decide, and reconstruct and amplify. With a degraded signal as input, it produces an output signal that is (almost) the equal of the signal present at the beginning of the transmission chain.

(2) Jitter and Timing

In general applications, the signal train will be irregular and will not contain a pulse in each timeslot. To generate a regular repetitive timing signal, the incoming signal is used to excite a resonant circuit that is tuned to the pulse repetition rate. The output of the resonant circuit is a sinusoidal signal synchronized with the fundamental frequency of the signal train. It is used to generate a square wave from which timing signals can be derived; they control the sampling of the received signal.

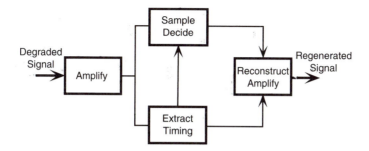

Figure 14.8 Functional Model of Regenerator
*With a degraded signal as input, the regenerator produces an output sig-
nal that is (almost) the equal of the signal present at the beginning of the
transmission chain. Only one direction of transmission is shown. A complete circuit will have a similar channel in which signals flow from right to
left.*

Because of the nature of the operations involved, the timing signal will
not exactly match the timing used to generate the pulse train for the first
time. The difference between the actual timing signal and an ideal signal is
known as *jitter*. It is produced by the various conversion processes occurring
in the regenerator. In a string, the resonant circuit in each regenerator
reduces incoming jitter from its upstream neighbor and passes the remainder
along with a component of its own to its downstream neighbor. Jitter can be
divided in two parts.[3] High-frequency jitter due to resonant circuit misalignment and other effects is attenuated rapidly so that, in a long string, the peak
jitter from these causes does not exceed four times the high-frequency jitter
in a single unit. Low-frequency jitter due to threshold detector misalignment
and other effects is accumulated so that, in a string of n units, the final value
will be n times the low-frequency jitter in a single unit.

(3) Propagation of Errors

Assuming identical repeaters and additive white Gaussian noise, the probability of error at each repeater is [see equation (10.14)]

$$P_\varepsilon = \mathrm{Erfc}(A/2\sigma)$$

where A is the signal amplitude and σ^2 is the mean square value of the noise.
For a string of m repeaters, an error is present at the receiver only if an odd
number of incorrect decisions is made along the transmission path. The prob-

---

[3]P. Bylanski and D. G. W. Ingram, *Digital Transmission Systems* (Stevenage, UK: Peter
Peregrinus, 1976), pp. 205–15.

ability of making k errors in m trials is given by the binomial distribution, so that the probability of an error at the receiver is

$$P_\varepsilon = \sum_{k \text{ odd}}^{m} C_k^m p^k (1-p)^{m-1}$$

and, for $p \ll 1$

$$P_\varepsilon \cong mp$$

Hence

$$P_\varepsilon \cong m \operatorname{Erfc}(A/2\sigma) \tag{14.11}$$

i.e., the probability of error in a chain of regenerative repeaters is directly proportional to the number of units. Through the use of such regenerators, while the average, voice-grade, twisted-pair telephone line (with equalizers) may support a signaling rate of 2400 baud, a twisted-pair line with regenerators every 6000 feet will support a T-1 channel with a signaling rate of 1.544 Mbits/s using AMI.

REVIEW QUESTIONS

14.1 Define *intersymbol interference.*

14.2 Name three ways of combatting intersymbol interference.

14.3 Explain how/why a Sinc pulse produces zero intersymbol interference.

14.4 Describe the shape of the passband of a lowpass filter with a raised cosine spectrum.

14.5 Using the model in Figure 14.6, explain the principle of *equalization.*

14.6 What is an *eye diagram?* Why is it used?

14.7 Why is jitter present in a repeatered line? What generalizations can be made about its magnitude at the end of a long string?

14.8 Comment on the statement "An error is present at the receiver only if an odd number of incorrect decisions is made along the transmission path."

15

Error Detection

If the recipient is to have confidence in the integrity of the messages received, the transmission process must include means for error control. Arguably, error control—the detection and correction of errors—is the most important function performed by terminal equipment. In this chapter, we deal with error detection.

- **Error detection:** a cooperative activity between sender and receiver in which the sender adds information to the character or frame to assist the receiver to determine whether an error has occurred.

A concept of this activity appears in **Figure 15.1**. Before sending, the sender and receiver must agree on the information that will be added, and how it will be used.

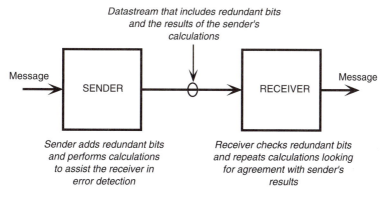

Figure 15.1 Error Detection Is a Cooperative Activity between Sender and Receiver

From Telecommunications Primer *by E. Bryan Carne, © 1995. Reprinted by permission of Prentice Hall, Inc., Upper Saddle River, NJ.*

15.1 PARITY CHECKING

A weak technique for the detection of errors adds bits to a message. Called parity checking, it is applied in series to each word or to a group of words by bit position.

(1) Vertical and Longitudinal Checking

One method of error detection adds parity bits. *Parity checking* is a process that adds a bit (0 or 1) to a sequence of bits representing the user's data to make the total number of 1s in the sequence even or odd:

- **Even parity check:** the parity bit makes (or leaves) the number of 1s even.
- **Odd parity check:** the parity bit makes (or leaves) the number of 1s odd.

Parity bits may be added to individual characters, in which case the process is known as *vertical redundancy checking* (VRC), or they may be added to bit sequences formed from bits assigned to specific positions in a block of characters, in which case the process is known as *longitudinal redundancy checking* (LRC). When operating with start (0s) and stop (1s) bits, they frame the entire set of character bits and parity bit, i.e.,

$$\leftarrow 0_s\, xxxxxxxx(0/1)_p 1_s 1_s.$$

Adding a parity bit to ASCII characters sent asynchronously reduces the transmitted coding efficiency to 70%. Thus, to send 7 information bits requires 1 start bit + 7 character bits + 1 parity bit + 1 stop bit, i.e., a total of 10 bits. If two stop bits are employed, the efficiency drops to 64%. Employing parity bits with ASCII in a standard, 136-octet frame (5 header, 128 text, and 3 trailer) sent synchronously produces an efficiency of 82%.

Parity is a relatively weak error detection technique; for one thing, it can detect only the presence of an odd number of errors. Nevertheless, that parity checking can be productive is shown by the following example.

EXAMPLE 15.1 Parity Checking with Random Errors

Given In a very long datastream, transmission errors occur at an average rate of one bit in error in one thousand bits. The errors are randomly distributed and independent of each other. The stream is made up of 8-bit words, seven of which are message bits and one of which is a parity check bit.

Problem Find the ratio of *errored words found* to *errored words not found*.

Solution   Because the transmission errors occur randomly and independently, the probability of k errors occurring in an n-bit word is given by the binomial distribution, i.e.,

$$P(k,n) = C_k^n p^k (1-p)^{n-k}$$

where p is the probability of error in one bit. If $p \ll 1$, this becomes

$$P(k,n) \cong C_k^n p^k = \frac{n! \, p^k}{k!(n-k)!}$$

For an 8-bit word, the probability of errors that are detected is the probability of the occurrence of an odd number of errors

$$P_D = P(1,8) + P(3,8) + P(5,8) + P(7,8)$$

If $p \ll 1$, (we are given $p = 0.001$)

$$P_D \cong P(1,8) = 0.008$$

Similarly, the probability of errors that are *not* detected is the probability of the occurrence of an even number of errors

$$P_{ND} = P(2,8) + P(4,8) + P(6,8) + P(8,8)$$

i.e.,

$$P_{ND} \cong P(2,8) = 0.000028$$

Hence,

$$P_D / P_{ND} = 0.008/0.000028 = 285.7$$

Extension   Evaluating P_D and P_{ND} completely gives

$$P_D = 8p + 56p^3 + 56p^5 + 8p^7$$

and

$$P_{ND} = 28p^2 + 70p^4 + 28p^6 + p^8$$

Figure 15.2 shows the values of P_D, P_{ND}, and P_D / P_{ND} for a range of probabilities of error from 0.1 to 1×10^{-6}. While parity may be a weak form of error detection, it can be effective when transmission errors are random and independent. The efficiency of parity checking increases as the probability of bit error decreases.

---

(2) VRC and LRC Combined

LRC adds parity bits to bit sequences formed from bits assigned to specific positions in a sequence of characters. The process produces an additional character, known as the *block check character* (BCC), that is transmitted in

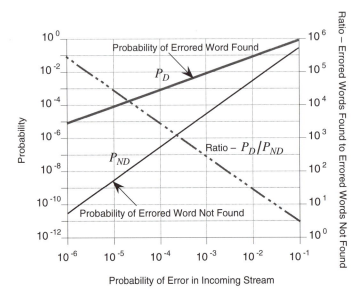

Figure 15.2 Performance of Parity Checking
If errors are randomly distributed and independent, parity checking detects all errored words with an odd number of errors. Parity checking does not detect errored words with an even number of errors. As the error rate in the incoming stream decreases, more errored words contain only one error, and the effectiveness of parity checking increases.

the trailer of the frame. As with VRC, the use of LRC detects only the presence of odd numbers of errors.

Figure 15.3 shows vertical and longitudinal redundancy checks on a sequence of five ASCII characters. Bits added to the characters provide even parity checking. Bits added to the rows provide even parity checking of each row of bits; they form the block check character. At the receiver, the VRC bits and the BCC are recalculated. If agreement is found with the values sent, the sequence of data is assumed to have been received without transmission error(s).

In certain circumstances, the combination of VRC and LRC can locate the exact position of an error. The bottom half of Figure 15.2 shows the determination of the location of a specific bit error. However, the technique fails to uncover many multiple errors. For instance, the second example at the bottom of Figure 15.2 shows a combination of four errors that will not be detected.

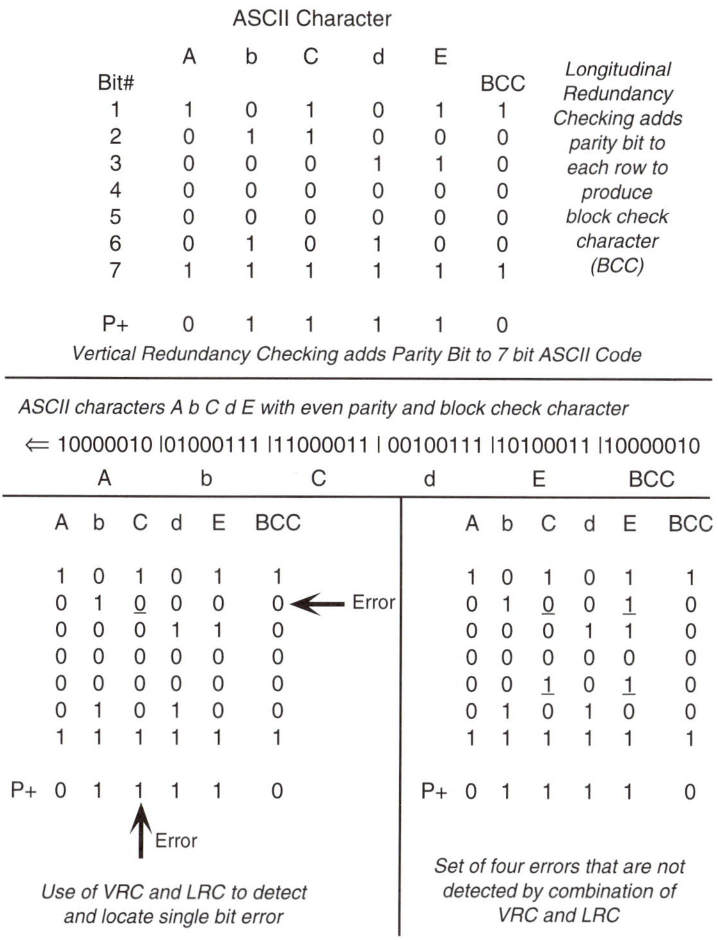

ASCII Character

Bit#	A	b	C	d	E	BCC	
1	1	0	1	0	1	1	Longitudinal Redundancy Checking adds parity bit to each row to produce block check character (BCC)
2	0	1	1	0	0	0	
3	0	0	0	1	1	0	
4	0	0	0	0	0	0	
5	0	0	0	0	0	0	
6	0	1	0	1	0	0	
7	1	1	1	1	1	1	
P+	0	1	1	1	1	0	

Vertical Redundancy Checking adds Parity Bit to 7 bit ASCII Code

ASCII characters A b C d E with even parity and block check character

⇐ 10000010 |01000111 |11000011 | 00100111 |10100011 |10000010

	A		b		C		d		E		BCC

	A	b	C	d	E	BCC			A	b	C	d	E	BCC
	1	0	1	0	1	1			1	0	1	0	1	1
	0	1	0	0	0	0 ← Error			0	1	0	0	1	0
	0	0	0	1	1	0			0	0	0	1	1	0
	0	0	0	0	0	0			0	0	0	0	0	0
	0	0	0	0	0	0			0	0	1	0	1	0
	0	1	0	1	0	0			0	1	0	1	0	0
	1	1	1	1	1	1			1	1	1	1	1	1
P+	0	1	1	1	1	0		P+	0	1	1	1	1	0

↑ Error

Use of VRC and LRC to detect and locate single bit error

Set of four errors that are not detected by combination of VRC and LRC

Figure 15.3 Use of Vertical Redundancy, Longitudinal Redundancy, and Block Check Character to Check the Integrity of ASCII Coding of the Five Character Sequence AbCdE

From Telecommunications Primer *by E. Bryan Carne, © 1995. Reprinted by permission of Prentice-Hall, Inc., Upper Saddle River, NJ.*

15.2 TECHNIQUES THAT EMPLOY CALCULATIONS

Stronger methods for the detection of errors perform mathematical operations on the message stream.

(1) Checksum

Another technique that the transmitter can use to assist the receiver in detecting the presence of errors employs addition. Called a *checksum,* the process treats the characters as binary numbers, adds them together, and expresses their sum as a 16-bit binary number to produce a 2-octet checksum.

(2) Cyclic Redundancy Checking

Another mathematical technique treats the datastream as one continuous binary number. Called *cyclic redundancy check* (CRC), the process is illustrated in **Figure 15.4**. Given a k-bit message M_k, the sender divides it by an $n+1$ bit prime number G_{n+1} to produce an n-bit remainder F_n [known as the *frame check sequence* (FCS)]. A prime number divisor is necessary to ensure that the remainders are unique. G_{n+1} is called the *generator function;* it is $n+1$ bits long, and is known to both the sender and receiver. The value of the FCS calculated by the sender is placed in the trailer of the frame. In the incoming

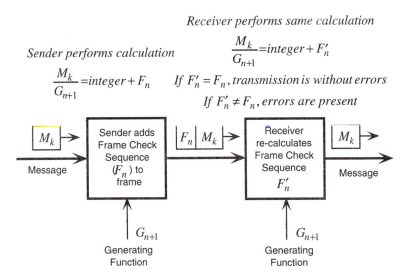

Figure 15.4 Generation and Use of Frame Check Sequence to Check Integrity of Message
The sender divides the message stream by a prime number (generating function) and attaches the remainder (frame check sequence) to the message. The receiver repeats the division and compares the remainder to the FCS. If a match occurs, the message has been transmitted without error.

frame, the receiver divides the k-bit message by G_{n+1} to produce the frame check sequence F'_n. If $F'_n = F_n$, M_k has almost certainly been received without error.

For convenience, the generating function can be expressed as a polynomial in a dummy variable (x, say) with coefficients that may be 1 or 0. Thus, the sequence 10100011 can be written

$$1x^7 + 0x^6 + 1x^5 + 0x^4 + 0x^3 + 0x^2 + 1x^1 + 1 = x^7 + x^5 + x + 1$$

For common use, several generating functions have been standardized. They are:

- **CRC-10:** $x^{10} + x^9 + x^5 + x^4 + x + 1$. Generates a 10-bit FCS for streams of 8-bit characters.
- **CRC-12:** $x^{12} + x^{11} + x^3 + x^2 + x + 1$. Generates a 12-bit FCS for streams of 6-bit characters.
- **CRC-16:** $x^{16} + x^{15} + x^2 + 1$. Generates a 16-bit FCS for streams of 8-bit characters.
- **CRC-CCITT:** $x^{16} + x^{12} + x^5 + 1$. Generates 16-bit FCS for streams of 8-bit characters.
- **CRC-32:** $x^{32} + x^{26} + x^{23} + x^{22} + x^{16} + x^{12} + x^{11} + x^{10} + x^8 + x^7 + x^5 + x^4 + x^2 + x + 1$. Generates 32-bit FCS for streams of 8-bit characters.

Each function is a binary prime number; it produces FCSs that are unique for all practical datastreams. It detects all single-bit errors, all double-bit errors, any odd number of errors, any burst errors for which the number of bits in error is less than the number of bits in the FCS, and most of the errors caused by larger bursts. Cyclic redundancy checking is by far the strongest of the common error-checking procedures. For randomly distributed errors, some estimates place the likelihood of CRC-16 not detecting an error at 1 in 10^{14} bits. For a system transmitting continuously at 1.544 Mbits/s, this amounts to approximately one undetected error every 2 years.

15.3 CODING

Error detection can be accomplished through the use of relatively complex coding schemes. Because they contribute to both error detection and error correction, we treat them in § 16.2.

REVIEW QUESTIONS

15.1 Define *error correction*.

15.2 Distinguish between odd and even parity checking.

15.3 A character is encoded as *xxxxxxx*. It is sent to a terminal with one start bit, two stop bits, and a parity check bit. How will the entire word look?

15.4 Explain how vertical and longitudinal checks can be combined to locate a specific bit in error.

15.5 What is a *checksum?*

15.6 What is *cyclic redundancy checking?*

16

Error Correction

Once detected, an error must be corrected. Two basic approaches to error correction are

- **Automatic-repeat-request** (ARQ): upon request from the receiver, the transmitter re-sends portions of the exchange in which errors have been detected.
- **Forward error correction** (FEC): employs special codes that allow the receiver to detect and correct a limited number of errors without referring to the transmitter.

16.1 ARQ TECHNIQUES

Automatic-repeat-request relies on the cooperation of sender and receiver to continue to send and to evaluate errored portions of the message until an error is no longer detected.

(1) Three Techniques

Three basic ARQ techniques are

- **Stop-and-wait:** the sender sends a frame and waits for acknowledgment from the receiver. If no error is detected, the receiver sends a positive acknowledgment (ACK), which may distinguish between even- and odd-numbered frames (ACK0 and ACK1). The sender responds with the next frame. If an error is detected, the receiver returns a negative acknowledgment (NAK) and the sender repeats the frame.
- **Go-back-n:** the sender sends a sequence of frames and receives an acknowledgment from the receiver. On detecting an error in the sequence it is receiving, the receiver discards the corrupted frame and ignores all further frames in the sequence. The receiver notifies the sender of the number of the first frame in error. On receipt of this

information, the sender begins re-sending a sequence starting with that frame. Because replacement of a corrupted frame depends on both receiver and sender sharing the same frame number, the sender must stop sending after an agreed upon number of frames (1 through 7 and 1 through 127 are common ranges). When the receiver acknowledges that all frames in the block have been received without detecting an error, the sender can reset the frame counter and begin to send another block.

- **Selective-repeat:** used on duplex connections, the sender repeats only those frames for which negative acknowledgments are received from the receiver. The sender must not send any more frames than numbers exist to identify them before receiving an acknowledgment from the receiver. The procedure puts a repeated frame out of sequence and may complicate data transfer for multi-frame messages.

(2) Throughput

To the network designer, the rate at which each link transports data is of vital importance. However, to the sender, the internal network speeds are much less important than the time it takes to deliver a message. Along with message information must go overhead to alert the receiver that a message is coming, to carry the address, to manage the data link, and to check for and correct errors. Accordingly, we define

- **Throughput** (S): the effective rate at which user's data bits are delivered, i.e.,

$$S = \frac{\text{number of user's data bits delivered in } m \text{ frames}}{\text{time to deliver all bits in } m \text{ frames}} \text{ bits/s} \qquad (16.1)$$

- **Normalized throughput** (S'): the ratio of the time to deliver only the user's data bits (i.e., mk/r) to the time to deliver all bits in a sequence of m frames

$$S' = S/r = \frac{\text{time to deliver only user's data bits in } m \text{ frames at bit rate employed}}{\text{time to deliver all bits in } m \text{ frames}}$$

$$(16.2)$$

where
- *all bits* includes the number of user's data bits (mk), the number of overhead bits in m frames [$m(n-k)$, where n is the number of bits in a frame], any additional bits included in signaling required by the protocol in use, and additional bits used to correct any errors detected
- *time to deliver all bits* includes the time to deliver m frames, propagation time(s), decision time(s), time for whatever additional signaling the protocol requires, and time for error correction, if needed.

In an error-free environment, the normalized throughput is a measure of protocol efficiency.

(3) Estimating Efficiency of ARQ Techniques

Figure 16.1 shows examples of stop-and-wait procedures. In Figure 16.1A, over a duplex connection, the sender transmits frames at a rate of r bits/s and they are acknowledged at the same rate by the receiver. In Figure 16.1B, frame F1 is found to be in error, a NAK is sent by the receiver, and the sender repeats F1. In Figure 16.1C, the receiver fails to acknowledge frame F0. Because no acknowledgment has been received by timeout, the sender repeats frame F0. This time it is acknowledged.

The sender's frames are n bits long and contain k information bits. The propagation time, i.e., the time for signal energy to cross the distance between sender and receiver, is designated τ seconds. The time to transmit each frame is designated t_F seconds, the time for the receiver to determine the condition of the received frame (the processing time) is μ seconds, and the time to transmit ACK (acknowledge, i.e., received without error) or NAK (negative acknowledgment, i.e., received, but error detected) is $t_{A/N}$ seconds. To simplify things, we assume the acknowledgment frames to be $n-k$ bits and the time for the sender to determine the next step in the procedure after receiving an acknowledgment is μ seconds. If the operation is error-free, the protocol efficiency is

$$\eta_{ARQ} = \frac{\text{Time to send user's bits}}{\text{Time to send all bits}} \qquad (16.3)$$

i.e.,

$$\eta_{ARQ} = \frac{kt_F/n}{\tau + t_F + \mu + \tau + t_{A/N} + \mu} = \frac{k}{(2n-k) + 2r(\tau + \mu)} \qquad (16.4)$$

If a protocol is employed that sends a block of m frames before pausing for acknowledgment and resetting the frame numbers, equation (16.4) becomes

$$\eta_{ARQ(m)} = \frac{mk}{\{(m+1)n-k\} + 2r(\tau + \mu)} \qquad (16.5)$$

EXAMPLE 16.1   Upper Bound of Protocol Efficiency

Given   The sender sends frames of 1088 bits (1024 information bits, 64 overhead bits) to the receiver. The receiver responds with acknowledgment frames of 64 bits. Propagation time and decision time are negligible (i.e., $\tau = \mu = 0$), and transmission is error-free.

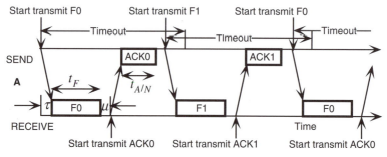

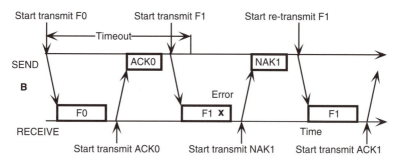

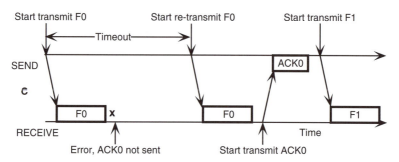

Figure 16.1 Stop-and-Wait ARQ Procedures
In Figure 16.1A, over a duplex connection, the sender transmits frames, and they are acknowledged by the receiver. In Figure 16.1 B, Frame F1 is found to be in error, a NAK is sent, and the sender repeats the frame. In Figure 16.1C, the receiver fails to acknowledge Frame F0; because no acknowledgment is received before timeout, the sender repeats Frame F0.

Problem Determine the upper bound of protocol efficiency when $m=1, 7,$ and 127 frames.

Solution   Under the conditions given, equation (16.3) becomes

$$\eta_{ARQ(m)} = mk/\{(m+1)n-k\}$$

Substituting $m=1, 7$, and 127 frames, $k=1024$ bits and $n=1088$ bits gives

$$\eta_{ARQ(1)} = 0.88889$$

$$\eta_{ARQ(7)} = 0.93333$$

$$\eta_{ARQ(127)} = 0.94074$$

Comment   In an error free environment, the only performance-degrading factor is the overhead to signal the sender when 1, 7, or 127 frames have been received. The values of η_{ARQ} calculated here, then, are the best that can be achieved.

EXAMPLE 16.2   Effect of Distance and Bit Rate on Protocol Efficiency

Given   The sender sends frames of 1088 bits (1024 information bits, 64 overhead bits) to the receiver. The receiver responds with acknowledgment frames of 64 bits. Duplex, error-free transmission occurs over a 1000-mile optical fiber facility for which the propagation speed is 1.26×10^5 miles/s (see Figure 22.5) so that the propagation time is 0.008 seconds (i.e., $\tau = 0.008$). The processing time to begin to send an acknowledgment or to decide to send the next frame is negligible (i.e., $\mu = 0$).

Problem   Determine the efficiency of error-free protocol procedures for bit rates of 4.8 kbits/s, 56 kbits/s, and 1.544 Mbits/s.

Solution   Under the conditions given, equation (16.3) becomes

$$\eta_{ARQ(m)} = \frac{mk}{\{(m+1)n-k\}+2r\left(10^3/1.26 \times 10^5\right)}$$

Substituting $k=1024$ bits, $n=1088$ bits, $r=4.8$ kbits/s, 56 kbits/s, and 1.544 Mbits/s, and $1 \leq m \leq 127$, gives the curves shown in **Figure 16.2**.

Comment   In an error-free environment, the greater the number of frames in a block (called a sequence above), the greater the protocol efficiency. For this reason, frame relay[1] is a popular transmission technique for moving large files at high speed.

---

[1]Frame relay is a technique in which a complete file can be sent as a single block over a permanent (virtual) circuit.

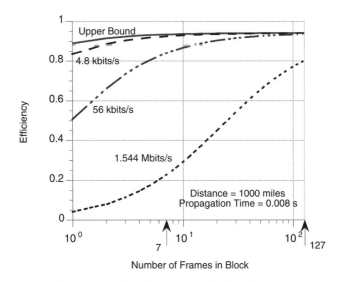

Figure 16.2 Effect of Number of Frames in Block and Bit Rate on Error-Free Protocol Efficiency

(4) Throughput of Stop-and-Wait ARQ with Errors

In stop-and-wait ARQ, the sender sends a frame and waits for acknowledgment from the receiver ($m=1$). If no error is detected, the receiver sends a positive acknowledgment. The sender responds with the next frame. If an error is detected, the receiver returns a negative acknowledgment, and the sender repeats the frame.

If the probability of bit error is P_ε, the errors are randomly distributed, and if each error requires that a frame be repeated, the average number of frames per frame that must be repeated is nP_ε. If errors occur in the repeated frames, the number of frames that must be repeated a second time is $n^2P_\varepsilon^2$, etc. Adding the total number of original and repeated frames and the number of ACKs or NAKs, the number of *all bits* per good frame is

$$(2n-k)\left[1+nP_\varepsilon+n^2P_\varepsilon^2+...\right]$$

and the time to deliver all bits is the time to deliver these bits plus propagation and decision times scaled by the number of frames per frame, i.e.,

$$\left\{(2n-k)/r+2(\tau+\mu)\right\}\left[1+nP_\varepsilon+n^2P_\varepsilon^2+...\right]$$

The corresponding number of user's bits is k; also, the series in [ ] brackets is a geometric series (see Table A.7), so that

$$S'_{S\&W(1)} = \frac{k(1-nP_\varepsilon)}{(2n-k)+2r(\tau+\mu)} \qquad (16.6)$$

Figure 16.3 shows normalized throughput for $k=1024$ bits, $n=1088$ bits, $\tau=0.008$ seconds, $\mu=0.005$ seconds, $r=4.8$ kbits/s, 56 kbits/s, and 1.544 Mbits/s, and $1\times10^{-8} \le P_\varepsilon \le 1\times10^{-2}$. The curves reflect the conditions used to derive equation (16.6). An important point is that we have assumed that errors are uniformly distributed and each error causes a frame to be repeated. (The model does not take into account the probability of multiple errors in one frame.) Thus, on average, with frames of around 1000 bits, data transfer ceases at levels of probability of error around 1×10^{-3}.

(5) Throughput of Go-Back-N ARQ with Errors

In go-back-n ARQ, the sender sends frames in a sequence, and receives acknowledgments from the receiver. On detecting an error, the receiver discards the corrupted frame, and ignores any further frames. The receiver notifies the sender of the number of the first frame in error. On receipt of this information, the sender begins re-sending the data sequence starting from that frame.

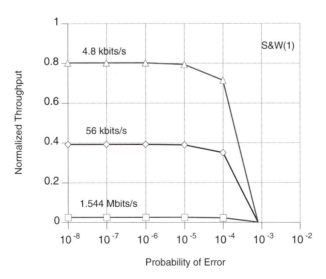

Figure 16.3 Normalized Throughput for Stop-and-Wait ARQ in which the Sender Sends 1 Frame Before Pausing for Acknowledgment

If the probability of error is P_ε, the errors are randomly distributed, and, on average, an error in a sequence of m frames causes $m/2$ frames to be repeated, the average number of frames per m frame sequence that are repeated the first time is $m^2 n P_\varepsilon/2$. When errors occur in the repeated frames, on average, $m/4$ frames must be repeated. Thus, the average number of frames per m frame sequence that must be repeated the second time is $m^3 n^2 P_\varepsilon^2/8$; and so on. Adding the total number of original and repeated frames, the number of bits in *all bits* per good sequence of m frames due to the frames is

$$mn\left[1+mnP_\varepsilon/2+m^2n^2P_\varepsilon^2/8+m^3n^3P_\varepsilon^3/64+...\right]$$

The number of bits due to NAKs and the final ACK is calculated in a similar fashion. Each repeat requires a NAK, so that the number of NAKs/ACK per m frame sequence is

$$1+mnP_\varepsilon+m^2n^2P_\varepsilon^2/2+...$$

and the number of bits per m frame sequence is

$$(n-k)\left[1+mnP_\varepsilon+m^2n^2P_\varepsilon^2/2+...\right]$$

The time to deliver all bits is the time to deliver these bits plus propagation and decision times scaled by the number of repeats per m frame sequence, i.e.,

$$mn/r\left[1+mnP_\varepsilon/2+m^2n^2P_\varepsilon^2/8+...\right]+\left\{(n-k)/r+2(\tau+\mu)\right\}\left[1+mnP_\varepsilon+m^2n^2P_\varepsilon^2/2+...\right]$$

The corresponding number of user's bits is mk, so that

$$S'_{GbN(m)}=\frac{mk}{mn\left[1+mnP_\varepsilon/2+m^2n^2P_\varepsilon^2/8+...\right]+\left\{(n-k)+2r(\tau+\mu)\right\}\left[1+mnP_\varepsilon+m^2n^2P_\varepsilon^2/2+...\right]}$$

$$(16.8)$$

Figures 16.4 and **16.5** show normalized throughput for $k=1024$ bits, $n=1088$ bits, $m=7$ or 127 frames, $\tau=0.008$ seconds, $\mu=0.005$ seconds, $r=4.8$ kbits/s, 56 kbits/s, and 1.544 Mbits/s, and $1\times10^{-8}\le P_\varepsilon\le1\times10^{-2}$. Because errored sequences occur more frequently, the performance of GbN(127) lags GbN(7) in the mid-range of probabilities of error. However, for low values of error probability, performance at 1.544 Mbits/s improves by a factor of 4 when $m=127$.

As before, I want to emphasize that the curves reflect the assumptions of the model. For GbN(127) at 56 kbits/s, **Figure 16.6** shows the effect of the sum of propagation and decision times; normalized throughput increases as the sum decreases. In the model, $\tau=0.008$ seconds and $\mu=0.005$ seconds.

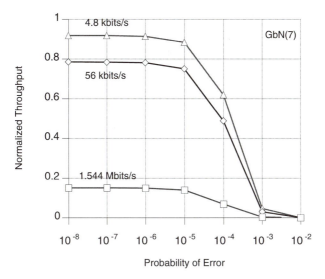

Figure 16.4 Normalized Throughput for Go-Back-N ARQ in which the Sender Sends 7 Frames Before Pausing for Acknowledgment

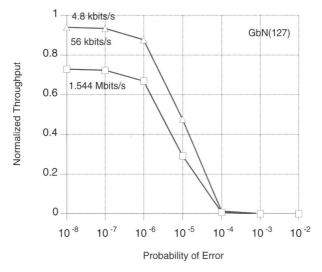

Figure 16.5 Normalized Throughput for Go-Back-N ARQ in which the Sender Sends 127 Frames Before Pausing for Acknowledgment

(6) Selective Repeat ARQ

In selective-repeat ARQ, the sender repeats only those frames for which negative acknowledgments are received from the receiver. Without disturbing the flow of frames from the sender, NAKs are sent by the receiver over the

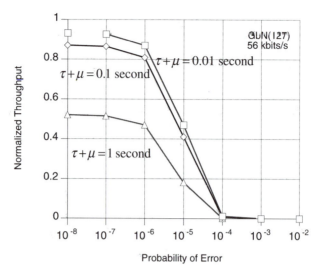

Figure 16.6 Effect of Delay Time on Go-Back-N ARQ
Delay is produced by the sum of propagation time and decision time

return channel. The sender repeats the frames as they can be inserted in the message flow. The sender must clear all requests for repeat before beginning a new sequence of frames.

If the probability of error is P_ε, the errors are randomly distributed, and each error causes a frame to be repeated, then the average number of frames per m frame sequence that are repeated the first time is mnP_ε. If an error occurs in a repeated frame, it is repeated again; thus, the average number of frames per m frame sequence that must be repeated the second time is $mn^2 P_\varepsilon^2$; and so on. Adding the total number of original and repeated frames, the number of bits in *all bits* per good sequence of m frames due to the frames is

$$mn\left[1+nP_\varepsilon+n^2 P_\varepsilon^2+...\right]=mn/(1-nP_\varepsilon)$$

When all frames have been received correctly, one ACK is sent to notify the sender to reset the frame counter and begin sending again. Thus, the time to send all bits is

$$mn/r(1-nP_\varepsilon)+(n-k)/r+2(\tau+\mu)$$

The corresponding number of user's bits is mk so that the time to send the user's data bits is mk/r, and

$$S'_{SelR(m)}=\frac{mk}{\left[mn/(1-nP_\varepsilon)\right]+(n-k)+2r(\tau+\mu)} \tag{16.9}$$

Figure 16.7 shows normalized throughput for $k=1024$ bits, $n=1088$ bits, $m=127$ frames, $\tau=0.008$ seconds, $\mu=0.005$ seconds, $r=4.8$ kbits/s, 56 kbits/s, and 1.544 Mbits/s, and $1\times10^{-8}\le P_\varepsilon\le1\times10^{-2}$. With selective repeat, the fact that NAKs are sent on a return channel without interrupting forward data flow reduces overhead delays significantly and, as might be expected, throughput performance improves rapidly as error probability decreases. However, corrected frames must be stuffed into the data stream, leading to frames out of sequence. In a multi-frame message, the receiver must reconstruct the correct sequence before the file can be used. Further, the sender and receiver must clear all corrupted frames before proceeding to the next 7 or 127 frame sequence.

(7) Comparison of ARQ Techniques

Figures 16.8, 16.9, and 16.10 compare the performance of the different ARQ techniques on the basis of normalized throughput for bit rates of 4.8 kbits/s, 56 kbits/s, and 1.544 Mbits/s and error probabilities from 1×10^{-8} to 1×10^{-2}. For convenience, I divide the discussion into: *low* levels of error probabilities, i.e., 1×10^{-8} to 1×10^{-6}; *moderate* levels of error probabilities, i.e., 1×10^{-6} to 1×10^{-4}; and *high* levels of error probabilities, i.e., 1×10^{-4} to 1×10^{-2}.

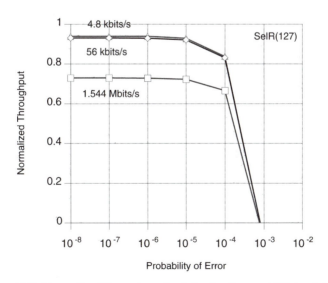

Figure 16.7 Normalized Throughput for Selective Repeat ARQ in which the Sender Sends 127 Frames Before Pausing for Acknowledgment

- **Figure 16.8:** comparison of performance at 4.8 kbits/s

  low levels of error probabilities: SelR(127) is the most efficient protocol; it is followed closely by GbN(127) and GbN(7).

  – *moderate* levels of error probabilities: The performance of GbN(127) drops off rapidly, and S&W(1) becomes a competitor in the range 1×10^{-5} to 1×10^{-4}.

  – *high* levels of error probabilities: SelR(127) and GbN(7) perform better than S&W(1) and GbN(127).

- **Figure 16.9:** comparison of performance at 56 kbits/s

  – *low* levels of error probabilities: SelR(127) is the most efficient protocol; it is followed closely by GbN(127).

  – *moderate* levels of error probabilities: In this range, the performance of GbN(127) drops off rapidly, and SelR(127) is without a close competitor.

  – *high* levels of error probabilities: SelR(127) and GbN(7) perform much better than S&W(1) and GbN(127).

- **Figure 16.10:** comparison of performance at 1.544 Mbits/s

  – *all* levels of error probabilities: Throughput performance is diminished (over Figures 16.7 and 16.8) due to the increasing fraction of time required for propagation and processing delays as the frame time decreases with increased bit rate.

  – *low* levels of error probabilities: SelR(127) is the most efficient protocol; it is followed closely by GbN(127).

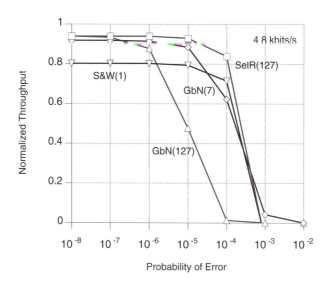

Figure 16.8 Throughput of ARQ Techniques at 4.8 kbits/s

 – *moderate* levels of error probabilities: SelR(127) is supreme.
 – *high* levels of error probabilities: SelR(127) performs better than
    the other three.

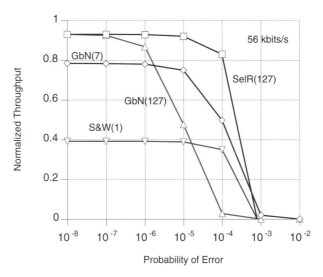

Figure 16.9 Throughput of ARQ Techniques at 56 kbits/s

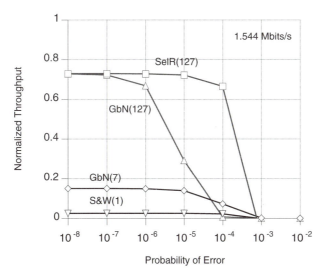

Figure 16.10 Throughput of ARQ Techniques at 1.544 Mbits/s

Obviously, the choice of a protocol is very dependent on the parameters of the operating environment. If the receiver can handle frames out of order and the connection is duplex, SelR(127) performs well under moderate- and low-probability-of-error conditions. If the receiver cannot handle frames out of order or the connection is half-duplex, GbN(7 or 127) is an appropriate choice. In a high-probability-of-error environment, if the receiver cannot handle frames out of order, low-speed S&W(1) may be the only ARQ technique that can be used. In the following section, I describe a way to achieve error correction that improves on ARQ techniques at high levels of error probabilities and permits operation at high speeds.

16.2 FORWARD ERROR CORRECTION

Forward error correction relies on the sender to encode message words in some way known to the receiver. For a limited number of errors, the receiver is able to estimate the original word. Several kinds of FEC are used.

(1) Repetition Code

Provided the number of errors is less than a value determined by the coding, the receiver can detect and correct errors without reference to the sender. This convenience is bought at the expense of adding bits. For instance, if you have only two messages to send, you can use an 8-bit repetition code to represent them, i.e., to send 1, send a block of 8-1s, and to send 0, send a block of 8-0s. If the receiver knows that the message is A or B and no other, and it is provided with the ability to determine the logical distance[2] between each incoming message and the two known messages, this strategy will allow the receiver to correct for up to three bits in error.

- Suppose the received word A′ is in error by 1 bit. The logical distance between A′ and A is 1, and the logical distance between A′ and B is 7; thus, A′ is likely to be A.
- Suppose A′ is in error by 3 bits. The logical distance from A is 3, and the logical distance from B is 5; thus A′ is likely to be A.
- Suppose A′ is in error by 4 bits. The logical distance from A is 4, and the logical distance from B is 4; thus A′ is equally likely to be A or B.

---

[2]The logical distance between two binary sequences is determined by finding the number of bits that must be changed to make one sequence into the other. This measure is also known as the Hamming distance.

Continuing the sequence to higher levels of error makes A′ more likely to be B than A. For this particular case, then, the limit of correction is 3 bits in error. In Example 16.3, I compute the probability of error for an 8-bit repetition code of this sort.

EXAMPLE 16.3   Probability of Error for an 8-Bit Repetition Code

Given   In a very long datastream, transmission errors occur at an average rate of one bit in error in ten bits. The errors are randomly distributed and independent of each other. An 8-bit repetition code is used to overcome the effect of the high error rate.

Problem   Find the probability of error using the 8-bit code.

Solution   Because the transmission errors occur randomly and independently, the probability of k errors occurring in an n-bit word is given by the binomial distribution, i.e.,

$$P(k,n) = C_k^n \, p^k (1-p)^{n-k}$$

where p is the probability of error in one bit. If $p \ll 1$, this becomes

$$P(k,n) \cong C_k^n \, p^k = \frac{n! \, p^k}{k!(n-k)!}$$

The coding will correct up to three errors. Hence, the remaining probability of error is

$$P_{\varepsilon 8BRC} = P(4,8) + P(5,8) + P(6,8) + P(7,8) + P(8,8)$$

Evaluating the RHS,

$$P_{\varepsilon 8BRC} = p^4 \left(70 + 56p + 28p^2 + 8p^3 + p^4\right)$$

For $p = 0.1$, this becomes

$$P_{\varepsilon 8BRC} = 0.0076$$

i.e., in a very long datastream, in which randomly distributed and independent transmission errors occur at an average rate of one bit in error in ten bits, an 8-bit repetition code has a probability of error of 0.0076. Of course, the coding efficiency is only 12.5%.

Extension   If the bit error rate is reduced to one bit in error in 100 bits, an 8-bit repetition code has a probability of error of less than 1 in 10^6.

---

(2) Practical Codes

Codes used to provide forward error correction (FEC) are more sophisticated than our example.[3] They are divided into two types:

- **Linear block codes:** in linear block coding, a block of k information bits is presented to the encoder; the encoder responds with a unique code block (codeword) of n bits $(n > k)$. A set of codewords is selected so that the logical distance between each codeword and its neighbors is approximately the same.
- **Convolutional codes:** convolutional codes are generated by continuously performing logic operations on a moving, limited sequence of bits (m bits, say) contained in the message stream. For each bit in the message stream, the encoder produces a fixed number (two or more) of bits whose values depend on the value of the present message bit and the values of the preceding $m-1$ bits.

Example 16.3 concerns a block code in which 1 information bit is represented by a block of 8 bits ($k = 1$ and $n = 8$) that corrects any combination of 3, or fewer, errors. Parity checking is a special case of a block code; it is a single error-detecting code—it cannot correct errors. Optimal encoding and decoding strategies have been determined by exhaustive computer searching of possible implementations. If the number of errors is larger than the correction span, the receiver must resort to an ARQ technique to achieve retransmission of the corrupted message.

(3) Linear Block Codes

By adding redundant bits to information bits in a disciplined way, linear block codes build on the principle of parity checking. Pioneered by R. W. Hamming,[4] a 20th-century American mathematician, they allow us to reduce the information bit error rate while maintaining a fixed transmission rate. If an efficient message source generates M equally likely messages from a k bit code so that $M = 2^k$, and r redundant bits are added to each codeword, the messages are expanded into codewords of length n bits, where $n = k + r$. With this expansion, $2^n - 2^k$ redundant codewords exist. Block codes of this sort

---

[3]For a comprehensive discussion of coding and codes, see Herbert Taub and Donald L. Schilling, *Principles of Communication Systems,* 2nd edition (McGraw-Hill: New York, 1986), pp. 529–78.

[4]R. W. Hamming, "Error detecting and correcting codes," *Bell System Technical Journal,* 29 (1950), 147–160.

are known as *systematic* and called (n,k) codes. Further, they are said to have a *rate* of k/n. This quantity reflects the fact that the information bit rate is k/n times the bit rate of the (n,k) word. If the speed at which information is to be delivered is not to be changed, the n bit word must be sent at a higher speed. In fact, the bit rate of the (n,k) word must be n/k times the bit rate of the uncoded word.

If the logical distance between two codewords is represented by d_{ij}, the value will vary with the combination i, j. Called the *Hamming* distance, the minimum distance d_{min} between words sets a limit to the effectiveness of the code. If ε errors are in a received codeword, you shall be able to detect that the word is errored provided

$$\varepsilon \leq d_{min} - 1 \tag{16.10}$$

Also, if

$$\varepsilon \leq \frac{d_{min} - 1}{2}, \ d_{min} \text{ odd; and } \varepsilon \leq \frac{d_{min}}{2} - 1, \ d_{min} \text{ even} \tag{16.11}$$

we can correct the errors by substituting the closest valid codeword.

In Example 16.3, we have only two valid codewords. The distance between them is 8; because we have only two, this is the Hamming distance also, i.e., $d_{min} = 8$. From equation (16.10), the maximum number of errors that can exist in a word detected to be in error is $\varepsilon = d_{min} - 1 = 7$. Anything less than 8 errors produces an unknown codeword. From equation (16.11), the maximum number of errors in a word that can be corrected is $\varepsilon = \frac{d_{min}}{2} - 1 = 3$. With three or less errors, the received word is closer to the word it is supposed to be than the word it is not supposed to be. These statements are illustrated in **Figure 16.11**.

(4) Cyclic Codes

A subclass of linear block codes, *cyclic* codes have a structure that gives them an efficient code rate and makes them relatively easy to implement. Suppose I represent an n bit codeword as

$$C = \left(x_{n-1}, x_{n-2}, \dots x_1, x_0 \right)$$

where the members are either 0 or 1. By shifting the terms one position to the left in a cyclic manner, we obtain

$$C' = \left(x_{n-2}, x_{n-3}, \dots x_1, x_0, x_{n-1} \right)$$

A second shift produces

$$C'' = \left(x_{n-3}, x_{n-4}, \dots x_1, x_0, x_{n-1}, x_{n-2} \right)$$

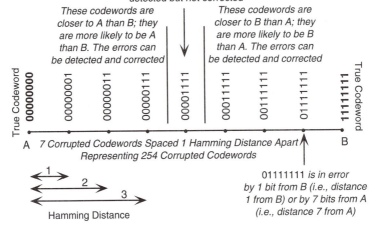

Figure 16.11 Illustration of Hamming Distances with Respect to Error Detection and Error Correction in an 8-Bit Repetitive Code

and so on. The vectors *C*, *C'*, *C''*, etc., are members of a cyclic code.

EXAMPLE 16.4   Eight-Member Cyclic Code

Given The codeword 01001101.

Problem

a) Find the other members of a cyclic code that includes this member.
b) Determine the minimum distance between members.
c) What are the error-detecting and correcting capabilities of the code set?

Solution

a) By moving the bits in order one position to the right, you can form a set of eight cyclic codewords, viz.,

$$C1 = 0\,1\,0\,0\,1\,1\,0\,1$$

$$C2 = 1\,0\,1\,0\,0\,1\,1\,0$$

$$C3 = 0\,1\,0\,1\,0\,0\,1\,1$$

$$C4 = 1\,0\,1\,0\,1\,0\,0\,1$$
$$C5 = 1\,1\,0\,1\,0\,1\,0\,0$$
$$C6 = 0\,1\,1\,0\,1\,0\,1\,0$$
$$C7 = 0\,0\,1\,1\,0\,1\,0\,1$$
$$C8 = 1\,0\,0\,1\,1\,0\,1\,0$$

If the process is pursued, the set of codewords is repeated.

b) By inspection, the logical distance between contiguous codewords, i.e., C2 – C1, C3 – C2, C4 – C3, etc., is 6. Also, the distance between a codeword and all codewords other than those that are contiguous is 4. Thus, C3 – C1 = 4, C4 – C1 = 4, C5 – C1 = 4, etc. Hence $d_{min} = 4$.

c) With $d_{min} = 4$, equations (16.10) and (16.11) tell us that up to 3 errors can be detected and 1 error can be corrected.

To treat the code members analytically, a customary approach is to associate each code vector with a polynomial in a real variable p. Thus, C becomes

$$C(p) = x_{n-1}p^{n-1} + x_{n-2}p^{n-2} + \ldots + x_1 p + x_0 \qquad (16.12)$$

Because the polynomial coefficients are either 0 or 1, the sum of two polynomials is obtained by *modulo*-2 addition. Multiplying equation (16.12) by p, we obtain

$$pC(p) = x_{n-1}p^n + x_{n-2}p^{n-1} + \ldots + x_1 p^2 + x_0 p \qquad (16.13)$$

and the polynomial associated with C' is

$$C'(p) = x_{n-2}p^{n-1} + x_{n-3}p^{n-2} + \ldots + x_0 p + x_{n-1} \qquad (16.14)$$

Summing equations (16.13) and (16.14), and remembering that

$$x_{m-1}p^m \oplus x_{m-1}p^m = 0$$

gives

$$C'(p) = pC(p) + x_{n-1}(p^n + 1) \qquad (16.15)$$

For an (n,k) cyclic code, the variable p forms a *generator* polynomial

$$G(p) = p^q + g_{q-1}p^{q-1} + \ldots + g_1 p + 1 \qquad (16.16)$$

where $q = n - k$, and the coefficients g_m are chosen so that $G(p)$ is a factor of $(p^n + 1)$. Under these conditions, each codeword corresponds to the polynomial product

$$C(p) = Q_M(p)G(p) \qquad (16.17)$$

where $Q_M(p)$ is a block of k message bits.

In **Table 16.1** I list generator polynomials for cyclic Hamming; Bose, Chaudhuri, and Hocquenghem (BCH);[5] and Golay[6] codes that have $d_{min} = 3$, 5, and 7. The Golay code (23, 12) is the only known code with $n = 23$ that can detect up to 7 errors and correct up to 3 errors. However, the rate is 0.52. A significantly better rate (0.71) for the same error performance is attained by BCH (63, 45). However, the codeword is 63 bits long, and the polynomial contains 11 terms. Because of the complexity, the design of codes and the evaluation of their performance is best left to a computer.

(5) Burst Error Correction

The parity bits (r) in the block codes of Table 16.1 allow correction of a limited number of errors. The techniques listed are most effective if the average bit error rate is small and the errors are evenly distributed within the codeword sequence. However, in many situations, the errors are closely clustered together—they occur in *bursts*—and affect entire codewords and short sequences of codewords. Two techniques that are effective in overcoming error bursts are

- **Block interleaving:** information bits are arranged in sequential rows to form columns of bits to which forward error correction parity bits are added [this action is analogous to longitudinal parity checking in § 15.1(1) and Figure 15.3]. The sequence of information bits is transmit-

Type	$G(p)$	n	k	$R = k/n$	d_{min}	#errors corrected
Hamming	$p^5 + p^2 + 1$	15	11	0.73	3	1
BCH	$p^{10} + p^9 + p^8 + p^6 + p^5 + p^3 + 1$	31	21	0.68	5	2
BCH	$p^{18} + p^{17} + p^{16} + p^{15} + p^9 + p^7 + p^6$ $+ p^3 + p^2 + p + 1$	63	45	0.71	7	3
Golay	$p^{11} + p^9 + p^7 + p^6 + p^5 + p + 1$	23	12	0.52	7	3

Table 16.1 Some Cyclic Codes
Codewords are formed from the product of a block of k message bits and the generator polynomial G(p). Golay (23,12) is the only known code with n = 23 that can detect up to 7 errors and correct up to 3 errors.

[5]R. C. Bose and D. K. Ray-Chaudhuri, "On a Class of Error Correcting Binary Group Codes," *Information and Control,* 3 (1960), 68–70.
[6]M. J. E. Golay, "Notes on Digital Coding," *Proceedings of the IRE,* 37 (1949), 657.

ted; it is followed by the sequence of parity bits. During transmission, a limited error burst will affect a sequence of bits in one or more rows. At the receiver, if the errors in a column are within the capabilities of the coding, the errors are corrected and the information bits can be read out in the original sequence. The principle of interleaving appears in **Figure 16.12**.

- **Reed-Solomon (RS) code:** a block code that employs groups of bits, not single bits. Known as *symbols,* the RS code has k' information symbols, r' parity symbols, and codewords of length $n' = k' + r'$ symbols. The number of symbols in a codeword is arranged to be $n' = 2^m - 1$, where m is the number of bits in a symbol (the symbol length). Thus, when $m = 8$, 255 symbols are in each codeword (2040 bits). The code is able to correct errors in $r'/2$ symbols; thus, if we wish to correct errors up to bursts that affect 16 consecutive symbols, the number of parity symbols must be 32. In this case, $n' = 255 = 223 + 32$, i.e., 233 information symbols and 32 parity symbols are in every 255-symbol codeword of 2040 bits. The code rate (k'/n') is 223/255 or 0.8745. Reed-Solomon codes are not efficient codes for correcting random errors.

When the datastream is subject to both random errors and error bursts, you can cascade coding techniques so that random errors are handled by one and

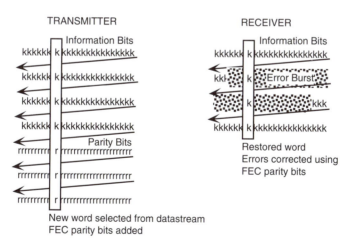

Figure 16.12 Principle of Interleaving Applied to FEC in a Bursty Error Environment

To protect data from bursts of errors, codewords are formed from bits selected periodically from the stream. To each word, FEC parity bits are added. An error burst affects a sequence of bits that are located in separate words so that they can be corrected at the receiver to give the original data sequence.

error bursts are handled by another. Sometimes this is called *concatenated coding*.

(6) Convolutional Codes

Convolutional codes employ relatively simple encoders whose span may extend over many bits; in contrast, the decoders are quite complex. Fundamentally different from block codes, information bits are not collected in a discrete sequence to which check bits are added to form a block or codeword. Instead, a structured mapping technique is used to convert the sequence of information input bits (k) to encoder output bits. Decoding requires the receiver to undo the mapping and determine the most likely known codeword for an errored word.

Consider a sequence of message bits of which the final bit is m_j and the leading bit is m_{j-L}. The output bit x_j is encoded from the sequence $m_{j-L} \rightarrow m_j$ according to the relationship

$$x_j = g_L m_{j-L} \oplus g_{L-1} m_{j-L-1} \oplus \ldots \oplus g_1 m_{j-1} \oplus g_0 m_j \qquad (16.18)$$

i.e.,

$$x_j = \text{modulo-}2 \sum_{i=0}^{L} g_i m_{j-i} \qquad (16.19)$$

where the coefficients g_n have the value 0 or 1. As shown in the top half of **Figure 16.13**, equation (16.19) is implemented readily by a tapped shift register.

To provide one or more redundant bits that can be used for error control, form additional output bits from the same sequence, so that

$$x_j' = \text{modulo-}2 \sum_{i=1}^{L} g_i' m_{j-i}$$

$$x_j'' = \text{modulo-}2 \sum_{i=2}^{L} g_i'' m_{j-i}$$

and so forth. An arrangement that produces 3 output (encoded) bits ($n=3$) for every input (message) bit ($k=1$) from a sequence of 4 message bits ($L=3$) is shown in the bottom half of Figure 16.13. Because the output is three bits for every 1 bit in the message sequence, the encoder has a rate of 0.33. Note that all message bits (m_{j-i}s) need not be connected to all the modulo-2 summers; I have shown only those connections for which the coefficients (g_{L-i}s) are 1s.

Decoding is a matter of reversing the logical operations that produced the encoded bits (x_j, x_j', x_j'') to determe the most likely message bit (m_j). In

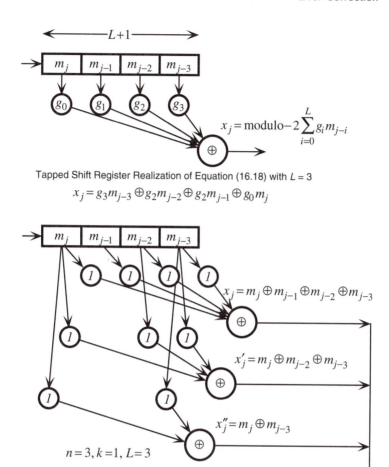

Tapped Shift Register Realization of Equation (16.18) with $L = 3$

$$x_j = g_3 m_{j-3} \oplus g_2 m_{j-2} \oplus g_2 m_{j-1} \oplus g_0 m_j$$

Realization of Convolutional Encoder

Figure 16.13 Principle of Convolutional Encoder

effect, the decoder estimates the path through the logical tree that was followed in the encoding process. Because complexity increases exponentially with the number of components, a common thread in decoder strategies is to reduce the number of paths to be searched. Three techniques have received attention:

- **Sequential decoding:**[7] searches for the most likely logical path by examining one path at a time. The algorithm evaluates probabilities from node to node and abandons likely incorrect paths as the search proceeds.

- **Viterbi algorithm:**[8] implements maximum-likelihood decoding. Employs an algorithm that limits searching the possible logical paths to a manageable number by creating a group of minimum (surviving) paths.
- **Feedback decoding:** searches the possible logical paths through a limited number of nodes. Selects the most likely and discards the remainder, and then searches the section that fans out from this path . . . and so on.

Of the three methods, the Viterbi algorithm is the most complex and requires the most computation. However, it is robust, and is optimum for many applications.

(7) Throughput Using Forward Error Correction

If we assume that a forward error correcting code of rate R ($=k/n$) corrects all errors, then $S = rR$, so that $\eta_S = R$. In **Figure 16.14**, I plot throughput for BCH (63, 45) and Golay (23, 12) codes, and a rate 0.33 convolution encoder

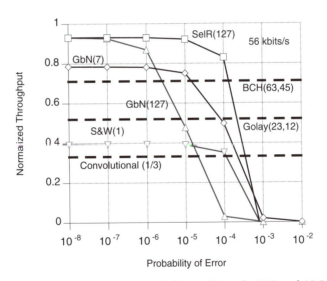

Figure 16.14 Throughput *versus* Probability of Error for FEC and ARQ Error Correction at 56 kbits/s

---

[7]R. M. Fano, "A Heuristic Discussion of Probabilistic Coding," *IEEE Transactions on Information Theory,* IT-9 (1963), 64–74.

[8]A. J. Viterbi, "Error Bounds for Convolutional Codes and an Asymptotically Optimum Decoding Algorithm," *IEEE Transactions on Information Theory,* IT-13 (1967), 260–69.

on the ARQ curves from Figure 16.8. For this particular set of circumstances, a conclusion we can draw is: when $P_\varepsilon > 1 \times 10^{-4}$, FEC should be considered.

REVIEW QUESTIONS

16.1   Describe two basic approaches to error correction.

16.2   What is *stop-and-wait ARQ?*

16.3   What is *go-back-n ARQ?*

16.4   What is *selective repeat ARQ?*

16.5   Define *protocol efficiency.*

16.6   Why is frame relay a popular transmission technique for moving large files at high speed?

16.7   Define *throughput.*

16.8   Why does a high-speed (i.e., *circa* 1.544 Mbits/s) data stream in an error-prone environment in which stop-and-wait ARQ is employed, perform so badly (see Figure 16.3)?

16.9   Why does a high-speed (i.e., *circa* 1.544 Mbits/s) data stream in an error-prone environment in which go-back-n ARQ is employed, perform better when blocks of 127 frames are used than when blocks of 7 frames are used (see Figures 16.4 and 16.5)?

16.10   Why is selective repeat (127) ARQ significantly better than go-back-n (127) ARQ in the mid-range of probabilities of error (see Figures 16.5 and 16.6)?

16.11   Why, at 4.8 kbits/s, with probabilities of error greater than 1×10^{-5}, is go-back-n (127) ARQ worse than go-back-n (7) ARQ (see Figure 16.8)?

16.12   Why, at 56 kbits/s, does stop-and-wait (1) ARQ perform worse than it did at 4.8 kbits/s with respect to the other protocols (see Figure 16.8 and 16.9)?

16.13   Why, at 1.544 Mbits/s, do go-back-n (127) ARQ and selective repeat (127) ARQ outperform go-back-n (7) ARQ and stop-and-wait (1) ARQ (see Figure 16.10)?

16.14   What is the Hamming distance between two binary words?

16.15   What is a *linear block code?* What is the relation between the Hamming distances among the codewords?

16.16   If $d_{min} = 5$, what is the greatest number of errors a word may contain and be detected as an errored word? What is the greatest number of errors the word may contain and be corrected?

16.17   Define a *cyclic code.*

16.18   What is special about a (23,12) Golay code?

16.19   Explain the principle of interleaving applied to coding in a bursty error environment.

16.20   What is a *convolutional code?*

16.21   Comment on Figure 16.14.

17

Access to Shared Media

Not all users can have dedicated access to the external resources they are likely to need. In fact, in most situations, because the time at which they are ready to send will depend on internal priorities and the length of time each sends will vary, sharing communications access is feasible. To avoid collisions among data packets from many sources, several levels of discipline are needed. Which one to invoke depends on the number of users, the length and frequency of transmissions from each user, the capacity of the shared medium, and the cost of repeating a message because part or all of it was destroyed by a collision. Access procedures are classified according to the technique they employ to share the common connection, and the degree of freedom given the sender.

17.1 PRE-ASSIGNED ACCESS

Frequency division, time division, and code division techniques are used to create channels that are assigned permanently to the users. Called *multiplexing,* the main features are

- **FDM** (Frequency Division Multiplexing): a technique that employs specific frequencies and bandwidths to create individual message channels on a cable or radio connection. Each user station is assigned a particular channel whether it has anything to send or not. [An example of FDM of voice channels is given in § 18.8.]
- **TDM** (Time Division Multiplexing): a technique that assigns fixed timeslots to users; during its timeslot, the full capacity of the connection is available to each station. Each station has a fixed allocation of time whether it has anything to send or not. The user's channel is defined by a succession of timeslots, one in every frame.
- **CDM** (Code Division Multiplexing): a spread spectrum technique [described in § 21.4(3)] in which all stations in the network transmit at

the same chip rate and carrier frequency with approximately equal power. Each station is assigned a code that is orthogonal to the codes used by the other stations. Each receiver sees the sum of the individual spread spectrum signals as uncorrelated noise. It can demodulate an individual signal if it has knowledge of the spreading code and the carrier frequency. With CDM, collisions are of no consequence, and the full bandwidth of the connection is available to the users at all times.

Figure 17.1 contrasts the basic natures of the three approaches. In the top figure, the bandwidth of the bearer is divided into 8 equal channels that are reserved for 8 users. In the middle figure, the full bandwidth of the bearer is assigned to one of 8 users at regular intervals. In the bottom figure, the full bandwidth of the bearer is available to any user at any time. The maximum number of users depends upon the number of spreading codes available and the ability to manage the output powers of the user group. FDM is used on radio circuits, coaxial cables, and some older wire systems; TDM is used on wire, cable, radio, and optical fiber circuits; CDM is principally a radio technique.

17.2 DEMAND ASSIGNED ACCESS

To obtain access to a shared medium, a straightforward way is to ask for it. Such a demand requires a system managing entity that can grant the request, allocate resources, and deny those resources to others until the activity is complete. In this Section, I describe some ways of accomplishing demand access to a common facility.

(1) Statistical Multiplexing

In many situations, a user's need to communicate is irregular so that, if assigned the exclusive use of a channel, it will be idle some of the time. A more efficient arrangement is to assign the channel for the transport of traffic between fixed points, and permit sharing by those that have sporadic needs to communicate between the points. This is the principle of *statistical multiplexing*.

Consider the case of n users that share a communication facility. Let the average traffic arriving at the facility be λ bytes/s and the capacity of the communication facility be μ bytes/s (λ must be less than μ). Further, suppose bytes arriving at the facility are Poisson-distributed, and that, after waiting in line, the bytes are served in the same time and sent on to their destination. This is a description of an M/D/1 queueing system [see § 9.2(2)]. The average waiting time in the system is

$$\overline{W}_{M/D/1} = 1/\{2(\mu - \lambda)\} \tag{17.1}$$

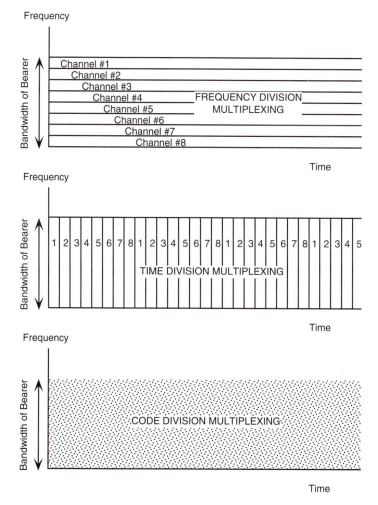

Figure 17.1 Frequency Division, Time Division, and Code Division Multiplexing
Eight users share a bearer. In the top diagram, the bandwidth of the bearer is divided into 8 separate channels. In the middle diagram, the bearer is used for a short time by 8 users in turn. In the bottom channel, 8 users employ different codes so that the signals do not interfere with one another.

and the average number of bytes in the system is

$$\overline{L}_{M/D/1} = \lambda / \{2(\mu - \lambda)\} \qquad (17.2)$$

Figure 17.2 shows relative values for these quantities for $\lambda/\mu < 1$. For values of $\lambda/\mu > 0.7$, the waiting time increases rapidly. If operation is based on con-

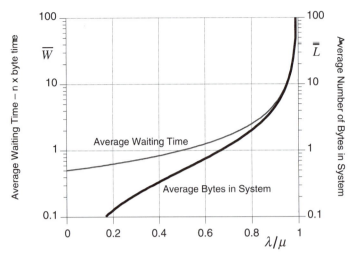

Figure 17.2 Performance of a Statistical Multiplexer When Service Times Are Constant

The multiplexer is modeled as an M/D/1 queue.

stant length packets, Figure 17.2 will give performance data in packet units rather than bytes.

However, if operation is based on variable data units that require custom service times, and they can be assumed to be exponentially distributed, use the results for an M/M/1 queueing system [see § 9.2(1)] for which

$$\overline{W}_{M/M/1} = 1/(\mu - \lambda) \tag{17.3}$$

and

$$\overline{L}_{M/M/1} = \lambda/(\mu - \lambda) \tag{17.4}$$

Alternatively, if operation is based on variable data units that require service times that are Erlangian, use the results for an $M/E_k/1$ queueing system for which

$$\overline{W}_{M/E_k/1} = \frac{(1+k)}{2k} \frac{\lambda}{\mu(\mu - \lambda)} + 1/\mu \tag{17.5}$$

and

$$\overline{L}_{M/E_k/1} = \frac{(1+k)}{2k} \frac{\lambda^2}{\mu(\mu - \lambda)} + \lambda/\mu \tag{17.6}$$

Figure 17.3 shows relative values of waiting time for $\lambda/\mu < 1$. As expected, the average waiting time increases as the service time distributions progress from constant through Erlang to exponential [see § 9.2(4)].

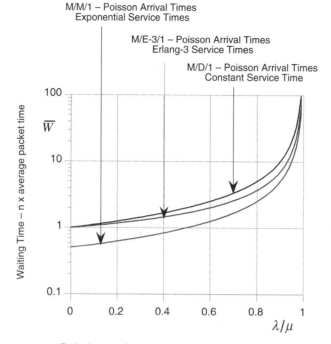

Figure 17.3 Waiting Time in a Statistical Multiplexer
Waiting time depends on the assumption made as to the distribution of service times.

(2) Polling

One of the older techniques for providing access to a network, *polling* can be implemented among peers, or between a primary (master) station and secondary (slave) stations. In both cases, the station with something to send waits until it is interrogated by another entity.

In *peer* polling, interrogation is done by a station of equal status. Each station in the network has an opportunity to send and then queries another station as to its need to send. A sequence is established so that all stations in the network are granted the opportunity to send a message. If some stations have more traffic than others (a server, for instance), the polling pattern can be modified to give them more opportunities than the average station. In normal operation, peer stations continually poll one another. In the case of a station failure, procedures must be in place to restart the polling sequence.

Exclusively, in a network in which one station is primary (a host, for instance) and all other stations are secondary, interrogation is done by the primary station. In a sequence that recognizes the needs of the individual secondary stations, it polls them to find out whether they have anything to send. All messages are sent to the primary station; if they are intended for a secondary station, the primary forwards them. Under normal operation, the primary continually interrogates the secondaries. In the case of a secondary failure, the primary station can continue to poll the others. In the case of a primary station failure, the network collapses.

(3) Token Access

Token access[1] employs a special datablock that is circulated among the stations on a network. If a station has something to send, it seizes the token, adds the message text to it, and sends the frame. Upon receiving it, the destination station copies the message and returns the frame with an acknowledgment to the sender. The sender regenerates the token block and sends it on to the next station. By adjusting the priority levels of the stations in the network, some stations can be given greater opportunities to seize the token. In this way, they have an opportunity to send more often than others.

(4) Transfer Delay

When using polling and token access methods, individual packets may be delayed up to a full cycle time. In addition, they will be lost if the queue at the entry node is full. The performance of polling and token access methods of operation is paced by *waiting time* and *transmission time.*

- **Waiting time** (t_W): time between the arrival of a packet for transmission and the time it is transmitted
- **Packet transmission time** (t_P): time required to send the packet; $t_P = N/r$ where N is the number of bits in a packet and r is the rate at which bits are sent.

To obtain an expression for the average transfer delay, define t_c as the average cycle time or the average time to serve all stations, and assume that, when permitted to send, a station sends all of its waiting packets. If a packet arrives at a station at random relative to time in the cycle and all stations are equally important and are receiving random traffic, on average, the packet must wait $t_c/2$ seconds before being transmitted. If an average λ packets per second arrive at each station, then λt_c packets arrive at each station during a cycle. Further, if M stations are in the system, all of the packets present at the sta-

---

[1]W. Bux, "Token Ring Local-Area Networks and Their Performance," *Proceedings IEEE*, 77 (1989), 238–56.

tion are sent when the station is permitted to send, and Δt is the average time to transfer use of the channel from one station to another, then we can write

$$t_c = M\left[\lambda t_c t_P + \Delta t\right] \tag{17.7}$$

so that,

$$t_c = \frac{M\Delta t}{1 - M\lambda t_P} \tag{17.8}$$

Thus, the average waiting time between the time of arrival of a packet for transmission and the time it is transmitted is

$$\overline{t_W} = t_c/2 = \frac{M\Delta t}{2\left(1 - M\lambda \frac{N}{r}\right)} \tag{17.9}$$

For equations (17.8) and (17.9) to hold, $M\lambda t_P < 1$. Thus, average rate of arrival of packets at each station, packet size, and serving data rate all interact to establish the maximum number of stations that can operate together. Their values are restricted to combinations that satisfy the inequality. In **Figure 17.4** I show the maximum arrival rate of 1088-bit packets at each station as a function of the number of stations and service data rate. The higher the service data rate, the greater the packet arrival intensity and the greater the number of stations that can operate in the system.

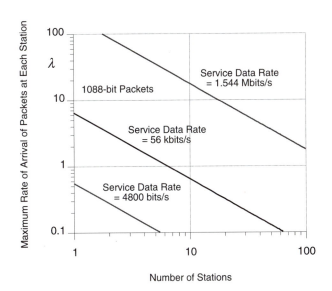

Figure 17.4 Maximum Permissible Rate of Packet Arrival at Each Station on a Token Ring
The rate depends on the number of stations and the service data rate.

(5) *Request Channel*

A direct approach to demand access employs a common channel on which stations request assignment of a private channel for their use. The request channel may operate as a fixed allocation channel in which each station is assigned a portion of the channel, or it may operate as a contention channel in which each station may send its request when ready. With a fixed allocation strategy, the capacity of the request channel limits the absolute size of the network. With a random access strategy, the capacity of the request channel limits only the rate at which requests are made.

17.3 RANDOM ACCESS

Another way of accomplishing access to a shared facility is for stations to send messages as they are ready. This procedure eliminates the need for a system managing entity. However, it runs the risk that messages interfere (collide) with each other, and must be repeated.

(1) *Basic Aloha*

Invented for a microwave radio system linking the Hawaiian islands, Aloha is a technique in which stations send data packets over a common channel as they are ready.[2] **Figure 17.5** shows the concept. On a common channel, stations transmit packets to a transponder that sends them on another common channel back to the receive ports of all stations. The technique is used widely with communication satellites and local-area networks.

Suppose k users each transmit packets of duration t_P seconds every nt_P seconds (where $n >> k$), so that the rate of packet presentation is low. The average rate at which packets are transmitted is

$$\lambda = k/nt_P \tag{17.10}$$

The full-load rate for the transponder is $\lambda = 1/t_P$ and occurs when $n = k$. If packets are generated randomly, the probability that m packets will be transmitted in time t seconds is approximated by a Poisson distribution, i.e.,

$$P(m,t) = \frac{(\lambda t)^m \exp(-\lambda t)}{m!} \tag{17.11}$$

---

[2]N. M. Abramson, "The ALOHA System, Another Alternative for Computer Communications," *AFIPS Conference Proceedings*, 37 (1970) 281–5. Also, L. Kleinrock and S. S. Lam, "Packet Switching in a Multiaccess Broadcast Channel: Performance Evaluation," *IEEE Transactions on Communications*, COM-23 (1975) 410–23.

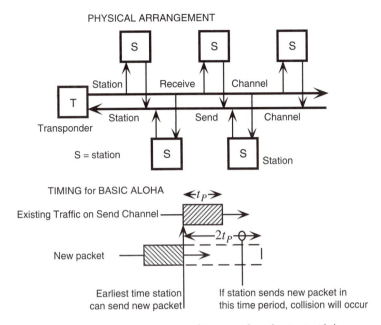

Figure 17.5 Physical Arrangement and Timing Chart for Basic Aloha

From the timing information in Figure 17.5, for a packet to be transmitted safely, no other packet can be transmitted in an interval $2t_P$ seconds. The probability that no other packet is sent over this period is found from equation (17.11) by setting $m=0$, $t=2t_P$, and $k=k-1$

$$P_0 = P(0, 2t_P) = \frac{\{2(k-1)/n\}^0 \exp\{-2(k-1)/n\}}{0!} \cong \exp(-2k/n) \quad (17.12)$$

On average, if the number of packets transmitted is k/nt_P per second, and the probability of successful transmission (i.e., no collisions) is $\exp(-2k/n)$, then the throughput in packets per second is

$$S = \frac{k}{nt_P} \exp(-2k/n) \quad (17.13)$$

If ρ denotes the input density or offered traffic as a fraction of full-load traffic, then $\rho = k/n$. To find the maximum throughput, differentiate equation (17.13) by ρ and equate to zero to give

$$S_{max} = 1/2et_P \text{ when } \rho = k/n = 1/2. \quad (17.14)$$

Figure 17.6 shows the relationship between the normalized throughput St_P (i.e., $S/(1/t_P)$ or throughput divided by full-load packet rate) and the input density $\rho = k/n$; clearly a maximum occurs at $St_p = 1/2e$ when $n = 2k$ and each user offers packets at an average rate of $1/2kt_p$ packets/s. Further:

- If the offered traffic from all the stations is full-load ($n=k$), approximately 13% of the packets will get through on the first attempt.
- If the offered traffic from all the stations is 50% of full-load ($n=2k$), throughput will be approximately 18% of full-load, or approximately 36% of the offered packets will get through on the first attempt.
- If the offered traffic from all the stations is 10% of full-load ($n=10k$), throughput will be approximately 8% of full-load, or approximately 80% of the offered packets will get through on the first attempt.

On the other hand:

- If the offered traffic from all the stations is 200% of full-load ($n=k/2$), throughput will be approximately 3% of full-load, or approximately 1.5% of the offered packets will get through on the first attempt.

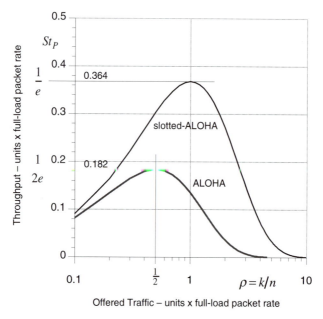

Figure 17.6 Throughput of Basic Aloha and Slotted-Aloha
In basic Aloha, if the offered traffic from all stations is 50% of full-load, throughput will be approximately 18% of full-load, or approximately 36% of the offered packets will get through on the first attempt. In slotted-Aloha, if the offered traffic from all stations is full-load, throughput will be approximately 36% of full-load; this is twice the capacity of basic Aloha.

Thus, so long as the total offered traffic from all stations is not a very large fraction of the capacity of the transponder, many packets have a reasonable chance of finding the transponder unoccupied and can be relayed without interference. Attempting to operate the system at an average rate greater than full-load will cause queueing at stations and overloading, and eventually packets will have to be abandoned.

(2) Slotted and Reservation Aloha

By modifying the process so that packets are sent to arrive at the transponder at the beginning of agreed upon timeslots, the throughput can be doubled. Called slotted-Aloha, the technique requires that all stations be synchronized so that their packets only arrive at the transponder at a slot start time. Under this circumstance, the period of vulnerability to a collision is reduced to one packet time, and equation (17.12) is modified to

$$P_0 = P(0, t_P) = \exp(\rho) \tag{17.15}$$

Equation (17.13) becomes

$$S = \frac{\rho}{t_P} \exp(-\rho) \tag{17.16}$$

and throughput is a maximum when $S = 1$. The maximum value is

$$S_{max} = 1/et_P \tag{17.17}$$

which is twice the value for simple Aloha. In Figure 17.6, I compare throughput for Aloha and slotted-Aloha as a function of offered traffic.

An extension of this technique allows stations to reserve time slots for their exclusive use. Called reservation-Aloha, it guarantees that messages will arrive one at a time at the transponder and will be relayed without collisions. Maximum throughput is the full-load packet rate, and the average delay is the propagation time plus send time. Thus, throughput is improved at the expense of imposing discipline on the senders and limiting the offered traffic to no more than the capacity of the transponder.

EXAMPLE 17.1   Local-Area Network

Given On a campus, a number of terminals are connected by twisted pair cables to a central hub. They employ 1088 bit packets at a speed of 56 kbits/s.

Problem

a) What is the full-load capacity of the hub?
b) What is the packet transmit time?

c) If the network is operated as an Aloha network, what is the expected maximum throughput when all terminals are active and the utilization factor is 50%?

d) If the network is operated as a slotted-Aloha network, what is the expected maximum throughput when all terminals are active and the utilization factor is 50%?

e) If the network is operated as a reservation-Aloha network, what is the expected maximum throughput? If this level of operation is achieved when all terminals are active and the utilization factor is 10%, how many terminals exist?

Solution

a) If each packet contains 1088 bits, and the hub operates at 56 kbits/s, the number of packets that can be relayed per second is

$$56000/1088 = 51.47 \text{ packets/s}$$

b) The bit rate is 56 kbits/s. Hence the time to transmit a 1088 bit packet is

$$1088/56000 = 0.0194 \text{ seconds}$$

c) From equation (17.14), $S_{max} = 1/2et_P$ when $\rho = k/n = 1/2$. Thus,

$$S_{max} = 1/(2e \times 0.0194) = 11.832 \text{ packets/s}$$

d) From equation (17.16), $S = \frac{\rho}{t_P}\exp(-\rho)$. Thus,

$$S = \frac{1/2}{0.0194}\exp(-1/2) = 15.632 \text{ packets/s}$$

e) In reservation-Aloha, transponder slots are reserved for each terminal. From a) the transponder capacity is 51.47 slots per second. Allowing some capacity for guard time, the maximum throughput might be 50 packets/s. From equation (17.10), the average rate at which a terminal transmits packets is

$$\lambda = k/nt_P$$

so that

$$\lambda = 1/(10 \times 0.0194) = 5.155 \text{ packets/s/terminal}$$

Hence, the number of terminals is $50/5.155 = 9.700$, i.e., nine terminals operating at a utilization factor of approximately 10% will drive the network to full capacity.

(3) Collisions

For stations operating under Aloha or slotted-Aloha, the occurrence of a collision can have a range of outcomes. At best, the addresses (originating and terminating) and packet identifier information will survive so that the terminating station can send a NAK. At worst, all of the identifying information will be destroyed so that no station accepts the garbled packet; consequently, no station responds. In between these limits, partial destruction of the information can occur, leading to wrongly addressed, wrongly identified, and otherwise garbled packets. If the sender receives an ACK, most likely, all is well. If the sender receives a NAK, or, after a fixed time has no information as to the status of a packet, it will repeat the packet. If, after resending a packet a number of times, the sender still receives no acknowledgment, the sender will cease to repeat the packet and may seek operator intervention. In this event, the most likely explanation is that the receiver is out of service. Other scenarios are possible, particularly if the sender sends several packets at a time.

Using the parameters of the stop-and-wait model of Figure 16.1, let t_P be the time to send a packet, t_N be the time to send a NAK, τ be the propagation time, μ be the processing time before the receiving station determines there is an error and sends a NAK, and t_W be the time the sender waits before sending again. To simplify the situation, assume that the NAK message is short enough so that it has a high likelihood of getting through without colliding with other traffic. Then the time Δ between the reception of the beginning of the original packet (which is found to be in error) and the reception of the beginning of a repetition of the packet is

$$\Delta = t_P + \mu + t_N + 2\tau + t_W \qquad (17.18)$$

When a collision does not occur, the delay between beginning to send a packet and beginning to receive it is τ, and the probability of such an occurence is P_0. When a collision occurs, the packet is re-sent, the delay is $\tau + \Delta$, and the probability of the packet getting through on the second attempt is $(1 - P_0)P_0$. If a second collision occurs, the procedure is repeated. In general, when the packet is received without errors after $q + 1$ sends, the delay is $\tau + q\Delta$, and the probability of this event is $(1 - P_0)^q P_0$. Thus, the average delay time between beginning to send a packet and receiving a packet without errors is

$$t_D = \tau P_0 + (\tau + \Delta)(1 - P_0)P_0 + \dots + (\tau + q\Delta)(1 - P_0)^q P_0 + \dots \qquad (17.19)$$

Simplifying, if τ is negligible with respect to t_D and Δ

$$t_D \cong \Delta P_0 (1 - P_0) \left[1 + \dots + q(1 - P_0)^{q-1} + \dots \right]$$

The quantity in the [ ] brackets is an infinite geometric series. Simplifying,

$$t_D \cong \Delta \left[(1 - P_0)/P_0 \right]$$

For basic Aloha, substituting the value of P_0 from equation (17.12),

$$t_D \cong \Delta \left[\exp(2\rho) - 1 \right] \qquad (17.20)$$

For slotted-Aloha, substituting the value of P_0 from equation (17.15),

$$t_D \cong \Delta\left[\exp(\rho)-1\right] \qquad (17.21)$$

In **Figure 17.7**, I show values of equations (17.20) and (17.21) in which the delay before the packet is received without error is plotted against offered traffic. In drawing the curves, I have assumed that $\Delta = t_P$, $2t_P$, and $11t_P$. When $\Delta = t_P$, all other quantities in equation (17.15) become zero, so that this curve represents the minimum time delay. When $\Delta = 2t_P$, the sum of the other quantities in equation (17.15) is one packet time. Finally, I plot curves for $\Delta = 11t_P$, on the assumption that they represent the performance for the case when the sender has received no information on the packet at timeout (assumed to be $10t_P$) and sends again. Loosely, these curves correspond to the average delay under light, medium, and heavy collision conditions.

EXAMPLE 17.2   Satellite Network

Given A number of earth terminals are connected to a central hub by a geostationary satellite. They employ 1088 bit packets at a speed of 56 kbits/s. The one-way propagation delay is 0.25 seconds.

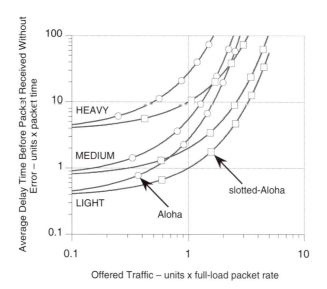

Figure 17.7 Time Delay Before Packet Is Received without Error As a Function of Offered Traffic and Repeat Time
The three sets of curves correspond loosely to the repeat times associated with the intensity of the collision environment.

Problem If the system operates as an Aloha network,

a) when all terminals are active and the utilization factor is 50%, what is the expected maximum throughput?
b) when all terminals are active and the utilization factor is 50%, what is the expected time delay between an earth station begining to send a packet and the hub receiving a packet without errors? Assume $\Delta = \tau + t_P$.
c) what is the expected time delay between an earth station begining to send a packet and receiving a reply packet without errors? Assume $\Delta = \tau + t_P$.

Solution

a) This is the same as c) above, i.e., from equation (17.14) $S_{max} = 1/2et_P$ when $\rho = k/n = 1/2$; thus,

$$S_{max} = 1/(2e \times 0.0194) = 11.832 \text{ packets/s}$$

b) When the propagation delay is not negligible, equation (17.20) becomes

$$t_D \cong \tau + \Delta \left[\exp(2\rho) - 1 \right] = \tau + (\tau + t_P) \left[\exp(2\rho) - 1 \right].$$

Thus when $\tau = 0.25$ seconds, $t_P = 0.0194$ seconds and $\rho = 1/2$ full-load packet rate

$$t_D = 0.25 + (0.2694 \times 1.7183) = 0.7129 \text{ seconds.}$$

c) The expected time delay between an earth station begining to send a packet and receiving a reply packet without errors under the assumption that $\Delta = \tau + t_P$ is

$$2 \times 0.7129 = 1.426 \text{ seconds.}$$

Comment The large propagation delay slows the completion of transactions over the satellite network. It has no effect on the maximum throughput (which is the same as Example 17.1).

---

(4) Carrier Sense Multiple Access

A procedure to alleviate the effect of collisions is known as Carrier Sense Multiple Access (CSMA) protocol; it requires that each station

- monitors activity on the shared medium
- can send only at a time determined by a random interval timer at each station
- does not send unless the shared medium is idle.

In effect, CSMA requires that stations in the network monitor what other stations are doing. Each of them is afforded opportunities to send at times specified by a local random interval timer. If the shared medium is occupied at the time designated, they do not send but wait for the next opportunity provided by the timer.

Suppose a receiver receives packets of duration t_P and the interarrival time t_I [see § 7.2(4)] is Poisson-distributed so that

$$P[t_I] = \tfrac{\rho}{t_P} \exp(-\rho t_I/t_P) \tag{17.22}$$

where, as before, ρ is the input ratio. The ratio of interarrival time to packet duration is a random process with average value

$$E[t_I/t_P] = \int_0^\infty \frac{t_I \rho}{t_P^2} \exp(-\rho t_I/t_P) dt_I = 1/\rho \tag{17.23}$$

Now throughput S is given by

$$S = 1/(t_P + t_I) = \tfrac{1}{t_P}\left[1/(1 + t_I/t_P)\right] \tag{17.24}$$

Substituting the average value of t_I/t_P from equation (17.23), equation (17.24) becomes

$$S = \tfrac{1}{t_P}\left[\rho/(1+\rho)\right] \tag{17.25}$$

so that the normalized throughput is

$$S t_P = \rho/(1+\rho) \tag{17.26}$$

In **Figure 17.8**, compare the normalized throughput of CSMA with Aloha and slotted-Aloha from Figure 17.6. It shows the performance that can be achieved, provided, before sending, each station is able to determine unequivocally that the shared medium is idle. Two factors influence this determination:

- **Detection delay time:** the time required for the terminal to sense when the shared medium is in use
- **Propagation delay time:** the time required for the signal to travel to other stations connected by the shared medium

These quantities affect the ability of stations to detect activity when another station on the shared medium has just started sending. Just after a station begins to send, and before the activity is sensed at a distant terminal, if it has just timed out, the distant terminal may begin to send. Thus, in a short time period equal to the propagation delay time plus the detection delay time, actions may legitimately occur that will result in collisions. As a result, CSMA has been modified to include a collision detection mechanism.

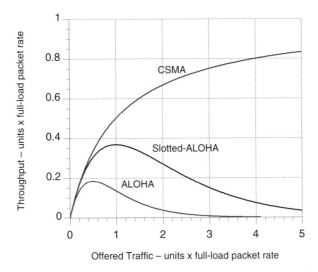

Figure 17.8 Normalized Throughput Performance of Aloha, Slotted-Aloha, and CSMA Protocols
In carrier sense multiple access (CSMA) protocol, each station monitors activity on the common medium and can only send if the medium is idle and a random interval timer at the station has timed out.

EXAMPLE 17.3   Local-Area Network (Continued)

Given As in Example 17.1, a number of terminals are connected by twisted pair cables to a central hub. They employ 1088 bit packets at a speed of 1.54 Mbits/s. The speed of propagation is 2×10^4 km/s, and the time required by the receiver to detect activity on the shared medium is 1 μs.

Problem

   a) If the system is operated as a CSMA network, what is the full-load capacity of the hub? What is the transmission time for an individual packet?
   b) If terminals A and B are each 1 km from the hub, and A begins to send a packet to terminal C at instant T_0, what is the magnitude of the timeslot in which B's local clock may timeout and B begins to transmit?
   c) If the header (which contains addresses and identifier) is 24 bits long, and B's transmissions are randomly distributed over the timeslot, what is the probability that A's packet will be identified by station C (also 1 km from the hub)?

Solution

a) If each packet contains 1088 bits and the hub operates at 1.54 Mbits/s, the number of packets that can be relayed per second is

$$1.54 \times 10^6 / 1088 = 1419 \text{ packets/s}$$

The time to transmit a single packet is

$$1088 / 1.54 \times 10^6 = 706.49 \text{ μs}$$

b) Over 2 kms, the propagation time is

$$2 / 2 \times 10^4 = 100 \text{ μs}$$

With an electronic delay of 1 μs to detect incoming activity, the time-slot is 101 μs.

c) Transmitting the header takes

$$24 / 1.54 \times 10^6 = 15.584 \text{ μs}.$$

B's interfering packet may arrive at C any time between T_0 and $T_0 + 101$ μs; hence the probability of a collision after the header has been received is

$$(101 - 15.584) / 101 = 0.8457$$

This is the probability of the header surviving so that C may return a NAK to A.

Extension If N terminals are located 1 km from the hub, and each has a packet to send when their local clocks time out, the number that send in a 100 μs interval will depend on the statistics associated with the clocks. If, in a 1000 μs period, the clocks provide a randomly distributed 10 μs window in which, provided the shared medium is unoccupied, the terminals may begin to transmit, then the probability of transmission by any terminal will be 0.01. For the same size window and clock periods of 100 μs and 10,000 μs, the probabilities of transmission are 0.1 and 0.001. Obviously, longer clock periods result in fewer collisions; however, longer clock periods slow the overall rate of packet transmission.

EXAMPLE 17.4   Local-Area Network (Continued)

Given A local-area network as in Example 17.3, except the terminals are connected to the hub with optical fibers. The speed of propagation is 2×10^5 km/s, and the time required by the receiver to detect activity on the shared medium is 1 μs.

Problem If the system is operated as a CSMA network, and A begins to send a packet to terminal C at instant T_0, what is the magnitude of the timeslot in which B's local clock may timeout and B begins to transmit?

Solution Over 2 kms, the propagation time is

$$2/2 \times 10^5 = 10 \ \mu s$$

With an electronic delay of 1 μs to detect incoming activity, the timeslot is 11 μs.

Extension With clocks that provide a randomly distributed 10 μs window and have periods of 100, 1000, and 10,000 μs, the probabilities of transmission by another terminal are are 0.01, 0.0001, and 0.00001. The shorter propagation delay of optical fiber has significantly decreased the probability of a collision.

---

(5) Carrier Sense Multiple Access with Collision Detection

Carrier Sense Multiple Access with Collision Detection (CSMA/CD) protocol[3] requires that each station

- monitors activity on the shared medium
- sends only at times designated by the local timer when the shared medium appears to be idle
- continues to monitor the medium throughout the packet transmission so as to detect any collision with another station's packet
- stop sending if a collision occurs, and, before trying to send again, wait for an interval designated by the local timer

If the time a user's packet must wait before being placed on the shared medium is t_W, the total delay time between the packet arriving at the sending station and arriving at the receiving station is

$$t_\Delta = t_W + \tau \tag{17.27}$$

If the probability of the shared medium being idle is P_{idle}, and the probability of the shared medium being busy is P_{busy}, then the probability of the medium being busy the first time activity was sensed and idle the second time activity

---

[3]F. A. Tobagi and V. B. Hunt, "Performance Analysis of Carrier Sense Multiple Access with Collision Detection," *Computer Networks,* 5 (1980), 245–59.

was sensed is $P_{idle}P_{busy}$. The probability of the medium being busy the first and second times activity was sensed and idle the third time activity was sensed is $P_{idle}P_{busy}^2$, and so on. Hence, the waiting time to enter the system is

$$t_W = \bar{t}_s P_{idle} P_{busy} + 2\bar{t}_s P_{idle} P_{busy}^2 + \ldots \qquad (17.28)$$

where $\bar{t}_s$ is the average time interval (determined by the local random timer) until sending is permitted. Now,

$$P_{idle} = \bar{t}_I / (t_P + \bar{t}_I)$$

and

$$P_{busy} = (1 - P_{idle}) = t_P / (t_P + \bar{t}_I)$$

so that, substituting in equation (17.28)

$$t_W = t_P \bar{t}_s / \bar{t}_I = \rho \bar{t}_s \qquad (17.29)$$

and substituting in equation (17.27)

$$t_\Delta = \rho \bar{t}_s + t_p \qquad (17.30)$$

If we assume that the propagation time τ is small and constant, equation (17.30) shows that delay will be minimized as $\bar{t}_s \to 0$ (i.e., the local timer produces very small intervals). However, as $\bar{t}_s \to \tau$, the other stations can no longer sense the activity, and the probability of collisions will increase rapidly and throughput will decrease equally rapidly.

To find the magnitude of the effect of collisions on throughput, assume that the times at which packets are ready to enter the system are Poisson-distributed. After user i sends, the probability that no packets enter the system for an interval equal to propagation delay τ is $P[0, \tau]$ so that the probability of no collision is

$$P_{no} = P[0, \tau] = \exp(-\rho \tau / t_P) \qquad (17.31)$$

and the probability of a collision is

$$P_{yes} = 1 - P[0, \tau] = 1 - \exp(-\rho \tau / t_P) \qquad (17.32)$$

If $\bar{T}$ is the average time between successfully received packets (i.e., no collision), and $\bar{i}$ is the average time, measured from the time user i begins to send a packet, at which a collision occurs $(\bar{i} < t_P)$

$$\bar{T} = (t_P + \bar{t}_I)P_{no} + \left[2(t_P + \bar{t}_I) + \bar{i}\right]P_{no}P_{yes} + \left[3(t_P + \bar{t}_I) + 2\bar{i}\right]P_{no}P_{yes}^2 + \ldots$$

where $(t_P + \bar{t}_I)$ is the average time between correctly received packets when no collisions occur, $2(t_P + \bar{t}_I) + \bar{i}$ is the average time between correctly received packets when a collision occurs during the first transmission but no collision occurs during the second packet transmission, and so on. Simplifying equation (17.33), we obtain

$$\overline{T} = \left(t_P + \overline{t_I}\right)\frac{1}{P_{no}} + \overline{t}\left(P_{yes}/P_{no}\right) \qquad (17.33)$$

Now $\overline{t_I} = t_P/\rho$, so that

$$\overline{T} = t_P\left(\frac{1+\rho}{\rho}\right)\frac{1}{P_{no}} + \overline{t}\left(P_{yes}/P_{no}\right) \qquad (17.34)$$

Because $0 < \overline{t} < \tau$, we can determine upper and lower bounds for $\overline{T}$; they are

$$t_P\left(\frac{1+\rho}{\rho}\right)\frac{1}{P_{no}} < \overline{T} < t_P\left(\frac{1+\rho}{\rho}\right)\frac{1}{P_{no}} + \tau\left(P_{yes}/P_{no}\right) \qquad (17.35)$$

and because throughput $S = 1/\overline{T}$, we obtain upper and lower bounds for the normalized throughput St_P. Substituting equations (17.31) and (17.32) in (17.35), they are

$$\frac{\rho\exp(-\rho\tau/t_P)}{1+\rho+(\rho\tau/t_P)\left[1-\exp(-\rho\tau/t_P)\right]} < St_P < \frac{\rho}{1+\rho}\exp(-\rho\tau/t_P) \qquad (17.36)$$

In **Figure 17.9**, I plot the upper and lower bounds of normalized throughput for a range of input intensities (ρ), and propagation times that are 100%, 10%, and 1% of packet time (i.e., $\tau/t_P = $ 1, 0.1, and 0.01). As $\tau \to 0$, $St_P \to \rho/(1+\rho)$, i.e., CSMA/CD is equivalent to CSMA. In Figure 17.9, I include the CSMA throughput performance curve from Figure 17.8 to emphasize the point. However, these values will be achieved only when propagation and detection times are vanishingly small compared to packet times. One message of Figure 17.9 is that, as propagation and detection times increase, so should packet time. This increase can be achieved by sending longer packets at the same speed, or sending the same length packets at slower speeds.

EXAMPLE 17.5   Local-Area Network (Continued)

Given A number of terminals are connected by twisted pair cables to a central hub. They employ 1088 bit packets at a speed of 56 kbits/s, and each terminal is 1 km from the hub. The speed of propagation is 2×10^4 km/s, and the time required by the receiver to detect activity on the shared medium is 1 μs.

Problem If the network is operated as a CSMA/CD network, what is the expected maximum throughput (upper bound) when all terminals are active and the utilization factor is 50%?

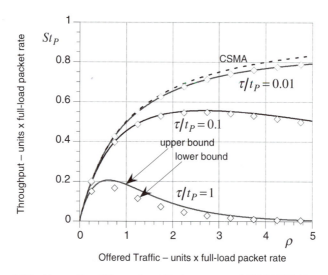

Figure 17.9 Normalized Throughput Performance of CSMA/CD Protocol for Propagation Times of 100%, 10%, and 1% of Packet Times
In carrier sense multiple access, collision detection (CSMA/CD) protocol, each station monitors the medium for activity, sends only when the common medium is idle and a local timer has timed out, continues to monitor the medium throughout transmission so as to detect any collision, and ceases transmission if a collision occurs. The station waits a random time before attempting to resend.

Solution The bit rate is 56 kbits/s. Hence the time to transmit a 1088 bit packet is

$$t_P = 1088/56000 = 0.0194 \text{ seconds}$$

Also, the distance between any two terminals is 2 km, so that the propagation time is

$$\tau = 2/2 \times 10^4 = 1 \times 10^{-4} \text{ seconds}$$

Thus,

$$S_{max} = 1/(2e \times 0.0194) = 11.832 \text{ packets/s}$$

Taking the upper bound in equation (17.36),

$$S < \frac{\rho}{t_P(1+\rho)} \exp(-\rho\tau/t_P)$$

Thus, when $\rho = k/n = 1/2$, $t_P = 0.0194$ seconds, and $\tau = 1 \times 10^{-4}$

$$S < \frac{1}{0.0194 \times 3} \exp(-1 \times 10^{-4}/2 \times 0.0194) \cong \frac{1}{0.0194 \times 3} = 17.18 \text{ packets/s}$$

Because the exponential term is equal to 1, this is also the value for CSMA [see equation (17.26)].

Comments Comparing this result to those in Example 17.1, for $\rho = k/n = 1/2$, we obtain

$$
\begin{aligned}
\text{Aloha} \quad & S = 11.83 \text{ packets/s} \\
\text{slotted-Aloha} \quad & S = 15.63 \\
\text{reservation-Aloha} \quad & S = 51.47 \text{ (i.e., transponder capacity)} \\
\text{CSMA} \quad & S = 17.18 \\
\text{CSMA/CD} \quad & S = 17.18
\end{aligned}
$$

REVIEW QUESTIONS

17.1   Distinguish among *frequency, time,* and *code division multiplexing.*

17.2   Describe *statistical multiplexing.* What difference does the service time make to the delay encountered by a message?

17.3   How is throughput of a statistical multiplexer affected by the average rate of traffic arriving at the input and its traffic handling capacity?

17.4   Distinguish between *peer polling* and *primary polling.*

17.5   What is a token?

17.6   In a polling or token operating environment, on what factors does the maximum number of nodes depend?

17.7   Define the random access method known as Aloha.

17.8   In an Aloha system, what percentage of offered packets are transmitted without collisions when the offered traffic from all stations is 50% of full load? What percentage of offered packets is transmitted without collisions when the offered traffic from all stations is 10% of full load?

17.9   Define *slotted-* and *reservation-Aloha.* What improvements in throughput are achieved with these operating procedures?

17.10   List the procedures employed when operating under a carrier sense multiple access protocol. Comment on Figure 17.8.

17.11   List the procedures employed when operating under a carrier sense multiple access, collision detection protocol. Comment on Figure 17.9.

18

Amplitude Modulation

- **Modulation:** a signal processing technique in which one signal (the *modulating* or message signal) modifies a property (or properties) of another signal (the *modulated* or carrier signal) to create a composite wave (the *modulated* wave); as a result, message signal energy is distributed at frequencies around the carrier signal frequency
- **Demodulation:** the action of recovering the modulating signal from the modulated wave
- **Baseband:** the frequency band (bandwidth) occupied by the modulating (message) signal

Modulation facilitates the long-distance transmission of messages. Because of the higher frequencies (and shorter wavelengths) it enjoys, the modulated wave can be launched from, and received by, practical sizes of antennas, or conducted by moderate sizes of cables or waveguides. With efficient coupling to the communication medium, modulated waves travel over greater distances than the original message signal can travel unaided.

18.1 MESSAGE SIGNAL

The message signal $m(t)$ contains the information to be transmitted by the modulated wave. It is convenient to define two forms:

- **Zero-mean message signal:** denoted by $m_0(t)$, the signal contains no dc-term so that $M_0(0)=0$. If $m(t)$ contains a dc-term, $m_0(t)$ is constructed by subtracting the time-average of $m(t)$ from $m(t)$, i.e.,

$$m_0(t)=m(t)-\langle m(t)\rangle \tag{18.1}$$

- **Normalized message signal:** denoted by $m_1(t)$, the zero-mean message signal has a maximum magnitude of 1. Normalized message signals are a convenience in the analysis of modulation processes. $m_1(t)$ is constructed by dividing $m_0(t)$ by its maximum magnitude, i.e.,

$$m_1(t) = m_0(t) / |m_0(t)|_{\max} \tag{18.2}$$

A sinusoidal signal can be modulated by a message signal in three ways. Consider the carrier signal $c(t)$ that is modeled as a cosine function, i.e.,

$$c(t) = A\cos(2\pi f t + \phi)$$

It may be modulated by varying the angle (i.e., $2\pi f t + \phi$) of the cosine function; in this case, we have *angle* modulation. Angle modulation (θM) is achieved in two ways: varying the frequency f produces *frequency* modulation (FM), and varying the phase angle ϕ produces *phase* modulation (ΦM).

Also, $c(t)$ may be modulated by varying the amplitude A in sympathy with the message function; in this case we have *amplitude* modulation (AM). Amplitude modulated waves can take several forms—large-carrier AM (LCAM), double-sideband, suppressed-carrier AM (DSBSC), single-sideband AM (SSB), and vestigial-sideband AM (VSB).

18.2 LARGE-CARRIER AM (LCAM)

LCAM is the modulation technique employed by all radio stations broadcasting information and entertainment in the frequency range 535 to 1605 kHz (called the AM-band). It has been specially crafted to make possible receiving the message signal with inexpensive circuits—a prerequisite for the development of large-scale, consumer-oriented, broadcast services. Because the envelope of the modulated wave mimics the message signal, demodulation can be accomplished with a relatively simple envelope detector.

An LCAM wave $s_{LCAM}(t)$ is produced by multiplying the carrier function by $[1 + \mu m_1(t)]$ to give

$$s_{LCAM}(t) = [1 + \mu m_1(t)] A\cos(2\pi f_c t) \tag{18.3}$$

$m_1(t)$ is a normalized, zero-mean message function of bandwidth W, f_c is the frequency of the carrier function, and μ is a constant ($0 < \mu < 1$) known as the *modulation index;* it controls the *depth* of modulation. If $m(t) \Leftrightarrow M(f)$, in the frequency domain, equation (18.3) can be expressed as

$$S_{LCAM}(f) = \frac{A}{2}[\delta(f - f_c) + \delta(f + f_c)] + \frac{\mu A}{2}[M(f - f_c) + M(f + f_c)] \tag{18.4}$$

Figure 18.1 shows the frequency domain representation of an LCAM wave for which $M(f)$ is given the arbitrary shape shown at the top of the diagram. The modulated function is *double-sideband* (i.e., the power associated with the message function is distributed on both sides of the carrier frequency), and a large impulse function exists at the carrier frequency. If the

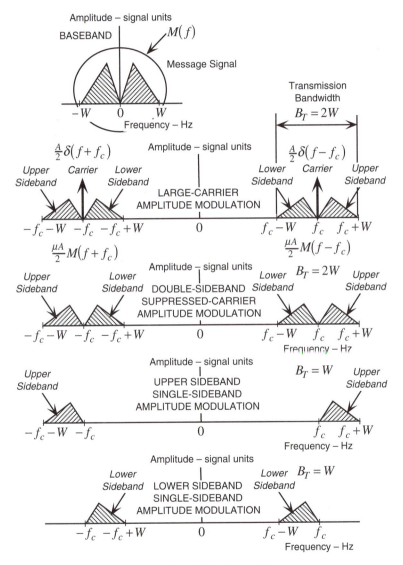

Figure 18.1 Spectrums of Amplitude Modulated Waves
Amplitude modulation is used to produce double-sideband and single-sideband waves.

message function (baseband) is limited to the frequency interval $-W \leq f \leq W$, the modulated wave is limited to the band $f_c \pm W$, and the transmission bandwidth is $2W$ Hz.

The frequency spectrum consists of impulses of area $A/2$ situated at f_c and $-f_c$ together with spectrums of the message function centered on f_c and $-f_c$. The portions of the spectrums of the message function lying above f_c and below $-f_c$ are referred to as *upper sidebands;* in a similar fashion, the portions of the spectrums of the message function lying immediately below f_c and immediately above $-f_c$ are referred to as *lower sidebands.* $M(f - f_c)$ and $M(f + f_c)$ are mirror images of one another.

EXAMPLE 18.1   LCAM with Sinusoidal Message Function

Given   A sinusoidal carrier $c(t) = A\cos(2\pi f_c t)$ and a sinusoidal message function $m_1(t) = \cos(2\pi f_m t)$ produce a large-carrier amplitude modulated wave for which the modulation index is μ.

Problem   Find $s_{LCAM}(t)$ and $S_{LCAM}(f)$.

Solution   From equation (18.3)

$$s_{LCAM}(t) = \left[1 + \mu\cos(2\pi f_m t) \right] A\cos(2\pi f_c t)$$

Multiplying out and simplifying gives

$$s_{LCAM}(t) = A\cos(2\pi f_c t) + \tfrac{\mu A}{2}\cos\left[2\pi(f_c + f_m)t \right] + \tfrac{\mu A}{2}\cos\left[2\pi(f_c - f_m)t \right] \quad (18.5)$$

The Fourier transform of equation (18.5) gives $S_{LCAM}(f)$

$$S_{LCAM}(f) = \tfrac{A}{2}\left[\delta(f \pm f_c) \right] + \tfrac{\mu A}{4}\left[\delta\{ f \pm (f_c + f_m) \} + \delta\{ f \pm (f_c - f_m) \} \right] \quad (18.6)$$

where

$$\delta(f \pm f_c) = \delta(f + f_c) + \delta(f - f_c)$$

and

$$\delta\{ f \pm (f_c + f_m) \} = \delta\{ f + (f_c + f_m) \} + \delta\{ f - (f_c + f_m) \}$$

Comment   To give form to these equations, suppose $A = 1.5$, $f_c = 6.3$, $f_m = 1$, and $\mu = 0.8$. In **Figure 18.2** I draw $m(t)$, $c(t)$, $s_{LCAM}(t)$, and $S_{LCAM}(f)$. In real life, the carrier frequency is considerably higher than the message frequency; a ratio of 6.3:1 is used to make it possible to draw the diagram.

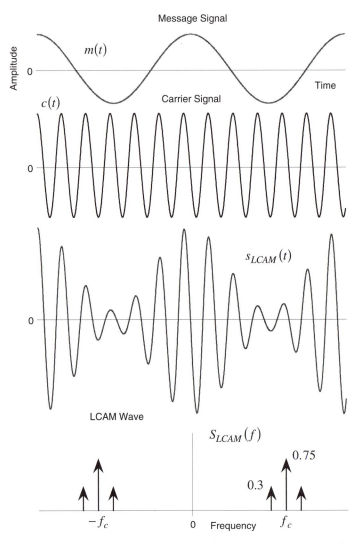

Figure 18.2 Time and Frequency Diagrams for Large-Carrier Amplitude Modulated Wave in Example 18.1

EXAMPLE 18.2   Ratio of Power in Carrier to Power in Sinusoidal Message Function

Given A sinusoidal carrier $c(t) = A\cos(2\pi f_c t)$ and a sinusoidal message function $m_1(t) = \cos(2\pi f_m t)$ produce a large-carrier amplitude modulated wave for which the modulation index is μ.

Problem Find the ratio of the total power in the carrier components to the total power in the message components.

Solution The power (in signal watts) in the carrier function is

$$\overline{P_c}[\infty] = \tfrac{A^2}{4}\big[\delta(f \pm f_c)\big]$$

i.e.,

$$\overline{P_c}[\infty] = A^2/2$$

and, the power (in signal watts) in the message side frequencies is

$$\overline{P_m}[\infty] = \left(\mu^2 A^2/16\right)\big[\delta\{f \pm (f_c + f_m)\} + \delta\{f \pm (f_c - f_m)\}\big]$$

i.e.,

$$\overline{P_m}[\infty] = \mu^2 A^2/4$$

Thus

$$\overline{P_c}[\infty]/\overline{P_m}[\infty] = 2/\mu^2 \tag{18.7}$$

Comment Because the maximum value of μ is 1, the power in the carrier components will always be greater than twice the power in the message components. Put another way, no more than one-third of the power in an LCAM wave is devoted to the message.

Additional Comment For $\mu = 0.8$, the value assumed in Figure 18.2, approximately 76% of the total power is found in the carrier components, and 24% is found in the message components. LCAM is a relatively inefficient way of producing a modulated wave.

18.3 DOUBLE-SIDEBAND SUPPRESSED-CARRIER AM (DSBSC)

DSBSC is used as the first step in the production of frequency-multiplexed signals, as the first step in producing single-sideband signals, and in FM stereo broadcasting systems (the channel that carries the difference signal uses DSBSC).

DSBSC waves are produced by multiplying the carrier signal by the message signal. Thus

$$s_{DSBSC}(t) = m_1(t)\,A\cos(2\pi f_c t) \tag{18.8}$$

In the frequency domain, equation (18.8) becomes

$$s_{DSBSC}(f) = \tfrac{A}{2}\big[M(f + f_c) + M(f - f_c)\big] \tag{18.9}$$

Equation (18.9) is shown in Figure 18.1. Unlike LCAM, no power exists at the carrier frequency; appropriately it is called *suppressed* carrier AM. If the message function is limited to the frequency interval $-W \le f \le W$, the modulated wave is limited to the band $f_c \pm W$, and the transmission bandwidth is $2W$ Hertz, the same as LCAM.

Without the powerful carrier component of LCAM, detection is much more difficult. Usually, DSBSC (and other modulation schemes that eliminate the carrier component) employs pilot tones, or other signals, from which the carrier signal can be derived at the receiver. All of the power in the wave is associated with the message-bearing sidebands.

EXAMPLE 18.3   DSBSC with Sinusoidal Message Function

Given A sinusoidal carrier $c(t) = A\cos(2\pi f_c t)$ and a sinusoidal message function $m_1(t) = \cos(2\pi f_m t)$ produce a double-sideband suppressed-carrier wave.

Problem Find $s_{DSBSC}(t)$ and $S_{DSBSC}(f)$.

Solution From equation (18.8)

$$s_{DSBSC}(t) = A\cos(2\pi f_m t)\cos(2\pi f_c t)$$

Multiplying out and simplifying gives

$$s_{DSBSC}(t) = \tfrac{A}{2}\left\{\cos\left[2\pi(f_c + f_m)t\right] + \cos\left[2\pi(f_c - f_m)t\right]\right\} \tag{18.10}$$

The Fourier transform of equation (18.10) is

$$S_{DSBSC}(f) = \tfrac{A}{4}\left[\delta\left(f \pm (f_c + f_m)\right) + \delta\left\{f \pm (f_c - f_m)\right\}\right] \tag{18.11}$$

Comment To give form to these equations, suppose $A = 1.5$, $f_c = 6.3$, and $f_m = 1$. In **Figure 18.3** I draw $s_{LCAM}(t)$ and $S_{LCAM}(f)$. In the time domain, the DSBSC wave suffers a phase reversal as the message function passes through zero; in the frequency domain, power impulses occur only at frequencies that are spaced $\pm f_m$ about f_c and $-f_c$.

18.4   QUADRATURE-CARRIER AMPLITUDE MODULATION (QAM)

QAM is a bandwidth conservation procedure. It is used in radio systems designed to carry digital signals and in voiceband data modems, and is employed to combine the in-phase and quadrature components of the luminance signal in color television.

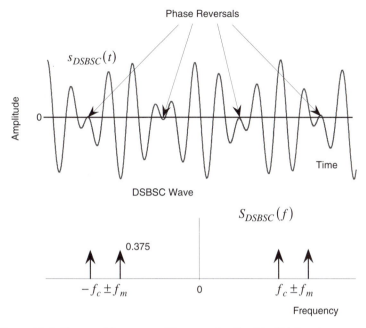

Figure 18.3 Time and Frequency Diagrams for Double-Sideband Suppressed-Carrier Wave in Example 18.3

QAM is a special case of DSBSC modulation in which sine and cosine functions are used to produce two DSBSC waves. Because they are orthogonal,[1] they occupy the same transmission bandwidth without interference. The procedure employs two independent message signals [$m'(t)$ and $m''(t)$] that modulate two carrier signals that differ in phase by 90° but have identical frequencies. Thus

$$s_{QAM}(t) = A\left[m'(t)\cos(2\pi f_c t) + m''(t)\sin(2\pi f_c t)\right] \qquad (18.12)$$

In the frequency domain, equation (18.12) becomes

$$S_{QAM}(f) = \tfrac{A}{2}\left[M'(f \pm f_c) + jM''(f + f_c) - jM''(f - f_c)\right] \qquad (18.13)$$

In **Figure 18.4**, we sketch the spectrums of equation (18.13) on real and imaginary planes. Because they are separated by 90° on the positive and negative frequency axis, $M'(f)$ and $M''(f)$ cannot interfere with each other. If each

---

[1]Real functions are orthogonal if

$$\int_b^a f_n(x) f_m(x)\,dx = 0, \quad m \neq n$$

The integral is called the inner product of f_n and f_m.

message signal is limited to the frequency interval $-W \le f \le W$, the transmission bandwidth of QAM is $2W$ Hz, the same as LCAM and DSBSC—but two message signals are transmitted.

18.5 SINGLE-SIDEBAND AM (SSB)

In LCAM and DSBSC-AM, the message power is carried in two sidebands that are mirror images of one another. If we are given details of the ampli-

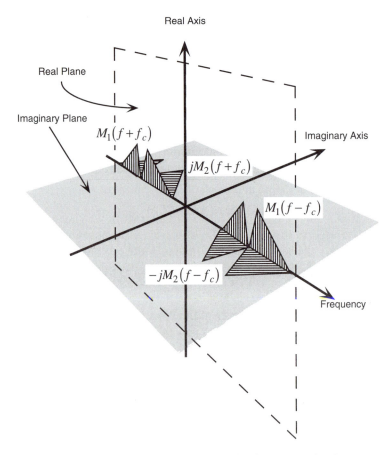

Figure 18.4 Representation of Spectrum of Quadrature Amplitude Modulated Wave
Two carriers with identical frequencies but out of phase by 90° are modulated by two independent signals. At the instant shown, one modulated carrier has energy distributed in the real plane, and the other has energy distributed in the imaginary plane. Using QAM, two message signals are carried in the bandwidth normally required by one message signal.

tude and phase spectrums of one, we can determine the other. This suggests that a further improvement in efficiency can be achieved by sending only one sideband so that all of the transmitted power is devoted to sending one copy of the message. Called single-sideband (SSB) AM, the precise description depends on which sideband is employed. When the higher frequency sideband is employed, the modulated wave is described as upper-sideband (USB) SSB; when the lower frequency sideband is employed, the modulated wave is described as lower-sideband (LSB) SSB. SSB is an efficient, bandwidth-mini-mizing, modulation technique that is used extensively in voice multiplexing systems (see § 18.8). SSB is not recommended for digital signals or signals (such as television) that contain significant energy at frequencies close to 0 Hz.

(1) Representation of SSB Wave

In Figure 18.1, comparison of the spectrums of DSBSC-AM and SSB-AM suggests that SSB functions may be derived from a DSBSC function through the use of a bandpass filter; thus

$$S_{USB}(f) = \tfrac{A}{2}\big[M(f \pm f_c)\big]\Pi\left\{\tfrac{f \pm (f_c + W)}{W}\right\} \qquad (18.14)$$

and

$$S_{LSB}(f) = \tfrac{A}{2}\big[M(f \pm f_c)\big]\Pi\left\{\tfrac{f \pm (f_c - W)}{W}\right\} \qquad (18.15)$$

For the arrangement to work, the filters must have a very sharp transition from bandpass to bandstop so as to divide the double sideband exactly. Alternatively, the function must not contain significant power at low frequencies so that a less than perfect transition will perform adequately. For this reason, the use of SSB is restricted to well behaved, zero-mean, message signals such as are generated in voice applications. To describe SSB functions in the time-domain requires some specialized theoretical development.

(2) Hilbert Transform

The Hilbert transform[2] (named for its inventor David Hilbert, a late-19th- and early-20th-century German mathematician) is a computational device that produces a phase shift of −90° for all positive frequency components of the signal $s(t)$ and a phase shift of +90° for all negative frequency components of the signal $s(t)$. $\hat{s}(t)$, the Hilbert transform of $s(t)$, is defined as

$$\hat{s}(t) = \frac{1}{\pi}\int_{-\infty}^{\infty}\frac{s(\tau)}{t - \tau}d\tau = \frac{1}{\pi t}\otimes s(t) \qquad (18.16)$$

---

[2]The Hilbert transform is the quadrature function of $s(t)$. See Ronald N. Bracewell, *The Fourier Transform and Its Applications,* 2nd Edition, (New York: McGraw-Hill, 1986), pp. 267 *et seq.*

and the inverse Hilbert transform is defined as

$$s(t) = -\frac{1}{\pi} \int_{-\infty}^{\infty} \frac{\hat{s}(\tau)}{t-\tau} d\tau = -\frac{1}{\pi t} \otimes \hat{s}(t) \qquad (18.17)$$

To avoid the singularity at the point $t=\tau$, integration is performed piecewise (Cauchy's principal value). If $s(t) \Leftrightarrow S(f)$ and $\hat{s}(t) \Leftrightarrow \hat{S}(f)$, the Fourier transform of equation (18.16) gives

$$\hat{S}(f) = -jS(f) \operatorname{sgn}(f) \qquad (18.18)$$

EXAMPLE 18.4 Hilbert Transforms of Sine and Cosine Functions

Given $s(t) = \cos(2\pi f_c t)$.

Problem Find the Hilbert transform.

Solution We know that

$$s(t) = \cos(2\pi f_c t) \Leftrightarrow \frac{1}{2}\{\delta(f+f_c) + \delta(f-f_c)\} = S(f)$$

Substituting in equation (18.18)

$$\hat{S}(f) = -j\operatorname{sgn}(f)S(f) = -\frac{j}{2}\operatorname{sgn}(f)\{\delta(f+f_c) + \delta(f-f_c)\}$$

Implementing the signum function

$$\hat{S}(f) = \frac{1}{2j}\{\delta(f-f_c) - \delta(f+f_c)\} \Leftrightarrow \sin(2\pi f_c t) = \hat{s}(t)$$

Thus, the Hilbert transform of $\cos(2\pi f_c t)$ is $\sin(2\pi f_c t)$.

Extension In the same way we can show that the Hilbert transform of $\sin(2\pi f_c t)$ is $-\cos(2\pi f_c t)$.

---

(3) Analytic Signal

A real bandpass signal can be represented as a complex function $[z(t)]$ as follows

$$z(t) = s(t) + j\hat{s}(t) \qquad (18.19)$$

Known as an *analytic* signal,[3] the real part $s(t)$ is the original bandpass function, the imaginary part $\hat{s}(t)$ is the Hilbert transform of $s(t)$, and the whole

[3]D. Gabor, "Theory of communication," *Journal IEE*, 93, part III (1946), 429–441.

has a one-sided Fourier transform $Z(f)$. In the frequency domain

$$Z(f) = S(f) + j\hat{S}(f) \qquad (18.20)$$

and

$$\hat{S}(f) = \begin{cases} -jS(f), & f > 0 \\ jS(f), & f < 0 \end{cases} = -j\hat{S}(f)\,\mathrm{sgn}(f) \qquad (18.21)$$

so that

$$Z(f) = \begin{cases} 2S(f), & f > 0 \\ 0, & f < 0 \end{cases} \qquad (18.22)$$

Thus, an analytic signal is a complex-valued signal whose real part consists of a real-valued signal $s(t)$, whose imaginary part consists of the Hilbert transform of $s(t)$, and whose spectral density is one-sided.

(4) Spectral Density of General SSB Wave

Analytically, you create an SSB wave from an arbitrary, band-limited message signal and a sinusoidal carrier signal as follows:

- Form an analytical signal whose real part is $m(t)$

$$z(t) = m(t) + j\hat{m}(t) \Leftrightarrow M(f) + j\hat{M}(f) = Z(f) \qquad (18.23)$$

  where

$$Z(f) = \begin{cases} 2M(f), & f > 0 \\ 0, & f < 0 \end{cases}$$

- In the time domain, multiply $z(t)$ by $A\exp(j2\pi f_c t)$. In the frequency domain, this translates $Z(f)$ to $Z(f - f_c)$, i.e.,

$$z(t)A\exp(j2\pi f_c t) \Leftrightarrow AZ(f - f_c) = AM(f - f_c) + jA\hat{M}(f - f_c) \qquad (18.24)$$

- The real part of $Z(f - f_c)$, i.e., $M(f - f_c)$, produces spectral densities of magnitudes $M(f)/2$ at $\pm f_c$. It is $S_{SSB}(f)$.
- In the time domain, these actions result in

$$s_{SSB}(t) = \tfrac{A}{2}\big[m(t)\cos(2\pi f_c t) - \hat{m}(t)\sin(2\pi f_c t)\big] \qquad (18.25)$$

  Equation (18.25) describes the upper-sideband case.

- If we start the procedure by forming the complex conjugate of $z(t)$, i.e.,

$$z^*(t) = m(t) - j\hat{m}(t) \Leftrightarrow M(f) - j\hat{M}(f) = Z(-f) \qquad (18.26)$$

we will obtain

$$s_{SSB}(t) = \tfrac{A}{2}\left[m(t)\cos(2\pi f_c t) + \hat{m}(t)\sin(2\pi f_c t)\right] \tag{18.27}$$

which describes the lower-sideband case.

(5) Limitations of SSB Modulation

Because equations (18.25) and (18.27) contain the Hilbert transform of the message signal, they describe a complex function that produces sharp peak values when the message signal makes abrupt transitions. An extreme example is provided by modulation with a unit pulse $m(t) = \Pi[t]$ for which the Hilbert transform is

$$\hat{m}(t) = -\frac{1}{\pi}\log_e\left|(t-1/2)/(t+1/2)\right| \tag{18.28}$$

The expression becomes infinite at $t = \pm 1/2$. Thus, a very large peak is generated at each signal transition. For this reason, the use of SSB is restricted to well behaved message signals such as are generated in voice applications.

As mentioned earlier, an SSB wave can be created from a DSBSC wave through the use of a bandpass filter that passes only the upper sideband or the lower sideband. In theory, this is simple; in practice, it is difficult. Particularly, this is true if $m(t)$ is not zero-mean, or is zero-mean and contains significant energy at low frequencies. Under these circumstances, practical bandpass filters introduce distortion. For this reason, the use of SSB is restricted to zero-mean message signals with $f_m|_{min} >> 0$ Hz, such as are generated in voice applications. Television, a wideband message signal, with significant energy at low frequencies, cannot take advantage of the efficiency of SSB modulation.

EXAMPLE 18.5   SSB with Sinusoidal Message Function

Given A sinusoidal carrier $c(t) = A\cos(2\pi f_c t)$ and a sinusoidal message function $m_1(t) = \cos(2\pi f_m t)$ produce an upper-sideband, single-sideband wave. The Hilbert transform of $\cos(2\pi ft)$ is $\sin(2\pi ft)$.

Problem Find $s_{USB}(t)$ and $S_{USB}(f)$.

Solution From equation (18.25)

$$s_{USB}(t) = \tfrac{A}{2}\left[\cos(2\pi f_m t)\cos(2\pi f_c t) - \sin(2\pi f_m t)\sin(2\pi f_c t)\right]$$

Multiplying out and simplifying

$$s_{USB}(t) = \tfrac{A}{2}\left[\cos\left\{2\pi(f_c + f_m)t\right\}\right] \tag{18.29}$$

The time domain representation is a half-magnitude function of frequency $(f_c + f_m)$. Taking the Fourier transform of equation (18.29)

$$S_{USB}(f) = \tfrac{A}{4}\left[\delta\left\{f \pm (f_c + f_m)\right\}\right] \tag{18.30}$$

The frequency domain representation consists of delta functions of magnitude $A/4$ at $\pm(f_c + f_m)$.

18.6 VESTIGIAL-SIDEBAND AM (VSB)

Signals that are not zero-mean, or that have significant energy at low frequencies, can approach SSB performance. Called VSB modulation, one sideband is passed almost intact together with a *vestige* of the other sideband that compensates for the amount removed from the first sideband. In addition, a carrier component is present so as to simplify signal detection at the receiver. VSB is the modulation technique used for domestic television service broadcasting in the United States.

Approximating the bandwidth efficiency of SSB is achieved by employing a bandpass filter designed to have odd-symmetry about the carrier frequency and a relative response of 50% at the carrier frequency. **Figure 18.5** shows how vestigial sideband operation is applied to an LCAM signal. For equal total power, VSB provides less message power than SSB. However, the power at the carrier frequency makes signal detection less expensive and has made the development of consumer-oriented television services possible. Like SSB, VSB modulation is not suited to message signals that contain abrupt transitions.

18.7 SIGNAL-TO-NOISE RATIOS

The signal-to-noise ratio at the output of a receiver depends on the noise present in the transmission environment and the type of modulation employed. For the purpose of signal-to-noise analysis, frequency changes that only facilitate communication leave no lasting changes in the noise and message signals. Accordingly, a receiver can be represented by an equivalent circuit that includes a bandpass filter and a demodulator. The input signal consists of a modulated wave $[s(t)]$ and a white noise signal $[w(t)]$. This arrangement appears in **Figure 18.6**.

(1) Narrowband Noise Signal

Because the ratio of the transmission bandwidth to the center frequency of the bandpass filter is very small (< 0.001), the noise signal presented to the

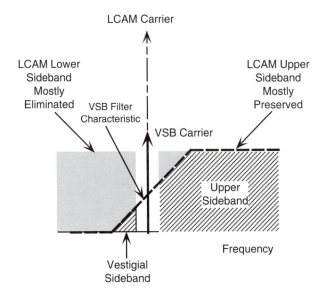

Figure 18.5 Spectrum of Vestigial-Sideband AM Wave
An asymmetrical filter is used to convert a double-sideband, large-carrier AM signal to one that contains most of one sideband, a reduced carrier component, and a vestige of the other sideband.

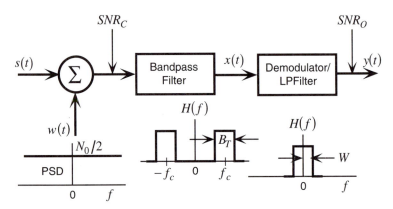

Figure 18.6 Equivalent Circuit of Receiver for SNR Calculations

demodulator is a band-limited, narrow band signal. The power spectral density is uniformly distributed across the passband to produce a total average power of $N_0 B_T$ signal watts. I represent this narrow band noise signal by $n(t)$ to distinguish it from the broadband, white noise signal $w(t)$ present at the input to the receiver.

In the actual receiver, a series of sharply tuned circuits perform the bandpass function. When these circuits are excited by random impulses (which are characteristic of the noise signal), they respond by oscillating at the frequency to which they are tuned. The narrow band noise signal becomes a band-limited, sinusoidal signal that can be represented as

$$n(t) = n_i(t)\cos(2\pi f_c t) - n_a(t)\sin(2\pi f_c t) \tag{18.31}$$

where $n_i(t)$ and $n_q(t)$ are the in-phase and quadrature components of the narrow band signal.[4]

(2) Signal-to-Noise Ratios

In Figure 18.6, I identify signal-to-noise ratios at the input and output terminals of the receiver. At the receiver output, SNR_O is given by

$$SNR_O = \frac{\text{average power of message signal at receiver output}}{\text{average power of noise at receiver output}}$$

At the input of the receiver, the definition is not so easy. While the white noise signal contains power at all frequencies, the modulated wave exhibits power only in a band of frequencies determined by the message signal and the modulation employed. To provide a consistent measure, define the SNR at the input of the receiver to be the SNR of the communications channel, i.e.,

$$SNR_I = SNR_C =$$
$$\frac{\text{average power of modulated wave at receiver input}}{\text{average power of noise at receiver input in message bandwidth}}$$

(3) Signal-to-Noise Ratios for LCAM with Envelope Detection

From equation (18.4), the power in $s_{LCAM}(t)$ is

$$P[s_{LCAM}(t)] = \frac{A^2}{2} + \mu^2 P[m_1(t)]$$

and the noise power at the receiver input in the message bandwidth is

$$P[w(t)] = N_0 B_T/2 = N_0 W$$

Thus,

$$SNR_C = A^2\{1 + \mu^2 P[m_1(t)]\}/2N_0 W$$

---

[4]A band-limited signal $s_B(t)$ can be expressed as

$$s_B(t) = \text{Re}[\gamma(t)\exp(j2\pi f_c t)].$$

Putting $\gamma(t) = \gamma_i(t) + j\gamma_q(t)$, $s_B(t)$ becomes

$$s_B(t) = \gamma_i(t)\cos(2\pi f_c t) + \text{Re}[\gamma_q(t)\exp\{j(2\pi f_c + \pi/2)t\}]$$
$$= \gamma_i(t)\cos(2\pi f_c t) - \gamma_q(t)\sin(2\pi f_c t)$$

Now, from Figure 18.6,

$$x(t) = s(t) + n(t) = \{A + A\mu m_1(t) + n_i(t)\}\cos(2\pi f_c t) - n_q(t)\sin(2\pi f_c t)$$

and, if the noise signal is small compared to the modulated signal, i.e.,

$$A + A\mu m_1(t) \gg n_i(t)$$

then the envelope $|x(t)|$ is approximately

$$|x(t)| \cong |A + A\mu m_1(t) + n_i(t)|$$

If the demodulator includes a low-pass filter that stops the carrier signal

$$y(t) = A + A\mu m_1(t) + n_i(t)$$

thus, the average power in the message signal at the output of the receiver is

$$A^2\mu^2 P[m_1(t)]$$

The average power in the noise signal at the output of the receiver is the same as the average power in the noise signal at the input of the envelope detector (demodulator), i.e., $2N_0W$ signal watts so that

$$SNR_O = A^2\mu^2 P[m_1(t)]/2N_0W$$

Taking the ratio of SNR_O and SNR_C, we find

$$[SNR_O/SNR_C]_{LCAM} = \mu^2 P[m_1(t)]/\{1 + \mu^2 P[m_1(t)]\} \qquad (18.32)$$

LCAM with envelope detection produces a lower value of SNR_O; it degrades the channel signal-to-noise ratio.

In the case of tone modulation, if $m_1(t) = \cos(2\pi f_m t)$, then $P[m_1(t)] = 1/2$ signal watts. Substituting in equation (18.32)

$$[SNR_O/SNR_C]_{LCAM} = \mu^2/2 + \mu^2 \qquad (18.33)$$

Because $0 < \mu < 1$, equation (18.33) has a maximum value of $1/3$. Thus, the signal-to-noise ratio at the output of the receiver is no more than one-third the value at the input to the receiver.

(4) Signal-to-Noise Ratios for DSBSC with Coherent Detection

From equation (18.8), the power in $s_{DSBSC}(t)$ is

$$P[s_{DSBSC}(t)] = \frac{A^2}{2}P[m_1(t)]$$

As before, the noise power at the receiver input in the message bandwidth is

$$P[w(t)] = N_0W$$

so that

$$SNR_C = A^2 P[m_1(t)]/2N_0W$$

In coherent detection, $x(t)$ is multiplied by the carrier signal to give

$$v(t) = x(t)\cos(2\pi f_c t) = \{s(t) + n(t)\}\cos(2\pi f_c t)$$

i.e.,

$$v(t) = \tfrac{1}{2}Am_1(t) + \tfrac{1}{2}n_i(t) + \tfrac{1}{2}[Am_1(t) + n_i(t)]\cos(4\pi f_c t) - \tfrac{1}{2}An_q(t)\sin(4\pi f_c t)$$

After the LPF, we obtain

$$y(t) = \tfrac{1}{2}Am_1(t) + \tfrac{1}{2}n_i(t)$$

so that the average power in the noise signal at the receiver output is $\tfrac{1}{4}n_i^2(t)$ signal watts, or $\tfrac{1}{2}N_0W$. Hence,

$$SNR_O = A^2 P[m_1(t)] / 2N_0W$$

and

$$[SNR_O / SNR_C]_{DSBSC} = 1 \tag{18.34}$$

If coherent detection is used to demodulate a DSBSC signal, the process neither degrades nor improves the SNR. It is equal to the value of the channel SNR.

(5) *Signal-to-Noise Ratios for SSB with Coherent Detection*

For SSB modulation and coherent detection, the SNRs follow directly from DSBSC. The average power at the receiver input due to the SSB signal is one-half of the DSBSC signal, and the average power in the noise signal in the message bandwidth at the receiver input is N_0W signal watts. Thus,

$$SNR_C = A^2 P[m_1(t)] / 4N_0W$$

From equation (18.25), the starting amplitude of $s_{SSB}(t)$ is one-half of $s_{DSBSC}(t)$ [see equation (18.8)] so that

$$y(t) = \tfrac{1}{4}Am_1(t) + \tfrac{1}{2}n_i(t)$$

Thus, the average power in the message signal at the receiver output is $\tfrac{1}{16}A^2 P[m_1(t)]$ signal watts. Because the SSB signal has one-half the bandwidth of a DSBSC signal, the bandpass filter has half the passband, and the noise power at the receiver output is one-half the DSBSC case, or $\tfrac{1}{4}N_0W$ signal watts. Hence,

$$SNR_O = A^2 P[m_1(t)] / 4N_0W$$

and

$$[SNR_O / SNR_C]_{SSB} = 1 \tag{18.35}$$

18.8 FREQUENCY-DIVISION MULTIPLEXING

For many years, frequency modulated (FM) microwave radio systems carried frequency-division multiplexed (FDM) voice signals in medium- and long-haul

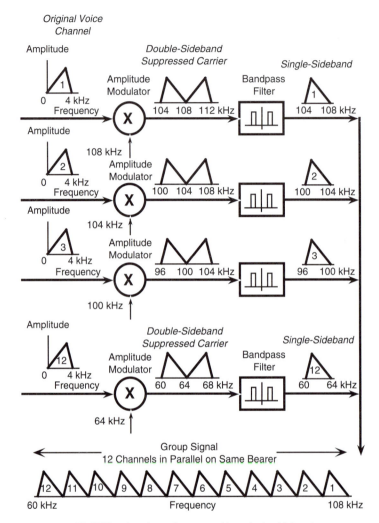

*12, 4 kHz voice channels arranged in a single-sideband,
amplitude-modulated, frequency-division multiplexed group*

Figure 18.7 Frequency-Division Multiplexing
*Using FDM, 12 voice channels are placed in parallel on the same bearer
to form a group. From* Telecommunications Primer *by E. Bryan Carne,
© 1995. Reprinted by permission of Prentice-Hall, Inc., Upper Saddle
River, NJ.*

telephone transmission applications. Later, to increase the capacity of long-haul systems, single-sideband (SSB), amplitude modulated (AM) microwave systems were employed. Now, the combination of optical fibers and digital techniques has displaced FDM in favor of TDM. Nevertheless, because of the amount of FDM equipment in place, a significant fraction of the world's voice traffic will continue to use these techniques into the next century.

Figure 18.7 shows how 12 analog voice channels are combined in a *channel bank* to form a composite, frequency-division multiplexed signal. To produce a double-sideband, suppressed-carrier (DSBSC) signal, each voice channel modulates a carrier chosen from the set 64, 68, 72, etc., up to 108 kHz. When passed through a filter, only the lower sideband remains. Because each LSB spans a different frequency range, all LSBs may be added as shown at the bottom of Figure 18.7. This arrangement of channels is known as a *group*. They are combined in steps to become part of an FDM signal containing 600 (or more) channels. In turn, these signals are used to modulate carriers in radio relay systems. Of course, a complete transmission system consists of two multiplexing and demultiplexing arrangements—one for each direction of transmission.

REVIEW QUESTIONS

18.1   Define *modulation, demodulation,* and *baseband.*

18.2   Describe a *zero-mean message signal* and a *normalized message signal.*

18.3   If $m(t) \Leftrightarrow M(f)$, sketch the frequency domain representation of an LCAM wave.

18.4   Define the *upper sidebands* and *lower sidebands* of an LCAM wave.

18.5   Why is LCAM regarded as an inefficient way to produce a modulated wave?

18.6   If $m(t) \Leftrightarrow M(f)$, sketch the frequency domain representation of a DSBSC-AM wave.

18.7   Describe a *QAM wave.* Why is QAM used?

18.8   If $m(t) \Leftrightarrow M(f)$, sketch the frequency domain representation of an SSB-AM wave.

18.9   Distinguish between *upper sideband* and *lower sideband* SSB-AM.

18.10   What is the purpose of the Hilbert transform?

18.11   What is an *analytic signal?* Why is it employed?

18.12   What limitations on the message signal are imposed by the use of SSB-AM?

18.13   Describe *VSB-AM.* Why is it used?

18.14   Describe the operations that are performed on a 4 kHz voice signal in order to fold it into an FDM signal.

19

Angle Modulation

Angle modulated (θM) waves exist in two forms—phase modulated (ΦM) and frequency modulated (FM). Frequency modulation is used to implement point-to-point, microwave-radio links, and cellular, mobile radio telephone service. In addition, it is employed by all radio stations broadcasting information and entertainment in the frequency range 88 to 108 MHz (FM-band). Further, FM is employed in the sound segment of NTSC television signals. Phase modulation is an important ingredient of digital modulation [see § 20.1(2)] and it is employed to convey color (chrominance) information in NTSC television signals.

19.1 RELATIONSHIP BETWEEN ΦM AND FM

The basic relationship between phase and frequency modulation is illustrated by examining the instantaneous angle.

(1) Instantaneous Angle

Consider the function

$$s(t) = A\cos\{2\pi f_c t + \phi(t)\} = A\cos\{\theta(t)\} \tag{19.1}$$

where $\theta(t)$ is the instantaneous angle produced by the modulation. It describes an angle-modulated wave. Restating equation (19.1) in exponential form,

$$s(t) = A\,\mathrm{Re}\big[\exp\{j\theta(t)\}\big] \tag{19.2}$$

so that the wave is an exponential function of the instantaneous angle. Thus, analysis is likely to be more difficult than for amplitude-modulated waves, and angle-modulated waves are unlikely to resemble the message waveforms.

(2) Phase Modulation

In phase modulation

$$\theta(t) = 2\pi f_c t + \phi_\Delta m_1(t) \tag{19.3}$$

where ϕ_Δ is the *phase deviation,* or phase modulation index. Remembering that $m_1(t)$ is a zero-mean, normalized message function, ϕ_Δ is the maximum phase shift produced in the modulation process. To prevent ambiguity, it is limited to $\leq \pi$ radians. Thus

$$s_{\Phi M}(t) = A\cos\left\{2\pi f_c t + \phi_\Delta m_1(t)\right\} \tag{19.4}$$

Varying the phase of the carrier to produce a phase modulated wave changes the argument of the cosine function; in effect, this changes the frequency, also.

(3) Frequency Modulation

In frequency modulation, $f_i(t)$, the instantaneous frequency of $s(t)$, can be expressed in a form analogous to equation (19.3)

$$f_i(t) = f_c + f_\Delta m_1(t) \tag{19.5}$$

where f_Δ is called the *frequency deviation.* It represents the maximum change in frequency produced by the zero-mean, normalized message function. In practice $f_\Delta \ll f_c$. Further,

$$f_i(t) = \frac{1}{2\pi}\frac{d\theta}{dt} = f_c + \frac{1}{2\pi}\frac{d\phi}{dt} \tag{19.6}$$

so that, from equations (19.5) and (19.6), we obtain

$$\phi(t) = 2\pi f_\Delta \int_{t_0}^{t} m_1(\lambda)d\lambda + \phi(t_0) \tag{19.7}$$

where λ is a dummy variable of integration. Choosing t_0 so that $\phi(t_0) = 0$, the equation for a frequency-modulated wave becomes

$$s_{FM}(t) = A\cos\left\{2\pi f_c t + 2\pi f_\Delta \int_t m_1(\lambda)d\lambda\right\} \tag{19.8}$$

Be sure to note that the integrals in equations (19.7) and (19.8) will not converge unless $m(t)$ is zero-mean; in practice, this condition corresponds to the usual state of the message signal. Varying the frequency of the carrier to produce a frequency-modulated wave changes the argument of the cosine function; in effect this changes the phase, also. In reality, both ΦM and FM exhibit time-varying phase and frequency. **Table 19.1** compares expressions for instantaneous phase and instantaneous frequency for FM with those for ΦM.

	Frequency Modulation	Phase Modulation
Instantaneous Phase	$\phi(t) = 2\pi f_\Delta \int_t m(\lambda)d\lambda$	$\phi(t) = \phi_\Delta m(t)$
Instantaneous Frequency	$f(t) = f_c + f_\Delta m(t)$	$f(t) = f_c + \frac{1}{2\pi}\phi_\Delta \frac{dm(t)}{dt}$

Table 19.1 Comparison of Instantaneous Frequency and Instantaneous Phase of Frequency-Modulated and Phase-Modulated Waves

(4) Constant Power

From equations (19.4) and (19.8), you see that the amplitude of an angle-modulated wave is constant. Consequently, because power is independent of phase, the wave exhibits constant power. As modulation occurs, the power is shared among the carrier and the side frequencies, but the total power remains constant at the value for the unmodulated carrier ($A^2/2$ signal watts).

19.2 NARROWBAND ANGLE MODULATION

In narrowband angle modulation, $\phi(t) << 1$ radian. Expanding the cosine term in equation (19.1) gives

$$s(t) = A\cos\{\phi(t)\}\cos(2\pi f_c t) - A\sin\{\phi(t)\}\sin(2\pi f_c t) \qquad (19.9)$$

Now, when $\phi(t) << 1$, $\cos\{\phi(t)\} \cong 1$ and $\sin\{\phi(t)\} \cong \phi(t)$, so that equation (19.9) becomes

$$s_{NB\theta M}(t) = A\cos(2\pi f_c t) - A\phi(t)\sin(2\pi f_c t) \qquad (19.10)$$

and, in the frequency domain,

$$S_{NB\theta M}(f) = \tfrac{A}{2}\delta(f \pm f_c) - \tfrac{jA}{2}\{\Phi(f + f_c) - \Phi(f - f_c)\} \qquad (19.11)$$

Thus, narrowband angle-modulated waves are characterized by impulses at the carrier frequency and double sidebands—the same as LCAM. If $m(t)$ has a bandwidth of W Hz, the transmission bandwidth of the modulated wave is $2W$.

EXAMPLE 19.1   Narrowband Angle Modulation with Sinusoidal Message Function

Given A sinusoidal carrier $c(t) = A\cos(2\pi f_c t)$ and a sinusoidal message function $m_1(t) = \cos(2\pi f_m t)$ are used to produce a narrow band phase-modulated wave and a narrowband frequency-modulated wave.

Problem Find $S_{NB\Phi M}(f)$ and $S_{NBFM}(f)$.

Solution For NBFM, equation (19.10) becomes

$$s_{NB\Phi M}(t) = A\cos(2\pi f_c t) - A\phi_\Delta \cos(2\pi f_m t)\sin(2\pi f_c t) \tag{19.12}$$

Expanding the cos×sin term (see Table A.6) gives

$$s_{NB\Phi M}(t) = A\cos(2\pi f_c t) - \tfrac{1}{2}A\phi_\Delta\left[\sin\{2\pi(f_c - f_m)t\} + \sin\{2\pi(f_c + f_m)t\}\right]$$

Taking the Fourier transform of both sides,

$$\begin{aligned}
S_{NB\Phi M}(f) = {} & \tfrac{1}{2}A\left[\delta(f \pm f_c)\right] - \tfrac{1}{4}jA\phi_\Delta\left[\delta(f - f_c + f_m) - \delta(f + f_c - f_m)\right] \\
& - \tfrac{1}{4}jA\phi_\Delta\left[\delta(f - f_c - f_m) - \delta(f + f_c + f_m)\right]
\end{aligned} \tag{19.13}$$

For NBFM, equation (19.10) becomes

$$s_{NBFM}(t) = A\cos(2\pi f_c t) - 2A\pi f_\Delta \int \cos(2\pi f_m t)\,dt\,\sin(2\pi f_c t)$$

Performing the integration (see Table A.3),

$$s_{NBFM}(t) = A\cos(2\pi f_c t) - \frac{Af_\Delta}{f_m}\sin(2\pi f_m t)\sin(2\pi f_c t) \tag{19.14}$$

Expanding the sin×sin and taking the Fourier transform of both sides,

$$\begin{aligned}
S_{NBFM}(f) = {} & \tfrac{1}{2}A\left[\delta(f \pm f_c)\right] - \frac{Af_\Delta}{4f_m}\left[\delta(f + f_c - f_m) - \delta(f - f_c + f_m)\right] \\
& - \frac{Af_\Delta}{4f_m}\left[\delta(f - f_c - f_m) - \delta(f + f_c + f_m)\right]
\end{aligned} \tag{19.15}$$

Comment **Figure 19.1** compares the positive frequency plane configuration of the sidebands of LCAM, NBΦM, and NBFM when modulated by a sinusoidal message signal. In LCAM, the sidebands are real and even; in NBΦM, the sidebands are imaginary and even; and in NBFM, the sidebands are real and odd.

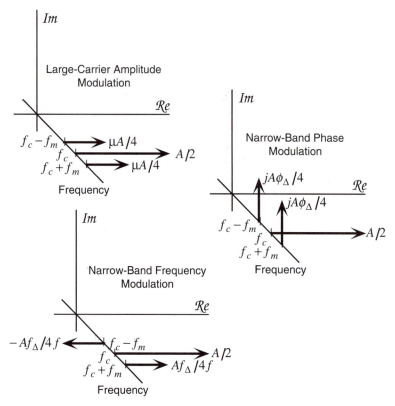

Figure 19.1 Comparison of Spectrums Produced by Single-Tone Modulation of Narrow-Band Angle-Modulated Signals and Large-Carrier Amplitude-Modulated Signal

Extension Impulse magnitude squaring equation (19.13) or (19.15) shows that $\overline{P}[\infty] = A^2/2$ signal watts. The power is constant and equal to the power of the unmodulated carrier.

19.3 ESTIMATING THE BANDWIDTH OF WIDEBAND ANGLE MODULATION

For the general case of angle modulation that involves a sinusoidal carrier and a zero-mean Gaussian process, we can obtain an estimate of the power spectral density of the wave using the autocorrelation function. For convenience, so as to recover the symbol ϕ for the autocorrelation function, rewrite equation (19.1) as

$$s(t) = A\cos\{2\pi f_c t + \mu(t)\} \tag{19.16}$$

where $\mu(t)$ is a zero-mean Gaussian process. The autocorrelation function, i.e.,

$$\phi(\tau) = \langle s(t)s(t+\tau) \rangle$$

is[1]

$$\phi(\tau) = \frac{A^2}{2} \langle \cos\{\mu(t+\tau) - \mu(t)\} \rangle \cos(2\pi f_c \tau) \tag{19.17}$$

The time average of $\cos\{\mu(t+\tau) - \mu(t)\}$ is the time average of the characteristic function, i.e.,

$$\langle \cos\{\mu(t+\tau) - \mu(t)\} \rangle = \exp\left[-\frac{1}{2}\langle \{\mu(t+\tau) - \mu(t)\}^2 \rangle\right] \tag{19.18}$$

so that, if $\phi_\mu(\tau)$ is the ACF of the modulating function $\mu(t)$, equation (19.17) becomes

$$\phi(\tau) = \frac{A^2}{2} \exp\left[-\{\phi_\mu(0) - \phi_\mu(\tau)\}\right] \cos(2\pi f_c \tau) \tag{19.19}$$

To estimate the power spectral density, we use the Wiener-Khinchine relationship

$$\Phi_X(f) = \int_{-\infty}^{\infty} \phi_X(\tau) \exp(-j2\pi f\tau) d\tau \tag{19.20}$$

Taking only the real part [because $\Phi_X(f)$ is always real],

$$\Phi_X(f) = \int_{-\infty}^{\infty} \phi_X(\tau) \cos(2\pi f\tau) d\tau \tag{19.21}$$

Substituting equation (19.19) in equation (19.21) gives the power spectral density

$$\Phi_X(f) = \frac{A^2}{2} \int_{-\infty}^{\infty} \exp\left[-\{\phi_\mu(0) - \phi_\mu(\tau)\}\right] \cos(2\pi f_c \tau) \cos(2\pi f\tau) d\tau \tag{19.22}$$

Multiplying out,

$$\Phi_X(f) = \frac{A^2}{2} \int_{-\infty}^{\infty} \exp\left[-\{\phi_\mu(0) - \phi_\mu(\tau)\}\right] \left[\cos\{2\pi(f+f_c)\tau\} + \cos\{2\pi(f-f_c)\tau\}\right] d\tau \tag{19.23}$$

---

[1]See J. H. Roberts, *Angle Modulation: the theory of system assessment* (Stevenage, UK: Peter Peregrinus Ltd, 1977), pp. 40–46.

If we restrict our interest to the region around f_c so that $f = f_c + \Delta f$, the second cosine term in equation (19.23) becomes the dominant term with respect to the frequency distribution of the power spectral density. Accordingly, an estimate of $\Phi_X(f)$ is

$$\Phi_{\Delta f}(f) = \frac{A^2}{2} \int\limits_{-\infty}^{\infty} \exp\left[-\left\{\phi_\mu(0) - \phi_\mu(\tau)\right\}\right] \cos(2\pi\Delta f\tau) d\tau \qquad (19.24)$$

In equation (19.24), the power spectral density of an angle-modulated wave is seen to depend on the ACF of the message function. When $\phi_\mu(0) < 1$ (so that $\phi_\mu(\tau) < 1$), the exponential term can be replaced by a power series and the value of $\Phi_{\Delta f}(f)$ determined by numerical methods.

In theory, $\Phi_{\Delta f}(f)$ extends to infinity; in practice, a bandwidth exists beyond which the value is so low as to be negligible. A widely recognized rule of thumb for determining its value is known as *Carson's* rule. For frequency modulation by a zero-mean Gaussian process for which the root-mean-square noiseband frequency deviation is η_Δ, the peak-to-rms factor is $\sqrt{10}$, and the maximum modulating frequency is $\hat{f}_m$, it states

$$B_T = 2\hat{f}_m\left[1 + \sqrt{10}\,\eta_\Delta / \hat{f}_m\right]$$

Carson's rule for a sinusoidal message function is described in § 19.4(5).

19.4 SINGLE-TONE ANGLE MODULATION

The study of angle modulated waves is simplified if the message signal is a sinusoidal function.

(1) Comparison of Single-Tone Modulated Waves

The time domain representation of a frequency-modulated wave formed from a single-tone message signal and a sinusoidal carrier signal is shown in the upper diagram of **Figure 19.2**. It differs from the phase modulation diagram (under it) in the location of the maximum frequency. To aid in visualization, I have included the outline of the message signal. In phase modulation, the maximum frequency of the wave occurs at the positive-going zero crossings of the message signal; in frequency modulation, the maximum frequency occurs at the positive peaks of the message signal.

Because of the complexity of the general analysis, I illustrate further properties of angle modulation with single-frequency, sinusoidal modulating

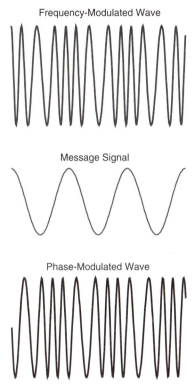

Figure 19.2 Comparison of Frequency-Modulated and Phase-Modulated Waves

With respect to the message signal, in frequency modulation, the maximum and minimum frequencies occur at the peaks; in phase modulation, the maximum and minimum frequencies occur at the zero crossings.

functions. Further, because phase modulation and frequency modulation are intimately related, I focus on FM waves. In doing so, I employ Bessel functions of the first kind.

- **Bessel function of the first kind:** a solution of Bessel's differential equation, the nth-order Bessel function of the first kind with argument β is defined as[2]

$$J_n(\beta) = \tfrac{1}{2\pi} \int_{-\pi}^{\pi} \exp j(\beta \sin x - nx)\, dx \qquad (19.25)$$

---

[2] G. N. Watson, *A Treatise on the Theory of Bessel Functions,* 2nd edition (Cambridge University Press: 1966).

Three properties have application to our studies:

(1) $J_n(\beta) = (-1)^n J_{-n}(\beta)$

(2) For small values of β, $J_0(\beta) \cong 1$, $J_1(\beta) \cong \beta/2$, $J_n(\beta) \cong 0$ for $n > 1$

(3) $\displaystyle\sum_{n=-\infty}^{\infty} J_n^2(\beta) = 1$

In **Figure 19.3**, I plot values of this function against the argument β for positive integer values of n.

(2) Single-Tone Frequency Modulation

In the time domain, a frequency-modulated wave is described by equation (19.8). If the single-tone modulating function $m(t) = \cos(2\pi f_m t)$, then

$$s(t) = A\cos\left\{2\pi f_c t + 2\pi f_\Delta \int_t m(\lambda)d\lambda\right\} = A\cos\left\{2\pi f_c t + \tfrac{f_\Delta}{f_m}\sin(2\pi f_m t)\right\} \quad (19.26)$$

f_Δ, the frequency deviation, is equal to the product of the frequency sensitivity of the modulator (k_f expressed in Hertz per volt) and the amplitude of the message function (A_m). A customary practice is to call the ratio of the frequency deviation and the modulation frequency the modulation index (β), so that

$$s(t) = A\cos\left\{2\pi f_c t + \beta\sin(2\pi f_m t)\right\} \quad (19.27)$$

where

$$\beta = f_\Delta/f_m = k_f A_m/f_m \quad (19.28)$$

Expressing equation (19.27) in exponential form,

$$s(t) = \mathrm{Re}\left[A\exp\left\{j2\pi f_c t + j\beta\sin(2\pi f_m t)\right\}\right] \quad (19.29)$$

Separating the right-hand side into two exponentials,

$$s(t) = \mathrm{Re}\left[A\exp(j2\pi f_c t)\exp\left\{j\beta\sin(2\pi f_m t)\right\}\right]$$

When multiplied by A, the second exponential is known as the complex envelope of $s(t)$. Denoted by $\tilde{s}(t)$, it can be expanded in a complex Fourier series, so that

$$\tilde{s}(t) = \mathrm{Re}\left[A\exp\left\{j\beta\sin(2\pi f_m t)\right\}\right] = \sum_{n=-\infty}^{\infty} C_n \exp(j2\pi n f_m t) \quad (19.30)$$

where

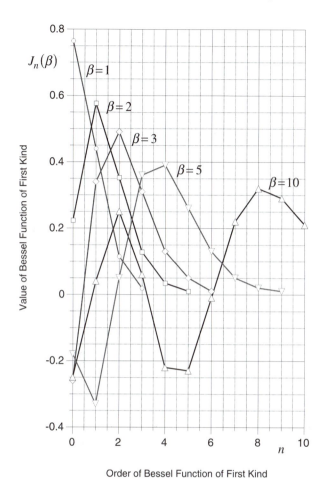

Figure 19.3 Values of Bessel Functions of the First Kind for Integer Orders

$$C_n = f_m \int_{-f_m/2}^{f_m/2} \tilde{s}(t)\exp(-j2\pi n f_m t)\,dt$$

i.e.,

$$C_n = f_m A \int_{-f_m/2}^{f_m/2} \exp\{j\beta\sin(2\pi f_m t) - j2\pi n f_m t\}\,dt \qquad (19.31)$$

For convenience, let $x = 2\pi f_m t$; then equation (19.31) becomes

$$C_n = \tfrac{A}{2\pi} \int_{-\pi}^{\pi} \exp j(\beta\sin x - nx)\,dx \qquad (19.32)$$

Now the nth-order Bessel function of the first kind with argument β is defined as

$$J_n(\beta) = \tfrac{1}{2\pi} \int\limits_{-\pi}^{\pi} \exp j(\beta \sin x - nx)\,dx$$

so that equation (19.32) can be written

$$C_n = AJ_n(\beta) \tag{19.33}$$

Substituting equation (19.33) in equation (19.30),

$$\tilde{s}(t) = A \sum_{n=-\infty}^{\infty} J_n(\beta)\exp(j2\pi n f_m t) \tag{19.34}$$

Next, substituting equation (19.34) in equation (19.29),

$$s(t) = A\,\mathrm{Re}\left[\sum_{n=-\infty}^{\infty} J_n(\beta)\exp\{j2\pi(f_c + nf_m)t\}\right]$$

Interchanging the order of summation and taking the real part gives

$$s(t) = A \sum_{n=-\infty}^{\infty} J_n(\beta)\cos\{j2\pi(f_c + nf_m)t\} \tag{19.35}$$

In the frequency domain, equation (19.35) becomes

$$S(f) = \tfrac{A}{2} \sum_{n=-\infty}^{\infty} J_n(\beta)\delta\{f \pm (f_c + nf_m)\} \tag{19.36}$$

Thus, the spectrum of an FM wave contains a carrier component and an infinite set of side frequencies separated by the modulating frequency.

(3) Effect of Varying the Range of the Modulating Frequency

Figure 19.4 shows spectral diagrams of single-tone modulated FM waves. In theory, the number of side frequencies is infinite. I show only those with amplitudes $\geq 1\%$ of the amplitude of the unmodulated carrier (they are called *significant* side frequencies). As the modulation index increases (i.e., the range of frequencies through which the modulated wave is changed increases), the bandwidth and the number of side-frequency components increases. In **Figure 19.5**, I plot the number of significant side frequencies against the modulation index.

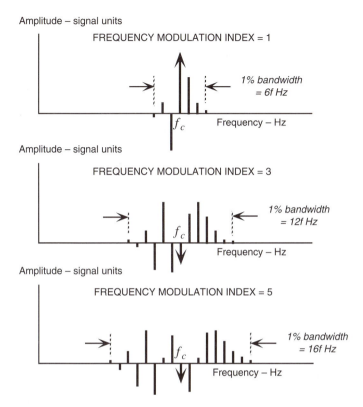

Figure 19.4 Spectrums of Frequency-Modulated Waves for a Range of Values of Modulation Index
The side frequencies shown are those that are ≥1% of the amplitude of the unmodulated carrier. From Telecommunications Primer *by E. Bryan Carne, © 1995. Reproduced by permission of Prentice-Hall, Inc., Upper Saddle River, NJ.*

(4) Bandwidth of Frequency-Modulated Wave

The interval that just contains components whose amplitudes are ≥ 1% of the amplitude of the unmodulated carrier is known as the *1% bandwidth*. It is an alternative to the bandwidth derived from *Carson's Rule*.

- **Carson's rule:** a *rule of thumb* for approximating the bandwidth of an FM wave produced by a single-tone modulating function. The transmission bandwidth is

$$B_T = 2f_\Delta + 2f_m = 2f_\Delta(1 + 1/\beta) \tag{19.37}$$

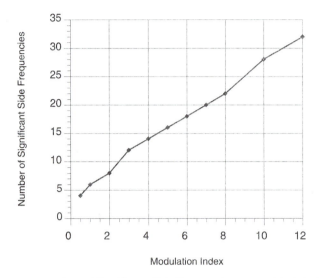

Figure 19.5 Number of Significant Side Frequencies *versus* Modulation Index

For small values of modulation index ($\beta < 1$), the bandwidth is $2f_m$ Hz. This result can be confirmed by substituting Bessel function property (2) in equation (19.36). Only $J_0(\beta)$ and $J_1(\beta)$ have significant values so that $S(f)$ is reduced to a carrier component and a single pair of side frequencies at $f_c \pm f_m$. For large values of modulation index, the bandwidth is approximately $2\Delta f$ Hz, where Δf is the frequency deviation of the modulated wave.

EXAMPLE 19.2   Power of Frequency-Modulated Wave

Given   A frequency-modulated wave produced by a single-tone modulating function.

Problem   Using the properties of Bessel functions listed in § 19.4(1), show that the power of the wave is independent of the modulation index and equal to $A^2/2$ signal watts.

Solution   In the frequency domain, the modulated wave is described as

$$S(f) = \tfrac{A}{2} \sum_{n=-\infty}^{\infty} J_n(\beta)\delta\{f \pm (f_c + nf_m)\}$$

The power is equal to the sum of the squares of the magnitudes of the impulses, i.e.,

$$\overline{P}[\infty] = \tfrac{1}{2}A^2 \sum_{n=-\infty}^{\infty} J_n^2(\beta)$$

From property (3), the sum is equal to 1. Hence the power in the wave is independent of the modulation index and equal to $A^2/2$ signal watts.

Extension As β varies, so the fraction of the power in the carrier component varies, and at some values of β (β = 2.4048, 5.5201, 8.6537, 11.7915, etc.), the carrier component disappears. However, at all times, the total power in the signal remains constant at $A_c^2/2$ signal watts.

19.5 SNR OF FREQUENCY MODULATION WITH DISCRIMINATOR DETECTION

The signal-to-noise ratio at the output of an FM receiver depends on the characteristics of the demodulation technique employed.

(1) Input SNR

The equivalent circuit of an FM receiver with discrimination detection appears in **Figure 19.6**. The average power in a frequency-modulated wave is $A^2/2$ signal watts, and the average power in the noise signal is $N_0 W$ signal watts, so that

$$SNR_C = A^2/2N_0 W$$

(2) Output SNR

To find the SNR at the output of the receiver, determine the average signal power at the receiver output in the absence of noise, and then determine the average noise power in the presence of the unmodulated carrier.

The average signal power at the output of the receiver is $P[v(t)]$. At the discriminator output $v(t) = k_f m(t)$, so that $P[v(t)]$ is $k_f^2 P[m(t)]$ signal watts. To find the average noise power in the presence of an unmodulated carrier, draw an Argand diagram to determine the instantaneous frequency; it is shown in **Figure 19.7**. A is the magnitude of the carrier, $n(t)$ is the narrowband noise signal [see equation (18.31)], and $\phi_n(t)$ is the time-varying phase angle relative to the carrier due to $n(t)$. If $A \gg n(t)$, from the diagram,

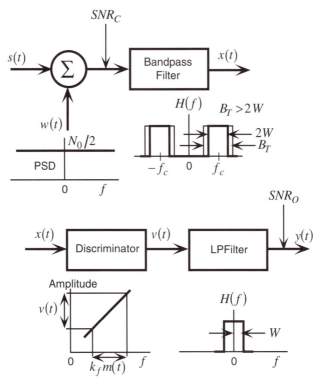

Figure 19.6 Equivalent Circuit of FM Receiver with Discriminator Detection

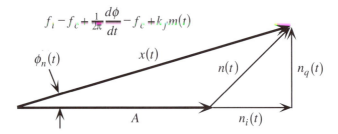

Figure 19.7 Argand Diagram Used to Determine Instantaneous Frequency Produced by Noise and Unmodulated Carrier Signal

$$\phi_n(t) \cong n_q(t)/A$$

so that

$$f_i = f_c + \frac{1}{2\pi A}\frac{d}{dt}n_q(t)$$

and, at the output of the discriminator, the signal due to noise is

$$v_n(t) = \frac{1}{2\pi A}\frac{d}{dt}n_q(t)$$

To find the noise power at the output of the receiver, use Property #9 in Table A.1 to find $V_n(f)$ and integrate the modulus squared over the passband of the LPF to give

$$P[y_n(t)] = \int_{-W}^{W} \frac{N_0 f^2}{A^2}\,df = 2N_0 W/3A$$

Thus, the output SNR is

$$SNR_O = 3A^2 k_f^2 P[m(t)]/2N_0 W^3$$

and

$$[SNR_O/SNR_C]_{FM} = 3k_f^2 P[m(t)]/W^2 \qquad (19.38)$$

Putting $D = k_f/W$, equation (19.38) becomes

$$[SNR_O/SNR_C]_{FM} = 3D^2 P[m(t)]$$

D is known as the *deviation* ratio. It is proportional to the frequency deviation of the modulated signal divided by the message signal bandwidth. Thus, D is proportional to the transmission bandwidth, and you can write

$$[SNR_O/SNR_C]_{FM} \propto B_T^2 \qquad (19.39)$$

When the carrier-to-noise ratio is high, an increase in the transmission bandwidth produces a quadratic increase in the ratio of the output and input SNRs. In the case of tone modulation, you can show that

$$[SNR_O/SNR_C]_{FMTone} \propto \beta^2 \qquad (19.40)$$

so that the ratio of the output to input SNRs improves with increasing modulation index.

(3) Pre-Emphasis and De-Emphasis

At the output of the discriminator, the power spectral density of the noise signal is

$$\Phi_n(f) = \begin{cases} f^2 N_0/A^2, & |f| \le B_T/2 \\ 0, & \text{elsewhere} \end{cases}$$

so that the noise signal energy is greater at the cut-off frequency of the LPF than at the center of the bandpass. To compensate for this imbalance, commercial broadcasting stations (and other operators) distort the message signal before transmission to emphasize the higher frequencies (pre-emphasis), and restore the signal to its original frequency content at the receiver (de-emphasis). In public broadcasting applications, the pre-emphasis and de-emphasis characteristic curves are standardized to produce a 13.4 db improvement in output SNR.

REVIEW QUESTIONS

19.1 Define *angle modulation, phase modulation,* and *frequency modulation.*

19.2 Explain why an angle-modulated wave is described as a constant power wave.

19.3 Explain why narrowband angle modulation is equivalent to amplitude modulation.

19.4 Describe the frequency spectrum of a frequency-modulated wave in which the modulation is produced by a single-tone message signal. How does it vary with modulation index?

19.5 State Carson's rule for approximating the bandwidth of a frequency-modulated wave in which the modulation is produced by a single-tone message signal.

19.6 Define the 1% bandwidth of a frequency-modulated wave in which the modulation is produced by a single-tone message signal.

20

Digital Modulation

By modulating a sinusoidal carrier with a stream of digital symbols, data are transported over radio channels or cable connections. Noise causes errors in the received message, and the fixed transmission bandwidth assigned to a particular channel sets an upper limit to the symbol rate.

In digital modulation, the modulated wave changes abruptly from one modulation state to another as the amplitude, frequency, and/or phase of the carrier is (are) changed in sympathy with symbols in the message stream. When each symbol corresponds to one bit, this process is known as *keying*, and the actions are known as BASK (binary amplitude-shift keying), BPSK (binary phase-shift keying), and BFSK (binary frequency-shift keying). When each symbol represents more than one bit, combinations of amplitudes, frequencies, and phases are used to create the set of modulation states needed to transmit a datastream.

Comparisons among modulation techniques may be based on two measures:

- **Power efficiency:** the ratio of the signal energy per bit (E_b) required at the receiver to achieve a specific probability of error, and the noise power (N)

$$\eta_P = E_b/N$$

- **Bandwidth efficiency:** the gross throughput data rate per unit of bandwidth, i.e., the ratio of the number of bits per second transmitted, and the bandwidth employed (in Hertz). For a binary datastream in which 1s and 0s are equally likely, you may use the Hartley-Shannon law [see § 11.4(3)] and express channel capacity in information bits/s or binits/s

$$\eta_{B_T} = C/B_T = \log_2(1 + AS/N)$$

where A is equal to 1 for a Gaussian channel, and to $12/k^2$ for a PCM channel [see § 12.5(3)].

20.1 BINARY KEYING

- **Binary keying:** modulating action that changes the amplitude, phase, or frequency of a carrier signal in sympathy with the bits in a stream of binary data.

Time-domain waveforms for binary amplitude, phase, and frequency keying are shown in **Figure 20.1**.

(1) Binary Amplitude-Shift Keying (BASK)

As shown at the top of Figure 20.1, in binary amplitude-shift keying, the transmitted signal is a sinusoid whose amplitude is changed by *on-off* keying (OOK) so that the presence of the carrier signal corresponds to a 1, and no carrier corresponds to a 0. For a unipolar data stream,

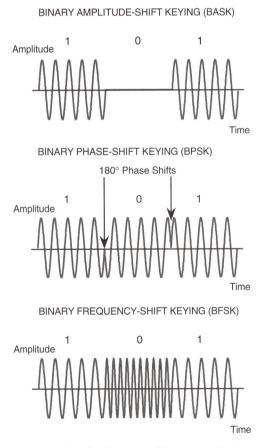

Figure 20.1 Binary Amplitude, Phase, and Frequency Keying

$$m(t) = \sum_{k=-\infty}^{\infty} A_k s_g\left(t - kT_b\right) \tag{20.1}$$

where

$$A_k = \begin{cases} 1, & \text{when signal one is present} \\ 0, & \text{when signal zero is present} \end{cases}$$

$s_g(t)$ is the normalized shaping pulse

T_b is the bit-duration (in seconds)

The modulated wave is produced by multiplying the carrier $A\cos(2\pi f_c t)$ by $m(t)$; thus

$$s_{BASK}(t) = A m(t)\cos(2\pi f_c t) \tag{20.2}$$

The amplitude-modulated wave is double-sideband, suppressed-carrier. If we assume that the digital shaping pulse is normalized and rectangular, when signal one is present, the modulated pulse is

$$p_1(t) = \begin{cases} A\cos(2\pi f_c t), & \text{when } -T_b/2 < t < T_b/2 \\ 0, & \text{otherwise} \end{cases} \tag{20.3}$$

and, when signal zero is present,

$$p_0(t) = 0.$$

These are the pulses shown in Figure 20.1.

E_b, the energy in $p_1(t)$, is equal to $A^2 T_b/2$ signal joules, so that, provided 1s and 0s are equally likely, the average signal energy per bit $\overline{E_b}$ is one-half, or $A^2 T_b/4$ signal joules. Assuming an eternal, periodic function, $\overline{E_b}/T_b$ is the signal power (i.e., $\overline{P[T]}$ or $\overline{P[\infty]}$) in $m(t)$; it is $A^2/4$ signal watts. Sometimes, expressing the coefficient A in terms of the average signal power is convenient. Denoting this quantity by P_s, for BASK we can write

$$P_s = A^2/4, \quad \text{i.e.,} \quad A = 2\sqrt{P_s}$$

so that equation (20.2) becomes

$$s_{BASK}(t) = 2\sqrt{P_s}\, m(t)\cos(2\pi f_c t) \tag{20.4}$$

In the presence of Gaussian-distributed white noise, and assuming synchronous detection [see § 18.7(4)] with $B_T = r$, the probability of error is

$$P_{\varepsilon_{BASK}} = \text{Erfc}\sqrt{S/2N} = \text{Erfc}\sqrt{P_s/2N} \tag{20.5}$$

Equation (20.5) is identical to equation (10.16) so that, from a signal-to-noise point of view, BASK is equivalent to a random unipolar baseband signal.

The power spectral density, i.e., the distribution of power density (in signal watts/Hz) about the carrier frequency, is shown in **Figure 20.2**. The impulse function at the carrier frequency represents power due to the unipolar binary data stream (see Example 13.5). Because the spectrum extends well beyond the transmission bandwidth, *spillover* is an important concern with digital modulation. To reduce interference (*interchannel* interference), bandpass filters are used. At the risk of degrading the signal by increasing *intersymbol* interference, a bandpass filter may be applied to limit the energy of the NRZ pulses that form the datastream. By restricting the baseband signal to a frequency range that just encompasses the primary lobe of the spectral density diagram, approximately 90% of the datastream energy is available to the modulation process. Another bandpass filter is placed at the output of the transmitter.

(2) Binary Phase-Shift Keying (BPSK)

As shown in Figure 20.1, in binary phase-shift keying, the transmitted signal is a sinusoid of constant amplitude whose presence in one phase condition corresponds to 1, and, in a state 180° out of phase with it, corresponds to 0.

$$s_{BPSK}(t) = A\cos\{2\pi f_c t + \pi m(t)\} \tag{20.6}$$

If we assume that the digital shaping pulse is normalized and rectangular, when signal one is present, the pulse can be described as

$$p_1(t) = \begin{cases} A\cos(2\pi f_c t), & \text{when } -T_b/2 < t \le T_b/2 \\ 0, & \text{otherwise} \end{cases} \tag{20.7}$$

and when signal zero is present, the pulse can be described as

$$p_0(t) = \begin{cases} -A\cos(2\pi f_c t), & \text{when } 0 < t \le T_b \\ 0, & \text{otherwise} \end{cases} \tag{20.8}$$

These are the pulses shown in Figure 20.1. For obvious reasons, BPSK is also known as *phase-reversal* keying (PRK).

The energy in $p_1(t)$ is equal to the energy in $p_0(t)$, i.e., $A^2 T_b/2$ signal joules, and the signal power in the modulated signal is $A^2/2$ signal watts. Thus, you can write equation (20.6) as

$$s_{BPSK}(t) = \sqrt{2P_s}\,\cos\{2\pi f_c t + \pi m(t)\} \tag{20.9}$$

In the presence of Gaussian-distributed white noise, the probability of error is

$$P_{\mathcal{E}_{BPSK}} = \text{Erfc}\sqrt{S/N} = \text{Erfc}\sqrt{P_s/N} \tag{20.10}$$

So long as 1s and 0s are equally likely, and the two phase states are equally disposed within 2π, the power spectral density (shown in Figure 20.1)

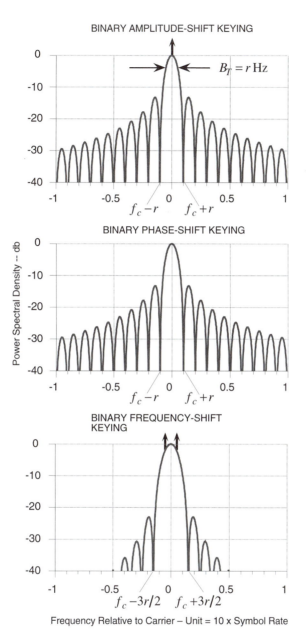

Figure 20.2 Power Spectral Densities
The PSD of amplitude-shift keying has first zeros at carrier frequency ± bit rate, slowly decaying sidelobes, and an impulse at the carrier frequency. The PSD of phase-shift keying is identical to the PSD of ASK, except no impulse occurs. Frequency-shift keying consists of two ASK functions whose carriers are the FSK carrier ± the frequency deviation. While the first zeros occur at the carrier frequency ±1.5 × bit rate, the impulses are generated by the two ASK components.

has the same form as BASK without the impulse at the carrier frequency. Thus, the percentage of the power devoted to signaling is greater in PSK than ASK and, as shown by equation (20.5) and equation (20.10), the error probability will be less for a given signal power level.

(3) Differential BPSK

Recovery of the datastream from a PSK-modulated wave requires *synchronous* demodulation, i.e., the receiver must reconstruct the carrier so as to be able to detect phase changes in the received signal. *Differential* phase-shift keying (DPSK) eliminates the need for the synchronous carrier in the demodulation process and simplifies the receiver. At the transmitter, the datastream is processed to produce a modulated wave in which the phase changes by $+\pi$ radians whenever a 1 appears in the stream and remains constant whenever a 0 appears in the stream (i.e., $1 \to 1$ and $0 \to 1$, increase by $+\pi$; $1 \to 0$ and $0 \to 0$, no change). Thus, the receiver need detect only phase changes, not search for specific phase values. **Figure 20.3** shows a segment of a datastream (original), the changes in phase angle using the rules given above, what the angles mean in terms of 1s and 0s in the processed datastream, and how the processed stream is interpreted at the receiver. To draw these sequences, I assume that the bit before the bit beginning the original stream is 0.

(4) Binary Frequency-Shift Keying (BFSK)

As shown in Figure 20.1, in binary frequency-shift keying, the modulated wave is a sinusoid of constant amplitude whose presence at one frequency $(f_c + f_d)$ corresponds to 1, and whose presence at another frequency $(f_c - f_d)$ corresponds to 0.

Original Datastream

0 1 0 0 1 1 0 0 0 1 1 1 0 0 0

Relative Phase Angle

0 $+\pi$ $+\pi$ $+\pi$ $+2\pi$ $+3\pi$ $+3\pi$ $+3\pi$ $+3\pi$ $+4\pi$ $+5\pi$ $+6\pi$ $+6\pi$ $+6\pi$ $+6\pi$

Processed Datastream

0 1 1 1 0 1 1 1 1 0 1 0 0 0 0

Demodulated Datastream

0 1 0 0 1 1 0 0 0 1 1 1 0 0 0

Figure 20.3 Datastream Processing for Differential BPSK
At the transmitter, the datastream is processed to produce a modulated wave in which the phase changes by $+\pi$ radians whenever a 1 appears in the stream and remains constant whenever a 0 appears in the stream (i.e., $1 \to 1$ and $0 \to 1$, increase by $+\pi$; $1 \to 0$ and $0 \to 0$, no change). Thus, the receiver need detect only phase changes, not search for specific phase values.

$$s_{BFSK}(t) = A \cos\left\{ 2\pi f_c t + 2\pi f_d \int_0^t m(t') dt' \right\} \tag{20.11}$$

If we assume that the digital shaping pulse is normalized and rectangular, when signal one is present, the pulse can be described as

$$p_1(t) = \begin{cases} A\cos\{2\pi(f_c + f_d)t\}, & \text{when } -T_b/2 < t \le T_b/2 \\ 0, & \text{otherwise} \end{cases} \tag{20.12}$$

When signal zero is present, the pulse can be described as

$$p_0(t) = \begin{cases} A\cos\{2\pi(f_c - f_d)t\}, & \text{when } -T_b/2 < t \le T_b/2 \\ 0, & \text{otherwise} \end{cases} \tag{20.13}$$

These are the pulses shown in Figure 20.1. Comparing equations (20.12) and (20.13) with equation (20.3) shows that the BFSK wave can be considered to be made up of two BASK waves of frequencies $(f_c + f_d)$ and $(f_c - f_d)$.

The energy in $p_1(t)$ is equal to the energy in $p_0(t)$, i.e., $A^2 T_b/2$ signal joules, and the signal power in the modulated signal is $A^2/2$ signal watts, so that

$$s_{BFSK}(t) = \sqrt{2P_s} \cos\left\{ 2\pi f_c t + 2\pi f_d \int_0^t m(t') dt' \right\} \tag{20.14}$$

In the presence of Gaussian-distributed white noise, the probability of error is

$$P_{\varepsilon_{BFSK}} = \text{Erfc}\sqrt{S/N} = \text{Erfc}\sqrt{P_s/N} \tag{20.15}$$

For binary-FSK created by a random datastream and $f_d = r/2$, the distribution of power density about the carrier frequency is shown in Figure 20.2. The impulse functions are associated with the unipolar binary signals in the BASK component waves. They occur at the frequencies employed to represent the 1 and 0 states $(f_c \pm r/2)$. The main lobe of the signal spans the frequency range $f_c - 3r/2$ to $f_c + 3r/2$ Hz, a range that is 50% greater than with BASK; however, the power in the sidelobes drops off more quickly. This particular version of BFSK is known as Sunde's FSK,[1] after its inventor E. D. Sunde, a 20th-century American engineer.

---

[1]E. D. Sunde, "Ideal pulses transmitted by AM and FM," *Bell System Technical Journal*, 38 (1959), 1357–1426.

How the symbol frequencies are generated has an important influence on the modulated signal. If the symbol frequencies are derived from two independent sources, discontinuities will occur when one symbol changes to another. Under this circumstance, the modulation is described as *discontinuous* FSK. If the symbol frequencies are derived from a single source by frequency modulation, discontinuities can be avoided and the modulation is described as *continuous*-phase FSK (CPFSK). The amplitude of the modulated wave remains constant, and interchannel interference effects are avoided. In order to achieve continuity of phase across the boundary between symbols, the phase of the frequency representing the next symbol must be adjusted to compensate for a phase shift introduced by the frequency keying process. Thus, CPFSK embodies both frequency shifts and phase shifts. In binary-CPFSK, the phase shift between symbols is $\pi/2$ radians.

(5) Minimum-Shift Keying (MSK)

Also known as fast FSK and binary CPFSK, MSK uses half-sinusoidal shaping pulses $[s_g(t)=\Pi(t/T_b)\sin(\pi t/T_b)]$ and $f_d=r/4$, i.e., one-half the spacing used in Sunde's FSK. The distribution of power density about the carrier frequency appears in **Figure 20.4**. The impulses present in BFSK are eliminated, and the sinusoidal datastream waveforms produce smaller sidelobes. The main lobe spans the frequency range $f_c-3r/4$ to $f_c+3r/4$ Hz, a range that is

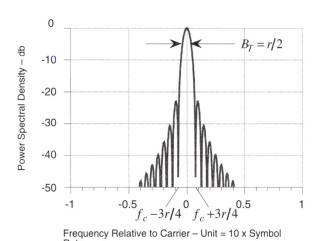

Figure 20.4 Power Spectral Density of Minimum-Shift Keying
From Telecommunications Primer by E. Bryan Carne, © 1995. Reprinted by permission of Prentice Hall, Inc., Upper Saddle River, NJ.

25% less than BASK and 50% less than BFSK. The power in the sidelobes drops off very quickly, more quickly than BFSK and much more quickly than BASK.

20.2 QUADRATURE BINARY MODULATION

Quadrature binary modulation is a bandwidth-conserving technique that transports two digital streams in a bandwidth that normally carries a single stream.

(1) Quadrature-Carrier AM (QAM)

In QAM, we transmit symbols at rate $2r$ in the bandwidth required by binary-ASK (r Hz). By dividing the datastream into two streams (even-numbered bits and odd-numbered bits) that are used to amplitude modulate orthogonal carriers [e.g., $\sin(2\pi f_c t)$ and $\cos(2\pi f_c t)$] derived from the same source, we produce two modulated waves containing r symbols/s that occupy the same bandwidth of r Hz. **Figure 20.5** compares signal space diagrams for binary-ASK and QAM. For binary-ASK, signal points (1 and 0) are arranged

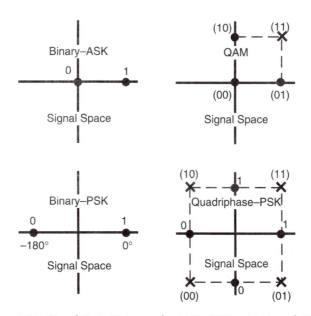

Figure 20.5 Signal State Diagrams for BASK, BPSK, QAM, and QPSK

on the horizontal axis. For QAM, signal points (1 and 0) occur on the horizontal and vertical axes. The set on the horizontal axis represents the signal states assumed by $\cos(2\pi f_c t)$ when amplitude-modulated by one of the datastreams, and the set on the vertical axis represents the signal states assumed by $\sin(2\pi f_c t)$ when amplitude-modulated by the other datastream. The four points 00, 01, 10, and 11 are the states that can be occupied by the composite signal.

(2) Quadrature PSK (QPSK)

QPSK (known as quadrature phase-shift keying, quaternary-PSK, or quadriphase PSK) employs orthogonal sinusoidal carriers derived from a common source. As before, the datastream is divided into two streams (of rate r symbols/s) that are applied to orthogonal carriers [$\sin(2\pi f_c t)$ and $\cos(2\pi f_c t)$]. Improved operation is achieved by delaying one of the datastreams by one *bit* period. Known as *offset*-QPSK (OQPSK), phase changes are limited to 90°; this restriction produces a beneficial effect. Figure 20.5 shows signal space diagrams for binary-PSK and QPSK. For binary-PSK, signal points (1 and 0) are arranged on the horizontal axis equidistant from the origin. The point representing 1 is used as the reference phase (0°), and the point representing 0 is 180° out of phase. For QPSK, signal points (1 and 0) are arranged on the horizontal and vertical axes. The set on the horizontal axis represents the signal states assumed by $\cos(2\pi f_c t)$ when phase-modulated by one of the datastreams, and the set on the vertical axis represents the signal states assumed by $\sin(2\pi f_c t)$ when phase-modulated by the other datastream. The four points 00, 01, 10, and 11 are the states that can be occupied by the composite signal.

Phase-shift keying produces abrupt changes in phase that translate into substantial changes in the amplitude of the modulated wave on the output side of the bandwidth-limiting transmit filter. In a communication system, when passed through equipment with non-linear characteristics (the power amplifier of a transponder in a communication satellite, for instance), these amplitude variations are suppressed—giving rise to additional spectral components outside the authorized bandwidth, and producing interchannel interference. Because the phase changes in OQPSK are limited to 90°, its performance in this sort of environment is better than QPSK, which relies on phase changes of 180°.

(3) Quadrature FSK (QFSK)

QFSK employs four frequencies spaced at equal intervals from one another. QFSK is likely to require a greater bandwidth than other quadrature modulation techniques; however, the probability of error decreases as the number of frequencies increases.

20.3 M-ARY MODULATION

The number of bits transported by a carier can be increased by using symbols that represent several bits at once.

(1) *M Unique Symbols*

To produce more complex symbols that represent groups of bits, the amplitude, phase, or frequency (alone, or in combination) of a carrier (either a single carrier, or a coherent quadrature-carrier arrangement) is changed to create M unique states. For instance in M-ary FSK, several frequencies are employed. Usually, they are spaced at equal intervals from one another and the carrier frequency. Thus, a set of frequencies can be described as $f_n = f_c + a_n f_d$, where f_n is the frequency corresponding to symbol n, f_c is the carrier frequency, $a_n = \pm 1, \pm 3, \ldots \pm(M-1)$ with M even, and f_d is the spacing between consecutive frequencies.

(2) *Spectral Efficiency*

When using M-ary signaling, each symbol represents $\log_2 M$ bits so that the binary data rate $r_b = r \log_2 M$, where r is the symbol rate (symbols/s), and $r_b/B_T = \log_2 M$. M-ary signaling schemes are used to transmit datastreams over bandpass channels when the desired data rate is more than can be achieved using binary-ASK, -PSK, or -FSK. With increasing amounts of power in a fixed bandwidth channel, you can increase the binary bit rate several fold. Thus: when $B_T = 1$ MHz and $M = 2$, $r = 1$ Msymbols/s and $r_b = 1$ Mbits/s; when $B_T = 1$ MHz and $M = 4$, $r = 1$ Msymbols/s and $r_b = 2$ Mbits/s; and, when $B_T = 1$ MHz and $M = 256$, $r = 1$ Msymbols/s and $r_b = 8$ Mbits/s. The spectral efficiencies for these examples are 1, 2, and 8 bits/s/Hz, respectively.

(3) *Comparisons*

The selection of a digital modulation scheme is a complex tradeoff among bandwidth, symbol rate, bit rate, and radiated power. Because noise affects the amplitude of the received signal, ASK techniques are inefficient in noisy environments. Because they are constant amplitude signals in which the information resides in parameters not greatly affected by amplitude perturbations, PSK and FSK perform better in noisy environments. As M increases, however, the transmitted power must increase to maintain the signal-to-noise ratio above an operating limit. In practice, QPSK is widely used because it offers attractive power/bandwidth performance for acceptable levels of error.

REVIEW QUESTIONS

20.1 Define the term *binary keying*.

20.2 Define two measures that can be used to compare modulation techniques.

20.3 Describe *binary amplitude-shift keying*.

20.4 Comment on the validity of the statement, "From a signal-to-noise point of view, BASK is equivalent to a random unipolar baseband signal."

20.5 Describe *binary phase-shift keying*.

20.6 What is the advantage of differential BPSK?

20.7 Describe *binary frequency-shift keying*.

20.8 Comment on the validity of the statement, "The BFSK wave is made up of two BASK waves...."

20.9 What is *minimum-shift keying?* What are its advantages?

20.10 Describe *QAM*.

20.11 Describe *quadrature PSK*.

20.12 Describe *M-ary modulation*. Why is it used?

20.13 Comment on the statement, "QPSK is widely used because it offers attractive power/bandwidth performance for acceptable levels of error."

21

Spread Spectrum Modulation

First used by the military because it is hard to detect and almost impossible to jam, spread spectrum modulation is employed in global positioning systems (GPSs), mobile telephones, personal communication systems (PCSs), and very small aperture terminal (VSAT) satellite communication systems.

- **Spread spectrum modulation** (SSM): a modulation technique in which the modulated message bearing signal occupies a much greater bandwidth than the minimum required to transmit the information it carries.

The spread spectrum signal has a low power density spectrum; as a consequence, it creates virtually no interference in other signals operating in the same frequency band. Further, at the receiver, demodulation spreads the energy of any interfering signal and reduces its effect.

21.1 CHANNEL CAPACITY

For a given signal-to-noise power ratio, the Hartley-Shannon law [see § 11.4(3)] states that a Gaussian channel of bandwidth B_T Hz can transport (a maximum of) C error-free information bits/s given by

$$C = B_T \log_2(1 + S/N) \tag{21.1}$$

When $0 < S/N << 1$, equation (21.1) becomes

$$B_T = 0.693 C(N/S) \tag{21.2}$$

The reasoning is as follows. For $-1 < x < 1$,

$$\log_e(1+x) = x - x^2/2 + x^3/3 - \dots$$

so that, when $0 < S/N << 1$,

$$\log_e(1 + S/N) \cong S/N$$

Changing bases,

$$\log_2(1+S/N) = \tfrac{1}{\log_e 2}\log_e(1+S/N) \cong 1.443\,S/N \qquad (21.3)$$

Substituting in equation (21.1),

$$C/B_T \cong 1.443\,S/N \qquad (21.4)$$

Thus, when operating under very noisy conditions, you can transport C error-free information bits/s by adjusting the channel bandwidth so that

$$B_T = 0.693C(N/S)$$

This is equation (21.2). As noted before, for a binary datastream in which 1s and 0s are equally likely, C is also the capacity in binits/s.

EXAMPLE 21.1 Channel Bandwidth

Given A 56 kbits/s datastream in which 1s and 0s are equally likely.

Problem Using spread spectrum modulation, find the channel bandwidth required when $S/N = 0.1, 0.01,$ and 0.001.

Solution Substitute in equation (21.2).

> When $S/N = 0.1$, use a modulation technique that occupies a bandwidth of approximately 0.39 MHz.
> When $S/N = 0.01$, use a modulation technique that occupies a bandwidth of approximately 3.88 MHz.
> When $S/N = 0.001$, use a modulation technique that occupies a bandwidth of approximately 38.81 MHz.

Comment From the curves in Figure 13.8, if you use 2B1Q or AMI coding, the bandwidth required to transmit 56 kbits/s is around 56 kHz. This is orders of magnitude less than the bandwidth range determined above. However, to transmit 56 kbits/s in a bandwidth of 56 kHz without appreciable errors requires a signal-to-noise ratio of some 30 db.

21.2 PROCESS GAIN AND JAMMING MARGIN

What equation (21.2) tells us is that we can achieve (virtually) error-free transmission of signals well below the noise level if we have a wideband channel and a modulation scheme that makes use of the bandwidth. **Figure 21.1**

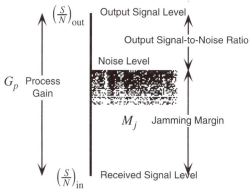

Figure 21.1 Showing the Relationships among Output Signal Level, Jamming Margin, Process Gain, and Received Signal Level

shows the relationships among output signal level, noise level, and received signal level in spread spectrum modulation. Defining two additional quantities, process gain and jamming margin, is convenient.

- **Process gain** (G_p): the ratio of the output signal-to-noise power ratio to the input signal-to-noise power ratio, i.e.,

$$G_p = (S/N)_{\text{out}} / (S/N)_{\text{in}}. \qquad (21.5)$$

Expressed in decibels, equation (21.5) becomes

$$G_p = (S/N)_{\text{out}} - (S/N)_{\text{in}} \text{ db.} \qquad (21.6)$$

If n_0 is the average noise power density in watts/Hz, then, at the output

$$(S/N)_{\text{out}} = S/n_0 C$$

and, at the input

$$(S/N)_{\text{in}} = S/n_0 B_T$$

Hence

$$G_p = B_T/C \qquad (21.7)$$

[At the beginning of Chapter 20, I define the ratio C/B_T to be bandwidth efficiency (i.e., the number of bits/s transmitted per Hertz of bandwidth). Equation (21.7) shows the inverse ratio B_T/C to be process gain (i.e., the bandwidth in Hertz required to transmit 1 bit/s).]

- **Jamming margin** (M_j): the difference between the noise level at the receiver and the received signal level that will assure the desired value of $(S/N)_{\text{out}}$

$$M_j = G_p - (S/N)_{\text{out}}\text{db} \qquad (21.8)$$

If G_p = 30 db and we require $(S/N)_{\text{out}}$ = 10 db, the jamming margin is 20 db, i.e., at the receiver, the noise power can be 20 db greater than the signal level.

EXAMPLE 21.2 Process Gain

Given A 56 kbits/s datastream in which 1s and 0s are equally likely.

Problem Using spread spectrum modulation, find the process gain required when $S/N = 0.1, 0.01$, and 0.001.

Solution Substitute in equation (21.7).

When $S/N = 0.1$, $B_T = 0.39$ MHz, so that

$$G_p = 0.39 \text{ MHz}/56 \text{ kbits /s} = 6.96 \cong 8.5 \text{ db}$$

When $S/N = 0.01$, $B_T = 3.88$ MHz, so that

$$G_p = 69.28 \cong 18.4 \text{ db}$$

When $S/N = 0.001$, $B_T = 38.81$ MHz, so that

$$G_p = 693 \cong 28.4 \text{ db}$$

EXAMPLE 21.3 Combined Parameters

Given A spread spectrum communication system operates in an environment in which the noise power is 100 times the received signal power. With a 56 kbits/s datastream in which 1s and 0s are equally likely, a 10 db output signal-to-noise power ratio is achieved at the receiver.

Problem Find the jamming margin, process gain, and RF bandwidth of the system.

Solution In Figure 21.1, if the received signal power is –20 db with respect to the noise level, the noise margin is 20 db, and the output signal-to-noise ratio is 10 db. Thus, the process gain is 30 db, or 1000. Also, $M_j = 20$ db, and, for C = 56 kbits/s, and $G_p = 1000$, $B_T = 56$ MHz.

21.3 PSEUDO-NOISE SEQUENCE

As you see in § 21.4, generating a spread spectrum signal requires a semi-random or noise-like sequence of 1s and 0s. Called *pseudo-noise sequences,* they are generated by deterministic strategies so that the same sequence can be available to both transmitter and receiver.

- **Pseudo-noise (*pn*) sequence:** a periodic binary sequence with a noise-like distribution of 1s and 0s in which the number of 1s is approximately equal to the number of 0s
- **Chip:** the name given to each element (0 or 1) of a *pn*-sequence; the output of a pseudo-noise sequence generator during one clock period
- **Maximum-length (*m*) sequence:** a *pn*-sequence produced by an *n*-stage feedback shift-register whose repetition period is $2^n - 1$ chips; when generated by a feedback shift-register arrangement, the parameters of the sequence are determined by the following:
  - *length.* Depends on the number of stages in the shift register; commonly expressed in *chips*
  - *composition.* Depends on the initial state of the register; the nature of the feedback connection(s) and the logic in the feedback loop(s)
  - *period.* Depends on the length (in chips) and the rate at which the register is driven (in chips per second)

Figure 21.2 shows the use of a 3-stage shift-register to generate a *pn*-sequence. In accordance with the truth table, the state of the second stage is combined with the state of the third stage to provide an input state. In the particular instance detailed, the initial state of the registers is 011. Stepping the register produces the 7-chip sequence

$$0001011 \rightarrow$$

It contains three 1s and four 0s. Because the initial state of the register and the state after seven shifts are the same, continuing to step the register repeats the sequence.

Different initial conditions produce different sequences. In Figure 21.2, for the eight possible initial settings, I list the sequences they will produce. Ignoring the all-1s case, I have seven sequences that contain three 1s and four 0s. Emphasizing that they are unique to the feedback arrangement shown is important. Changing the logic to an exclusive-*or* produces a set of seven 7-chip sequences in which there are four 1s and three 0s, and an all-0s sequence (which is discarded). These examples mimic the properties of the general *m*-sequence in which a shift-register of *n*-stages produces

- **Length:** $2^n - 1$ chips per sequence
- **Number:** $2^n - 1$ different sequences

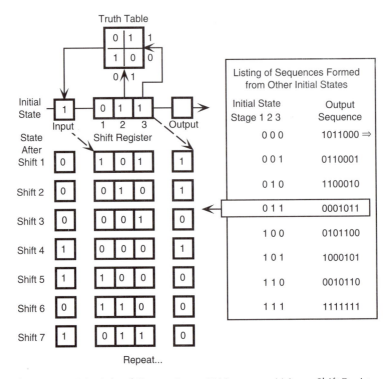

Figure 21.2 Principle of Generating a *PN*-Sequence Using a Shift Register and Feedback Logic

- **Composition:** (number of 1s) = (number of 0s) ± 1

For practical applications, an immense range of register sizes and feedback connections has been explored to find useful codes of almost any length. (Codes that are long enough so that they repeat only every few days are not uncommon.) Further, more sophisticated means of generating sequences are employed to achieve families that have low correlations.[1]

21.4 SPREADING TECHNIQUES

To create spread spectrum signals, *pn*-sequences are employed to convert narrowband information-bearing signals to wideband signals. Two basic techniques are used.

---

[1]D. V. Sarwate and M. B. Pursley, "Crosscorrelation properties of pseudorandom and related sequences," *Proceedings IEEE,* 68 (1980), 593–619.

(1) Direct-Sequence Spread Spectrum

To create a direct-sequence (DS) spread spectrum signal $[W_{DS}(t)]$, the binary data message signal $[m(t)$, as in equation (20.1)] modulates a very high speed *pn*-sequence $[p(t)]$ to produce a wideband signal $[m(t)p(t)]$. It is used to modulate a sinusoidal carrier $[\cos(2\pi f_c t)]$ to create a spread spectrum radio frequency wave [say BPSK, $s_{BPSK}(t)$, as in § 20.1(2)]. Thus

$$W_{DS}(t) = A\cos\{2\pi f_c t + \pi m(t)p(t)\} \qquad (21.9)$$

$m(t)$, $p(t)$, and $W_{DS}(t)$ are shown in **Figure 21.3**. The speed of $p(t)$ may range from 1 Mchips/s to 100s of Mchips/s so as to produce frequency spectrums that extend up to several hundred megahertz. The result is a signal

- that extends over the wide spectrum of $p(t)$
- whose power density spectrum is very low
- whose appearance is noiselike
- whose level is significantly below ambient noise

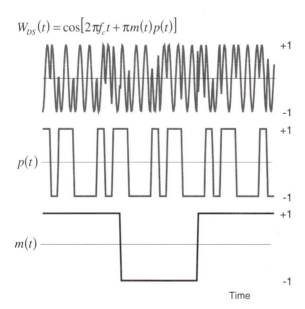

$$W_{DS}(t) = \cos[2\pi f_c t + \pi m(t)p(t)]$$

Figure 21.3 A DS Spread Spectrum Signal and the Message and Chipping Components
A binary data message signal and a very high-speed pn-*sequence are used to phase modulate a carrier to produce a spread spectrum signal.*

At the receiver, the spread spectrum signal is demodulated to a wideband signal and demodulated a second time by the pn-sequence to produce $m(t)p(t)^2$. Provided that the chipping signal $p(t)$ used in the transmitter and the chipping signal used in the receiver are exact mimics, and the chipping signal in the receiver is exactly synchronized with the chipping signal arriving at the receiver, $p^2(t) = 1$ and the message signal is recaptured. This state is shown in **Figure 21.4**. Fortuitously, the second use of $p(t)$ for demodulation spreads any interfering signals and reduces their effect on the wanted signal.

By restricting knowledge of the *pn*-sequence to the transmitter and receiver, the communicators can be assured of a significant level of privacy. Also, by choosing the *pn* sequences carefully so that they are orthogonal, several users can communicate in the same spectrum space [see § 21.4(3)].

EXAMPLE 21.4 Power Density in DS Spread Spectrum Signal

Given Using a direct sequence technique, a 1-watt, information-bearing signal of bandwidth 10 kHz is spread by a 10 Mchip/s *pn*-sequence in a bandwidth of 10 MHz.

Problem Determine the average power density of the spread signal.

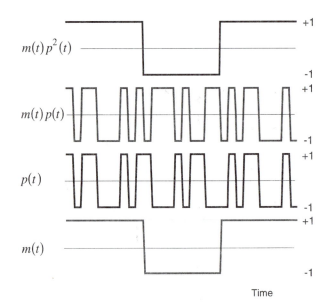

Figure 21.4 Demonstration that $m(t)p^2(t) = m(t)$, i.e., $p^2(t) = 1$

Solution Employing a 10 Mchip/s *pn*-sequence to spread the signal over 10 MHz creates a signal with an average power per unit bandwidth (i.e., power density) of

$$1 \text{ watt}/10 \text{ MHz} = 0.1 \text{ } \mu\text{watts/Hz}$$

Comment The average power in 10 kHz will be 0.1% of the transmitted power (a reduction of 30 db).

---

(2) Frequency Hopping Spread Spectrum

To create a frequency hopping (FH) spread spectrum signal $[W_{FH}(t)]$, the binary data message signal $[m(t)]$ modulates a sinusoidal carrier $[\cos(2\pi f_c t)]$ to create an *M*-ary FSK modulated wave $[s_{MFSK}(t)]$. $s_{MFSK}(t)$ is then translated to a sequence of pseudo-random frequencies produced by mixing the signal with the output of a frequency synthesizer controlled by a *pn*-code generator $[p(t)]$. Thus

$$W_{FH}(t) = s_{MFSK}(t)\sum \Pi(t - iT_D)\cos(2\pi f_i t) \qquad (21.10)$$

where $\Pi(t - iT_D)\cos(2\pi f_i t)$ represents a signal generated by the synthesizer at frequency f_i in response to the *i*th code. Each signal is present for a dwell time T_D. As a result, the transmitted signal *hops* from one frequency to another as the code changes. Hopping rates may be as high as 100 khops/s. The FH signal

- occupies a narrow *slice* of spectrum for a very short time before shifting to another slice in the spectrum
- has a low average power density across the operating spectrum
- radiates standard power above the noise level in the slice that happens to be occupied

To reclaim the information signal, the receiver must know the frequency-hopping sequence of the transmitter. **Figure 21.5** shows the process of creating an FH spread spectrum signal at the transmitter and reconstructing the message signal at the receiver. The MFSK signal is mixed with the output of the frequency synthesizer to produce a short burst of signal at frequency f_i in response to the *i*th code. The synthesizer then responds to a change in the *pn*-sequence and produces another frequency, etc. At the receiver, the procedure is reversed.

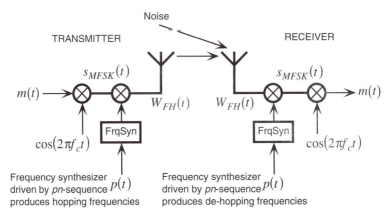

Figure 21.5 Creating an FH Spread Spectrum Signal and Reconstructing the Message Signal

At the transmitter, the pn-sequence controls the frequency output of the frequency synthesizer so as to cause the transmitted frequency to hop from one frequency to another in a random manner. At the receiver, the pn-sequence controls the frequency output of the frequency synthesizer so as to change the received frequencies to the original carrier frequency.

When several transmitters are operating, multiple signals may appear at the same time in the same spectrum slice. This occurrence creates interference and errors that set a limit to the number of channels that may be operated simultaneously. As with DS operation, by restricting knowledge of the *pn*-code to the transmitter and receiver, the communicators can be assured of a significant level of privacy.

EXAMPLE 21.5 Power Density in FH Spread Spectrum Signal

Given A 1-watt, information-bearing signal of bandwidth 10 kHz is spread in a bandwidth of 100 MHz using a frequency-hopping technique that makes 10,000 hops/s.

Problem Determine for what percentage of time the signal is present in a slice, and the average power density of the spread signal.

Solution Without overlap, the signal can occupy any one of

$$100 \text{ MHz}/10 \text{ kHz} = 10,000 \text{ slices}$$

If the hopping rate is 10,000 times per second, on average, a slice will be selected once per second. Each time it is selected, the signal will dwell on it for 100 μs, i.e., the signal is present for 0.01% of the time. The average power density is

$$1 \text{ watt}/100 \text{ MHz} = 0.01 \text{ μwatts}/\text{Hz}$$

Comment   The average power density is less than that of the DS example above; however, the power is concentrated in a slice at a time, and is 1 watt in the occupied slice.

---

(3) Code Division Multiple Access

Using unique spreading sequences that are orthogonal to each other, more than one user at a time can transmit in the same bandwidth. Called *code division multiple access* (CDMA),[2] cooperating stations transmit at the same carrier frequency with approximately equal power and the same chip rate using DS or FH techniques. Each transmitting station employs a code that is uncorrelated with the codes used by the other stations. In practice, the spreading sequences in the set will have a cross-correlation with the rest of around –20 db. The receivers see the sum of the individual signals as uncorrelated noise. Each station can demodulate one or more signals if they have been provided with the associated spreading code(s) and the carrier frequency and means to synchronize with them. For reliable performance in a multi-user environment, controlling the power level of competing signals is important in order to prevent higher power signals from capturing the receiver.

Figure 21.6 shows the principle of CDMA in a direct spreading environment. It shows several transmitters operating with orthogonal chipping codes. At the receiver, the correlator responds only to $p(t)$.

REVIEW QUESTIONS

21.1 Define *spread spectrum modulation.*

21.2 When $S/N \ll 1$, what is the bandwidth of a Gaussian channel that transports C bits/s without error?

21.3 Define *process gain* and *jamming margin.*

---

[2]Andrew J. Viterbi, *CDMA: Principles of Spread Spectrum Communication,* (Reading, MA: Addison Wesley Longman, 1995).

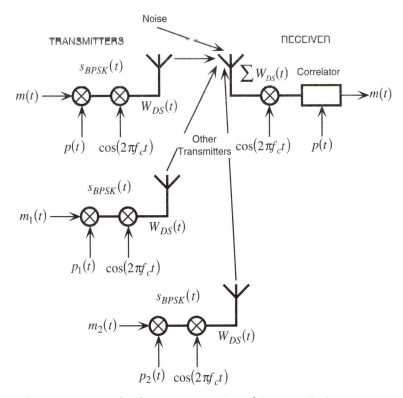

Figure 21.6 Principle of CDMA in a DS Spread Spectrum Environment
*As well as the wanted signal, the receiver receives signals from other
transmitters and noise. Using the known chipping code, the correlator
selects the wanted signal and spreads the spectrums of the other transmitters and noise.*

21.4 What is the relation among process gain, jamming margin, and output
signal-to-noise ratio?

21.5 Define *pseudo noise sequence* and *chip*.

21.6 Distinguish between *direct-sequence* and *frequency-hopping* spectrum
spreading techniques.

21.7 Explain how code division multiple access schemes work.

22

Transmission Media

By effecting *action at a distance,* electromagnetic energy carries messages between communicating entities to make real-time telecommunication possible. Described by *Maxwell's* equations (after James Clerk Maxwell, a 19th-century Scottish physicist), signal energy propagates under the influence of oscillating electrical and magnetic fields. Carried by *photons* at relatively low frequencies, it is manifest as electric currents and potentials in copper wire cables; at higher frequencies, it is manifest as electromagnetic waves in coaxial cables and free-space; and, at very high frequencies, it is manifest as infrared light in optical fibers.

22.1 WIRE CABLES

In 1843, copper wire was used to carry the first telegraph signals between Washington, DC, and Baltimore, MD. Twisted into cables, it became the dominant transmission medium in the first century of telephony. Today, twisted-pair cables still make up all but a small fraction of local loop transmission facilities. Thus, knowing some of the characteristics of wire cables is important.

(1) Ideal Transmission Line

An ideal transmission line is one without losses—losses due to the resistance of the conducting members and the leakance (conductance) between them. It is characterized by inductance (L, Henries per unit length) and capacitance (C, Farads per unit length). **Figure 22.1** shows an equivalent circuit representation in which the line consists of an infinite ladder network of identical small sections (length Δx) containing incremental capacitors ($C\Delta x$) and inductors ($L\Delta x$). The line begins at the left and extends to infinity on the right. The impedance looking at terminals a_n, and a_n' is Z_n. Z_1 is equal to the impedance of $L\Delta x$ in series with the parallel combination of Z_2 and the impedance of $C\Delta x$, i.e.,

IDEAL TRANSMISSION LINE

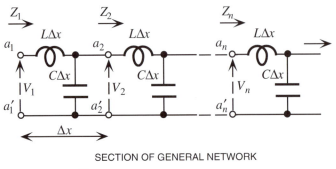

SECTION OF GENERAL NETWORK

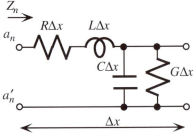

Figure 22.1 Equivalent Circuits for Transmission Lines
An ideal transmission line has no loss. The general network includes resistances that produce losses.

$$Z_1 = j\omega L\Delta x + \frac{Z_2}{j\omega C\Delta x}\bigg/\left(Z_2 + \frac{1}{j\omega C\Delta x}\right) \qquad (22.1)$$

where $\omega = 2\pi f$. Now, because the line is infinitely long, you see the same impedance whether you look at terminals a_1 and a_1' , a_2 and a_2' , or a_n and a_n', so that $Z_1 = Z_2 = Z_n$. Substituting in equation (22.1) and simplifying,

$$Z_1^2 - Z_1 j\omega L\Delta x = L/C \qquad (22.2)$$

and, in limit $\Delta x = 0$

$$Z_1 = \sqrt{L/C} \qquad (22.3)$$

Thus, the input impedance of an ideal transmission line is a pure resistance of value $\sqrt{L/C}$ ohms. If the line is broken at any point and terminated with a resistor of this value, from the point of view of a signal at the input, distinguishing between the truncated line and an infinite line is impossible. Z_1 is known as the *characteristic impedance* of the line and is denoted by Z_0.

Returning to Figure 22.1, straighforward manipulation shows that

$$V_1/V_2 = 1 - \omega^2 LC(\Delta x)^2 + j\omega\Delta x\sqrt{LC} \qquad (22.4)$$

Because the ratio is a complex quantity, express it in terms of magnitude and phase. Thus, the magnitude is the square-root of the sum of the squares of the real and imaginary parts

$$|V_1|/|V_2| = \sqrt{\left(1 - \omega^2 LC(\Delta x)^2\right)^2 + \omega^2 LC(\Delta x)^2} \qquad (22.5)$$

and the tangent of the phase angle is the imaginary part divided by the real part

$$\tan\beta_{\Delta x} = \omega \Delta x \sqrt{LC} \Big/ \left\{1 - \omega^2 LC(\Delta x)^2\right\} \qquad (22.6)$$

If Δx is small, $(\Delta x)^2$ is very small, so that, provided ω is not too large (i.e., the frequency is not too high), $\omega^2 LC(\Delta x)^2 \ll 1$. Under this condition,

$$|V_1|/|V_2| \cong 1$$

i.e., for a loss-free line, no attenuation of the signal occurs along the line. Also, if Δx is small, $\tan\beta_{\Delta x} \cong \beta_{\Delta x}$, so that

$$\beta = \beta_{\Delta x}/\Delta x = \omega\sqrt{LC} \qquad (22.7)$$

For a loss-free line, the phase angle of the signal varies along the line, depending on the signal frequency and the square-root of the product of L and C. The phase shift along the line is a manifestation of a wave that propagates with velocity

$$\vartheta = \omega/\beta = 1/\sqrt{LC} \qquad (22.8)$$

EXAMPLE 22.1 Propagation Velocities

Given At 1 kHz, the inductance and capacitance of 24 AWG air core, twisted pair cable are 18 mH/kfoot and 16 nF/kfoot. At 1.5 MHz, the values are 14 mH/kfoot and 16 nF/kfoot.

Problem Determine the velocity with which 1 kHz and 1.5 MHz signals propagate over the cable.

Solution Convert parameters to quantity per mile and substitute in equation (22.8)

$$\vartheta_{1kHz} = 1/\sqrt{18 \times 5.28 \times 10^{-3} \times 16 \times 5.28 \times 10^{-9}} = 1.1160 \times 10^4 \text{ miles/s}$$

$$\vartheta_{1.5MHz} = 1/\sqrt{14 \times 5.28 \times 10^{-3} \times 16 \times 5.28 \times 10^{-9}} = 1.2654 \times 10^4 \text{ miles/s}$$

Comment The velocities of signals on twisted pairs are relatively slow compared to their velocities over optical fibers (approximately $\frac{2}{3}c$, see Figure 22.5).

---

(2) Lossy Transmission Line

In the lower diagram of Figure 22.1, I show a section of transmission line that includes the resistance (R Ohms per unit length) of the conducting members and the conductance (G Siemens per unit length) between them. R, L, C, and G are the fundamental electrical parameters of the line, and are called *primary* constants.

At a point x, if a current i (Amps) flows in the conductors under the influence of a voltage v (Volts) between them, then

$$\partial v/\partial x = -\{Ri + L\,\partial i/\partial t\} \tag{22.9}$$

and

$$\partial i/\partial x = -\{Gv + C\,\partial v/\partial t\} \tag{22.10}$$

For sinusoidal voltages, the solutions are

$$V = A\exp(-px) + B\exp(px) \tag{22.11}$$

and

$$I = \tfrac{1}{Z_0}\{A\exp(-px) - B\exp(px)\} \tag{22.12}$$

where

$$p = \sqrt{(R + j\omega L)(G + j\omega C)}$$

and

$$Z_0 = \sqrt{(R + j\omega L)/(G + j\omega C)}$$

p is called the *propagation* constant. It consists of a real part—the *attenuation* constant, α, and an imaginary part—the *phase delay* constant, β. Thus

$$p = \alpha + j\beta \tag{22.13}$$

Z_0 is the characteristic impedance. (Because p and Z_0 are derived from primary constants, they are called *secondary* constants.)

Both p and Z_0 are independent of length. In equations (22.11) and (22.12), the negative exponential terms [i.e., $A\exp(-px)$] represent a wave traveling from source to load, and the positive exponential terms [i.e., $B\exp(px)$] represent a wave traveling from load to source. The constants A and B are determined by the values of the terminations at each end of the line.

(3) Reflected Waves

If the source impedance is Z_S and the load impedance is Z_L, on a line of length l, at the load terminals, the ratio of the reflected (i.e., directed to the source) voltage wave to the forward (i.e., directed to the load) voltage wave is

$$\vartheta_R = B\exp(pl)/A\exp(-pl) = (Z_L - Z_0)/(Z_L + Z_0) \qquad (22.14)$$

and the ratio of the reflected current wave to the forward current wave is

$$\iota_R = -B\exp(pl)/A\exp(-pl) = (Z_0 - Z_L)/(Z_0 + Z_L) \qquad (22.15)$$

Thus, to prevent reflections at the load end, the load impedance must equal the characteristic impedance of the line. A similar result can be developed for the source end. For reflection-free operation, then, both the source impedance and the load impedance must be equal to the characteristic impedance.

(4) Skin Effect

At dc, and low frequencies, the current flowing in a conductor is distributed uniformly over the cross-sectional area. However, due to self-induction, as the operating frequency is increased, the distribution is disturbed, and the current density falls at the center and begins to increase in the outer layer of the conductor. The result is that the apparent resistance of the conductor increases. Eventually, if the frequency is high enough, almost all of the current flows along the skin. Although the exact physical mechanism is difficult to describe, most analyses conclude that

$$R_{HF}/R_{dc} \propto \sqrt{f_{HF}} \qquad (22.16)$$

i.e., the ratio of the resistance at a high frequency to the resistance at dc is proportional to the square-root of the frequency.

(5) Twisted Pair Metallic Cables

Telephone wires are pervasive. Often called *twisted pairs* because of the way in which they are constructed, they carry voice and data signals from customers' premises to local switching and routing centers.

The primary constants, R, L, C, and G, depend on the gauge of the wire, the materials employed to construct the cable, and the operating frequency. C is close to constant; it exhibits a value of around 16×10^{-9} F/kft from 1 kHz to 1.5 MHz. L exhibits a value of around 0.20 mH/kft at 1kHz to 0.16 mH/kft at 1.5 MHz. G exhibits a value of around 0.04 µS/kft at 1 kHz to 60 µS at 1.5 MHz. R depends on the gauge of wire; it exhibits values ranging from about 16 to 80 ohms/kft at 1 kHz, and 100 to 230 ohms/kft at 1.5 MHz.[1]

The characteristic impedance of the medium varies with wire size and operating frequency. Because Z_0 is complex, we consider magnitude and

---

[1]For a comprehensive discussion, see Chapter 7 of Whitham D. Reeve, *Subscriber Loop Signaling and Transmission Handbook—Digital* (New York: IEEE Press, 1995). Also, see Whitham D. Reeve, *Subscriber Loop Signaling and Transmission Handbook—Analog* (New York: IEEE Press, 1992).

phase. At 1 kHz, for a range from 19 through 26 AWG, $|Z_0|$ exhibits values from 400 to 900 ohms, and $\angle Z_0$ exhibits values between -0.75 and -0.80 radians. At 1 MHz, for a range from 19 through 26 AWG, $|Z_0|$ exhibits a value of approximately 100 ohms, and $\angle Z_0$ exhibits values between -0.05 and -0.10 radians. Because of these wide variations, you cannot terminate the lines exactly. Compromises have been reached that use 900 ohms on predominantly fine gauge (i.e., 26 and 24 AWG) circuits, and 600 ohms on predominantly coarse gauge (i.e., 22 and 19 AWG) circuits. For private lines intended for high-speed data transport, terminations of 135 to 150 ohms are used. **Figure 22.2** shows values of the primary constants and the characteristic impedance of 19, 22, 24, and 26 AWG air core twisted pair cables at approximately 20°C.

Figure 22.3 shows estimated values of attenuation and phase delay for the medium. Both parameters are frequency-dependent. In addition, attenuation depends on the gauge of the wire employed, but phase delay is independent of wire size. Although twisted pair is thought of as a voiceband transmission medium, the diagram shows that it can be used at much higher frequencies over short distances, particularly if processing is used to restore the level and compensate for signal distortion.

(6) Crosstalk

Electromagnetic coupling between isolated circuits causes a signal in one to produce a disturbance in the other. Called *crosstalk,* it contributes to the noise background in the receiving circuit. **Figure 22.4** shows two parallel transmission paths that are terminated by their characteristic impedances. If the distances are small compared to the wavelengths of the operating frequencies, their interaction can be characterized by the inductive and capacitative elements shown. A voltage v_0 from the source will produce a current i_0 around a,a'. In turn, through the action of the inductive coupling represented by the mutual inductance M, i_0 will induce a voltage v_m and drive a current i_m around the circuit b,b'. Also, v_0 will drive a current i_c around the shunt circuit a,b,b',a' through the two halves of the capacitance C_U. Between b and b' the current divides into equal parts that traverse the circuit in different directions.

For the sake of illustration, assume that the characteristic impedance is much less than the impedance of the two halves of C_U in series, and that the effects of capacitance and inductance can be treated separately. Then $v_0 = i_0 Z_0$, and

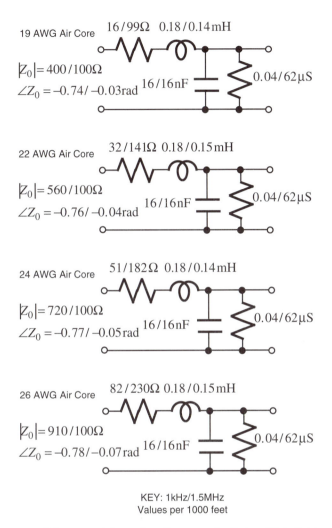

Figure 22.2 Primary and Secondary Constants for Twisted Pair Cable
*Values are per 1000 feet at 1 kHz and 1.5 MHz for 19, 22, 24, and 26
AWG air core cable.*

$$i_c = \tfrac{1}{4} i_0 Z_0 j\omega C_U \tag{22.17}$$

In circuit b,b' the current i_0 produces a voltage $v_m = i_0 j\omega M$ and

$$i_m = -i_0 j\omega M / 2 Z_0 \tag{22.18}$$

I define two forms of interference, or crosstalk:

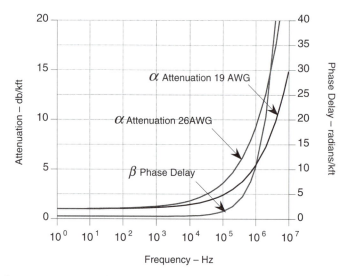

Figure 22.3 Estimated Secondary Parameters for Twisted Pair Cable *Attenuation depends on wire size. Phase delay is independent of wire size.*

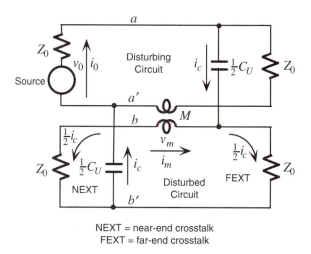

NEXT = near-end crosstalk
FEXT = far-end crosstalk

Figure 22.4 Inductive and Capacitive Contributions to Crosstalk between Parallel Cables

- **Near-End crosstalk** (NEXT): crosstalk whose energy flows in the opposite direction to that of the signal in the disturbing channel. If the NEXT current is i_{NEXT},

$$i_{NEXT} = i_c - i_m = \tfrac{1}{4} i_0 Z_0 j \omega C_U + i_0 j \omega M / 2 Z_0 \qquad (22.19)$$

- **Far-End crosstalk** (FEXT): crosstalk whose energy flows in the same direction to that of the signal in the disturbing channel. If the FEXT current is i_{FEXT},

$$i_{FEXT} = i_c + i_m = \tfrac{1}{4} i_0 Z_0 j \omega C_U - i_0 j \omega M / 2 Z_0 \qquad (22.20)$$

In NEXT, the effects of mutual inductance and unbalanced capacitance reinforce each other, and in FEXT, they oppose each other. Thus, NEXT is usually a more serious problem than FEXT where parallel transmission paths are concerned—such as pairs of conductors in the same cable.

(7) Other Impairments

In a telephone plant, many existing arrangements impair transmission integrity. Often, they represent compromises made many years ago that have become embedded in operations, and are too expensive to remove. I mention three items briefly:

- **Multiple wire sizes:** because space is limited in well developed wire centers, cables connected to the distributing frames employ 26 AWG twisted pairs. On longer loops, a necessary action is to increase the wire size once or twice to keep the loop resistance within prescribed limits (usually 110 to 1300 ohms). The change in wire gauge produces an imbalance that can create reflections.
- **Two-to-four wire conversion:** most subscribers are served by a single twisted pair that carries signals in two directions. At a point in the loop, or at the serving office, the channels are separated on two twisted pairs, one for each direction of signaling. This conversion is achieved through the use of a *hybrid* transformer, or an electronic device. Almost impossible to balance exactly, these arrangements create reflections.
- **Bridged taps:** in the outside plant, some loops may have been used to connect to several terminating points at different times. In the process of rearrangment, the pair may have been left connected to some of them so that there are unused, open-circuited, connected cable pairs. They represent a shunting impedance at the point of connection and cause reflections.

22.2 OPTICAL FIBER

An optical fiber is a strand of exceptionally pure glass about the diameter of a human hair (125 micron = 0.005 inches). The refractive index varies from the center to the outside in such a way as to guide optical energy along its length. Although several types of fibers are recognized in the scientific and engineering communities, the predominant design in communications applications is *single-mode fiber*. Shown in **Figure 22.5**, in such a fiber the central core of elevated refractive index glass is < 10 microns in diameter; because this is less than an order of magnitude greater than the wavelength of the optical energy, propagation is governed by classical wave equations. Although a significant (and essential) fraction of the optical energy travels in the cladding, most energy is contained within the core. Because its velocity is slightly slower than the energy in the cladding, conditions are right to support single-mode propagation. With the refractive index of 1.475 shown, the velocity of energy in the core is approximately 2×10^8 meters per second (i.e., approximately two-thirds the velocity of light in free space).

(1) Comparison with Copper Wire

Transmission of information over an optical fiber has several fundamental advantages over transmission along a copper wire:

- Optical fibers are insulators; they provide electrical isolation between transmitter and receiver
- Optical energy is not affected by lower-frequency electromagnetic radiation; optical communication can occur in noisy electrical environments without degradation
- When launched properly, all of the optical energy is guided along the fiber; adjacent fibers do not interfere with one another; no fiber exists to fiber crosstalk
- Optical frequencies are very high compared to any conceivable message bandwidth; they can be used to transport wideband signals

Optical fibers have disadvantages, too:

- Optical fibers are a one-way transmission medium; unlike electrical energy in a twisted pair cable, when injected in a fiber, optical energy propagates in one direction only
- Optical fibers do not conduct electricity; thus powering equipment (such as telephones or repeaters) down the fiber is impossible; if required, it is accomplished over wires placed in the cable sheath
- Optical fibers are subject to increased losses due to microbending and other mechanical insults

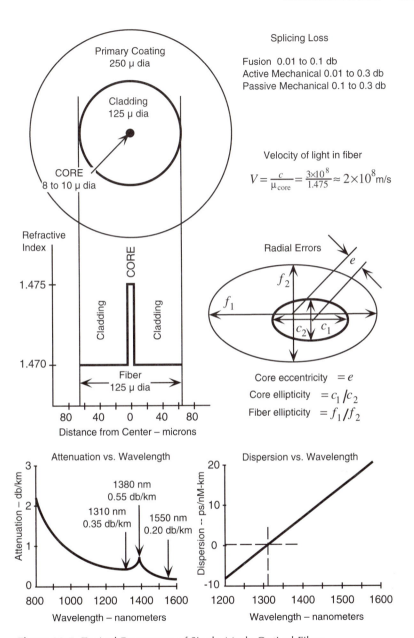

Figure 22.5 Typical Parameters of Single-Mode Optical Fiber

(2) Mechanical Properties, Cables, and Connections

The mechanical properties of optical fiber are difficult to measure. Glass is a brittle solid that does not deform plastically to relieve local stress—it shatters. The strength of a given length of optical fiber is determined by the most severe flaw present. For a 1 μm flaw, it is estimated that failure occurs around 80,000 pounds per square inch (psi). This is somewhat higher than the tensile strength of hard drawn copper wire (60,000 to 70,000 psi). During manufacture, the fiber is subjected to a proof test, coated with a polymer coating (primary coating) to provide abrasion protection, and coated with a second polymer coating (buffer coating) to provide additional protection. Cables are produced with as few as 2 fibers up to 200 fibers. As noted above, most cables contain a few wire pairs for service use, and a tensile strength member to minimize fiber stress. Cables are finished off with a plastic jacket and, if environmental conditions require it, steel armor.

The success with which fiber-to-fiber connections are made depends critically on the radial geometry of each fiber and on the preparation given to the fiber ends. Radial geometry must be built into the fiber during manufacturing. The closer the fibers conform to same-size concentric circles, the better the connection that can be made. Referring to Figure 22.5, for each fiber, the core eccentricity (e) must be minimized, and the core and fiber ellipticity measures (c_1/c_2 and f_1/f_2) must equal 1. For the two fibers, the diameters must be equal. As for the mating fiber ends, they must be cleaved square or polished, so that they make intimate contact with each other.

Permanent connections are called *splices*. They are made by:

- Aligning the fibers with reference to their outer surfaces and employing a coaxial connector structure to hold them in place. With some connectors, adjustments can be made to obtain the lowest splice loss after installation is made.

- Aligning the fibers with reference to their outer surfaces and cementing them into a structure to hold them in place.

- Employing a gaseous arc to fuse the ends together. During fusion, surface tension assists in aligning the fiber ends.

Figure 22.5 lists representative values of losses due to these techniques. Although fusion splicing produces the lowest loss connection, it cannot be used in hazardous environments, and requires a skilled operator and significant capital investment. In the twenty years since the first trial installations, the cost of optical fibers has reached parity with wire systems. The greatest concern of those engineering contemporary installations is with the joints and connections, not with the glass medium itself.

(3) Optical Properties

Single-mode fibers are used with solid-state lasers and photodetectors and operated at wavelengths of 1310 or 1550 nanometers. The lasers are switched on and off to produce unipolar pulses of intense infrared energy. Figure 22.5 shows the attenuation and dispersion per kilometer of a single-mode fiber as a function of the optical wavelength employed. At 1550 nanometers, the fiber has a minimum attenuation of around 0.2 db/km, and a dispersion of some 17 ps/nm-km. At 1310 nm, the fiber has zero dispersion and an attenuation of 0.35 db/km. The great majority of current installations were designed to operate in the minimum dispersion region. Their capacity is limited by a combination of fiber loss and the bandwidth of electronic repeaters.

At the attenuation minimum of 1550 nm, optical amplification is possible (amplification but no regeneration) using erbium-doped fiber. However, intersymbol interference produced by dispersion becomes a limiting factor. To make full use of the enormous bandwidth, an increasingly popular strategy is to transmit several optical carriers simultaneously in the same fiber. Called *wavelength division multiplexing* (WDM), the carriers can be spaced about 2 nm apart. Current practice may employ up to 8 carriers, and research organizations have demonstrated the use of many more.[2] Crosstalk is a major concern in WDM. Interference is produced by fiber non-linearities that scatter the optical energy of the carriers, and by imperfections in network equipment.

22.3 CELLULAR RADIO

Cellular radio has shown how to uncouple terminals from fixed service points. No longer must a telephone or data terminal be attached to a cable to send and receive information. Exploiting frequency reuse for mobile telephone service, cellular radio makes possible serving a large population with a relatively small band of frequencies.

(1) Frequency Reuse

- **Frequency reuse:** the principle of using a set of frequencies to communicate in one location, and simultaneously using the same frequencies to support independent communication in another location

For fifty years, *frequency reuse* has been employed in terrestrial microwave radio relay systems in which focused beams of microwave energy distributed about a specific carrier frequency are sent from one station to another. The same carrier frequency can be used in other beams so long as individual receivers only *see* their transmitter. Successful operation depends on providing sufficient physical separation between entities that use the same frequencies so as to prevent one from receiving signals from another. With mobile units (mobiles), we use a band of frequencies in a limited area (known as a *cell*) and arrange that other areas in which the same frequencies may be in use are separated by enough distance so that the energy is properly attenuated. **Figure 22.6** shows a cellular radio building block. Seven hexagonal cells are arranged in a *cluster*. At the center of each cell is a transmitter and

---

[2]See IEEE Special Issue on Multiwavelength Optical Technology and Networks, *Journal of Lightwave Technology,* 14 (1996).

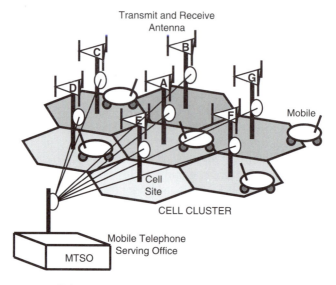

Figure 22.6 Cellular Radio Building Block
The cell cluster consists of seven cells that operate with different radio fre-
quencies. Provided the cells are large enough to produce sufficient attenu-
ation, similar cell clusters are positioned to give continuous coverage over
a large area, and the frequencies are reused.

receiver combination that uses a specific set of frequencies to communicate
with mobiles moving within the cell. Signals transmitted to, or received from,
the mobiles by the cell site are relayed to a mobile telephone switching office
(MTSO) over line-of-sight radio links, or by other means. In all, seven sets of
frequencies (designated A, B, ... G in Figure 22.6) are employed so that the
cells within the cluster do not interfere with one another. Signals from
mobiles are picked up by the cell site antennas and relayed to a mobile telephone serving office (MTSO). Signals originating elsewhere are sent by the
MTSO to the cell site transmitter for relay to mobiles.

Outside the 7-cell cluster, the frequencies are reused. **Figure 22.7** shows
how the cluster fits with other clusters so that the same sets of frequencies
are separated by cells operating at different frequencies. Common clusters
contain 4, 7, or 12 cells. If N is the number of cells in a cluster, B_{CL} is the
radio bandwidth assigned to the cluster, b_{ch} is the bandwidth required for
each two-way communication channel, and the radio resources are shared
equally among all cells. The number of channels k_{CE} that are available in
each cell is

$$k_{CE} = B_{CL}/Nb_{ch} \qquad (22.21)$$

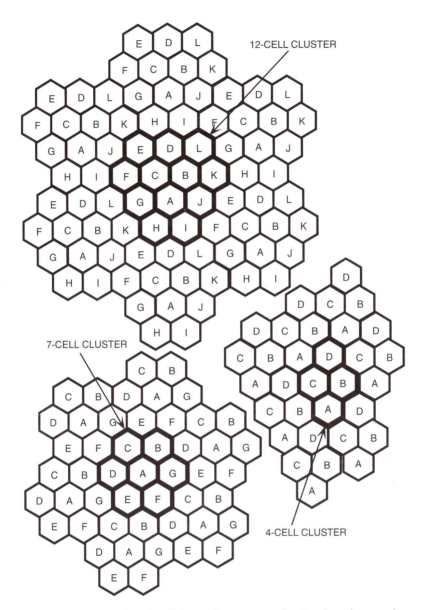

Figure 22.7 Examples of Cellular Radio Coverage that Employ Clusters of 4, 7, and 12 Cells

This is the number of simultaneous communications that can be supported in a cell. In the cluster, it is

$$k_{CL} = B_{CL}/b_{ch} \qquad (22.22)$$

and, if the cluster is duplicated M times to form a system, the number of simultaneous communications that it can support is

$$k_{SYS} = MB_{CL}/b_{ch} \qquad\qquad (22.23)$$

Thus, the number of simultaneous communications that a system can support is proportional to the number of clusters and the total allocated bandwidth, and inversely proportional to the bandwidth required by each communication.

The quantity $1/N$ is known as the *frequency reuse factor*. Denoting it by c_{FR}, we can write equation (22.21) as

$$k_{CE} = c_{FR}k_{CL}$$

and substituting equation (22.22) in equation (22.23),

$$k_{SYS} = Mk_{CL}$$

EXAMPLE 22.2   Influence of Cluster Size on System Capacity

Given Four separate cellular radio systems operate in two frequency bands: 825 to 845 MHz mobile to cell site, and 870 to 890 MHz cell site to mobile. A duplex circuit consists of a pair of 30 kHz simplex channels: one in the 825 to 845 MHz band, and the other separated by 45 MHz in the 870 to 890 MHz band. The systems are distinguished by cluster size; in system A it is 4, in system B it is 7, in system C it is 12, and in system D it is 19. In all systems the cluster is duplicated 16 times.

Problem

  a) Find the number of simultaneous communications that can be supported by the systems.
  b) Find the number of simultaneous communications that can be supported by single cells in each system.

Solution

  a) From the information given, $M=16$, $B_{CL} = 40\,\text{MHz}$, and $b_{ch} = 60\,\text{kHz}$. Hence, substituting in $k_{SYS} = MB_{CL}/b_{ch}$ gives

$$k_{SYS} = \frac{16 \times 40 \times 10^3\ \text{kHz}}{60\ \text{kHz}} = 10,666\ \text{channels}$$

  b) Substituting in $k_{CE} = B_{CL}/Nb_{ch}$ gives

$$k_{CE} = \frac{40 \times 10^3\ \text{kHz}}{N60\ \text{kHz}}$$

so that, for system A

$$N = 4, \ k_{CE} = 166 \text{ channels}$$

for B

$$N = 7, \ k_{CE} = 95 \text{ channels}$$

for C

$$N = 12, \ k_{CE} = 55 \text{ channels}$$

and, for D

$$N = 19, \ k_{CE} = 35 \text{ channels}$$

Comment If the cell size is the same in each system, A covers an area equal to 64 cells, B covers an area equal to 112 cells, C covers an area of 192 cells, and D covers an area of 304 cells.

Extension Depending on the cluster size, if the cell size is the same in each system, and you wish to cover a fixed area—say 100 cells—the cluster will be duplicated a different number of times. Under these circumstances,

in system A

$$N = 4, \ M = 25, \text{ and } k_{SYS} = 16{,}666 \text{ channels}$$

in system B

$$N = 7, \ M \cong 14, \text{ and } k_{SYS} = 9{,}332 \text{ channels}$$

in system C

$$N = 12, \ M \cong 8, \text{ and } k_{SYS} = 5{,}333 \text{ channels}$$

and, in system D

$$N = 19, \ M \cong 5, \text{ and } k_{SYS} = 3{,}333 \text{ channels}$$

Thus, the smaller the cluster size, the greater the number of simultaneous communications that can be served in a fixed area.

---

(2) Signal-to-Interference Noise Power Ratio

In free space, the energy in radio waves drops off as the square of distance. In engineering terms, the energy decreases by 20 db for every tenfold increase in distance traveled. Thus, a wave at one mile from the antenna in free space with a power level of 0 db will have a power level of –20 db at 10 miles. In an urban environment, buildings, structures, and other impediments square the attenuation over the free space value. Measurements in the range 30 MHz to 3 GHz show that the average path loss is proportional to the fourth power of distance. Thus, a wave at one mile from the antenna in an urban environment with a power level of 0 db, will have a power level of 40 db at 10 miles.

Figure 22.8 shows the distance between a mobile at point *a* on the periphery of one cell and the transmitter in the closest interfering cell. If the

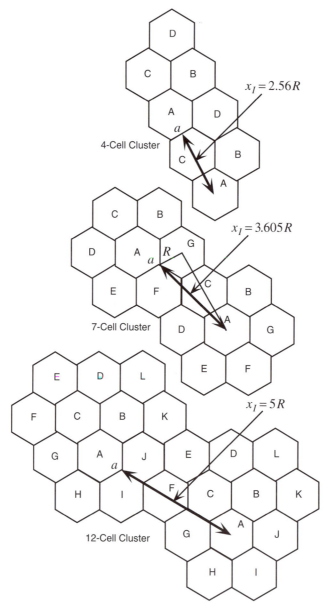

Figure 22.8 Interference Geometry for Clusters of 4, 7, and 12 Cells

mobile is distant x_0 from the primary transmitter and x_I from the interfering transmitter, and is receiving from the transmitter in its cell (known as receiving on the forward channel) the ratio of the primary power received to the co-channel interfering power for the worst interferor is

$$P_0/P_I = S/I = (x_0/x_I)^4 \qquad (22.24)$$

If the radius of the circle that just contains each hexagon is R,

for a 4-cell cluster,

$$x_0 = \sqrt{3R}/2 \text{ and } x_I = 2.56R \text{ so that } S/I = 18.7 \text{ db}$$

for a 7-cell cluster,

$$x_0 = R \text{ and } x_I = 3.605R \text{ so that } S/I = 22.3 \text{ db}$$

for a 12-cell cluster,

$$x_0 = R \text{ and } x_I = 5R \text{ so that } S/I = 28 \text{ db}$$

for a 19-cell cluster (not shown),

$$S/I \cong 35 \text{ db}$$

In actual fact, six first-layer interferors exist in the case of seven- and twelve-cell clusters (check Figure 22.7), and many more that are two, or more, layers away.

Equation (22.24) can be generalized to give an approximation to the signal-to-interference (SIR) noise power. Suppose that n co-channel interfering cells exist and that the distance between the primary receiver and the ith interferor is D_i; then

$$S/I = R^{-\alpha} \bigg/ \sum_{i=1}^{n} D_i^{-\alpha} \qquad (22.25)$$

Now, if you consider only the first layer of co-channel interferors (n_1), let the interfering stations be equidistant from the primary receiver, let the distances between the interfering transmitters and the primary station be the same and equal to D, and assume all n_1 interferors are active,

$$S/I = \frac{1}{n_1}(D/R)^\alpha = \frac{1}{n_1}\left(\sqrt{3N}\right)^\alpha \qquad (22.26)$$

The quantity $\sqrt{3N}$ is sometimes known as the *co-channel reuse ratio* and written Q, so that equation (22.26) becomes

$$S/I = Q^\alpha / n_1 \qquad (22.27)$$

In equation (22.26), if $\alpha = 4$, and $n_1 = 6$, then, when

$$N = 4, \quad S/I = 24 = 13.8 \text{db}$$

$$N = 7, \ S/I = 73.5 = 18.7 \text{db}$$

$$N = 12, \ S/I = 216 = 23.3 \text{db}$$

$$N = 19, \ S/I = 541 = 34.6 \text{db}$$

These results, and the capacity of a 100-cell system (from Example 22.1) are plotted in **Figure 22.9**. Obviously, a significant tradeoff occurs between system capacity and SIR.

If all the interferors are not active full-time, the values of SIR will improve. Suppose activity is randomly distributed among six interferors; the number of simultaneously active interferors is given by the binomial relationship, i.e.,

$$p(m) = \frac{6!}{m!(6-m)!} p^m (1-p)^{6-m} \tag{22.28}$$

where the average probability of the interferors being active is p and $p(m)$ is the probability of m interferors ($m \leq 6$) being active simultaneously. **Figure 22.10** plots $p(m)$ versus p. As the probability of activity drops below 1, fewer interferors are active simultaneously, and the SIR values decrease. In the general case, they will be between values derived from equation (22.26) and those derived from equation (22.24).

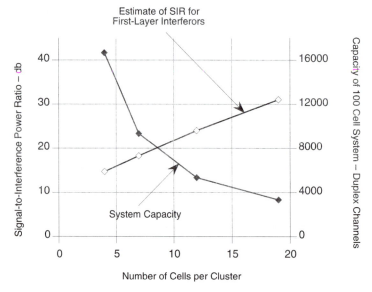

Figure 22.9 Signal-to-Interference Power Ratio and Capacity of 100 Cell System *versus* Number of Cells in Cluster

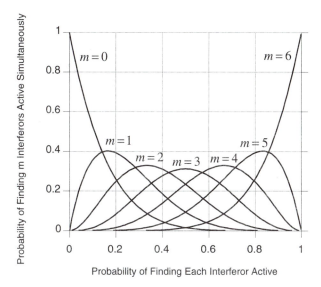

Figure 22.10 Number of Simultaneously Active Interferors When Activity Is Randomly Distributed Among Six Interferors

(3) Number of Simultaneous Users

You may be tempted to assume that the maximum number of simultaneous users is equal to the number of channels available in the cellular system. Only under very special circumstances is this true. If call requests are generated at random and holding times vary widely, as the number of call attempts increases, a level is reached at which calls are blocked. When blocking occurs, the call attempt may be cleared from the system, or it may be held in the system until resources are available to handle it. As described in § 9.3, the first strategy is *blocked calls cleared* (BCC); the second strategy is *blocked calls delayed* (BCD). If call attempts are Poisson-distributed and holding times (service times) are Erlangian, the performance of a system operating under BCC is calculated using the Erlang-B formula, and the performance of a system operating under BCD is calculated using the Erlang-C formula. In addition, the latter case may be investigated using an $M/E_k/n$ queueing model.

(4) Cell Size

For convenience, in Figures 22.6, 22.7, and 22.8, I have pictured a cellular radio system consisting of regular hexagonal cells. In the real world, the cell boundary is defined by

- the continuum of points at which the power received from the transmitter in the cell is just equal to the minimum level at which satisfactory detection is achieved by the mobile receiver (i.e., the probability of detection is $\geq 90\%$, say). This level will depend on the signal-to-noise ratio due to interferors (and possibly other disturbers).
- the maximum power of the mobile transmitter. It must be heard at the cell site receiver.
- the maximum speed at which mobiles traverse cells. If they are too small, the rate of handoff between cells will burden the system management functions. For systems intended to serve fast-moving vehicles in which relatively robust transmitters and receivers can be mounted, cells may be several miles across. For systems intended to serve pedestrians that employ small hand-held units, cells may be a few hundred feet across.
- any anomalous propagation conditions or geographical features.

Establishing a cell size is a complicated process that requires full definition of the purpose, location, environment, and equipment employed.

(5) Digital Operation

Cellular radio systems may be analog or digital. Whereas the original systems employed FDMA techniques in each cell, new systems employ TDMA and CDMA techniques.[3] As a result, they can service more subscribers. For instance, coding can be employed to provide error correction so as to improve performance in noisy environments. Considerable success has been achieved using Reed-Solomon coding to offset bursty noise [see § 16.2(5)]. Effective increases in SNR of around 3 to 6 db can be obtained.

22.4 COMMUNICATION SATELLITES

Communication satellites orbit the Earth; most perform as *repeaters in the sky.* They communicate with units on the ground, and may communicate among themselves (satellite-to-satellite). In the simplest case, the spacecraft electronics capture the signals from earth, convert them to lower frequencies,

---

[3]For an excellent description of these new systems, see Chapters 3 through 7 of David J. Goodman, *Wireless Personal Communication Systems* (Reading, MA: Addison Wesley Longman, 1997).

and transmit them back to earth so that a message originating from a single point in the field of view of the satellite can be received at a second point on the earth's surface in the field of view of the satellite. The relayed message can be broadcast to the entire area, in which case it can be received by everyone, or dispatched to a specific area by a spot-beam antenna on the spacecraft. In this case, only those within the illuminated area can receive it.

(1) Orbits

The most useful orbit for communications is one that is concentric with the earth's circumference in a plane that intersects its center. For a mass m (the satellite) that moves in a circular orbit R about a mass M (the earth) with angular velocity ω radians/s, the centripetal force is balanced by the gravitational force, so that

$$mR\omega^2 = GmM/R^2 \qquad (2.29)$$

where G is the gravitational constant. At the surface of the earth, the gravitational force on a body of mass m is mg where g is the acceleration due to gravity. If R_E is the radius of the earth,

$$mg = GmM/R_E^2 \qquad (2.30)$$

Substituting equation (2.30) in equation (2.29) and solving for R,

$$R^3 = gR_E/\omega \qquad (2.31)$$

If f is the rate at which the satellite rotates about the earth (in revolutions/s), we can write $f = \omega/2\pi$, and

$$R^3 = gR_E/4\pi^2 f^2 \qquad (2.32)$$

or

$$R^3 f^2 = \text{constant} \qquad (2.33)$$

If the height above earth is H_E and the radius of earth is 3,963 miles, equation (2.33) becomes

$$(H_E + 3963)^3 f^2 = \text{constant} \qquad (2.34)$$

Figure 22.11 plots height above the surface of the earth against the number of revolutions per day required to sustain the orbit. In an orbit 22,300 miles above the surface, a satellite completes one revolution a day; when in the equatorial plane, it appears stationary from a fixed point on earth.

- **Geostationary orbit** (GEO): some 22,300 miles above the equator, a path on the equatorial plane in which satellites rotate about the earth's axis at the same angular velocity as the earth (i.e., 360° or 2π radians in 24 hours)

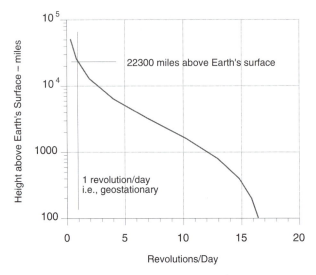

Figure 22.11 Revolutions per Day Required to Maintain Orbital Height above the Surface of the Earth
At 22,300 miles above the surface of the earth in the equatorial plane, a satellite completes one revolution per day and appears stationary relative to a fixed point on earth.

The satellite travels at approximately 6900 miles/hr, appears stationary from any point on earth, and can illuminate a circular area approximately 7,000 miles in diameter. When positioned between approximately 90° and 130° West longitude, GEO satellites can cover the 50 states of the United States, and when positioned between approximately 60° and 90° West longitude, they can cover the 48 continental states. GEO satellites deployed by major communications organizations for intercontinental and intracontinental trunking applications are spaced at intervals of arc of ≥ 2°. When used to provide direct television broadcasting, they are spaced ≥ 9° apart.

Other orbits are

- **Polar earth orbit:** a path on the polar plane—one that passes through the center of the earth and contains the poles
- **Inclined earth orbit:** a path on a plane that passes through the center of the earth situated between the polar and equatorial planes
- **Medium earth orbit** (MEO): a path about 6000 miles above the surface of the earth; the orbit is likely to be polar or inclined, and the footprint is around 5 to 6000 miles in diameter

- **Low earth orbit** (LEO): a path a few hundred miles (400 to 900 miles) above the earth's surface; the orbit is likely to be polar or inclined, and the footprint is around 1000 to 3500 miles in diameter

(2) Transmission Delay

A delay occurs between transmission and reception back on earth due to the distance the energy must travel. For GEOs, it is approximately 0.25 seconds round-trip; for MEOs, it is 60 to 70 milliseconds round-trip; and for LEOs it is less than 10 milliseconds round-trip. **Figure 22.12** shows values of the minimum round-trip transmission delay as a function of height above the surface of the earth.

(3) Interference among Satellites

To avoid signal interference among satellites, both space and frequency diversity are employed:

- **Space diversity:** satellites are physically separated from each other. Either they occupy different orbits, or, if in the same orbit, they occupy different positions in the orbit.

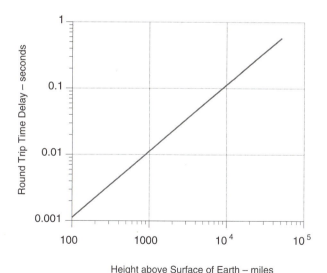

Figure 22.12 Transmission Delay and Height above the Surface of the Earth

- **Frequency diversity:** satellites assigned to the same orbit positions operate in different frequency bands

(4) Frequency Allocations

- **GEO satellites:** employ frequencies in C band (around 4 GHz, space to earth, and 6 GHz, earth to space) and Ku band (around 11 and 12 GHz, space to earth, and 14 GHz, earth to space). Additional frequencies are allocated in Ka band (around 18 GHz, space to earth, and 30 GHz, earth to space).

At 4 and 6 GHz, interference with terrestrial microwave networks that share the same frequency bands may restrict opportunities for siting earth stations. At the higher frequencies (Ku and Ka bands), the absence of terrestrial facilities competing for the same frequencies makes the selection of sites for earth stations easier. However, atmospheric attenuation due to rainfall, ice crystals, and fog can be significant. During heavy rainstorms in the eastern and southern regions of the United States, pairs or triplets of stations may be required to assure reliable signal reception. On the basis of statistics describing the extent of local storms, they are sited so that at least one station is likely to be in the clear at all times.

- **MEO and LEO satellites:** allocations exist from as low as 150 MHz (149.9 to 150.05 MHz) to as high as 30 GHz. Important bands for satellite-provided personal communication services exist at 312 to 315 MHz (downlink) and 387 to 390 MHz (uplink), 1610 to 1622.5 MHz (downlink) and 2483.5 to 2520 MHz (uplink), and 1980 to 2010 MHz (downlink) and 2170 to 2200 MHz (uplink). Other allocations exist in S, C, and Ka bands.

EXAMPLE 22.3   Received Power

Given A satellite orbits the earth at a distance of 40,000 kms and operates in the 4/6 GHz bands. The gain of the transmitting earth station antenna is 70 db; the gains of the receive and transmit antennas on the spacecraft are 30 db; the gain of the spacecraft amplifier is 80 db; and the gain of the receiving earth station antenna is 50 db. The transmitted power is 50 dbm.

Problem Find the power at the receiver.

Solution The free-space loss on a line-of-sight path is given by

$$L = (4\pi l/\lambda)^2 = (4\pi f/c)^2 \text{ watts}$$

where l is the path length, λ is the wavelength of the radio energy, f is the frequency of the radio energy, and c is the velocity of light. Restating the relationship with L in db, l in kms, and f in GHz gives

$$L_{db} = 92.4 + 20\log_{10} f_{GHz} + 20\log_{10} l_{km}$$

For the uplink, $f_{GHz} = 6$ and $l_{km} = 4 \times 10^4$; hence

$$L_{db} = 92.4 + 20\log_{10} 6 + 20\log_{10} 4 \times 10^4 = 200 \text{ db}$$

For the downlink, $f_{GHz} = 4$ and $l_{km} = 4 \times 10^4$; hence

$$L_{db} = 92.4 + 20\log_{10} 4 + 20\log_{10} 4 \times 10^4 = 196 \text{ db}$$

Since all gains (losses) are in decibels, you may add (subtract) them to determine the power at the receiver, i.e.,

Transmitted power =	+ 50 dbm
Gain of transmitting antenna =	+ 70 db
Loss in uplink =	− 200 db
Gain of spacecraft receive antenna =	+ 30 db
Gain in spacecraft amplifier =	+ 80 db
Gain of spacecraft transmit antenna =	+ 30 db
Loss in downlink =	− 196 db
Gain of receiving antenna =	+ 50 db
Received power =	− 86 dbm

The received power is approximately 2.5×10^{-9} milliwatts or 2.5×10^{-12} watts.

REVIEW QUESTIONS

22.1 What is an *ideal transmission line?*

22.2 What are the primary constants of a lossy transmission line?

22.3 What are the secondary constants of a lossy transmission line?

22.4 What conditions must be met for reflection-free operation of a lossy transmission line?

22.5 Explain skin effect on a current carrying conductor.

22.6 Define *near-end* and *far-end crosstalk* between parallel twisted pair cables.

22.7 Name three items that impair transmission over twisted pair telephone plant.

22.8 What is an *optical fiber?*

22.9 List the advantages optical fibers have over copper wires.

22.10 List the advantages copper wires have over optical fibers.

22.11 What is *wavelength division multiplexing?* Why is it used on optical fibers?

22.12 Define *frequency reuse.*

22.13 In a cellular radio system, how does cluster size influence system capacity?

22.14 Describe the influence of interferors on the capacity of a cellular radio system.

22.15 Describe the effect of the number of simultaneous users on the capacity of a cellular radio system.

22.16 What factors influence cell size in a cellular radio system?

22.17 Define *geostationary orbit, polar orbit, inclined earth orbit, medium earth orbit,* and *low earth orbit.*

List of Acronyms

θM: angle modulation
ΦM: phase modulation
2B1Q: two binary, one quaternary (signal format)

ACF: autocorrelation function
ACK: acknowlege
ADPCM: adaptive differential pulse code modulation
AM: amplitude modulation
AMI: alternate mark inversion (signal format)
ANSI-T1:   American National Standards Institute, Committee T1
ARQ: automatic repeat request
ASCII: American Standard Code for Information Exchange
AWG: American Wire Gauge

BASK: binary amplitude-shift keying
BCC: block check character
BCC: blocked call cleared
BCD: blocked call delayed
BCH: Bose, Chaudhuri, and Hocquenghem (code)
BFSK: binary frequency-shift keying
BPSK: binary phase-shift keying

CCF: cross-correlation function
CDF: cumulative (probability) distribution function
CDM: code-division multiplexing
CDMA: code division, multiple access
CPFSK: continuous phase frequency-shift keying
CRC: cyclic redundancy check
CSMA: carrier sense, multiple access
CSMA/CD: carrier sense, multiple access with collision detection

DPCM: differential pulse code modulation
DPSK: differential (binary) phase-shift keying
DS: direct sequence (spread spectrum)

DSBSC: double-sideband suppressed-carrier (AM)
DSP: digital signal processor

EBCDIC: Extended Binary Coded Decimal Interchange Code
ESD: energy spectral density

FCS: frame check sequence
FDM: frequency-division multiplexing
FDMA: frequency-division, multiple access
FEC: forward error correction
FEXT: far-end crosstalk
FH: frequency hopping (spread spectrum)
FIFO: first in, first out
FM: frequency modulation
FSK: frequency-shift keying

GbN: go-back-n (ARQ)
GEO: geostationary orbit
GPS: global positioning system

ISI: intersymbol interference
ITU–T: International Telecommunications Union, Telecommunication Sector

LAN: local-area network
LCAM: large-carrier amplitude modulation
LCC: lost call cleared
LCD: lost call delayed
LEO: low earth orbit
LIFO: last in, first out
LRC: longitudinal redundancy checking
LSB: lower-sideband (SSB-AM)
LTI: linear, time-invariant (system)

M/D/1:   a queueing system with a Poisson arrival process, a constant service time distribution, and a single server
MEO: medium earth orbit
M/E$_k$/1:   a queueing system with a Poisson arrival process, an Erlang-k service time distribution, and a single server
M/M/1:   a queueing system with a Poisson arrival process, an exponential service time distribution, and a single server
MSK: minimum-shift keying
MTSO: mobile telephone switching office

NAK: negative acknowledgment
NB ΦM: narrow-band phase modulation
NBFM: narrow-band frequency modulation
NEXT: near-end crosstalk
NRZ: non-return to zero

OOK: on-off keying
OQPSK: offset-QPSK

PCM: pulse code modulation
PCS: personal communication system
PDF: probability density function
pn: pseudo-noise (sequence)
PRK: phase-reversal keying
PSD: power spectral density

QAM: quadrature-carrier amplitude modulation
QFSK: quadrature FSK
QPSK: quadrature PSK, quaternary PSK, quadriphase PSK

ROM: read-only memory
RHS: right-hand side
RS: Reed-Solomon (code)
RZ: return to zero

S&W: stop and wait (ARQ)
SelR: selective repeat (ARQ)
SNR: signal-to-noise (power) ratio
SSB: single-sideband (AM)
SSM: spread spectrum modulation

TDM: time-division multiplexing
TDMA: time-division, multiple access

USB: upper-sideband (SSB-AM)

VRC: vertical redundancy checking
VSAT: very small aperture terminal
VSB: vestigial-sideband (AM)

List of Symbols

SPECIAL SYMBOLS

$\langle\ \rangle$: time-averaging

$\Leftrightarrow$: Fourier transform pair

$\leftrightarrow$: Laplace transform pair

$\mathbf{F}[\ ]$: Fourier transform operator

$\mathbf{F}^{-1}[\ ]$: inverse Fourier transform operator

$\mathbf{L}[\ ]$: Laplace transform operator

$f[\ ]$: rule that relates system input to system output

$\oplus$: modulo-2 addition

$\otimes$: convolution operator

$|\ |^{2_i}$: impulse magnitude squaring operator

$\varnothing$: null set

$Q \subseteq R$: all elements of Q are contained in R

$Q \cup R$: union of Q and R; set of all elements contained in either Q or R, or both

$Q \cap R$: intersection of Q and R; set whose elements are common to both Q and R

ENGLISH ALPHABET

A: constant

A: mean value (in Erlangs) of offered traffic

A: signal amplitude

A_k: binary valued function

A_m: amplitude of message function

b_{ch}: radio bandwidth required for each two-way communication channel

392

$b(k{:}n,p)$: binomial probability density distribution; probability that n Bernoulli trials with $P(S)=p$ and $P(F)=q=(1-p)$ result in k successes

B_{CL}: radio bandwidth assigned to a cluster of cells

B_T: transmission bandwidth

c: velocity of light in free space

c_{FR}: frequency reuse factor

$c(t)$: carrier signal

C: capacitance (in Farads)

C: channel capacity (in information bits/s)

C: number of service channels in trunking system

C, C', C'' etc.: members of a cyclic code

C_k^n: binomial coefficient—the number of ways in which k thing can be taken from n things, $= n!/k!(n-k)!$

$d_{\min}$: minimum logical distance (Hamming distance) among codewords

D: average delay for all calls held in system operating under BCD

D: deviation ratio of frequency-modulated signal

D_i: distance between primary (mobile) receiver and ith interferor

E_b: signal energy per bit

E_n: system state (Markov process)

$E[X]$: signal energy in time T

$E[X]$: the mean or expected value of the random variable $X(s_i)$

$E[X]$: total signal energy

f: frequency

f: rate at which satellite orbits earth (in revolutions/s)

f_1: fundamental frequency

f_c: carrier frequency

f_{co}: cut-off frequency of a low-pass filter

f_d: frequency offset

$f_i(t)$: instantaneous frequency

f_n: nth harmonic frequency

f_s: sampling rate in samples/s, $= 1/T_s$

f_Δ: frequency deviation

$f_X(x)$: probability density distribution function, the probability of occurrence of event x

$f_{X,Y}(x,y)$: joint probability density function of X and Y

$f_{Y|X}(y|x)$: conditional probability density function of Y given X

F_n: frame check sequence number

$F_X(x)$: cumulative probability distribution function, the probability of the event $(X \leq x)$

g: acceleration due to gravity

g_n: binary coefficients with the value 0 or 1

G: conductance (in Siemens)

G: gravitational constant

G_{n+1}: $n+1$-bit binary prime number

G_p: process gain

$G(p)$: generator polynomial

$h(t)$: impulse response of (LTI) system, $h(t) = f[\delta(t)]$

H: holding time, average duration of a call

$H(f)$: transfer function of (LTI) system

$H(X)$: source entropy, the average value (expected value) of information per symbol (in bits) emitted by a source

$H(Y|X)$: equivocation, measure of the average uncertainty of received sample Y given that sample X was transmitted

i: current (in Amps)

i_{FEXT}: far-end crosstalk current

i_{NEXT}: near-end crosstalk current

I: current (in Amps)

I_j: self information of symbol x_j

j: imaginary operator, $\sqrt{-1}$

$J_n(\beta)$: nth order Bessel function of the first kind with argument β

k: number of information bits (user's data bits) in a frame

k: number of information bits in block codeword

k: number of successes in n Bernouilli trials

k_{CE}: number of channels available in each radio cell

k_f: frequency sensitivity of modulator (in Hertz per volt)

L: expected line length (in calling units) in queueing system

L: inductance (in Henries)

L: length of bearer

L: number of quantization levels

L_q: expected queue length (in calling units)

m: any integer (positive, negative, or zero)

m: mass (of satellite)

m: number of frames sent before pause for response

$m(t)$: message signal

$\hat{m}(t)$: Hilbert transform of $m(t)$

$m_0(t)$: zero-mean message signal

$m_1(t)$: zero-mean, normalized message signal

M: mass (of earth)

M: mutual inductance

M: number of stations in system

M: number of times cluster is duplicated to form a cellular system

M: number of unique states

M_j: jamming margin

M_k: k-bit message

$M[\ ]$: set (of elements)

$M(f)$: frequency spectrum (Fourier transform) of $m(t)$

$M_X(t)$: moment generating function of random variable X

n: number of co-channel interfering cells

n: number of trials

n: order of Bessel function of first kind

n: number of bits in a frame

n: number of users that share a communication facility

n: total number of bits in codeword, $n = k + r$

$n(t)$: narrowband noise signal

$n(t)$: band-limited white noise signal

$n(t)$: sample function (of a wide-sense stationary random process)

$n_i(t)$: in-phase component of narrow-band signal

$n_q(t)$: quadrature component of narrow-band signal

N: noise power

N: number of bits in codeword

N: number of cells in a cluster (in cellular radio)

N: number of intervals examined in Poisson process

N: number of terminals

N_D: decoding noise power

N_Q: quantization noise power

N_T: total noise power

$N(t)$:   wide-sense stationary random process

p: propagation constant

p: average probability of interferors being active

p: probability of success (in Bernouilli trials)

p_0:   probability of sending a 0

p_1:   probability of sending a 1

$p(t)$: pn-sequence signal

$p_0(t)$:   normalized rectangular pulse signal representing 0

$p_1(t)$:   normalized rectangular pulse signal representing 1

$p(k;\lambda)$:   Poisson probability density function, $= \left[\lambda^k/k!\right]\exp(-\lambda)$

$p(k;\lambda\tau)$:   Poisson process, probability of exactly k events occurring in the interval $(t, t+\tau)$ where $\lambda\tau$ is the average of the number of events observed in all intervals, $= \left[(\lambda\tau)^k/k!\right]\exp(-\lambda\tau)$

$p(m)$:   probability of m interferors being active simultaneously

$p(\tau)$:   interarrival time (Poisson process)

P_{busy}:   probability of shared medium being busy

P_D: probability of detecting errored words

P_ε: probability of error

P_{idle}:   probability of shared medium being idle

P_I: power received by mobile from worst co-channel interferor

P_{ND}:   probability of not detecting errored words

P_{no}:   probability of no collision

P_0:   power received by mobile from cell site transmitter

P_s: average signal power

P_{yes}:   probability of collision

$P(A)$:   probability of event A

$P(A \cap B)$:   joint probability of events A and B

$P(A \cup B)$:   joint probability of event A or B

$P(A|B)$:   the probability of event A occurring when event B has occurred

$P(F)$:   probability of failure (in Bernouilli trials)

$P(R_0)$:   the probability of receiving 0

$P(R_1)$:   the probability of receiving 1

$P(S)$:   probability of success (in Bernouilli trials)

$P(S_0)$:   the probability of sending 0

$P(S_1)$:   the probability of sending 1

$\overline{P}[T]$:   power averaged over time T

$\overline{P}[T_1]$:   power averaged over one period of a periodic function

$\overline{P}[\infty]$:   power averaged over all time

q: probability of failure (in Bernouilli trials)

q: step size (in linear quantizer)

Q: co-channel reuse ratio

r: bit rate

r: number of redundant (parity) bits in block codeword

r: symbol rate

R: radius of circle just enclosing hexagonal cell

R: radius of circular orbit about center of earth

R: resistance (in Ohms)

R: source information rate

R_E: radius of earth

$s(t)$:   signal

$\hat{s}(t)$:   Hilbert transform of $s(t)$

$\tilde{s}(t)$:   complex envelope of $s(t)$

$s_\delta(t)$:   sampled wave in time domain, $= s(t)\delta_{T_s}(t)$

$s_g(t)$:   normalized shaping pulse

$s_g(t)$:   generating function, specifically one period of periodic function,
$= s_p(t), \ -T_1/2 \le t \le T_1/2, \ = 0, \ \text{elsewhere}$

s_i: sample point, outcome of ith trial

$s_p(t)$:   periodic signal

$s_t(t)$:   transient signal, $s_t(t) = s(t), \ t_1 \le t \le t_2, \ = 0$, elsewhere

$\text{sgn}(t)$:   signum (sign) function, $= 1, \ t > 0, \ = -1, \ t < 0$

S: set of all elements, sample space, universal set

S: throughput, effective rate at which information (user's data) bits are delivered

S':   normalized throughput, ratio of the time to deliver user's data bits to the time to deliver all bits

$S(f)$: frequency spectrum (Fourier transform) of $s(t)$

$\hat{S}(f)$: frequency spectrum (Fourier transform) of $\hat{s}(t)$

$S_\delta(f)$: sampled wave in frequency domain, $= S(f) \otimes \delta_{f_s}(f)$

S_n: complex Fourier coefficient

S_n^*: conjugate complex Fourier coefficient

S_n: normally distributed random variable

SNR_C: channel signal-to-noise power ratio

SNR_O: signal-to-noise power ratio at output of receiver

t: time

t: real number

Δt: average time to transfer use of a channel from one station to another

t_Δ: total delay time

$t_{A/N}$: time to transmit ACK or NAK

t_c: average cycle time, average time to serve all stations

t_F: time to send a frame

t_N: time to send a NAK

t_P: time required to send a packet

$\overline{t_s}$: average time interval until sending is permitted

t_w: pulse width (duration)

t_W: time between arrival of a packet for transmission and its transmission

T: time

T: total number of events observed in Poisson process

T: width (duration) of timeslot

T_1: time period

T_B: dwell time

T_b: bit duration

T_s: period of sampling function

$U(t)$: unit step function, $= 0$, $t < 0$, $= 1$, $t > 0$

v: voltage (in Volts)

V: voltage (in Volts)

$\text{Var}(X)$: variance of random variable $X(s)$

$w(t)$: white noise signal

W: expected waiting time in queueing system

W: message signal bandwidth

$W_{DS}(t)$:   direct-sequence spread spectrum signal

$W_{FH}(t)$:   frequency-hopping spread spectrum signal

W_q: expected waiting time in queue

x: number of calls in progress simultaneously

x_0:   distance from mobile to primary transmitter

x_I: distance from mobile to interfering transmitter

$x(t)$:   system (LTI) input signal

$\mathrm{x}(t, s)$:   sample function, a component of a random process

$\overline{X}$:   mean or expected value of the random variable $X(s)$

X^*:   standardized random variable, $X^* = (X - \mu)/\sigma$

$X(s_i)$:   random variable that assigns a real number to every sample point s_i

$X(t, s)$:   random process, a family of time-dependent, sample functions

$y(t)$:   observed waveform

$y(t)$:   system (LTI) output signal

$z(t)$:   analytic signal

$z(x)$:   compressor output signal

Z: impedance (in ohms)

Z_0:   characteristic impedance (in ohms)

Z_n: impedance looking at terminals a_n, a'_n

$Z(f)$:   frequency spectrum (one-sided Fourier transform) of $z(t)$

GREEK ALPHABET

α: arbitrary instant in time

α: attenuation constant (of a transmission line)

β: argument of Bessel function of first kind

β: modulation index—ratio of the frequency deviation and the modulation frequency

β: phase angle of signal on transmission line

β: phase delay constant (of a transmission line)

$\beta_{\Delta x}$:   incremental phase shift in signal along transmission line

δ_{nk}:   Kronecker's delta function, $= 1$, $n = k$, $= 0$, $n \neq k$

$\delta(t)$:   Dirac's improper function or unit impulse function,

$$\delta(t)=0,\ t\neq 0\ ;\ \int_{-\infty}^{\infty}\delta(t)dt=1$$

$\delta_{f_1}(f)$:   Dirac comb, $\delta_{f_1}(f)=\sum_k\delta(f-kf_1)$

$\delta_{T_1}(t)$:   ideal sampling function, $\delta_{T_1}(t)=\sum_k\delta(t-kT_1)$

Δx: small length

ε: number of errors in received codeword

ε_k: quantization error

η: efficiency

η_{ARQ}:   efficiency of ARQ technique

η_{B_T}:   bandwidth efficiency

η_P: power efficiency

$\theta(t)$:   instantaneous angle produced by the modulation

ι_R: ratio of the reflected (i.e., directed to the source) current wave to the forward (i.e., directed to the load) current wave (on a transmission line)

λ: average rate at which units arrive at queue

λ: average traffic arriving at a facility in bytes/s

λ: mean value of Poisson distribution

$\Lambda(x)$:   unit triangle function, $=1-|x|$, when $|x|<1$, $=0$, when $|x|>1$

μ: average rate at which units are serviced and leave queueing system

μ: capacity of a communication facility in bytes/s

μ: decision threshold

μ: mean value of random variable

μ: modulation index

μ: positive real number (Erlang-distributed random variable)

μ: processing time

μ_r: the rth moment of a random variable X about the mean, $\mu_r=\text{E}\!\left[(X-\mu)^r\right]$

$\mu(t)$:   a zero-mean Gaussian process

ν: traffic intensity ratio

π: universal constant, 3.14159

Π_{ch}:   channel filter characteristic

Π_{eq}:   equalizer characteristic

Π_{rx}:   receive filter characteristic

Π_{tx}:   transmit filter characteristic

$\Pi(t)$:   unit rectangle function, $= 1, |t| < 1/2, \ = 0, |t| > 1/2$

ρ: rolloff factor

ρ: input traffic density, offered traffic as a fraction of full-load traffic

σ: standard deviation of random variable, positive square root of $\mathrm{Var}(X)$

τ: time difference, scanning or searching parameter

τ: propagation time

ϑ: velocity of signal (in unit of length per unit of time)

ϑ_R: ratio of the reflected voltage wave to the forward voltage wave (on a transmission line)

ϕ: phase angle

ϕ_Δ: phase deviation, or phase-modulation index

$\phi(t)$:   time-varying phase angle

$\phi(\tau)$:   power-type autocorrelation function of $s(t)$

$\phi_{ij}(\tau)$:   power-type cross-correlation function of two deterministic power-type functions $s_i(t)$ and $s_j(t)$

$\phi_\mu(\tau)$:   autocorrelation function of $\mu(t)$

$\phi_N(\tau)$:   autocorrelation function of sample function of white noise process

$\phi_p(\tau)$:   periodic component of auto-correlation function

$\phi_R(\tau)$:   random component of auto-correlation function

$\phi_X(t)$:   characteristic function of random variable X

$\Phi(f)$:   power spectral density of $s(t)$

$\Phi_{\Delta f}(f)$:   estimate of power spectral density of wideband angle-modulated signal

$\Phi_n(f)$:   power spectral density of noise signal

$\Phi_p(f)$:   periodic component of power spectral density function

$\Phi_R(f)$:   random component of power spectral density function

$\psi(\tau)$:   energy-type autocorrelation function of the deterministic energy-type function $s(t)$

$\psi_g(\tau)$:   energy-type autocorrelation function of generating function $s_g(t)$

$\psi_{ij}(\tau)$:   energy-type cross-correlation function of two deterministic energy-type functions $s_i(t)$ and $s_j(t)$

$\Psi(f)$:   energy spectral density of $s(t)$

ω: angular velocity, $= 2\pi f$

Appendix

Table A.1   Properties of the Fourier Transform

Property #1 Linearity

If $s_1(t) \Leftrightarrow S_1(f)$. and $s_2(t) \Leftrightarrow S_2(f)$, for all values of A and B

$As_1(t) + Bs_2(t) \Leftrightarrow AS_1(f) + BS_2(f)$

Property #2 Duality

If $s(t) \Leftrightarrow S(f)$, then $S(t) \Leftrightarrow s(-f)$

Property #3 Conjugate Function

If $s(t) \Leftrightarrow S(f)$, then $s^*(t) \Leftrightarrow S^*(-f)$

Property #4 Time Scaling

If $s(t) \Leftrightarrow S(f)$, then $s(at) \Leftrightarrow \frac{1}{|a|} S\left(\frac{f}{a}\right)$. Compressing $s(t)$ by a factor a, where $|a| > 1$,

expands and diminishes $S(f)$ by a. Likewise, expanding $s(t)$ by a factor a, where $|a| < 1$,

contracts and increases $S(f)$ by a.

Property #5 Time Shifting

If $s(t) \Leftrightarrow S(f)$, then $s(t - t_0) \Leftrightarrow S(f) \exp(-j2\pi f t_0)$. If $s(t)$ is shifted in the positive direction

by an amount t_0, the phase of its Fourier transform $S(f)$ is changed by an amount $-2\pi f t_0$.

Property #6 Frequency Shifting

If $s(t) \Leftrightarrow S(f)$, then $s(t)\exp(j2\pi f_0 t) \Leftrightarrow S(f - f_0)$. Multiplication of $s(t)$ by an exponential

(sinusoidal) function of frequency f_0 is equivalent to shifting its Fourier transform in the

positive direction by an amount f_0. This action is known as *modulation*.

Property #7 Area Under $s(t)$

If $s(t) \Leftrightarrow S(f)$, then $\int\limits_{-\infty}^{\infty} s(t)\, dt \Leftrightarrow S(0)$. The area under $s(t)$ is equal to its Fourier

transform when $f = 0$.

Property #8 Area Under $S(f)$

If $s(t) \Leftrightarrow S(f)$, then $s(0) \Leftrightarrow \int\limits_{-\infty}^{\infty} S(f)\,df$. The area under its Fourier transform is equal to $s(t)$

when $t=0$.

Property #9 Differentiation in the Time Domain
If $s(t) \Leftrightarrow S(f)$, then $ds(t)/dt \Leftrightarrow j2\pi f S(f)$. Differentiating $s(t)$ multiplies its Fourier transform by $j2\pi f$.

Property #10 Integration in the Time Domain
If $s(t) \Leftrightarrow S(f)$, then $\int\limits_{-\infty}^{t} s(\tau)\,d\tau \Leftrightarrow \dfrac{1}{j2\pi f} S(f) + \dfrac{S(0)}{2}\,\delta(f)$. If $S(0)$ is zero, integrating $s(t)$

divides its Fourier transform by $j2\pi f$. If $S(0)$ is not zero, integrating $s(t)$ divides its
Fourier transform by $j2\pi f$ and adds an impulse function of area equal to one-half the value
of $S(0)$.

Property #11 Multiplication in the Time Domain
If $s_1(t) \Leftrightarrow S_1(f)$, and $s_2(t) \Leftrightarrow S_2(f)$, then

$$s_1(t)s_2(t) \Leftrightarrow \int\limits_{-\infty}^{\infty} S_1(\lambda)S_2(f-\lambda)\,d\lambda = S_1(f) \otimes S_2(f)$$

Multiplication of two signals in the time domain results in the convolution of their individual
Fourier transforms in the frequency domain.

Property #12 Convolution in the Time Domain
If $s_1(t) \Leftrightarrow S_1(f)$, and $s_2(t) \Leftrightarrow S_2(f)$, then

$$\int\limits_{-\infty}^{\infty} s_1(\tau)s_2(t-\tau)\,d\tau = s_1(t) \otimes s_2(t) \Leftrightarrow S_1(f)S_2(f)$$

Convolution of two signals in the time domain results in the multiplication of their individual
Fourier transforms in the frequency domain.

Note on Convolution
The convolution $[\,h(x)\,]$ of two functions $f(x)$ and $g(x)$ is defined by

$$h(x) = \int\limits_{-\infty}^{\infty} f(u)g(x-u)\,du = \int\limits_{-\infty}^{\infty} f(x-u)g(x)\,du$$

or, in shorthand, $h(x) = f(x) \otimes g(x)$.
Convolution is:
- commutative $\left[f \otimes g = g \otimes f \right]$
- associative $\left[f \otimes (g \otimes h) = (f \otimes g) \otimes h \right]$
- distributive $\left[f \otimes (g+h) = (f \otimes g) + (f \otimes h) \right]$

Table A.2   Some Common Fourier Transform Pairs

Time Function	Fourier Transform
Rectangle $\Pi\!\left(\frac{t}{T}\right)$	$T\mathrm{Sinc}(fT)$
Triangle $\Lambda\!\left(\frac{t}{T}\right)$	$T\mathrm{Sinc}^2(fT)$
Sinc $\;\mathrm{Sinc}(2Wt)$	$\frac{1}{2W}\Pi\!\left(\frac{f}{2W}\right)$
$\mathrm{Sinc}^2(2Wt)$	$\frac{1}{2W}\Lambda\!\left(\frac{f}{2W}\right)$
Exponential $\exp(-\alpha t)U(t),\ \ \alpha>0$	$\frac{1}{\alpha+j2\pi f}$
$\exp(-\alpha\lvert t\rvert),\ \ \alpha>0$	$\frac{2\alpha}{\alpha^2+(2\pi f)^2}$
$\exp\!\left(-\pi t^2\right)$	$\exp\!\left(-\pi f^2\right)$
$\exp\!\left(-j2\pi f_0 t\right)$	$\delta(f-f_0)$
Impulse $\delta(t)$	1
1	$\delta(f)$
$\delta(t-t_0)$	$\exp(-j2\pi f t_0)$
$\delta_{T_0}(t)$	$f_0\delta_{f_0}(f)$
Sinusoidal $\cos(2\pi f_0 t)$	$\frac{1}{2}\left[\delta(f+f_0)+\delta(f-f_0)\right]$
$\sin(2\pi f_0 t)$	$\frac{1}{2j}\left[\delta(f-f_0)-\delta(f+f_0)\right]$ or
	$\frac{j}{2}\left[\delta(f+f_0)-\delta(f-f_0)\right]$
Signum $\mathrm{sgn}(t)$	$1/j\pi f$
$1/\pi t$	$-j\,\mathrm{sgn}(f)$
Step $U(t)$	$\frac{1}{2}\delta(f)+\frac{1}{j2\pi f}$
Sampling $\displaystyle\sum_{k=-\infty}^{k=\infty}\delta(t-kT_0)$	$\displaystyle\frac{1}{T_0}\sum_{n=-\infty}^{n=\infty}\delta(f-nf_0)$

Table A.3   Some Common Integrals

Indefinite Integrals

$$\int x^n\, dx = x^{n+1}/(n+1)$$

$$\int \tfrac{1}{x}\, dx = \log_e|x|$$

$$\int \exp(ax)\, dx = \tfrac{1}{a}\exp(ax)$$

$$\int \cos(ax)\, dx = \tfrac{1}{a}\sin(ax)$$

$$\int \sin(ax)\, dx = -\tfrac{1}{a}\cos(ax)$$

$$\int x\exp(ax)\, dx = \tfrac{1}{a^2}(ax-1)\exp(ax)$$

$$\int x\exp\!\left(ax^2\right) dx = \tfrac{1}{2a}\exp\!\left(ax^2\right)$$

Definite Integrals

$$\int_0^\infty \operatorname{Sinc}(x)\, dx = \tfrac{1}{2}$$

$$\int_0^\infty \operatorname{Sinc}^2(x)\, dx = \tfrac{1}{2}$$

$$\int_0^\infty \exp(-ax)\, dx = 1/a$$

$$\int_0^\infty x\exp\!\left(-x^2\right) dx = \tfrac{1}{2}$$

$$\int_0^\infty \exp\!\left(-ax^2\right) dx = \tfrac{1}{2}\sqrt{\pi/a}\,,\ a>0$$

$$\int_0^\infty x^2\exp\!\left(-ax^2\right) dx = \tfrac{1}{4a}\sqrt{\tfrac{\pi}{a}}\,,\ a>0$$

Table A.4   Power Ratios Expressed in Decibels

DECIBELS	POWER RATIO	
	Gain	Loss
0.0	1.000	1.0000
0.1	1.023	0.9772
0.2	1.047	0.9550
0.3	1.072	0.9333
0.4	1.097	0.9120
0.5	1.112	0.8913
0.6	1.148	0.8710
0.7	1.175	0.8511
0.8	1.202	0.8318
0.9	1.230	0.8128
1.0	1.259	0.7943
2.0	1.585	0.6310
3.0	1.995	0.5012
4.0	2.512	0.3981
5.0	3.162	0.3162
6.0	3.981	0.2512
7.0	5.012	0.1995
8.0	6.310	0.1585
9.0	7.943	0.1259
10	10	0.1000
20	100	0.0100
30	1×10^3	1×10^{-3}
40	1×10^4	1×10^{-4}
50	1×10^5	1×10^{-5}
60	1×106	1×10^{-6}

Table A.5   Some ASCII and EBCDIC Characters

Alphas	ASCII	EBCDIC	Alphas	ASCII	EBCDIC
a	1100001	10000001	A	1000001	11000001
b	1100010	10000010	B	1000010	11000010
c	1100011	10000011	C	1000011	11000011
d	1100100	10000100	D	1000100	11000100
e	1100101	10000101	E	1000101	11000101
f	1100110	10000110	F	1000110	11000110
g	1100111	10000111	G	1000111	11000111
h	1101000	10001000	H	1001000	11001000
i	1101001	10001001	I	1001001	11001001
j	1101010	10001010	J	1001010	11001010
k	1101011	10001011	K	1001011	11001011
l	1101100	10001100	L	1001100	11001100
m	1101101	10001101	M	1001101	11001101
n	1101110	10001110	N	1001110	11001110
o	1101111	10001111	O	1001111	11001111
p	1110000	10010000	P	1010000	11010000
q	1110001	10010001	Q	1010001	11010001
r	1110010	10010010	R	1010010	11010010
s	1110011	10010011	S	1010011	11010011
t	1110100	10010100	T	1010100	11010100
u	1110101	10010101	U	1010101	11010101
v	1110110	10010110	V	1010110	11010110
w	1110111	10010111	W	1010111	11010111
x	1111000	10011000	X	1011000	11011000
y	1111001	10011001	Y	1011001	11011001
z	1111010	10011010	Z	1011010	11011010

Control	ASCII	EBCDIC	Numerics	ASCII	EBCDIC
SYN	0010110	00110010	0	0110000	11110000
SOH	0000001	00000001	1	0110001	11110001
STX	0000010	00000010	2	0110010	11110010
ETX	0000011	00000011	3	0110011	11110011
EOT	0000100	00110111	4	0110100	11110100
ENQ	0000101	00101101	5	0110101	11110101
ACK	0000110	00101110	6	0110110	11110110
NAK	0010101	00111101	7	0110111	11110111
DLE	0010000	00010000	8	0111000	11111000
ETB	0010111	00100110	9	0111001	11111001

The least significant (i.e., the leading) bit is on the right of codeword. Characters representing punctuation marks and special characters are not listed.

Table A.6   Trigonometric Functions

$$\cos(\theta) = \tfrac{1}{2}\big[\exp(j\theta) + \exp(-j\theta)\big]$$

$$\sin(\theta) = \tfrac{1}{2j}\big[\exp(j\theta) - \exp(-j\theta)\big]$$

$$\exp(\pm j\theta) = \cos(\theta) \pm j\sin(\theta)$$

$$\cos^2(\theta) = \tfrac{1}{2}\big[1 + \cos(2\theta)\big]$$

$$\sin^2(\theta) = \tfrac{1}{2}\big[1 - \cos(2\theta)\big]$$

$$\cos^2(\theta) + \sin^2(\theta) = 1$$

$$\cos(A \pm B) = \cos(A)\cos(B) \mp \sin(A)\sin(B)$$

$$\sin(A \pm B) = \sin(A)\cos(B) \pm \cos(A)\sin(B)$$

$$\cos(A)\cos(B) = \tfrac{1}{2}\big[\cos(A-B) + \cos(A+B)\big]$$

$$\sin(A)\sin(B) = \tfrac{1}{2}\big[\cos(A-B) - \cos(A+B)\big]$$

$$\sin(A)\cos(B) = \tfrac{1}{2}\big[\sin(A-B) + \sin(A+B)\big]$$

Table A.7   Some Useful Constants and Other Information

$$e = 2.71828$$

$$\pi = 3.14159$$

Logarithms

$$\log_e(x) = 2.3026\log_{10}(x)$$

$$\log_2(x) = 3.3219\log_{10}(x)$$

$$\log_b(x) = \log_c(x)\log_b(c)$$

Geometric Series

$$1 + a + a^2 + ... + a^n + ... = 1/(1-a),\ |a| < 1,\ n \to \infty$$

Exponential Series

$$\exp(x) = 1 + x + x^2/2! + ... + x^n/n! + ...,\ n \to \infty$$

Taylor's Series

$$f(x) = f(a) + \frac{f'(a)}{1!}(x-a) + \frac{f''(a)}{2!}(x-a)^2 + ... + \frac{f^{(n)}(a)}{n!} + ...$$

$$\text{where } f^{(n)}(a) = \frac{d^n f(x)}{dx^n}\bigg|_{x=a}$$

Binomial Series

$$(1+x)^n = 1 + nx + \frac{n(n-1)}{2!}x^2 + ...,\ |nx| < 1$$

Index

JOHN LUDLOW

The Autobiography of a

Christian Socialist

Edited and Introduced by

A. D. MURRAY

FRANK CASS

First published in 1981 in Great Britain by
FRANK CASS AND COMPANY LIMITED
Gainsborough House, Gainsborough Road,
London, E11 1RS, England

and in the United States of America by
FRANK CASS AND COMPANY LIMITED
c/o Biblio Distribution Centre
81 Adams Drive, P.O. Box, 327, Totowa, N.J. 07511

British Library Cataloguing in Publication data

Ludlow, John
  John Ludlow
  1. Ludlow, John   2. Socialists – Great Britain
  – Biography
  335'.7'0924   HX243

ISBN 0 7146 3085 3

Photoset in Baskerville by Saildean Limited
Printed and Bound in Great Britain by
Robert Hartnoll Limited, Bodmin, Cornwall

To my parents

CONTENTS

EDITORIAL NOTE

The 1,000 page manuscript of the autobiography is among the Ludlow papers left to the University Library at Cambridge by Charles des Graz, great-nephew of Ludlow, in 1953, (catalogue no. Add. 7348.) Des Graz also left a typescript of much of the manuscript, though this is very inaccurate and has many omissions, and a most useful transcription of Ludlow's own bibliography of his books, pamphlets, lectures and contributions to journals. This bibliography has unfortunately, due to exigencies of space, had to be omitted from the present edition, but it can of course be consulted among the Ludlow Papers.

Ludlow's own footnotes are printed in the text within square brackets; the Editor's footnotes are gathered together at the end of the book. The latter, due to shortage of space, have been kept to an absolute minimum. On the few occasions where the omission of a passage has made an editorial insertion necessary, this has been enclosed in double round brackets, to avoid confusion with Ludlow's own brackets. Ludlow's punctuation was sometimes erratic, and there were also occasional minor mistakes in wording; for the sake of clarity these have been corrected, but the corrections have not been acknowledged, as in no case was the meaning of a passage changed by a correction.

Considerable alterations have been made to Ludlow's original chapter list, as some chapters have been run together and a few omitted entirely. But, as far as possible, Ludlow's chapter-headings have been preserved in the new arrangement.

ACKNOWLEDGMENTS

My initial debt is to Dr. A. R. Vidler, former Dean of King's College, Cambridge, and to the Revd. Simon Barrington-Ward, former Dean and Chaplain of Magdalene College, Cambridge. In their lectures and supervisions my interest in Christian Socialism and in its founder, John Ludlow, was first awakened. I should also like to thank Professors Emile Delavenay and Philippe Séjourné, who were my mentors in the preparation of a *maîtrise* dissertation for the University of Nice on the influence of French socialists on Ludlow and the Christian Socialists. I am also grateful to Mr. N. C. Masterman, author of a fascinating biography of Ludlow, for so willingly answering the questions I put to him.

The preparation of this edition of Ludlow's autobiography would not have been possible, given the fact that my own teaching, first in Nice and then in Oran, has kept me away from Cambridge for all but two months or so of the year, without the assistance, extending far beyond that normally due to a researcher, of Mr. A. E. B. Owen, Under-Librarian of the Cambridge University Library and in charge of the Manuscripts Department. The Ludlow papers are in the possession of the Library, and were catalogued by Mr. Owen, whose knowledge of the collection is unrivalled. My only regret is that, despite Mr. Owen's wishes, the exigencies of modern publishing conditions have forced me to edit the manuscript to less than two-thirds of its original length. Nevertheless, I feel sure that Ludlow himself, with his severely practical mind, would have carried out at least a fair amount of hatchet-work on his very lengthy manuscript, had he not renounced the idea of publication in his lifetime.

To the staffs of Cambridge University Library and the

British Museum Reading Room, and to Mr David Muspratt, Curator of Muniments and Assistant Librarian of the Working Men's College, I would like to express my thanks for their unfailing courtesy and kindness. To Mrs. Alison Ingram, typist and friend, who had to contend not only with a much-amended manuscript but also with an absent author, I owe a great deal. And finally I would like to acknowledge with gratitude the comments and encouragement of Pat Parrinder of the University of Reading, George Morgan, lecturer at the University of Nice, and, most constantly, my wife and colleague at the University of Oran, Anne.

EDITOR'S INTRODUCTION

John Ludlow's life (1821-1911) spanned the century in which the Labour movement in Britain grew from unorganised and sporadic beginnings to a mass movement with its own political party.

Brought up in Paris and educated in the heady atmosphere of the *Collège Bourbon* at a time when the students were enthusiastically discussing the ideas of the French social thinkers, John Ludlow completed his studies and was promised a brilliant future in France. But he chose to move to England to read for the Bar, and there began to apply the socialist ideas he had learned in Paris. As founder and central figure of the Christian Socialist Movement after 1848, he devoted immense energy to the organisation of the first Working Men's Associations and to the planning and co-ordination of the nascent Co-operative movement. As editor of the movement's journal, *The Christian Socialist*, and in a large number of tracts, pamphlets and articles in national periodicals, Ludlow developed a cogent programme for Socialism through a social revolution. As *éminence grise* behind the early campaigns of the Amalgamated Society of Engineers, precursor of the 'New Model' Trade Unions, Ludlow made a major contribution to the development of Trade Unionism.

But the fact that the Christian Socialist Movement ceased to work as an organised group after 1854-5 has meant that most historians of the development of socialism in Britain, from the Webbs onward, have either ignored Ludlow and the Christian Socialists, or treated them as interesting, but insignificant, paternalist Christian ancestors of the true socialist movement.

That this assessment is inadequate, if not erroneous, can be argued from two separate sets of evidence. Firstly the assessment fails to take account of the valuable inauguratory role, both theoretical and practical, which Ludlow and his friends played in awakening the working class to a consciousness of the possibility of social, as zell as political, action as a prelude to the reconstruction of society. In terms of practical results, Ludlow was here more successful than either the Owenites or the Chartists. Secondly, the assessment, by treating the Christian Socialist Movement as an isolated phenomenon, ignores the wide-ranging activities of the Christian Socialists as individuals, after the movement itself had broken up.

Here again, Ludlow is the key figure, as leading advocate for the Trade Union movement to the Government and various Commissions, as a leader of the growing national Co-operative movement and then of the Labour Association, as secretary and president of the Congresses of these two organisations, and, finally, in high Public office. On his appointment in 1875 as the first Chief Registrar of Friendly Societies, Ludlow became the Government official solely responsible for the registration and supervision of all working-men's organisations, of which there existed by 1888 some twenty five thousand, including Trade Unions, Co-operative Stores and Associations. In this position he was able to exercise considerable influence on the growing Labour movement by advising and encouraging the development of working class organisations of all kinds. He retired in 1891, with a well-deserved C.B., and continued, in the twenty years before his death in 1911, to serve the causes he had always supported, though now in the role of elder statesman rather than activist.

*   *   *   *   *

Good Victorian autobiographies are thin on the ground. John Ludlow's personal record of a life spent in the service of the working class movement and in the attempt to develop a specifically English brand of socialism in the context of a strong, humane Christian commitment is a welcome addition to their number. But what is at least as fascinating as the record of his public life are his candid recollections of the private reflections, reactions and emotions with which he responded to the events in which he was concerned. As with Mill, Carlyle, Dr. Arnold and so many of the Victorians, beneath the confidence of Ludlow's public pronouncements lay a deep personal insecurity. This was of a less cosmic quality than that of a J. A. Froude or Sterling; with a more pragmatic and earthbound disposition, Ludlow's doubts were linked to the major events of his life, particularly in connection with the 'seven spiritual crises of (his) life.' In describing these Ludlow eschews histrionics, and writes, with remarkable penetration, of his own motives and emotions, giving us a notable insight into the forces which beset a Victorian social conscience.

In his public life, his greatest commitment was to the Christian Socialist movement in which he collaborated intimately with F. D. Maurice; in his inner life, the greatest of his crises was that generated by the collapse of the movement and its apparent abandonment by Maurice. In chapter 20, Ludlow relives the whole episode and reveals the inner conflicts he suffered. He never attempts to hide the embarrassments or failures which he experienced, whether dealing with his personal charitable visiting in the 1840s, or his public involvement in the Royal Commission on Friendly Societies in the seventies. And he actually begins his autobiography by summing up his achievement as that of a good second-rater: he felt that he had failed to realise the high promise of his early years. The reader suspects that a major motive in the writing of his autobiography was the attempt to discover the reasons for this failure.

If he is at his most serious when examining his own character, Ludlow is at his liveliest and most mordant when dealing with his contemporaries. For Ludlow was equally willing to subject the behaviour of others to his scrutiny. To those who excited his whole-hearted admiration – Maurice, Kingsley, John Bright, Proudhon and others – Ludlow was generous in his appreciation of their qualities, yet honest in his discussion of their limitations. Where his admiration was less than uncritical, his comments could be devastating, as when he pricks the balloon of Disraeli's reputation for political courage, or uncovers the hypocrisy masquerading as timidity of Lord Shaftesbury.

But if Ludlow's shrewd penetration into character sets him apart from some of his contemporaries, he remains very much a Victorian. His nicely-observed vignettes of the social life of the period are often accompanied by a moralising comment. It is with just a suspicion of relish that he recounts the peccadilloes of some of his working-men co-operators: the occasional case of 'levanting' with the funds, or tragedy due to drink or women. Again, he is not wholly free from Victorian sentimentality, notably concerning his mother, but also in referring to strong friendships and affairs of the heart. In his account of his epic courtship of his cousin, Maria Forbes, the attitudes of the two 'lovers' are almost a caricature of Victorian concepts of duty. His moral earnestness, his unrelenting driving of himself to hard work, his occasional dark night of the soul (chapter 8 contains a revealing discussion of his urges to suicide) are all emotional experiences shared by many of his generation.

Yet, throughout his life, and even by his colleagues at the Chief Registrar's office, he was noted for his logical, disciplined, *French* cast of mind. He could never abide British compromise and muddle. His upbringing and education had been completely French, and although he had left France, covered with prizes, immediately on graduating *bachelier* from the *Collège Bourbon* he continued to write and think in French for several years. It was not until he was caught up in the events of 1848 that he first began to feel himself an Englishman.

Ludlow's chapter on 'Paris in the Forties' is, unfortunately, one of the several chapters which were either never written or have since been lost (it was included in Ludlow's handwritten chapter list at the beginning of the Autobiography.) Nevertheless, throughout the book from the lively account of his youth in France onwards, Ludlow refers to men, events and ideas in France. In Ludlow's socialism we find clear evidence of his debt to (among others) Fourier, Proudhon and Louis Blanc. We find also the assumption, common to almost every French socialist of the period, that social regeneration must go hand in hand with some form of religious regeneration.

The introduction of French socialist ideas and, what is more important, their application in an English context were among Ludlow's major contributions to English socialism. And not only Ludlow's ideas, but also his character, owed much to France. It is difficult to imagine any of his English collaborators forging the cogent and effective force of Christian Socialism from the somewhat vague notions of social regeneration held by Maurice, the Tory Young Englandism of Charles Kingsley, and the various reforming interests in health, education, co-operation and sanitation of their friends. It was Ludlow's ability to force others to see the logical conclusions and practical possibilities of their own ideas which made Christian Socialism a reality.

The Christian Socialist movement ceased to work as an organised group in 1854, and its concrete achievements were hardly momentous. The Working Men's College, founded in 1854, inspired many imitators and has been a permanent and effective institution; the Industrial and Provident Societies Act of 1852 had justly been called 'the co-operators' charter'; Working Men's Associations, though they have never threatened the capitalist economy, have continued to flower and fade at intervals over the last hundred years. If these are regarded, however, as the sole results of the movement, then its later neglect by historians is perhaps comprehensible. But the Christian Socialists also had a major influence, even before 1854, on two developments of critical importance: the organisation of a national Co-operative movement, and the formation of the 'New Model' Trade Unions.

In the two decades after 1855, when Christian Socialism collapsed as a movement, there was relative industrial peace and national prosperity; in this period the foundations were being laid for the beginning of the labour movement proper.

During this 'dormant' period for working-class activity, the Christian Socialists, though now working as individuals, were, with the Positivists, among the few continuing to fight for working class progress. Even without mentioning Ludlow's work, the list of their activities is impressive: Maurice and Kingsley influenced a generation of undergraduates from pulpit and professorial chair at Cambridge, Tom Hughes and Lord Goderich (now the Marquis of Ripon) were in Parliament promoting working-class interests and Trade Union legislation, Neale and Lloyd Jones were at the head of the Co-operative movement, while a group of 'slum priests' were agitating on their parishioners' behalf. The Working Men's College brought many Londoners, teachers, including John Ruskin, William Morris and Sidney Webb, as well as students, into contact with the Christian Socialists. Later, in the East End, Canon Samuel Barnett's Toynbee Hall, another Christian Socialist foundation, attracted other important figures such as Octavia Hill, Bernard Shaw and Beatrice Webb.

But it was not their educational work alone which kept them in the public eye; Neale, Maurice, Kingsley, Hughes, Lloyd Jones and Furnivall all published books or pamphlets on social issues, while Ludlow remained the most effective propagandist for socialism. He continued to write, throughout these years, frequent articles for reviews and periodicals of all kinds, especially the *Spectator*. In 1867 he produced an influential book, *Progress of the Working Class 1832-67* (written with the help of Lloyd Jones). This was his most substantial work of propaganda for the working class movement. But although its content was both democratic and socialist, its tone was undoubtedly pitched to appeal to the middle classes and to overcome their fears and prejudices. It might be said, indeed, that one factor which prevented Ludlow's being remembered as a major theoretician of English socialism was his tendency to write with middle-class reactions constantly in mind. This was probably less true of the other group active on behalf of the working class, the Positivists, who have been paid the attention due to them, notably by Royden Harrison in *Before the Socialists* (1954). But ever since the Webbs relied on Frederic Harrison, the leading Positivist, for their account of this period in their *History of Trade Unionism* (1894), the Christian Socialists have been neglected by historians of socialism. But this is not the place for a revaluation of the Christian Socialists, nor indeed for the badly-needed re-assessment of Ludlow's socialism.

In any case, in his autobiography, Ludlow is more concerned with events than ideas. He does of course give an account of his opinions and their development at various important periods of his life, and is never slow to justify his actions when they excited controversy. But he does not attempt to give any detailed account of the principles of his socialism. He prefers to refer us to his writings. These, being voluminous and widely scattered, have to be sifted through with careful attention before the elements of his socialism become clear. What Ludlow gives us in the *Autobiography* is a candid and forthright record of the life, work and opinions, both public and private, of a Victorian socialist reformer.

A. D. MURRAY

John Ludlow

Reproduced by permission of the Chief Registrar of Friendly Societies

John Ruskin

J. M. Ludlow

Thos. Woolner

F. J. Furnivall

Frederick Denison Maurice

Tom Hughes

John Westlake

J. Llewelyn Davies

R. B. Litchfield

From "The Working Men's College – 1854-1904," Rev. J. Llewelyn Davies (ed.).

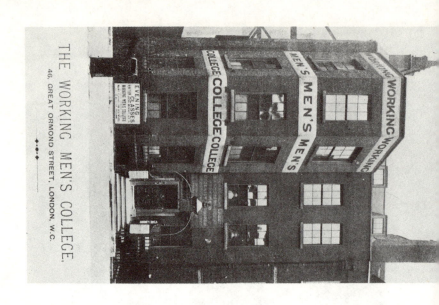

THE WORKING MEN'S COLLEGE.

46, GREAT ORMOND STREET, LONDON, W.C.

Frederick Denison Maurice
(1805-1872)
Principal Founder
of the Working Men's College.

xxii

Entrance Hall, Working Men's College, Great Ormond Street

Art Room, Working Men's College, Gt Ormond Street.

THE AUTOBIOGRAPHY

PRELIMINARY NOTES

I think I see, at last, what my calling and position have been in God's world, so far as my sphere of action is concerned. I have more than once, and in various directions, been for the time being virtually at the top, but never permanently. I seem now to understand that I have been one of God's odd men – the men who, in a great household, being neither steward nor cook, coachman nor groom, butler nor valet, yet are kept on because, in case of emergency, they are able to fill, pretty tolerably, any place that suddenly falls vacant, and are ready to do any odd jobs. I have never been permanently a first rate lawyer, a first rate writer, a first rate critic, a first rate economist, a first rate speaker, a first rate thinker, a first rate man of business, a first rate leader. But when the call was upon me, I have been able temporarily to fulfil all these various parts, as the odd man of the household, and it is only when my 79th year is more than half over that I have at last understood this. – 29th October 1899.

PREFACE

I have had a somewhat varied experience[1]. I kept for some months the ledger of a French colony. I practised for many years at the bar as a conveyancer, on my own account or as a devil, doing a somewhat large amount of Parliamentary drafting. I was secretary for four years to an important Royal Commission; I was for about 17 years at the head of one of the smaller government departments responsible both to the Treasury and the Home Office. I was afterwards for some years a member of a Committee of Inspection appointed by virtue of an Act of Parliament to look after a class of bodies of no inconsiderable social importance.

I have contributed to daily and weekly papers (including one Indian one), to monthly and quarterly periodicals (one American); I have edited weekly journals, both social and literary, and for a short time a monthly periodical. Brought up in France and being as familiar with French ,as with English I have contributed to two French (religious) journals. To say nothing of Blue-books, I have published some eleven volumes, on legal, historical, political, literary, religious and social subjects, besides one written in conjunction with a friend, and contributions to series of Tracts, Dictionaries, Transactions, etc. The work written in conjunction with a friend has been translated into German (two editions), as also has another paper by me.

I missed a fortune by declining to take Suez Canal shares when they were offered to me at a very low price by Ferdinand de Lesseps himself.

Though often preferring to act as a free-lance, I have been on the council of various institutions and bodies of a social, educational or philanthropic character, have taken part in some important social and political movements, and have been in contact of fellow-work with members of various religious bodies, churchmen and dissenters of several denominations (Quakers included), French Roman Catholics, Calvinists and Lutherans. Outside of England I have, or have had, friends in France, Germany, Denmark, the United States.

My range of travel has been a very narrow one according to present practice, but my foreign experience has included a stay of between 11 and 12 years in France, of seven months twice in the West Indies, a winter in Madeira, and travel in or at least visits to Italy, Belgium, Germany, Switzerland, Spain and Portugal. Besides French, I have known enough German, Italian and Spanish to read without difficulty works in those languages (I do not reckon Portuguese – have yet to find any book worth reading in it).

I have received a prize from the hands of Guizot. I have had cousin to Villemain among my examiners for a French degree. I have been in Paris during one revolution, and immediately after another. I was present at the last expiatory Mass solemnised at St. Denys for the soul of Louis Seize, the then Duc de Chartres (who took my boy's fancy prodigiously in his blue Hussar's uniform), being present with the King's sister, Mme. Adelaide, who, six months later was a fugitive, whilst the young Duke was heir to the throne. I have seen Louis Philippe beat time to the Marseillaise and Parisienne from his balcony, and Ledru-Rollin give his first audience as Minister of Justice. I have declined to dine in company with Mazzini, and to be introduced to the Comte de Paris. My friends have included clergymen, a Lutheran Minister, artists, politicians, literary men, commercial men, working men. I have been privileged to number among them one great man, – Maurice –, several eminent or distinguished men and women, Charles Kingsley, T. Hughes, W. E. Forster, Norman Macleod, C. B. Mansfield, Llewelyn Davies, Mrs. Gaskell, Arthur Stanley, Vansittart Neale, F. G. Penrose, Lloyd Jones, John Westlake, R. H. Hutton, Lowes Dickinson, Samuel Clark, the Lushington Twins. I have had the entrée of Tennyson's upstairs sanctum at Farringford. The first Lord Metcalfe, Lowell, Clough, Froude, and Mr.

Martin have sat at my table. Among my foreign friends have been or are V. A. Huber, Lujo Brentano, Adolf Held, Harald Westergaard, Martin Nadaud, George Thomson of Copenhagen. I have been in close contact with statesmen such as the present Lord Ripon, the first Lord Aberdare, the late Lord Shaftesbury, Henry Fawcett, W. H. Smith, G. I. Goschen, Lord Rosebery. I have dined in a small company with Brougham.

I have known, on terms often amounting to friendship, Archdeacon Hare, John Bright, A. J. Scott, Dean Plumptre, Lord Welby, John Taylor, Henry Kingsley, the first Lord Houghton, Harriett Martineau, Thomas Cooper, Huxley, Woolner, Richard Barwell, Charles Bennett, J. F. Maclennan, Louis Blanc, Chevalier Bunsen, Edgar Quinet and his wife, Amédée Thierry; to a less extent Maréchal Bertrand, Erskine of Trinlathen, Sir Charles Eastlake, Archbishop French, W. Lovett, Phillips Brooks. In various ways I have been a fellow worker with John Stuart Mill, Sir Fitzjames Stephen, Sir C. E. Pearson, Sir M. E. Grant Duff, Ruskin, Dante Rossetti, Herbert Spencer, Canon Scott Holland and Bishop Gore. I have spoken with Chopin. I have seen Paul Delaroche in his class, Ingres in his studio and Horace Vernet. I have given an introduction to Ferdinand de Lesseps for an English Under-Secretary of State. I have stayed as a fellow guest with the then future Lord Herschell at Dr. Norman Macleod's. I have received a visit from Cabet on his way from America to face voluntarily a criminal tribunal of the Second Empire; have seen Thackeray writing at the Reform Club; have heard Father Mathew speak from an open air pulpit in Whitechapel, Lacordaire in Notre Dame, Cobden and Bright in Drury Lane Theatre, Daniel O'Connell from more than one platform, Evanson in Exeter Hall. I have exchanged letters with Mr. Gladstone. I have heard Carlyle deliver his lectures. [On 'Revolutions' (not, as it is generally worded – 'On the French Revolution' – this being the subject of only one of them) – I have met him more than once, without caring to be introduced to him.] Though I have never laid myself out for an orator, I have spoken once in Exeter Hall, once in Hyde Park, once in the Albert Hall . . .

Something of all this experience should be worth reading,

– not on my account, whose utmost capacity has been that of good second-rate, but for the help of the few generations immediately following my own, so long as interest in the period to which I have belonged shall not have entirely died out. No one can know as well as myself how far beyond my deserving God has been gracious to me throughout life. I began to write these lines at 75, with dulled senses of hearing, smell and taste indeed, and failing memory for recent things, yet otherwise clear of thought, capable of enjoyment, love and loving; waiting for death so far without fear for myself, but only of that dread time when the wife whom I have loved for now 53 years shall be taken from me in her increasing weakness, and I shall be left alone on this earth, – God in His mercy grant, not for long!

CHAPTER I

Introductory

I am three parts English and one part Scotch. On my father's side I know only by his portrait my grandfather, painted as a boy with a bow in his hand by Sir Joshua Reynolds when quite young, and the colours of which are perfectly fresh to this day. It hung for years in my dining room, where it had been deposited whilst its owner of the elder branch, the late General Ludlow, had no home of his own, until eventually restored to him after his marriage. When Tom Taylor, who was then engaged on his biography of Reynolds, saw it his first exclamation was 'But that cannot be a Reynolds, whose is it?' And Tom Taylor was constrained to admit that there was no other painter of the day on whom it could possibly be fathered, and that it was no doubt painted by him on his first stay in London, and he inserted the picture accordingly in the catalogue at the end of his work.

[I was told by a gentleman from the Heralds' office, on the occasion of my nephew Edwin Liot's assuming the name of Ludlow, that since the death of Charles, my cousin General Ludlow, I have been 'Mr. Ludlow', the head of the whole Ludlow family. I have in my possession a pedigree once belonging to my uncle Ludlow, and the earlier part of which has long since been enrolled at the Heralds' office, carrying back the family to one William Ludlow of Hill Deverill, 'boteler' to King Henry IV, V, VI, 1399-1425.

I am, myself, profoundly indifferent to genealogy, holding with Plato that we must all have about the same proportion of decent folk and rascals among our ancestors.

Since the above was written I have been told by Mr. Ludlow Bruges (who is a true Ludlow) that William Ludlow the 'boteler' is undiscoverable in the records of the royal household under Henry IV and Henry V, but that he is described as 'Groom Trayer[2] to the King's cellar', 1451, and 'Yeoman of the Cellars', 1454, i.e. under Henry VI. There was also a Richard Ludlow who was 'Serjeant of the Cellars' in 1437, but there is nothing to show what was the relation between William and Richard.

Meanwhile a peerage of Ludlow has been revived in favour of Lord Justice Lopes (since dead), whose claim to belong at all to the true Ludlow family was made fun of in the *Saturday Review* and is, I believe, entirely unsupported by evidence.]

I may mention that my Uncle who owned the picture had also in his possession an unprinted sketch by Reynolds of his 'Perdita' far more graceful in outline, but altogether fady as compared with the portrait of my grandfather. Of the latter, all I know is that he was very fond of playing the violin and very extravagant, and that born heir to a good landed estate he virtually fiddled away the most of his property. He left two sons. The younger son was my father John, afterwards Lt. Colonel Ludlow, the elder, Edmund Ludlow, I shall hereafter designate as my Uncle Ludlow. Of my father's youth I know nothing, except that when quite young he is said to have fallen in love with 'the beautiful Lady Oxford', then a girl, and to have been sent out to India in the army to cure him. He took part in the Egyptian expedition under Sir R. Abercrombie, was aide-de-camp to Lord Wellesley when Governor General, and was by all accounts a first rate soldier, winning his C.B. – in days when the authorities were not prodigal of the honour, especially in the Indian Service — by marked gallantry in the Gurkha war. His last action in that war was, however, a failure – an attack upon a hill fort, when his men would not follow him, alleging afterwards as an excuse that their colours had been changed, and that they could not beat the enemy under the new ones! This repulse preyed upon his mind, and though ill of the disease which caused his death he insisted on taking his regiment into the Birdaneei war, but the fatigues of the expedition killed him. My mother, in her last illness, handed me over a batch of his letters, which I have somewhere. I read them once, – I could not bear to read them again. Those written during his last campaign, the enfeebled handwriting straggling crookedly down the page (and often blurred with what evidently were her tears), all breathing the simplest affection to her and to his children, are more touching than I can say. He was devoted to his profession, and used always to read some military work while dressing, in which he would become so engrossed that my mother has told me he actually worked through a perfectly good tooth by brushing at it, quite unconscious of the lapse of time. To encourage his men, he used always to lead them into action – as an old comrade of his, the late General Pitman, has told me was the custom with the best Indian officers of those days – with nothing but a walking stick in his hand.

My maternal grandfather, Murdoch Brown, I know also only by his portrait, though he knew me and was very fond of me as a baby. The face, without being handsome, is singularly alive and shrewd. His history was a remarkable one. He was the son, so it is reported, of a Scotch Minister in Edinburgh (but of this I have my doubts). Lame from a boy through some accident, and clever at his books, he was yet passionately fond of the sea, of sea-things, and whenever the time could be found, was down at Leith among the shipping. Here he scraped an acquaintance with a skipper, trading to Portugal on his own brig and who, having no children, used always to have his wife aboard with him. This skipper took so much to the lad that when the latter was sixteen he came to my great-grandfather and offered to take his son to Lisbon that he might see a little of the world, promising that he (the skipper) and his wife would take good care of the lad. At Lisbon, however, the skipper died of fever . . . and from thenceforth ((he)) maintained himself, never returning to the British Isles and, it is said, speaking English with a Scotch accent to the last. I wish I could tell the story of his subsequent wanderings – he wrote his memoirs, but they were burnt in some Indian insurrection . . . In Austria he obtained

with a Frenchman, a Baron Dineur, a grant of a silver mine
in Croatia, I think (which I have seen), and eventually
embarked for India as Consul of the Empress Maria Theresa
at Tellicherry, then a port of great importance on the
Malabar coast. Here he became one of the merchant princes
of India . . .

He married late in life, under very singular circumstances,
which I will not here detail, a girl much younger than
himself. They had several daughters and a son, the eldest of
the daughters being my mother. The children were sent home
to England to be educated and my grandmother, who had
been brought up in France and knew scarcely anyone in
England, placed them under the care of a Mrs. Spence,
who . . . appears to have been an Irishwoman of the worst
type, – speaking ill of persons behind their backs almost in
proportion as she fawned on them to their faces; virtually
without a word of truth in her; pretending to anybody that
she was keeping her charges out of charity, while my
grandfather was paying her a very handsome allowance for
them, – leaving their education – except as to music and
dancing – utterly neglected, so much so that my mother used
to read, and get her sisters to read with her by stealth,
Robertson's *History of America* and such other instruction
books as she could get hold of. Instead of educating the girls
indeed, Mrs. Spence used to employ them in writing out
lying puffs for some quack medicine, the recipe for which she
had got hold of and money out of. Nor is it possible to believe
that all this was mere neglect for the creature at one time
actually sent one of my aunts, then a very handsome girl, to
walk under the windows of the notorious Duke of Queens-
berry – fortunately without result. Yet thanks be to God, the
girls passed through the ordeal unscathed. Indeed, the
experience of Mrs. Spence's lying and duplicity inspired my
mother with a perfect horror of such vices, and though she
was simple-hearted as a child to the last herself, gave her a
singular instinct for seeing through falsehood and double-
dealing in others.

Apart however, even from this strange conflict between
good and evil in her bringing up, my mother's childhood
passed odd vicissitudes. During the peace of Amiens Mrs.

Spence went over to Boulogne with her charges, professedly to improve their French... Whilst they were at Boulogne war broke out again, and all the English at Boulogne... were sent into the interior – made *déternes* and *internes*... Eventually the women and children *déternes* were allowed to return to England, which they did via Rotterdam, and not long after my mother and one of my aunts were sent for to India, where at last they enlightened their father and mother as to Mrs. Spence's character. During the whole time of their stay with her not one single genuine letter had they been allowed to write, all being dictated or at least revised by Mrs. Spence...

My mother married not very long after her return to India – at eighteen, from the house of the first Sir Charles Forbes, who had not then received his baronetcy... Twelve years of married life were all that were vouchsafed to my mother. I was her youngest child of five who outlived birth. My elders being Eliza, Maria, Julia and my brother Edmund Ludlow, whom I never saw, but my sister Maria has often spoken to me about him...

I was born in India, but brought to England when two years old after my father's death... The period at which memory becomes retentive varies greatly. One friend of mine (the late Septimus Hansard) maintained that he had a clear recollection of a matter which took place when he was about eighteen months old. With me, memory is an absolute blank till I was over two years old. I have not the faintest recollection either of India, or of the voyage to Europe. But of my old Ayah – old Jane – who stayed with us for some months after our arrival, I have a dim, shadowy remembrance as of a dusky face bending over me, and I have still distinctly in my memory the tune of a lullaby which she used to sing over me, and also what seem to me have been its words:

Serotin peis a murchali minar peis an ghee,

only unfortunately no old Indian has been able to interpret them for me, except that they seem to speak of fish and butter. My mother told me that she always found Jane absolutely truthful and trustworthy, and would have been very glad to have kept her on, but Jane could not stand the climate, and had to be sent back...

I have a vague remembrance of lying in bed with typhoid – it was then called typhus-fever . . . but my first distinct recollection is that of falling into the fire, while on a visit to an old lady. I was trying to reach something on the mantel shelf above the kitchen fireplace, standing on the fender, when I lost my footing, and fell foremost into the grate, one of the high ones of those days. I was pulled out in a moment, and was not, I believe, much hurt, but I can see that grate now, and smell the horror of the embers and ashes (there was nothing else, I believe) about my face. Another still keen recollection of the same visit is of the first rocking-horse I had ever seen, a grand white steed with flowing mane, and of the delight of riding such a charger . . . - which was the means of fixing in my mind the first item of legal knowledge I ever possessed. A few years later, when we were living in France, Mrs. Marriott died. Kindly recollecting my pleasure in the rocking-horse, she had bequeathed it to me, . . . but it was, of course, not worth while to bring it over, and to quiet me I was told that legacies were only paid at 21. I gave in at once, never doubting but that at 21 I should have as high an appreciation of rocking-horses as I had at 7, and never from that time forgetting that legacies were only paid at 21.

. . . We sailed across from Ramsgate, I think, to Dieppe, my mother and her four children, and my uncle Ludlow. It was, I remember, a beautiful day, and my first recollection of being at sea is of that crossing, as well, alas! as my first recollection of personal dishonesty, though certainly not malicious. I was, as I considered, fishing, by means of a small Smyrna fig-basket at the end of a long string, into which I hoped that the little fishes would kindly enter for my benefit. On one occasion of my hauling it up, I distinctly saw, but pretended not to see, my Uncle put something into my basket. I hauled it up again as soon as possible to see what was in it, and found – a boiled shrimp, at sight of which I pretended great astonishment. I had two or three more hauls with the same result, and my Uncle always maintained afterwards that I had believed the Channel to be stocked with boiled shrimps.

From the time of our arriving at Dieppe memory is with

me fairly continuous. I remember my surprise at the construction of the Dieppe house, with its large inner court-yard,
over which I can in fancy now look out from the kitchen
window – the novelty of living in a flat . . . I remember the
going away of an English servant – Catherine – whom we had
brought with us. She was a bigoted dissenter, and believed
that all Church people must go to a bad place, and her horror
at the sayings and doings of French papists may be conceived.
Everything was a cause of offence. '*Vous avez quelque chose là*,'
said the French *douanier* as he slightly felt her when she first
disembarked, according to the then custom. 'Shoes!' she says,
'I am carrying shoes!' She came in a towering rage to
complain to her mistress. I remember in the early days of our
stay a scolding bestowed by my mother on the baker, who
had given some cause for complaint, and his astonished look
under it, which was explained when it turned out she had
been giving him his jobation in Portuguese! the only European language which she had of late years been accustomed
to speak when not speaking English.

I remember, though somewhat vaguely, an excursion to
Arques and my first donkey-ride. Also I remember, after the
lapse of a few months, the journey to Paris in a yellow chariot
which my mother had brought over, and the changing horses
and the long whips and jackboots of the French postilions. (I
will answer for it, that if anyone in the spirit of prophecy had
in those days foretold one of the 'Paris in $7\frac{1}{2}$ hour' advertise-
ments which are now in every newspaper and railway guide,
it would have been held to be overwhelming evidence against
him on a commission in lunacy).

The arrival in Paris I do not recollect – very likely it was in
the evening and I was asleep or sleepy.

CHAPTER II

First Residence in France – Trip to Italy –
The Revolution of 1830

With the exception of occasional short visits to London, a trip to Italy, and one seven months' stay in London, our residence in France extended over 12 years, and was followed by a stay in a French colony of seven months, and with the exception of a trip to Normandy and a winter in Tours, was confined to Paris and its neighbourhood, Fontenay aux Roses and Ville d'Orsay.

I was the youngest of the household, separated by several years in age from the youngest of my sisters, but I do not recollect the time when I was not treated as a member of the family, dining at table, and reading freely whatever books I chose. I have been told that I did not learn much till I was four years old, when I suddenly took to reading, and after that went on by myself.

I can see now before me the drawing room in our first apartment in the Rue Castiglione, where my novel reading began (at 5) with *The Last of the Mohicans* followed by Walter Scott's *Woodstock* and various others. I began reading poetry as soon as novels, and have still in my possession my first book-present, a French edition of *The Lady of the Lake* given to me in June 1826, when, consequently, I was five years and three months old. Of course at first I only cared for the story, but it was very delightful to set one's back against a chair as representing a rock, and repeat James Fitzjames's words, which appeared to me sublime:

8

Come one! come all, this rock shall fly
From its firm base as soon as I.

I read *Marmion* soon after, but somehow did not care for it
as much as for *The Lady of the Lake*. From Fenimore Cooper and
Walter Scott I went on to Byron, for whom I had for some
years a passionate admiration, counterbalanced fortunately
by an equally passionate one for Shakespeare. Strange to say,
however, the attractions of both were soon rivalled by those of
a book which for years proved a constant resource to me, and
out of which I really believe I drew the foundation of most
that I have since learned – a queer favourite for so small a
child – Nicholson's *Encyclopedia,* in six thick 8vo volumes. In
this I nibbled at most of the subjects, but I am bound to say
that the natural history articles were those to which I mostly
turned. And not only were the many Latin names of animals
which I unconsciously stored up in my mind from my dear
Encyclopedia of material help to me, I believe, when I began
to learn Latin, but they were perhaps of still greater moral
help by predisposing me in favour of a language which
seemed to belong to these beloved creatures. I can remember
now the thoughts to which those readings sometimes gave
rise, e.g., my surprise at finding that men and women-
people – formed a 'genus Homo' in the Linnaean scheme.
History also, to anticipate a little, soon began to interest me,
and I have in my possession a *Goldsmith's History of Greece,*
given to me when I was a little after six. But however eagerly
I might read the stories of Marathon and Thermopylae and
Salamis, I am bound to say they never rivalled *The Lady of the
Lake* in my affections. Geography also had great attraction for
me, and I could spend hours with delight lying face-
downward, poring over an atlas . . .

After a few months we removed to the Faubourg St.
Germain, to a house in the Rue de Regard, afterwards turned
into a convent. It was here I first began to talk French, with
the *propriétaire,* M. Dupuis, who was his own gardener . . . I
remember one serio-comic passage in my life at the Rue de
Regard, illustration of the odd fancies of childhood. There
was one plant in the garden which I had been forbidden to
touch – what it was I have quite forgotten – as being poisonous.

On one evening, without meaning it, I did touch it, and jumped at once to the conclusion that, being poisonous, it must kill me. So I went up to my own room, washed my hands several times, and prepared to die, and that – heaven knows why – at 12 o'clock at night. Furthermore, I considered myself as pestiferous, and when my mother or my sisters, missing me, searched me out, I would not let them touch me. Finally my mother succeeded in persuading me that if I was to die it would be better for me to die in bed. So I undressed and went to bed, and of course to sleep, and woke up, to my great surprise, the next morning, as much alive as ever . . .

My sole teacher, during all this time, after the very first, was my second and favourite sister Maria, seven years older than myself. I never had any other educator worthy of the name; from her alone I learnt to work, and to love working, and all the mis-teaching of other teachers was not sufficient to undo what she had done for me.

Of corporal punishment in the shape of pain inflicted I never knew anything. I remember one physical struggle with her, at the end of which I found myself on the floor tied on my back underneath a heavy arm chair, which I was not able to move, and the shame with which I looked up at its seat. Years after I told my sister how thoroughly she had beat me on that occasion, when she replied with a laugh that she had been at her wits end when she thought of the punishment, as she felt I was getting too strong for her, and if it had failed she must have asked my mother to get a master for me. But my usual punishments, and very painful they were to me – were to go to bed or come into the room in the morning without a kiss from her. Sometimes this was withheld for a whole day, and on one awful occasion for three days running. What the faults were which occasioned these punishments (which cost at least as much to the inflicter as to the victim) I have totally forgotten. But my sister and I were all in all to each other.

At the Rue de Regard we used to take our lessons in a little kind of boudoir which was given up to us – a narrow room looking out upon the garden, and here I read with her my first French book, Florian's *Guillaume Tell* . . . My third and youngest sister was an invalid, and on her account we had, for six months, a house in the country at Fontenay-aux-Roses.

There I was sent for three months to a school, the only one, properly so called, that I ever attended. I was there as a day scholar only – not even as a day-boarder, taking all my meals at home except on the first day when, the maid not having arrived to fetch me, I went in to the school mid-day dinner, and made my first acquaintance with *boeuf en salade*, of which I always remained very fond. My recollections of the school are very slight. I remember the difficulty my fellow-pupils had in pronouncing my name and the joy of one of them when he had transformed it into 'Lilas'; I remember that one of the bigger boys was named de Noé, and that his father, a royalist refugee, had served in India. Learning that I was English he came at first to speak with me, and put two or three questions to me in English, but I do not remember his taking any notice of me afterwards. Except that one of the subjects taught was *dessin linéaire* (line-drawing) I do not in the least recollect what I learnt there, or whether I learnt anything. My schooling was interrupted by a most pleasant cause. My mother's brother, Mr. Francis Carnac Brown ('Uncle Frank' in the family) had come over from India, and he proposed that we should take a trip to Italy. A young lady (who did not turn out very satisfactory) was left in charge of my invalid sister with her nurse; the old big-bellied yellow chariot was furbished up and off we started. The chariot was fortunately very roomy, as it had to seat three ladies and a small boy, the last on a small seat in front which could be let down if I preferred standing. My uncle sat on the box by the side of the driver, and to his amusement was, more than once, taken for a courier and offered a drink. We had intended to go as far as Rome, and perhaps Naples, but were detained three weeks at Turin through my mother falling ill. An Italian doctor was called in, who after the fashion of those days wanted to bleed her. She steadily refused to allow this, and when she was getting well he had the candour to admit that it would have killed her though maintaining still that it would have been the proper treatment for one of his own countrywomen.

My uncle having to return to India, we were thus not able to proceed beyond Florence ... My sisters both wrote their diaries and so did I (at seven). Strange as it may seem, I have never been to Italy since, but a large portion of the tour has

remained indelibly fixed in my memory . . . We returned to
Fontenay-aux-Roses in time for the vintage, which I remember as three days of great enjoyment. The children while
gathering the grapes (the small black grape of northern
France) were not grudged any number which they might eat,
and I never remember such delicious *pot-au-feus*, cooked in the
open air as we ate by the road-side. The peasant of the
banlieue was not under ordinary circumstances a very delightful character, but somehow at vintage-time he showed
himself at his best, freehanded and altogether jolly.

From Fontenay-aux-Roses we went for a year to a house in
the Allée-des-Veuves, now altogether in Paris, then virtually
country. Never before or since have I seen in any place such a
variety of butterflies as came into our garden – again a large
one. I began collecting them and had already a majority of the
species pictured in my book of French *papillons*. But I knew
nothing of the scientific way of killing them, simply pinning
them down. One day I was horrified to see a big peacock, which
I had pinned about three weeks before, walking over a distant
table with the pin sticking through him. This gave me such a
shock that I did not collect any more . . .

I rather think that it was at the Allée des Veuves that I had
my first master, a chirpy little old man named Monsieur Rey,
who took a great deal of snuff, author of a French grammar of
which I still possess a copy. Beyond French grammar, I do not
well recollect what he taught me.

It was also, I think, whilst at the Allée des Veuves that I sat
for the first (and only) time for my portrait, to a miniature
painter . . . He painted with great delicacy of touch, but his
drawing is not quite correct. I well remember the tediousness of
those sittings, and resolved within myself that I would never
submit to the like again, a vow which I have faithfully kept.

In the year 1829 we moved again into Paris proper, to the
Rue du Luxembourg, debouching into the Rue de Rivoli. It
was intended to send me to the *Collège* in a year or two, and I
had here a classical tutor . . . At the time I speak of social life
in Paris was very different from what it is now. When we first
came there the dinner hour was still often four to half past,
though this soon grew to five. Up to 1834 six was a very late
hour, for only grand occasions, as a wedding dinner. And no

wonder. The hours for all evening amusements were equally early. At the smaller theatres on the Boulevards the performances began at half past six, the Grand Opera at seven. The streets were, of course, only lighted with oil. Open side-gutters in the streets were the last improvement, and only existed, it may be said, in the Tuileries quarter. Elsewhere the old *ruisseau* became, in rainy weather, a torrent, and men would then make a good thing by carrying about planks to throw as bridges across it, receiving a *sou* generally from those who passed over, or in default discharging at least a franc's worth of abuse at the non-payer. Omnibuses were not introduced till, if I recollect aright, 1828 or 1829.

For many of the suburbs there was no public conveyance but the *coucou*, a hooded cart with two parallel benches looking in the same direction as the driver, dreadfully jolty, and the occupants of which had rather a bad time of it in wet weather, the hood being seldom water-tight. Within the city, beyond a few lumbering *fiacres*, the only public conveyance was the one-horse *cabriolet*, with straw-strewn floor in which the driver sat with his fare or fares (in the corresponding London conveyance the driver sat in the same position, but on an outside seat).

Wood was the only fuel, burnt on the hearth.

One of the great inconveniences was the use of silver for a standard. Gold was entirely driven out of circulation, 20F. pieces fetching a premium, and being never seen except in the hands of English travellers, though a small proportion of all official salaries was paid in gold. The necessity of keeping a stock of five franc pieces was a universal nuisance. In large transactions indeed, bags containing each 1000 francs in such pieces, sealed with the seal of the Bank of France, formed a necessary element to save the trouble of counting, and these sealed bags might remain current for years, without the seal being broken. [This led sometimes to singular results. My then future brother-in-law, then an *employé* in the *Ministère de la Marine*, was paid his salary in such thousand franc bags. On opening one of them, he found once that instead of containing 200 five franc pieces, the pieces were all old *écus de six livres*, so that he was overpaid by between £7 and £8. He took it back to the Bank of France, but after some consultation they returned it to him, telling him that it had been in circulation for over 20 years as representing 1000 fr. and that the tracing the mistake to its source and correcting all the entries in their books would cost them more than the difference in value.] Let it be observed that such a thing as a cheque was not recognised – if I recollect aright – cheques date in France only from the Second Empire – and some idea will be formed of the cumbrousness of money transactions in Paris in those days. My mother had always to go to Rothschilds to fetch her English remittances. [I used often to go with her, and to see the cashier, Mr. Browning, a very friendly elderly gentleman, whose son was to be the Robert Browning of English literature. Old Mr. Browning himself wrote a History of the Huguenots, a copy of which he gave to my mother, but I do not know what became of it eventually. On one occasion I met there young Browning, a few years my senior. We were put into relations with each other, but we were both shy, and the difference of age between us was considerable in those days. I felt afraid of him, he no doubt looked down on me, and we didn't seem to have anything to say to each other. So all I can say of Robert Browning (whom I never met in after life) is *Tantum vidi*.]

Seeing very little of French society we took, at this time, no interest in French politics. I received indeed in 1827 my first political impression through the death of Canning (8 August), from the grief expressed for this event by my mother and our other English friends. From this time I began to read diligently Galignani's *Messenger* and I became deeply interested in the struggle for Roman Catholic emancipation, following all the debates in Parliament through Galignani's summaries ... By instinct or family tradition or both I was a Liberal, as I trust to be till death. It needed a revolution to make me care for French politics.

At the same time something occurred in the early part of January 1830 which in a manner prepared me to do so. We attended the celebration of the expiatory service for Louis XVI (Jan. 21, 1830) at the Cathedral of St. Denys. We had very good seats in a side gallery closely overlooking the chancel, where sate the Royal Family. The Duke and Duchess of Angoulême were there – and I think the Duchess de Berry. And thither came, as the representative of the Orleans branch, the young Duc de Chartres, the future heir to a throne which he never lived either to ascend or lose. I can see him now as he entered the Chancel, looking very handsome in a light blue hussar uniform, the jacket hanging over his shoulder, and bowing gracefully to his cousins. I was greatly taken with his good looks and graceful bearing, and should always from that time have felt an interest in him. Little did we or any one present imagine that we were witnessing the last public solemnity of the kind.

We were living at this time in an apartment on the second floor of a house, No. 4, Rue Neuve-Luxembourg, exactly opposite the barracks of the corps known as the *Cent Suisses* – though there were very few Swiss in it – and fully overlooking the barrack yard . . . Notwithstanding our little interest in French politics, towards the end of July 1830 we could not fail to notice the famous *Ordonnances* of Charles X; nor could we but observe something quite unwonted taking place in the barracks opposite, which were usually extremely quiet when the drum was not beating. Till late at night on the 26th there were lights going to and fro, and a noise of clinking of glasses and hurrahing. On the other hand, that same evening down the street came a cry '*A bas les Ministres*'. On hearing it Marguerite, who was in the room with my mother, in her excitement – she was by no means agile – jumped upon a chair exclaiming '*Ah! Madame, c'est une révolution! C'est comme ça que la Révolution a commencé*'. Her guess was a true one on this occasion, though it would not have been so afterwards, when the same cry heralded only insurrections more or less bloodily suppressed. A revolution it was, and for three days we remained confined to our apartment, the *persiennes* always closed, but, I need hardly say, all within always at the windows. The noise of firing was constant, that of artillery more or less frequent, but only three shots were fired in our street, one indeed close to our windows. Those were certainly the three most exciting days of my life. Nothing happened to us, and even in the way of food we were fortunate, as there was a pastry-cook on the ground floor of the house, and he kindly kept us to a great extent supplied out of his own stores. Our great apprehension was on account of the barracks. The troops had been taken away by a gate opening onto another street on the Wednesday morning 28th July, but they were known to have left behind them several barrels of powder. On Thursday afternoon, the 29th, the victory of the people was complete, and with our heads out of window – the *persiennes* being by this time flung open – we were watching the retreat of the *Garde Royale* by the *Rue de Rivoli*. I have the scene before me now – the soldiers in disorder, not running, I am bound to say, but hurrying away, sometimes turning round to fire yet a shot at their pursuers. One poor fellow, I recollect, levelling his gun to aim, finding it was not loaded, and throwing up his arms in despair. And then the mob following them! – the most respectable in blouses, most of them nondescript vagabonds, many mere boys! I came, then and there, to a resolution which fixed to some extent the course of my whole life. Till then I had always looked forward to being a soldier, like my father. But if a soldier's fate nowadays could be ignominy like this I would never be a soldier.

Till then we had really been in no personal danger. But now, when the revolution was triumphant, an hour of real danger began. All over Paris the mob poured into the various barracks. They came amongst others to our *Caserne des Cent Suisses*, and swarmed into it, a motley lot, armed with all sorts of weapons, muskets, swords, pistols and what not, all with tricolour cockades, several of them half-seas over. Foremost among them was one drunken fellow, who before long began to cry out that they must '*mettre le feu à la caserne*'. The idea won favour, and they were already seeking means of kindling the fire, not knowing of the existence of the powder, which had it caught fire would have blown up our house as well as the barracks and their occupants. But there were no friction matches in those days for anybody to carry, the last invention being that of some liquid preparation of sulphur and phosphorous into which a long match was dipped and took fire on being drawn out, and this was only kept in the kitchen. The people of the house durst not remonstrate for fear of being treated as royalists, and we were all in a perfect agony of suspense when a new personage entered on the scene. This was a pupil of the Polytechnic School – the *Polytechniciens* had distinguished themselves at the barricades, and had to a great extent led the Revolution – in his uniform begrimed with powder, his sword by his side, a tricolour scarf round his waist, – a young fellow, say, of 20 or 21. Pushing himself into the midst of the rioters – for such they were now, breaking up whatever chairs and tables they could find, as materials for the intended conflagration – and they could not be far short of a hundred – he began appealing to their better feelings, stopping this man and that, sometimes putting his hand on the hilt of his sword to threaten, but never drawing it. We could only hear a few words now and then, and could not see him, except when in the street outside, or in the barrack yard, but were all watching with breathless expectancy, the result. A more magnificent triumph of moral over physical force, of good over evil, I have never witnessed. At the end of, perhaps, three quarters of an hour, this one young fellow had got the better of the whole unruly, roaring, drunken mob. He persuaded them to leave the barracks, all but two or three, and selecting the very man whom I have mentioned as a leader in the previous disorder, set him as a sentry, musket on shoulder, to guard the entrance to the barracks. None of us ever heard the name of that young student, – I never saw, in any of the numerous narratives of the Revolution, any account of what he had done, – but he saved our lives and those of hundreds of others at the risk, certainly, of his own, since half the mob whom he fronted alone so dauntlessly were better armed than himself.

But after the tragedy came the farce. The half-drunken rioter whom, with an instinct perhaps of prescient humour, the *polytechnicien* had posted as sentry, taking *au grand sérieux* his new duties, began to fulfil them with a ludicrous, though misplaced excess of zeal. To begin with, instead of pacing only up and down before the entrance to the barracks, he deemed it his duty to guard the whole frontage, and with unsteady gait paced backwards and forwards a distance of probably 100 to 150 yards, wholly forgetting that whilst he had his back turned at a little distance anyone might slip into the barracks. To prevent this indeed, he would not allow anyone to cross or go beyond the middle of the street, and two or three times we saw him level his musket at passers-by who might approach too near the now sacred walls entrusted to his keeping. Eventually a party of *gardes nationaux* relieved him of responsibility by taking charge of the barracks.

The excitement of the three 'glorious days' had been so great that I afterwards complained that 'it was so stupid now

there was no more revolution' and got, of course, a good deal
laughed at for so doing. Still, there was for some days a good
deal of interest in walking about the streets and boulevards,
looking at the barricades and other traces of the struggle. The
elms of the Boulevards had been splendid trees, and when cut
down formed of themselves a complete cover to a kneeling
shot, but as every boulevard and street was then paved with
big cubical paving stones, when a huge tree trunk was
strengthened with these, the obstacle formed ((a sound))
defence to all except artillery. The crowd were singularly
good humoured, and Paris was *en fête* altogether.

Rumours of royalist plots for a counter-revolution soon
brought back a little excitement, and I tried to add to this at
home, I am ashamed to say, by a little – or rather a good deal
of – fibbing on my own account. I used to go and play with
my hoop in the Tuileries garden, and as I was one day sitting
on a bench, the circumstance of two men sitting down on the
same bench and beginning to talk in a low voice first
suggested the idea that they might be conspirators, then that
of inventing their conspiracy and humbugging my people at
home with it. Which I did accordingly for a series of days
with much success, (old Marguerite, I recollect, was the
greatest sceptic), pretending that I saw the plotters repeatedly
(I never set eyes again on the two men I had at first noticed),
followed them and listened to them, and heard them say on
one occasion '*Nous avons les clefs du Château*', and asseverating
the truth of my assertions with all the earnestness I could
command. At last, when it came to be decided that the police
must be informed, I took fright, and without confessing my
lies asked for a day's respite, that I might watch the
conspirators again once more. Of course I never saw them
afterwards. It was no doubt very wicked of me to lie and not
confess my lies – which I did not do till many years afterwards – but the scare of imminent discovery was a sufficient
lesson to me – I never invented another conspiracy.

July 1830 was our only direct experience of political
disturbance. Subsequent insurrections all took place in distant quarters of Paris. One went afterwards to see the
mischief done – the torn up streets, the barricades, the
wrecked houses, but that was all . . .

CHAPTER III

Temporary Return to England 1831
– Second Stay in France 1831-7

We had now been five years in France, but a change was at hand. The various revolutions or revolutionary movements which on the Continent followed upon the French revolution of 1830 had alarmed our English friends, and they pressed my mother to return, – laying also stress upon the consideration that I was growing old enough to be sent before long to an English school. We came over, as far as I recollect, in the spring of 1831 . . .

I should now observe that during the period of our first stay in France I had several times come over with my mother to England, and we used always to stay with my father's elder brother, my favourite Uncle, Edmund Ludlow, who was also very fond of me . . . I used to enjoy my walks with my Uncle. He knew his London well, and took me to see magnificent St. Saviour's Church (where many of his family are buried), and the Roman stone, and the Exchange, and Whitehall and the old Houses of Parliament. Picturesque old London Bridge was at first still in existence, with its many arches widening as they approached the centre, and the shops upon it, the river forming actual rapids as it rushed underneath. I may recall the fact that in those days before river steamers and buses, a boat was a frequent means of locomotion. I have repeatedly crossed the river in one with my uncle. The Thames, if yellower than the Seine, was still greenish, and had not then

19

had all the sewage of London turned into it, whether above or below London. The Thames flounders were a delicacy taken in abundance above as well as below London Bridge, and I have eaten potted lampreys caught under the Bridge itself, whilst for many years even after this time the terrace at Somerset House was a favourite angling resort for roach, dace etc. Even salmon occasionally made their way up the river, tho' the capture of one was rare enough to be always chronicled in the newspapers.

When we came over, as it was thought definitely, in 1831, the question of my schooling was much discussed. Had it been decided in favour of a public school Rugby would have been chosen, and in after days when I read Arnold's life I had a queer sort of satisfaction in the thought that I might have been his pupil. But my uncle was a shareholder in University College London – then the most go-ahead educational establishment in the kingdom, and it was decided that after some preparation by an English tutor I should enter the school connected with it, a very go-ahead concern also. My mother consequently took a house in Gower Street, then comparatively a new quarter – one of those houses on the East side of the street, with strips of garden behind, in those days generally producing nothing but marigolds and harbouring innumerable cats; (since then trees have grown up in them, so that the houses tho' less sought after are really much more agreeable to live in than they were sixty years ago).

. . . One great show which I saw at this period was the opening of New London Bridge, August 1st. 1831. If my memory does not deceive me, part of the old bridge was still left standing at the time, probably by way of contrast.

Among other novelties of the time were the new police and omnibuses. I may here observe that I have a distinct recollection of the old watchmen whose loud-voiced observations as to the time of night and the weather served as the kindliest warnings to thieves and burglars to get out of their way. [It used to be a good story against Hallam the historian, as illustrating his craving for accuracy, that being a bad sleeper he would often throw up his window at night to engage in discussion with the watchman: 'Half past twelve o'clock – Fine night.' 'No, Watchman, don't you see that bank of cloud in the west? It will rain in ten minutes'.] One familiar sight of the time I am now speaking of has completely passed away, that of the portable gas carts, loaded with big iron tubes like elongated boilers, rounded at both ends. The meaning of the demand for 'portable gas' was, of course, that the vast network of gas mains and their connections, which now spreads through all London, was then only in its beginning, and that those mains which existed were only for street use. Gower Street itself, for instance, was lighted with gas, and yet the portable gas carts stopped at houses within a few doors on either side of our own – we ourselves using only candles and oil lamps.

Our stay in England on this occasion however, only lasted seven months. My sister Maria had an attack of pleurisy which kept her for weeks between life and death, and the doctors pronounced that the climate of England did not suit her. So my mother determined to return to Paris. Even while my sister was slowly recovering a new danger came upon her. The plaster of the ceiling in her room, just above her bed, which fortunately was an old, covered, fourposter, came down upon it and crushed the tester down, but formed an arch over her so that she was quite unhurt. She was destined in after life to meet with a far more terrible accident, yet somewhat resembling this.

We returned then to Paris in the early autumn of 1831, taking up our quarters in the first instance in the Rue de la Ferme des Mathurins, a street debouching into the Boulevard and afterwards in a parallel street, the Rue Godot de Mauroy, which for the first time my mother furnished as an apartment, it having been decided that I should be brought up in France. I here resumed under a master my Latin and Greek studies and began learning German, together with my eldest and second sisters . . .

I may mention that by this time we had got into the habit of attending the (French) services of the Reformed Church at the Oratoire. There was there a preacher of considerable rhetorical power, Athanase Coquerel, afterwards under the Republic of 1848 a somewhat prominent member of the National Assembly. My sister Maria and I became great admirers of his. He was the author, besides a volume or two of published sermons, of a fairly good Bible dictionary of biography. It was the first time that I had ever cared for sermons. Coquerel was a preacher of the so-called 'Liberal' school of French Protestants, and I am afraid his Christianity eventually drew nigh to the vanishing point. I heard him again years after, and his preaching appeared to me so hollow and his manner so artificial that I was amazed to think I had ever admired him. I am not, however, sorry to have done so. I think it is a great thing for a boy when he begins to take interest in *any* sermon. And if his bad taste and undeveloped judgment lead him astray in the first instance, the ripening of both is sure eventually to attract him to better models. One thought, indeed, I owe to M. Coquerel, which has influenced all my life. He mentioned in one of his sermons the astronomical commonplace, new to me at that time, that of all the stars of heaven not one in all probability can see our earth. If this be so, I said to myself, what is all earthly fame? And from that time to this I have never done anything for the sake of fame, and when I have sought publicity it has only been when responsibility had to be incurred, or, alas, money to be earned.

It was decided that before going to the *Collège* I was to take my *première communion* (answering to our confirmation) after a course of religious instruction, and M. Coquerel was the *pasteur* selected to give it. The course was given at his house; we had to write notes of his lessons, and this was the first occasion which I had to write French, otherwise than as a mere translation. Truth to say I did not like M. Coquerel in his house as well as in the pulpit. He was always pointedly civil to me, but I could see the man's vanity, and I found his lessons dry . . . M. Coquerel had also 'at homes', one of the avowed objects of which was to bring together young people of both sexes in the congregation. I found them dreadfully dull. M. Coquerel's younger brother Charles was a second-rate literary man of pleasant conversation, of whom we grew to see a good deal, in reference to the putting together of a new collection of hymns which was called for at the Oratoire, a partnership consisting of my elder sister, another lady and M. Charles Coquerel being formed to draw up a selection, and all the work being done at our house. This selection was sent in, but not accepted at the *Consistoire* . . .

The nearest *Collège* to our house was the *Collège Bourbon*, afterwards for some time *Lycée Condorcet*, and to this I was sent at the *rentrée des classes* in the autumn of 1832.

The French *Collège* of that day formed (I speak in the past as there have been changes since my time, though I believe inconsiderable ones on the whole) a combination of our school and college. The same *Collège* might contain in its different classes from boys of 7 to men of one and twenty . . . Bourbon was the college of *externes* [day-pupils] for the West end, as Charlemagne was of the East end, and there was a curious mixture of fellow-feeling and jealousy between the two – fellow-feeling as enjoying a freedom which enabled us to look down upon what we considered the serfage of the colleges of *internes*, but much jealousy too, Charlemagne priding itself on its educational successes, Bourbon belonging to the bright end of Paris, the end which had the Tuileries, and the Champs Elysées and the operas and the fashionable theatres and the chambers, and all that made life worth living, whilst Charlemagne was buried away in the Quartier Latin.

At the head of each college was a *Proviseur*, and next to him a *Censeur*, the latter having nothing to do with the studies, but only with the discipline of the *Collège* . . . The classes were of two hours each, twice a day, with a sufficient interval between the two to admit not only of a midday dinner but a couple of hours' preparation for the evening class as it was termed. These two hours were spent by me with a *répétiteur* or coach, a professor in the same college. The class rooms were large and whitewashed; three rows of benches and a boarding behind the highest, the professor's pulpit being at one end so as to overlook the whole. The rooms were certainly lofty and airy, but that is all the good that could be said about them. There were (except those of the professor) no desks or other conveniences for writing, which had to be done on one's knees, inducing a cramped attitude, especially for nearsighted boys, who had to bend double, which in course of time must have acted most prejudicially on the pupils' health and physical development, producing habitual stooping and narrowness of chest.

The curriculum was on the whole judicious and comprehensive. Latin and Greek of course formed the staple, but there were also history classes, geography, arithmetic, some exact science, and German and English were optional . . . Prizes were given each half year . . . But the great event of the college year was the *Grand Concours* or yearly examination, from the *sixième* upwards, of the representatives of the seven Paris *Collèges* and that of Versailles . . . No less than seven hours was allowed, for each examination, during all of which time one was not allowed to leave the building, nor was any book allowed beyond a dictionary. I never saw a paper given out that the dullest boy present could have taken more than four hours over, and the consequence was that the difficulty of killing time was simply frightful. I personally lost two prizes through the effort to do so . . .

The distribution of prizes at the *Grand Concours* was, of course a very grand affair. It was made by the Minister of Public Instruction himself, with the usual accompaniment of a kiss and wreath; the book prizes being, moreover, always really valuable ones, which was not always the case with the college prizes. I had to go up thus to M. Guizot, to M. de Salvandy, and I forget whether to anyone else. But I am anticipating.

I was considered 'strong' enough to enter at once into cinquième, though a couple of years below the limit of age allowed for competing at the *Grand Concours*, which virtually determined the class-age in all the colleges. And here (except for my three months' stay as a day scholar at the Fontenay-aux-Roses School some years before) I found myself for the first time thrown into the society of boys of my own age or thereabouts. I must say that, take them all in all, I found them a good lot . . . We were thoroughly democratic, though ranging from members of the old nobility to the son of a *faiencier* in the Rue Caumartin, *tutoiement* was universal and I never saw – in my own form at all events – the slightest difference of manner towards any boy by reason of his station in life.

There was a good deal of lying toward the masters, but except for fun now and then, we were altogether truthful among ourselves. There was virtually no bullying, or at least there was always a readiness to combine against anyone who tried to bully. On the whole there was a very large amount of mutual kindliness and fair dealing, and no jealousy of a good fellow's success. There was a good deal of foul talk and some precocious vice; and together with a perpetual readiness to combine against a master, not much pluck in facing him.

I tried to be good friends with all my school-fellows, and in the main succeeded. The only exception was a queer little West Indian, terribly spoilt, who lived in the same street as myself and was at first by way of making great friends with me. He was an *externe libre* like myself, and when I was confined to the house by measles and recovering from them, used to come and see me, the only one of my companions who did so. Then for some inscrutable reason he turned against me, and at last challenged me to fight. I had never learned to box, and had never struck a blow in anger in my life, and I am bound to say, though much smaller than me he floored me twice (my weak ankle failing me) once in wrestling, once in boxing; on the latter occasion, using the favourite negro trick so easily met when one knows how, of butting me in the stomach. We could only fight in the street or on the Boulevard and were soon separated both times. My only revenge came to me a couple of years afterwards. He became so unruly at home that he had to be put to school, but on one occasion when he had slept at home we met in the morning on the college steps. He tried some piece of impertinence with me, but I had grown much stronger by this time, and with a single push I sent him rattling down the steps. He never meddled with me again. I heard long afterwards that he had gone from bad to worse, and come to a bad end.

A more painful matter to me than this young scamp's persecution was the result of my own greenness – a matter which indeed belonged to a still earlier period and to the very beginnings of my college life. The class rooms, as I have said, consisted of tier upon tier of benches, with the professor's pulpit in the centre. A common practical joke was to *passer le coup de poing* – hit your next neighbour and tell him to pass it on. This happened when I had been a few weeks at college, but in the act of my passing on the blow the Professor saw me and I – really not knowing as yet but that it might be a

University regulation, and never having had to practise concealment – in answer to his calling to me by name, exclaimed 'It is Quélain (the name of my next neighbour) who told me to do so'. He did not punish either of us, but I was deservedly abused after class hours by my comrades and sent to coventry for some time. It was agony to me when I understood what I had done and that I was supposed to have shirked punishment by betraying another and I remember on my return home that day, when I was unable to eat my dinner and was asked by my sister what was the matter with me, bursting into tears as I replied (we usually talked French together) '*J'ai perdu l'estime de mes camarades*'. However I succeeded in showing that I could be as staunch as any, and by the end of my first year I believe I was at least as popular as any boy of my class. Quélain, the boy I had betrayed without meaning it, behaved remarkably well, excusing me as far as possible to our comrades on the ground of my inexperience. I intended to have made great friends with him, but he was taken away from the class before many months, and I never saw him again . . .

There was good enough instruction to be had at the College for those who chose to seek it, but virtually no education – no drawing out of the individual mind. Personally, I did not need it; I was, in a sense, educated when I came to the College, and all the mis-teaching of other so-called teachers was not sufficient to undo what my sister Maria had done for me. But at the College it was a fact that almost all the professors looked upon their pupils simply as *machines à succès*. The status of a college depended so entirely upon the distinctions won by it at the *Grand Concours* that as soon as a boy manifested any aptitude for a particular subject, the only object in view was for the most part to force him on in that subject regardless of his general education. There were two or three exceptions among the professors; one was M. Miraud, a very strong classicist in French literature, who would only admit Victor Hugo's *Odes, Ballads*, his first published poems, to be within the *tradition du grand siècle* but who, I verily believe, thought more of developing in us a true literary taste, according to his lights, than of any successes at the *Concours*. His brother professors, I am afraid, rather looked down upon him. But even in him there was no trace of any feeling towards the boys as individual men to be. The one great exception was a dear old history professor, Jarry de Mancy, one of the kindest souls I ever met with. He was a man perfectly saturated with his subject, full of the richest stores of historical anecdote and of kindly humour, and his classes were, to the very few I am bound to say who really appreciated him, simply delightful. But you must take him as he came; he had no method, and above all, no power of discipline. One of my friends, de Riancey (of whom more anon), a boy with a precocious gift for narration, used obviously to look down upon a man who could not relate events of a period consecutively without stopping to tell us a story, that one perhaps leading to another and another till the whole thread of the narrative was irretrievably broken. To the bulk of the boys de Mancy's good natured indulgence was simply a thing to be abused, and when they had got to that pitch of unrulyness which put him in a rage I must own that his rage was unspeakably comical. One intimate friend of mine and I were, I think, the only two who appreciated him, and we often remonstrated with our companions for behaving so badly to such a good fellow as de Mancy. More than once we got the reply '*C'est si amusant de le faire rager*'. And there can be no doubt that his classes were turbulent to a degree. M. de Mancy was the only one of my professors who ever took an interest in me individually, and the only one in whom I took an individual interest. I used to go and see him at his house, where he was always delightful, and continued to do so in after days . . .

Turning from the professors to boys, I may mention that during the five years of my stay at Bourbon there were at one time or another two English boys in my class. One, Angelo Hayter, was in *cinquième* with me. He was the son of Sir George Hayter, the Historical painter to the Queen; Hayter was a very good little fellow and we were great friends, but he left after that year. I met him later on when I returned to England, and dined at his father's. He was then intended also to be an artist, but he so frightened me by proposing to paint my portrait that I never went near him again . . .

None of my classmates turned out any great men, though

several distinguished themselves in after life. The one perhaps who came most prominently before the public was Cochery, repeatedly Minister under the present Republic, and who had once been envoy to England (but since dead); a plodding, hard working boy . . . Another, G—, became as the story was told me afterwards the hero of a somewhat pretty little idyll. His family were poor and he obtained a place as a private tutor to the son of a French ironmaster now of world wide reputation whose young daughter asked leave to be in the room during the lessons. After a time G— felt that he was falling in love with her, and it seemed to him also that the young lady was doing the same with him. Like a true gentleman, although the place was an excellent one, he went to his employer and gave him notice, and on the latter pressing him for his reasons, gave them. The employer commended the young man's frankness, told him that he had other views for his daughter and that they must part. He should not, however, be a loser, for he would give him a better place at a branch establishment of his, on condition, however, that he should have no communication with his daughter for two years. G— of course accepted, and as it happened was able to render essential service to his employer in his new position, detecting some serious frauds which had been going on. The young lady, meanwhile, had declared to her father that she would never marry anyone but G—; she remained staunch during the two years, and at the end of that period G— was allowed to marry her, and was admitted a partner in his father-in-law's business.

Of my own intimate school friends . . . the first, in point of affection, was Paul Nicod. He was the best of fellows, clever but of a cleverness which disdained all wile, so that whilst as a scholar he might be pronounced inveterately idle, he was yet always busy with something of his own. I shall never forget the amazement with which, having noticed him reading something very intently in class (of course a scholastic offence for which he was reproved and punished) I found it was a Basque grammar . . . [I asked him why he had not shown the book to the master, who of course supposed it to be a novel. His answer was that it had been lent to him, and that he could not run the risk of its being confiscated.] His own tendencies were essentially artistic . . . ((and, later)), it was an immense delight to me to visit my friend at the *Institut* where for the first time I found myself amongst artistic surroundings. A near neighbour, on the same floor if I recollect aright, was M. Ingres, the painter, a bright little old man whom I saw several times . . . Horace Vernet was another painter whom I met on several occasions. Different as his style was, he had a thorough appreciation of Ingres . . .

The first boy with whom I was intimate at all, was Charles de Riancey. Of good family – though he confided to me one day his suspicions that the de Rianceys owed their particle to the *savonnette à vilains*, i.e. that they must have bought their ennoblement – he was a strong legitimist and ardent Roman Catholic, the only one of our class . . . We fought out together the Roman Catholic question on the college benches, each without convincing the other, and I think it was a wholesome surprise for both of us that we did not. After leaving college Charles de Riancey, with an elder brother Henri, whom later on I also knew and liked, wrote an *Histoire du Monde* in four volumes which is really very creditable to men young as they were, and remarkable as being the first book of general history embracing countries which only in this century have been found to have a history, such as America. They both became writers in the legitimist and classical *Monde* of which Henri de Riancey was for many years the editor . . .

I only completed the latter half of my *rhétorique* at Bourbon, we having for family reasons spent the winter at Tours. I did not there go to the local *Collège*, but had lessons from one of the professors, M. Hatry, with Jarry de Mancy, the only one of my masters of whom I was really fond . . . Poor Hatry! I heard afterwards, from the lady who remembered him to my mother, of his pitiable end. He had a pretty little wife of whom he was passionately fond, but she could not bear to see him growing fatter and fatter. So to reduce his proportions he took to starving himself and drinking vinegar, and died of the process.

Although I always went on working to the best of my power while in Tours, my stay there threw me back a good deal in my studies, for on my return to Bourbon after Easter 1837 I no longer found myself taking such good places as I had been used to do. But in many respects that stay at Tours ripened and developed me. I was there no longer as a *collégien*, but as the son of the family; I had absolutely no male companion of my own age; two young girls in an excellent and charming French family were the barest substitute for such – but by a very long interval for, in truth I was horribly afraid of them, whilst they were perfectly at their ease with me.

Although I was strongly pressed by my *Rhétorique* professor to *redoubler*, i.e. pass a second year in the class – he dangling before me the bait of the prize for French Oration (*Discours Français*) which he told me I was sure to get the highest honour, save one, of a French *Collège* career (the very highest honour is that for Latin oration), I left the college itself without any regret, especially as none of my friends were staying on.

As I look back I cannot but feel that those were ten of the pleasantest years of my life; although there were some disagreeables in my first college year arising from my inexperience in the ways of boys, and schoolboys more especially, I cannot but wonder to recollect how popular I became. I must say I doubt greatly whether any foreigner in an English public school would ever meet with the same reception. At the yearly distribution of prizes it was always remarked that no other boy won such hearty applause, as I swept off prize after prize. I can scarcely account for this, except partly on such grounds as these: (1) that by ingrained democratic feeling I always made a point of speaking to everybody and making friends with whomever I could get the opportunity; (2) that my most intimate friendships with the exception of de Riancey were not taken from among the *forts* who generally formed an aristocratic clique of their own- ... Still I cannot but feel that I was treated by my French companions not only kindly but most generously ...

CHAPTER IV

Society in the Thirties – Decision as to My Future –
The French-West Indies, First Stay, 1837-8

I have purposely confined the last chapter to the subject of
my education. But to this period belongs a series of other
events, which have influenced all my life.

From the time of our return from England we lived mainly
in French society, and amongst ourselves my sisters and
myself generally talked French. We became specially intimate, as the consequence of a chance meeting in the street,
with the family of our old landlady of the Rue St. Maur, who
we found had migrated to a street not far from our
own . . . Old Mme. Liot, who had been thrice married,
belonged to one of the richest St. Domingo planter families,
but had been ruined by the revolution in the island . . . Her
marriage was not a happy one, and eventually they separated
by mutual consent, she coming to France with her children
whilst M. Liot obtained the place of treasurer of Martinique,
the most important of the French West Indian Islands after
the loss of St. Domingo . . .

The member of the Liot family however who was to be
most closely connected with our own was the son, Charles Liot.
I have never seen a handsomer young man than he was
then; then and then after, I have never known one more lovable. He had been strongly attracted by my sister Maria from
the time of his first meeting her in the Rue St. Maur;
with more frequent opportunities of meeting, the attraction

deepened into a lifelong attachment. In the year 1833 he proposed to her. My sister, before giving him a definite answer, consulted not only her mother; but even me, a lad of 12. I saw that it meant being deprived of her (one of the features of the matter was that it involved sooner or later my sister's removal to the West Indies, Liot having the promise of the succession to his father as Treasurer of Martinique) but, thank God! I was not selfish enough to oppose what seemed to be for her happiness. She told me afterwards – so deep was the affection she bore me – that if I opposed it, she would have declined M. Liot's proposal. The wedding, however, had to be postponed in consequence of the state of my invalid sister Julia's health . . . My sister Maria's wedding was ultimately fixed for July 22nd 1834. Some weeks before the wedding my mother, for the first and only time, gave a ball. I mention it on account of one particular recollection concerning it. I have said that Charles Liot was remarkably handsome. My sister Maria, without being absolutely beautiful, had a very finely shaped face and above all a noble brow, which the neo-paganism of our days had not taught women to hide by vulgar fringes and the like. She was, like him, perfectly graceful in all her movements. They waltzed together, and their movements were so perfectly attuned to each other that in its way that waltz was the loveliest thing of its kind I ever saw. And I was not alone to feel it, for after a time, as other couples fell out, instead of re-entering the waltz, they stood looking upon the couple, till at last they went round all alone, the whole ball room company being turned into mere spectators. They stopped, of course, as soon as they discovered the fact . . .

In the following year a baby appeared, and I held a niece in my arms within a few hours of her birth. I have a keen recollection of her having convulsions in teething, and of taking my turn in carrying her about and singing to her. Fortunately she recovered without any injury being caused to her system. She was a very intelligent child, and never having to do with children younger than myself, her development was a subject of great interest to me. My brother-in-law I soon came to love as a brother, and I never ceased to do so. Unfortunately these pleasant days could not last. The increased expenditure caused by the birth of his child rendered it necessary for Liot to increase his income, and just now some changes were being introduced into the system of colonial finance which his father, an old man, met with passive resistance, so that the old promise, or quasi-promise, that his son should be appointed to succeed him acquired special urgency ... The result was that my brother [5] got appointed to his father's place, and that he, his wife and child left for Martinique in the spring, I think, of 1836 ...

On all these occasions I was never treated as a child, but as a member of the family, always dining at table, and almost invariably invited out with my mother and sisters – at 15 I had attained my full height and had as much beard as I ever have had, so that if a boy still in years and on many points, in fact, I looked altogether the young man.

And indeed the time was at hand when I had to exercise a man's choice and determine for myself the course of my life. Before the year was out from her leaving us, my sister Maria had to return to Europe with her child, owing to a painful affliction which necessitated the care of a first rate surgeon, such as could not be found in the West Indies ... We had returned to Paris to meet her [6]; I also returning to my old college. I was now in the *rhétorique* class, and it became necessary to decide as to my future. I was by this time, I may say, practically a Frenchman. I habitually spoke French, thought in French. My interests were chiefly centred in France. My friends of my own age were French without exception, and I was very fond of them. I was particularly attached to my married sister and her child, and scarcely less so to her husband. I had had a distinguished career at the *Collège*; success, I may say, was assured to one if I chose to remain in France. Both my sisters wished me to do so. At sixteen the choice which should determine my life time was left to me. I may say I longed myself to remain (for so I really was) a Frenchman. My mother used no arguments to persuade to the contrary, but in the course of the discussion she said: 'I am sure that if your father had been alive he would not have wished you to be a Frenchman.' That one saying decided me. I chose for England.

I believe that decision – the immolation of all my own wishes to the desire of a dead father – aged me morally by ten years. Fortunately for me I did not know at the time what it would cost me – did not foresee the long years of spiritual solitude which I should have to endure. If I had done so I trust I should yet have had sufficient strength of mind to make the choice I did, for I know that it was a right one. But bitter was the misery through which I was to fight my way.

In case, however, of my ever changing my mind and returning to France, it was decided that I should take out my degree of *bachelier ès lettres*, virtually indispensable in every French profession, which I did as related in the last chapter. The earliest regular age for it is 16, but a *dispense d'âge* is sometimes granted. For this I applied, but it was refused, otherwise I should not have returned to the *Collège* for about a year. I obtained the degree, I may mention, without any special coaching, such as is usual. I thought myself I passed a very bad examination, but I was told afterwards that my repeating from memory (not being able to read a word or even find the page for sheer fright) the opening lines both of the Iliad and of the second Ode of Horace, and going on until stopped by my examiners in both cases had carried me through triumphantly. Had I been an examiner I should have plucked myself remorselessly. Among my examiners were Jouffroy for Philosophy and St. Mare Girardin for History.

Note.   I have lately (August 1902) come upon a batch of old exercise books and letters of my college period of the survival of which I was unaware. I do not know what French education is now, but I may safely say that at the period I am speaking of it was – apart from the absence of personal experimentation in the physical sciences – admirably thought out, and judging from what I saw afterwards of the results of English public school and college education especially in Mr. Ker's pupil room, immeasurably superior to English. Anyone who worked steadily through his classes in Paris must have come out an educated man. I may fairly say that until I met Kingsley and Charles Mansfield I never came across a young Englishman whom I could have called such, and most of their education had been worked out by themselves since leaving college.

* * * * *

Before my Mother and I should return definitively to England, my sister Maria having now quite recovered her health, and for the sake of spending one last year all together, it was decided that the whole family – my Mother, my elder sister and myself – should accompany her back, the intention being that my Mother and I should return to England via New Orleans and the United States so as to be back by the autumn of 1838 . . .

In Fort Royal I saw one of the most dreadful sights I ever came across in my life. We went to the *Maison des Fous*; such a term in France would mean a hospital or asylum. Nothing of the kind at Martinique at that time. It consisted of a long row of huts, or rather kennels, covered over but open on one side to a paved walk about four feet broad, some of the kennels barred like wild beasts' cages. Each kennel was about six feet broad by seven in depth. Anyone might walk through – at least as respects the male inmates; I was told that the treatment of the female inmates was just the same, only that men were not allowed to see them. In each of these kennels an inmate was chained – the more dangerous ones in the barred cages, with manacles or foot irons besides. No means of occupation; no privacy of any kind. All the inmates were negroes or men of colour; I do not remember whether white men would have been confined there as well. Those in the barred cages looked for the most part actual fiends, which is what such treatment would easily turn them into. We were most earnestly besought by one man of colour who spoke French and not Creole, to procure his release. 'All the others who are here' he told us 'no doubt are actual madmen, but I, gentlemen, I am Jesus Christ. I cannot do my work whilst I am here.' Let me say at once that this horrible state of things had ceased to be by the time of my second visit to the island . . .

As I am one of the few surviving who have seen slavery in the West Indies, it is but right that I should say something upon the subject. Although one heard occasional stories of cruelty I don't think the French slaves were as a rule unkindly treated – certainly not those of my brother-in-law who was constantly applied to by slaves to buy them (he was known to be in favour of abolition). I certainly don't think I have ever seen any sleeker, merrier set of servants anywhere than his slaves. Having said so much, I have to add that my experience of slavery has made me hate and loathe it. It degrades the slave; it degrades the master; it is an abomination before God. Purity of manners disappears necessarily before it; self-will, self-indulgence become the rule for the master – the slave follows his example as far as he can, and in turn corrupts the young generation of whites. Although the moral tone of the French West Indians was far superior to that, for instance, of certain parts of Brazil which simply cannot be described in print, I will just say this, that on my return voyage from the West Indies on this occasion a girl of eleven, daughter of a Martinique gentleman, who was being sent home for her education with an elder sister and a younger brother of nine, was found one day trying (success in the attempt being of course impossible) to teach him incest upon herself.

In order to dismiss this subject at once, I will say that when after the revolution of 1848 slavery was abolished in the French West Indies, notwithstanding the kindness with which my brother's slaves had been treated, all but three or four out of some forty left him. They wanted, they said, to see the world, which for them would mainly consist in loafing about the streets of St. Pierre. All professed great affection both to him and my sister, and admitted that the wages he offered were more than they could earn by huckstering as they mostly looked forward to doing. But they evidently wanted to *feel* themselves free. I cannot blame them.

CHAPTER V

A London Boarding House – A Conveyancer's Chambers –
My Early Impressions of London Society

On our return from our first voyage to the West Indies my
mother and I took up our quarters temporarily at a boarding
house in Weymouth Street, Portland Place, on the suggestion
of my Uncle Ludlow who lived there. Left a widower some
years before, and all his sons being in the Indian Army, his
only daughter in Australia, he did not care to keep house, and
boarding house life suited him. He was really as good as the
master of the house, without any trouble or responsibility.
The two Scots ladies who kept it – a widow, Mrs. Barton and
her sister Miss McIvor – were both devoted to him, and stood
in great awe of him. He had a very large room on the ground
floor, took the head of the table at all meals, and being a
capital carver (those were carving days) was a most valuable
guest on that account only, but still more so for himself. He
was as handsome and healthy looking an old man as you
could ever see, very rosy-cheeked, clean shaven except as to
his grey whiskers, comfortably stout, always clean and neat,
the very pattern of an elderly English gentleman. His
occupations, as I have said before, had been varied; he had
numerous relatives in the Indian Army, others in the Navy,
and he had brought a good deal of good and varied
experience from many points of contact with his fellows who
were guests at the dinner table. He was thoroughly upright,
kind hearted and genial to boot. He liked company at meals,

was indulgent to other people's weaknesses, and his presence at the breakfast and dinner table was undoubtedly a main attraction of the house. Into the drawing room on the first floor, on the other hand, he did not put his foot once in a twelvemonth, retiring after meals to his own room, in the evenings, to smoke over a single glass of spirit and water . . .

((Also boarding were three)) foreign counts, and of the genuineness of two I always had my doubts. The genuine one was a Count Salomes, a Greek from the Ionian Islands, half-brother to a noted Greek poet of the day, a very handsome young fellow with jet black hair and moustache; very good natured, very ignorant, and very empty headed. He had recently become engaged to an Irish lady, a Miss Taite, and the open way in which he spoke of his love affairs and descanted on his lady love's perfections at the boarding house dinner table was most amusing . . . Among the occasional boarders were an American couple of the name of Hyde, the husband an intelligent man with just a soupçon of Transatlantic twang. The lady was a very pretty young woman of about six or eight and twenty, and I was looking at her with great pleasure until she opened her lips for the first time in my hearing in tones so nasal that I cannot describe my feelings . . .

Before we had been long in the house we found that 1st all the elder ladies were setting their caps at my uncle, 2nd that Mrs. Weare, though nearing her 70th year, was openly professing her willingness to marry again (she as good as offered herself to my uncle), and that the two disengaged foreign counts were both either angling for her or screwing up their courage to do so . . . All this became soon extremely disgusting to us, as it was already to one of my uncle's sons, then home on furlough, and who had taken lodgings rather than go on 'feeding in a stall' as he phrased it, and we also took a house after about two months. Of the boarders who were there with us, none succeeded in making a prize of my uncle. But there was a dark horse, or rather a dark mare, in the running in the person of another old lady, a former boarder, a Mrs. Hooke, who had come two or three times to dinner while we were there to whom he became married, not, I am afraid, for his own happiness. But he survived her . . .

I have dined in Paris at a *pension* of a much humbler character than the Weymouth Street boarding house above spoken of, as the guest of an *employé* at 2,000 frs. a year, and I am bound to say that it offered none of the repellent features of the latter... There was throughout friendly feeling, mutual courtesy, pleasant if not brilliant talk. It was not 'stall-feeding' in any way.

It is indeed a curious thing how utterly unsuited the Englishman is for the casual society of a common table. I have heard as good converstaion at a French, a Belgian, a Swiss *table d'hôte* so long as there were no English people present or very few – as at almost any gentleman's table. But besides that, the Franco-German war has in a strange degree diminished much talkativeness; the always growing invasion of the continent by railways and Englishmen has pretty nearly ruined the old *table d'hôte*. An Englishman at such a table is either absolutely silent or else he talks loud to his own set, wholly ignoring the rest of the company. Now the essence of the *table d'hôte* is that you meet simply as fellow-creatures. You knew nothing of each other yesterday, you will know nothing tomorrow, but here and now you are fellows. In the old diligence days I have seen maid servants sit down to the inn dinner by the very side of their mistresses, as a matter of course, without giving the slightest offence either to them or to the other guests, behaving just as well at table, and showing themselves just as ready afterwards to wait upon the mistress and obey her as if they had been relegated for dinner to the furthest back kitchen. And to me the sight has been a delightful one...

* * * * *

I was entered of Lincoln's Inn in Michaelmas term, 1838, and as I did not belong to an English university had in those days to wait five years before I could be called. At the time I speak of there was absolutely no legal instruction provided for students. And yet the absurd form called 'Exercises' showed that in former days real disputations on legal questions used to be held before a bencher, a very practical form of education for the bar. There were required to be, if I recollect aright, eight of these exercises and they were performed as follows: In the lobby of Lincoln's Inn Hall, the steward or one of his assistants, put into your hands a pasteboard tablet on which a sentence was written. With this you presented yourself before a bencher, who only allowed you to read the first words, e.g. 'I hold that A will take the widow's estate' – 'That will do, Sir', – and the 'exercise' was performed. A more disgraceful swindle both of the student and of the public could not be imagined. For about £100 and the going through this ridiculous farce a certain number of times, a man obtained the right to share in the privilege of practising at the bar, and whatever might be his ineptitude, shielding from responsibility both himself and the most blundering solicitor who chose to employ him. Practically, however, the result was in almost all cases that he had to pay for the same thing twice over, viz., on the one hand for the right of admission to the bar, on the other hand for the acquisition of such knowledge as could alone fit him for practice, by entering himself as a pupil on the Equity side, to a conveyancer, usually followed by an Equity draftsman, on the common-law side, to a special pleader. I did not feel inclined for the common-law and as my uncle was well acquainted with one of the leading conveyancers of the day, Mr. Bellenden Ker, I was introduced to him and took my place in his pupil room, in a set on the first floor of No. 8, Old Square, Lincoln's Inn.

There is now, I believe, no such thing as a pure conveyancer, such as the Duvals, Kers, Cootes, Jarmans, and others of those days, the simplification of the law, the shortening of deeds and the better education of the solicitors, having both diminished conveyancers' fees and brought down much of their work within the solicitor's sphere of action. But in those days, with a well filled pupil-room at 100 guineas a year each, a conveyancer's income, while never reaching anything like the figure of a leader in the courts, might often amount to £2,000 a year, and in Duval's case was said to reach £3,000. There were, I think, from 8 to 10 pupils in Mr. Ker's rooms during most of the time I was there, though the number latterly fell off . . .

I soon found that I was the only Liberal (Radical in fact) in chambers, and had a good many political tussles with my fellow pupils, but held my own, and was at last pretty well left to myself till, as fate would have it, there came to the pupil room an ill-educated, stupid attorney from one of the Southern counties, who swore by Mr. Cobbett, could not call a clergyman anything but a black slug, and dropped and added his h's miscellaneously. Now I could fend for myself, but O ye gods and little fishes, how I suffered at having this would be political ally thrust upon me! I tried at first to help him as far as I could, but he understood me so little that I was obliged at last to leave him to be baited; though indeed, as he never knew when he was beaten, it didn't harm him. I remember on one occasion this M—, while munching his pen as he was in the habit of doing, breaking out dogmatically with the assertion: 'I don't think much of the manners of the female haristocracy.' This dogmatic proposition seemed so incongruous and surprising, coming from M—, that a dead silence ensued for a moment. Then Greenside, in the blandest tone, put the question, 'May I ask, Mr. M—, where you have had the opportunity of watching the manners of the class you speak of?' 'Oh, walking up and down Regent Street.' And then there was a roar.

A very curious personage was Mr. Ker's clerk, W. H. Comyns, reputed one of the two best conveyancer's clerks in Lincoln's Inn, (later on the author of *Comyns* on Abstracts of Title). If you can fancy Sancho Panza, the master, and Don Quixote the servant, but without the slightest fellow feeling between them, you have about the relation between Ker and Comyns. Ker was short, rotund, rubicund, quick, talkative, irascible; Comyns was tall, thin, whitefaced, deliberate, taciturn, saturnine – I might say, except with those he liked – spiteful. Neither understood, neither tolerated the other. Ker would say openly that he hated Comyns; Comyns had never a good word for Ker. And yet the two had been together for years, and kept together to the last. Ker absolutely trusted Comyns with one branch of his business, all drawing for the Ecclesiastical Commission, to which he was counsel; made him sign for all his orders for the Zoo! For myself, I always got on well with Comyns and learnt a great deal from him.

I soon found out indeed that Ker's chambers were a place where one was expected to learn without being taught, but where nevertheless one could learn a good deal by finding out how. If even after the most careful reading one felt a difficulty as to some legal point, there was no use in asking Mr. Ker; he would tell you sharply that it was your business to find out. But if you gave him his head, he might pour forth an unstinted flood of legal lore, so long as you simply listened, shutting up abruptly, however as soon as a question was asked. Otherwise the process of learning consisted in this: you were expected to read text books at home; in chambers you were set down to copy out in a book of your own the Ms. precedents used in chambers; after a time a draft was given you to draw, or an abstract to read and make marginal notes on in pencil, without the slightest explanation.

Then you were called into Mr. Ker's room when the time came for his settling the draft or perusing the abstract, or if he did either at home, it was handed back to you with his own corrections. I never cared much, I own, for ordinary drawing or abstract perusing. I was more interested in cases when, as one became a little more capable, these were handed over to the pupil to write upon. But the business that interested me most arose from Ker's position as conveyancing Counsel to the Board of Trade and to the Woods and Forests. In the former capacity – it being before the days of limited liability by general act of Parliament – he had to draw or settle a large number of deeds of settlement or charters for companies seeking incorporation from the Crown – then the only mode of obtaining limited liability, otherwise than by special act of Parliament. In the latter capacity – besides some interesting cases – he had frequently to draw Acts of Parliament, for which purpose he was also employed by other departments of the government, and somehow or other I developed a certain ability in drawing both deeds of settlement and Parliamentary bills which led to these being pretty generally handed over to me. And this again drew me into closer personal relations with Mr. Ker than, I believe, any other of my fellow-pupils. In short, I occupied with him for a longer period, I believe, than ever had been known before, the position of 'favourite pupil', and continued to 'devil' for him for years after being called to the bar – never paid by him directly for anything but, I believe, whenever he could get it for me, receiving pay from the Government for my services, or from private firms for drawing deeds of settlement when he was to draw the charter. [I remember, some years after I was called to the bar, a bit of very sharp practice on the part of a solicitor, head of a great firm, and bearing a high repute as a patron of art and legal reformer, who indeed for some time really befriended me – a very pleasant fellow to boot. On Mr. Ker's recommendation he instructed me to draw on my own responsibility, a deed of settlement and a charter for a company to be settled by Mr. Ker. In the deed I inserted of course all the provisions and none but the provisions which I knew would pass muster at Mr. Ker's hands. My client wished me to insert certain provisions which I knew would not be passed. I declined to do so. He insisted, saying I was his counsel, and bound to do what I was instructed. I inserted them, but appended an opinion stating that in adding the provisions as instructed, I felt sure that they would not be accepted by the government. And I wrote a private note to Mr. Ker telling him what I had done. I heard nothing more for about six weeks, when having to call upon Mr. Ker in reference to another matter, he told me that he had just found out that my client, once in possession of my draft, had gone to the Board of Trade with a fair copy of the deed and the draft charter under my signature (without the opinion) on the ground that the settling of the deed and charter by me were equivalent to the draft being settled by Mr. Ker himself and had all but got his charter sealed when by accident the latter heard of it, and was able to stop the affair. This dishonest trick was damaging to me in two ways. On the one hand that was the last piece of business but one that I ever got from the firm: on the other hand, warned by the Court, Ker ceased to recommend me for drawing deeds of settlement for his characters.]

I was for so many years in close connection with Mr. Ker that I may as well say here pretty nearly all that I shall have to say of him, though anticipating greatly on the date of various of these reminiscences. For good or for evil, he certainly greatly influenced my life; for a good many years he tried all he could to help me, and in one instance at least, many might think in two, I feel now that I did unwisely in not accepting the help he offered me. Remembering the sort of man he was, and the difference of age between us, I am surprised now when I look back to think how long he bore with me, considering how sturdily I stuck to my own way instead of following his, and how often in small matters I openly thwarted him. Yet when he did turn in a manner against me, the only matters which gave occasion for his doing so, as far as I can surmise, was not such as I could in anywise blame myself for I owed him much and I am very glad to have known him though, as a shrewd solicitor friend once told me, I am afraid on the whole he did me more harm (speaking in a worldly sense) than good.

Ker was a man who ought to have reached much higher than he did, and under given circumstances would have done so. Scarcely any of the cabinet ministers I have come across, I must say, were his equals in point of ability. He was in fact one of the three or four cleverest men I ever met with; for quickness of thought I don't think I ever knew his equal. His father's name had originally been Gawler (his uncle still retained that name), but the father had taken the name of Bellenden Ker by royal license, at or before the time of his becoming a claimant for the Duchy of Roxburgh. I never knew Mr. John Bellenden Ker, but from the son's account of him he must have been a most eccentric personage. Travelling in Spain as a young man, he had run a French valet through the body with his sword (fortunately without killing him) because he would say Toledó instead of Tolédo. Being a captain of the Guards, and silverstick in waiting to George the 3rd, but also a great friend of Horne Tooke's and of the extreme Whigs of the day, he had lost his office, and his commission, through the following extraordinary trick: It was in those days the custom, every time that the King passed under the archway at the Horse Guards, for four trumpeters, at the command of the officers on duty, to sound their trumpets. Captain Gawler being on duty on such an occasion, bought four penny rolls, took out the crumb from each, and with his own hands stuffed it into the trumpets of his four trumpeters. Then when the King's carriage came through, he gave the order to sound, when of course no sound came. Captain Gawler meanwhile commanding his men loudly to 'Blow, rascals, Blow!' The trick of course came out, with the result above mentioned. In his old age, Mr. John Bellenden Ker had two entirely different occupations. Passing through Egham or Windsor, I forget which, he noticed a flower which was new to him in a cottage window. He stopped, went in to inquire about it, bought it and took it home. It was an auricula, and from that time he grew by degrees to be one of the leading auricula fanciers in England, never caring for any other flower. His other hobby was a droller one still. He had travelled in Holland and spoke Dutch (he was a very good linguist), and he took it into his head that all our nursery rhymes are covert satires on priests, monks and friars, to be interpreted through Dutch, and published a work on the subject, which I believe is now very scarce. I have seen it; it is very ingenious, though I confess it did not impress me. But years after his son told that travelling by rail opposite to an evident Dutchman, and having by him a despatch box received in connexion with some Royal Commission of which he was a member, his name being printed in gold letters outside, he had observed the Dutchman his *vis-à-vis* on noticing the name became very excited, till at last he could contain himself no longer and asked, 'Are you the great Mr. Bellenden Ker?' On Ker's telling him he was only the son of the person probably referred to, he began to speak with enthusiasm of the book, saying that he had verified all the instances quoted, and that they were all quite true . . .

Ker in a good humour was certainly about as pleasant company as I have ever come across, his mind stored with anecdote, flashing both with wit and humour. [His stories were just as likely to be against himself as against other people. The following somewhat *risqué* one is by no means the most so of the former class. He used every week to go and see his mother, who never had a newspaper, and always depended upon him for news, which for her meant personal news only, gossip in short. One day he could think of absolutely nothing, and found himself reduced to draw the long bow, 'Well, the only news is, that the Pope has taken to keeping Fanny Elssler'. 'O! how terrible!' – At his next visit he found his mother in a state of towering rage – but not with him. 'I am going to turn away my confessor! I think I shall become a Protestant! What you told me was not true. I asked Abbé . . . whether it was so, and he flew into a passion with me, called me a wicked old fool, said it was probably a mortal sin for me to have believed such a thing, and that the least I could do was to perform some severe penance'. Ker must have been then nearly sixty, but no schoolboy who has escaped a whipping could chuckle over a successful trick with greater enjoyment than he did over this story.]

There was a great deal of kindliness in him in his better days, and in his way, of high purpose, though latterly I am afraid he became a good deal of a jobber. He had married a solicitor's daughter, a very handsome, clever, and at heart most kindly woman, of whom he was in fact very proud, but they certainly did not hit it off together. She was proud and reserved; not to say cold in manner; he was hot-tempered, outspoken, coarse in his language, so that he was perpetually wounding her, and she as perpetually putting him out. With another woman of sufficiently strong character, but who would have better known how to manage him, all the good in Ker might have been brought out; another husband, who could have appreciated her good qualities, Mrs. Ker might have made thoroughly happy. I was always on the most friendly terms with Mrs. Ker, whom I liked very much, and I have always thought that the first thing which began to set her husband against me was that Mrs. Ker consented at my request to join the Committee of a Needlewomen's Institution, after he had predicted that she never would. 'If I had asked her she would not have done it', he said angrily.

She had a younger sister, Miss Harriett C—, one of the first ladies who took up wood engraving on its revival in this country, the most sweet tempered creature, of whom Mr. Ker was really very fond and who was devoted to him. Yet I have known him to take a positive brutal pleasure in domineering over her and exhibiting her subserviency to him . . . ((Added here, and later half struck out, is the following anecdote:))

[I remember on one occasion at Cheshunt, where he had a charming country place, as we were sitting out on the lawn, his falling into a disquisition on the necessity of keeping women subservient, and by way of illustration untying his shoestring (he never wore boots) and calling to his sister-in-law, who was sitting on the lawn with us, and had heard all, 'Harriett, come and tie my shoestring for me.' I was indignant, and called to her, 'Don't do it, Miss Harriett!' The end was that she did it. 'He would never have left me quiet if I had not', she afterwards said to me. In my disgust I used words to him, which I ought not to have done, at all events to one so much my senior, and which I am surprised now that he bore with.]

At the time I first spoke of, however, and for many years after, nothing could exceed Ker's kindness to me. He was an old and intimate friend of Lord Brougham's, who was his wife's special antipathy (I have heard her say that her feeling toward him was what some people have for a cat), and he tried very hard to get me into the set of what I may call the Useful Knowledge Society Liberals. At an early period of my pupilship I was asked by him to meet Lord Brougham at dinner . . . I may here say that in the House of Lords he was a most undignified Lord Chancellor, perpetually changing one ungraceful attitude for another, making signs to this or that peer to come and speak to him, and in fact ceaselessly restless, as if the whole thing were an utter bore to him. (It is but fair to observe that the woolsack, having no backing, is a much less dignified seat than the Speaker's chair and renders specially prominent any ungraceful ways or tricks of manner in its occupant).

On the occasion I speak of we were eight at dinner, Mr. and Mrs. Ker, her father Mr. Clarke, a very clever man, but too highly educated in those days for his profession, whose great complaint as he grew older was that he could not get Greek authors in a sufficiently large type for his failing sight; then there was Bingham Baring afterwards the second Lord Ashenden, a Mr. Gower, and John Wrottesley afterwards the second Lord Wrottesley. Brougham was very pleasant and interesting. The conversation fell amongst other things on the French Revolution, and Brougham referred to the personal disinterestedness of Robespierre and his fellow members of the *Comité du Salut Public*, as evidenced by the fact that when arrested he and others were found in possession of just the number of days' salary (at 33 francs a day, if I remember aright, paid *in specie*) which would cover their current month of office.

As there can be but few comparatively now remaining who have seen Lord Brougham in the flesh, I may mention one facial peculiarity of his of which no portrait can give an idea. The tip of the nose was moveable, and could be directed – whether voluntarily or involuntarily I cannot say – up and down, from side to side, or even made to perform a complete gyration. It is said that a worthy Quaker who went up to him with a deputation as its spokesman was so amazed at the behaviour of Brougham's proboscis that he entirely lost the thread of his thoughts, and was obliged to sit down. This curious tic, however, was less noticeable in his old age when, as a member of the Council of the National Association for the Promotion of Social Science, of which he was President, I had pretty frequent opportunities of observing him. But at the dinner of which I spoke, the nose was scarcely at rest for a minute.

Brougham's memory was prodigious. His *Sketches of States-men of the reign of George the 3rd*, forming four volumes, if I recollect aright, of a series published by Bright, were dashed off *currente calamo* in the course of a few weeks, with scarcely a reference to any authority. He was quite unconscious himself of his extraordinary power in this respect, – on some occasion during the 30's justifying his statement of the vast dimensions (to within inches) of the cupola of St. Peter's on the ground that he and Playfair (the astronomer) had measured it in the year (I think) 1814! As bearing on this subject I may mention that Mr. Ker once showed me the notes of one of Brougham's great speeches on Education which had taken three hours in the delivery. The strip of paper was about a foot and a half long, but did not contain more than perhaps 25 or 30 words, never more than 5 or 6 together, these being separated by unequal spaces, corresponding, I presume, to the time intended to be taken in developing the ideas which they represented.

Of all the distinctions which had been bestowed upon Brougham, the one he was most proud of was his member-ship of the Institute of France. He had indeed serious thoughts of claiming French citizenship by virtue of it (though I doubt if this would have been admitted) and taking up his abode in France with a view to a second political career in that country. Unfortunately his knowledge of French was by no means on a par with his own estimate. I have in my autograph book one of his, given me by Ker, which is an excruciating specimen of French of 'Stratford-atte-Bowe'. It was addressed to the Librarian of the French Institute:

Ce 9 Oct, 1845.

Ayez la bonté de laisser avoir le porteur de cette lettre (M. Ker) le 5eme volume des mien de l'Institut pour l'annee 1821-2 pour moi

Brougham.

Few people know that in his latter years Brougham perpetrated an anonymous novel, *Albert Lunel*, on the subject of the persecution of Protestants in France in the last century. I have read it. It is very dull . . .

Although, as I said, Lord Brougham was perfectly good humoured on the occasion of the dinner I have mentioned and even addressed a few words to the young law student, I cannot say I felt inclined towards him, and never cared to claim acquaintance with him, declining even at a later date an invitation addressed to me from him through Mr. Ker to join the Council of the Useful Knowledge Society, which indeed broke up altogether after a few years more (it had latterly become, I believe, almost a purely Jewish body, so much so that, as Mr. Ker has told me, on the occasion of one of their dinners Dr. Mowbray Fen, who was in the chair and himself were the only two non-Jewish members present). Nor did I ever exchange a word with him in after years at the Council or other meetings of the Social Science Association. I have never cared to know those for whom I had neither regard nor respect and I cannot say that I felt either towards Lord Brougham . . .

Another interesting person whom I have also met and stayed with at the Ker's house at Cheshunt (a very charming place) and with whom I always remained on friendly terms, though I never met with her again after the last time we were

together at Cheshunt was Harriett Martineau. At the time I
speak of she had not fallen in with Mr. Atkinson, and was so
persuaded of her personal immortality that she had fixed on
the particular star which she was certain to inhabit after
death. She was always very pleasant, but her oracular
utterances were sometimes astounding. I have heard her say
that she considered Monckton Milnes 'The humblest minded
man in England', and again that 'three fifths of the women of
England smoked' . . .

Nor could one omit, in referring to Ker's Cheshunt life, to
notice his relations with his Haileybury friends. Haileybury
College, then the training institution for Indian civilians, was
within a drive, and the Haileybury Professors were the
nearest good company within reach. Several of them I have
met at his house, and I have also dined in company with him
at the College . . . There were several very able men among
the Professors, Mr. Ker's special friends being Empson,
Professor of Jurisprudence, who was also Editor of the
Edinburgh Review, Jeffrey's son-in-law; Jones *(Jones on Rent)*,
Professor of Political Economy, and Heaviside, Professor of
Mathematics, the two latter clergymen, of a somewhat – what
shall I call it? – easy type, Heaviside indeed afterwards dying
a canon. Both these were exceedingly clever and great
gourmets and the table certainly was first rate. There was a
story against Heaviside that, a dish of quails being handed
round, he had suddenly turned round to the neighbour who
would have been served before him, and exclaimed: 'I'll give
you half a guinea for your chance!' He had observed that the
number of birds was less than that of the intervening guests,
and had calculated the chances of any reaching him. At the
time I speak of, he was a very handsome man still. Empson I
did not meet at Haileybury, but first at Ker's and afterwards
pretty frequently at the Athenaeum Club; a thoroughly
kindly good man. In consequence of having stumbled upon
an anonymous work of mine, *Letters on the Criminal Code,*
which I had laid on the library table of the Club, Empson
asked me through Ker, to contribute to the *Edinburgh*, which I did on one occasion. But he had a habit of
what he called 'pumicing' his authors' contributions which I
could not stand, though he declared that he had altered my

article less than any other in the number in which it appeared. Another Haileybury notability whom I saw but only for a few minutes was the Professor of Persian, known as the Mirza Ibrahim, a very good looking and bright looking black-bearded man. By common consent he was one of the cleverest men there; his command of English was perfect, his familiarity with Shakespeare astounding – but still more the mastery over English slang and terms of abuse, by means of which he had been known to silence even a cabman. He was eventually invited to return to Persia in order to take a Professorship of English, on very handsome terms. He was desirous of accepting, but he had discovered an excellent English cook and further educated her to a level of admirableness. He felt he could not part with her and he wanted her to go with him, but the lady's one condition (she was no longer young, and anything but beautiful) was marriage. To the amazement of his brother professors, the marriage took place – in an English church be it observed, which would probably be no valid Mussulman Marriage. And Mr. & Mrs. Mirza Ibrahim departed for Teheran . . .

Though I have endeavoured to keep the balance equal, and have not told the worst I know of him, I almost fear that my judgement of a character so complex, so full of contrasts and anomalies as Ker's may have been too severe. Considering his upbringing, or rather his total want of upbringing, it may be said that his faults were due to others, his good qualities were his own. That he was fully capable of appreciating goodness and high purpose is clear from the fact that I never knew him so much struck by any book as by Stanley's *Life of Arnold*. He spoke of it for full six months. I am afraid that what most weakened the impression of the book with him was his meeting Chevalier Bunsen at dinner, and while ready with Arnold to 'drink wisdom from his lips', finding in him only a 'clever German' – a feeling which, I own, was very much my own in somewhat later days, when I in turn knew the Chevalier . . .

I have made a similar observation before, but I must here repeat that there is an extraordinary difference betwen the England of 1838 and the England of the present day. Scarcely an educated or merely well-to-do Englishman but is much less insular than he was then, much more cosmopolitan. The opportunities of foreign travel have placed the well-to-do artisan in point of knowledge of the world, well ahead of the bulk of the English gentry of those days. It was not only that people went much less abroad than now, they were far less capable of understanding what they saw there. The ignorance of the contemporary literature of the Continent for instance was something astounding. I happened to be thrown almost at once as will have been seen from a previous chapter into a society which prided itself on its cultivated tastes, a society of Edinburgh Reviewers and the like. They might be familiar with French, but the only French writer of that day whom they knew was Paul de Kock. [What revolted me most was that George Sand, although known by name, was considered unreadably immoral, whilst Paul de Kock, if not read by ladies, was laughingly tolerated by them as reading for their husbands, the difference between the two being that of a noble river too often eddying from rock to rock or made muddy by storms, and a common muck-pond.] Lamartine, Victor Hugo, Michelet were mere names to them. Of the race jealousies of the Continent, of the existence of a Hungarian question, a Bohemian question and the like, they were absolutely ignorant. The same with art. Coming after the French *Salon* and the works of Ingres, the Scheffers, Dela-roche, Delacroix, Léon Cogniet, I was amazed at the poverty of the Royal Academy, and at the praise lavished upon niggling elegance or crudeness of conception. So with society. Though we did not go out much, yet I had opportunities of judging the society of the day under various aspects – at Mr. Ker's, whose caustic humour would tolerate neither dullards nor fools; at the house of an East India Director, an old friend of my father's; at the house of a solicitor in large practice, my uncle L's old friend, and elsewhere. Yet I missed everywhere the easy freedom of French society, its unrestrained bright-ness. London society I found prevailingly dull, or when brilliant, it had no ease about it. The dullness of a whist evening at the H's in Russell Square was simply awful. I contracted such a horror of the game that I not only forgot it as soon as I was able, but have always shrunk from the idea of learning it again. The only source of interest there lay in our host who had contracted the habit of going to sleep at the whist table with his eyes open, his sleep being profound for all intents and purposes except those of the rubber. You might say anything you pleased on any other subject close to his ears – bring the falsest accusations against him; the doors might open or shut; guests come and go, sing, dance, shout – he was asleep to everything, but nothing that took place at the table in reference to the game escaped him, and he went on playing with just as much care and judgement as before dropping to sleep . . .

Although I was on good terms with my fellow pupils at Mr. Ker's, I did not for many years make any friends of my own age, and only began to do so when the strength of a common purpose bound me to others. At an early period I became a member of the Reform Club, to which both my uncle Ludlow, my father's brother, and my mother's brother my uncle Frank (Mr. F.C. Brown) belonged. I continued a member of it until I was elected to the Athenaeum, and I am bound to say have often regretted both its splendid Parliamentary library and the privilege of freely inviting a friend to dinner. Several friends of mine, as J. Bonham-Carter, joined it after I did. Without ever speaking to him, I used constantly to see Thackeray, generally busy writing at a particular table at the window. Cobden I have frequently seen there, both before and after his election to Parliament, and once exchanged a few words with him. Another member who could not fail to be known to every other, whether he knew them or not, was G.F. Muntz, the ironmaster, Radical M.P. for Birmingham. I do not know whether it was from having to speak amid the din of a foundry, but his voice was simply a roar. In whatever room he was, every word he uttered was heard by all present, and it was impossible to go on doing anything whilst he chose to speak. I remember hearing him shout from nearly the other end of the large drawing room in answer to an unheard but easily inferred question: 'My father was a Pole, and I dare say he was a Jew.' It is one of the strange 'revenges' of Time's whirligig that the 'Muntz M.P.' of a later day should have sat on the Conservative benches . . .

In company with my uncle Ludlow, I attended Carlyle's Course of Lectures on *Revolutions of Modern Europe,* (May

1839). I had read nothing of his, although *Sartor Resartus* had already appeared. It was then commonly supposed that he had been brought up in Germany, and that this accounted for the strangeness of his style. His delivery of the lectures was very painful – hammered out, as it were – though there were some very fine passages. The audience throughout was but small. My uncle, a dear old Philistine, expressed his opinion of him by saying that 'the man had some stuff in him, but he ought to learn the language better before attempting to lecture'. My own impression was that of a constant straining after effect, combined with difficulty of expression. And after hearing him on this occasion I did not feel tempted to attend any other course of his, nor, in fact, to put myself out of my way at any time to hear him. But indeed I was never a Carlylomaniac nor even a Carlylophil.

CHAPTER VI

A Few Impressions of Travel – The Martinique Earthquake –
British India Society and Anti-Corn Law League –
Second Visit to West Indies

During the three years after our first return from the West
Indies, my mother and I used every year to make a tour
together in the 'Long'. The first year (1839) crossing over to
Ostend, we went to Belgium, and did the Rhine, returning by
Brussels and Paris . . . In the following year (1841) our tour
was a home one. We went to Ireland, stopping on our way at
Birmingham, Manchester and Liverpool, with letters of
introduction from my Uncle, Mr. F. C. Brown, or from his
agents, Messrs Forbes, Forbes & Company, to the heads of
important commercial establishments. Thus at Birmingham
we saw the nail-factory of Joshua Scholefield, then M.P., vast
and airy, with an engine of 36 horse-power. We saw the iron
cut as if it were paper by immense shears, then made straight
off into nails, the head at first being quite brittle. The workers
were mostly men, the rest boys of from 8 to 12. We went next
to the Button factory of Hammond and Turner, where
everything was done by hand, the workers being mostly
women, fairly young, though one or two quite old, with a few
young men and children. From thence we went to Collis and
Sons, metal workers generally, and to Jenners and Betteridge,
the *papier maché* manufacturers, winding up with a copper
tube factory worked by a 90 horse-power engine. I found the
Birmingham working people, whether men, women or children, fine looking and healthy, and in general most cleanly;

the young girls often quite roguishly dressed, with pretty pink frocks, often coral necklaces, ear-rings almost always.

I shall tell in the next chapter our anti-Corn Law experiences at Manchester, which took up the whole of our first day. The second was devoted to visiting Manchester establishments. Our first visit was to Birley's cotton-mill, the largest in Manchester. It comprised both a spinning and weaving factory, but the former alone was at work. The two together employed 1400 hands, besides those employed in an India-rubber goods department. Every establishment was on a large scale, airy and well lighted; the work-people too, though sometimes pale, were healthy and cheerful . . .

During our first three years' stay in England we lived in the Portman Square quarter, first in York Street, afterwards in King Street . . . The house in York Street was owned by a Mrs. Popham, widow of the Admiral of that name, and was full of pictures, including several Loutherburgs.[7] I used to walk backwards and forwards to chambers, generally taking the Reform Club (when I had become a member) on my way coming back. We were comfortably off, not extravagant, and could indulge in an evening at the opera or a concert without feeling it. Perhaps our greatest pleasure was the choosing and sending out a monthly box of books to the West Indies, where those lived whom we loved best.

One day, a terrible piece of news for us was in the papers. There are three scourges which occasionally devastate the West Indian islands, hurricanes, fires, and earthquakes, the last being the least frequent. Now hurricanes blow down or blow away, fires burn down all wooden buildings; earthquakes shake down pretty well all that is not built of wood, the latter material being elastic, and yielding to the shocks. At the time of my first visit a hurricane had been the latest calamity, and in consequence of it an *ordonnance* had been promulgated, that no houses beyond a certain height should be built of wood. My brother-in-law had been building a new Treasury Office, of brick and stone with two stories of living rooms above the ground floor offices, the finest house in the town. And now we learnt that a frightful earthquake had ravaged Martinique on the 11th of January 1839 or 1840, I forget now which. Fort Royal was in ruins, it being particularly stated that the Treasury Office was razed to the ground. But no names of the dead or missing were mentioned. The more probable conclusion seemed to be that all our loved ones – my mother's two daughters, her son-in-law, her grandchild – must have been crushed in the fall of the building.

That was the great spiritual crisis of my life. Whatever faith I had had till then was intellectual only. The blow produced at first a furious rebellion against God. What had I done, that He should take away at a swoop all but one of those whom I loved best? Had I not sacrificed virtually all earthly happiness in renouncing France? Did I not feel the bitterness of the sacrifice daily more and more? Was not that trial enough already? I wrestled with faith, I wrestled with despair and doubt. Gradually – in about a week – the bitterness passed away, not the sorrow. Perhaps the sight of my dear mother's resignation, who lived virtually only by her affections, and whose loss must be far greater than mine (I had at least dearly loved friends of my own age remaining) shamed me out of my selfishly rebellious sorrow. I was able to say, 'Thy will be done, O God!' It was what the Evangelicals would call my conversion. I really turned to God, consciously, willingly. From that hour of submission I have been a Christian, however imperfectly so. [I say once more that I had been intellectually religious before. I have told of my fighting out the Roman Catholic controversy with de Riancey on the benches of the Collège Bourbon, of the pleasure I had felt in hearing Athanase Coquerel's sermons. Nay, more – prefiguring instinctively my work of later years, I remember, after leaving college, writing a page or two in French upon the Lord's Prayer, to show how it met all the aspirations of mankind, including those of socialism (though I had not then read a single socialist work). On the other hand, during my first stay in the West Indies I had faced and largely yielded to doubt and had edged away from my previous beliefs to doubts in the Bible, in the divinity of Christ, and had gone so far as to face the problem of atheism, shrinking back, however, from that bottomless pit. If there were no God, I felt that our life must be purposeless – and yet it must have some purpose. Still, I remained for a time with no belief beyond this, – that there must be a God, and that there had lived a pre-eminently good man named Jesus Christ about whom various contradictory tales had been preserved. We had never indeed when in Europe given up attending religious worship. But Church of England preaching was then dry to an extent almost inconceivable at the present day. After trying the Churches in the neighbourhood we used to go to the French Protestant Church – I forget whether it was then in St. Martin's le Grand – and when that became too far for my mother we used to attend in the morning the services in a congregational church.]

For ten days we remained crushed under suspense. Then came letters from my sisters, telling how all whom we loved had been marvellously preserved. The splendid new house had indeed been levelled to the ground. The oscillations of the earth had shaken away completely the wall and ceiling of the second floor on which was my married sister's bedroom, so that the roof closed over the bed, making all pitch dark and suffocating with dust. [The analogy of this disaster to the far less serious accident which had occurred to my sister in Gower Street 8 years before is curious.] My brother-in-law gave up all for lost. 'Prie Dieu, Marie, car tu es morte', he said to his wife. But just then she heard her little girl cry to her, who slept in an adjoining room, and maternal love saved them. Observing a chink in the roofing where one of the wooden tiles had got loose, she scrambled up, and tearing at the opening made it a little larger. Then her husband, coming to her assistance, ended by making the hole large enough for them to get out. Marvellous as it may seem, no one was killed in the house, though my elder sister was much hurt by a beam falling across her and pinioning her down, and one negro girl, caught between two beams, hung suspended head downwards for more than a quarter of an hour, uttering the most piercing shrieks all the time – and I would back a black girl against any white for screaming. My brother-in-law's clerks worked without sparing themselves in saving the cash at the Treasury (the special reserve was kept in vaults below, the flooring of which was split up by the earthquake). With a generosity perhaps quixotic (as he had one day cause to feel), when asked whether he would have an exact official account taken of the cash recovered, in order to show his complete confidence in his clerks declined the offer, saying that he would hold himself responsible for the balance of the 10th January. (He had in like manner some years before, in order not to hurt the old gentleman's feelings, taken on without verifying the balance stated by his father on succeeding to his place).

It would be impossible to describe our sensations of thankfulness, relief and joy on receiving this news. Our loved ones seemed to have come back to us from the dead.

I had come to Christ, if I may use the expression, by the Evangelical door. I was surprised to find after this how teaching and preaching which before had simply passed over me now came home to me. When I went to Paris again and heard again M. Coquerel I felt amazed at the hollowness of the man's preaching. I simply could not stand him. On the other hand Frédéric Monod (whose delivery became greatly improved in course of time), began by his earnestness to impress me more than any other minister. To anticipate somewhat on the order of events, I may say that I found myself now drawn nearer to my elder sister, who was and remained for life a strong Evangelical, not being unfortunately able to grow into a larger faith. I was still at heart a Frenchman, and perhaps the chief religious influence which acted upon me was that of Alexandre Vinet, to whose works I was introduced by my elder sister. In him at all events I found a breadth of literary and philosophical culture which I always instinctively required, and which I missed, for instance, in the very earnest and eloquent discourses of Adolphe Monod. I was indeed always conscious of a narrowness in the Evangelical School which prevented me from doing more than sympathise with it through one side of my spiritual nature. And fortunately, besides the very practical character of my conveyancing studies, and that bracing contact with Ker's sharp-eyed, sharp-tongued, yet at the bottom not unkindly cynicism, I was also in contact with other influences which kept up the breadth of my sympathies and prevented me from falling into mere pietism.

The first object of interest which I found in England to take me out of myself was the British India Society, a body which, though shortlived and greatly forgotten, was the real pioneer of all subsequent movements for Indian Reform. This I came into contact with through my Uncle, Mr. F. Carnac Brown, who if not its actual founder became really its guiding spirit and the pivot of its existence. His return from India in July 1838 (which had hastened our first return from the West Indies) had had for its principal motive to obtain redress for severities which he had seen perpetrated in the province of Canara (Madras Residency) in the putting down of local disturbances which were magnified into rebellion, the whole thing being the result of sheer misunderstanding and ignorance of native feeling. Mr. F. C. Brown was well qualified to judge, as his and my mother's father, Mr. Murdoch Brown, had been the first English landowner by legal title in all India . . .

A terrible famine was raging at the time in Northern India. At the same time the attention of the anti-slavery party had been directed to the existence of slavery in India as well as to the endeavours which were being made for introducing Indian coolies into some of our late slave colonies, under seven year contracts, perhaps in some respects the worst of all forms of slavery. A portion of the abolitionists began to be struck with the magnitude and urgency of the Indian question in itself. Mr. Joseph Pease and the Earl of Darlington and others put themselves in communication with Sir Charles Forbes and Mr. Montgomery Martin, writer then of note on Colonial and Indian subjects, and with the Aborigines Protection Society, and the services of Mr. George Thompson, the most eloquent of anti-slavery lecturers, were secured. O'Connell entered heartily into the subject, and the advent of Mr. F. C. Brown supplied what was most wanted – an intimate knowledge of all ranks of the native population in two provinces, particularly with the languages of both these, besides a general knowledge of the country and of Hindustani. Another relative of mine on the father's side, General John Briggs, author amongst other works of one on the Land-Law of India (a memoir of him by Major Evans Bell was published in 1885), also joined the band of Indian Reformers. On the 27th May 1839 a provisional committee was formed for the establishment of a London British India

Society, including amongst others the names of Lord Brougham, Sir Charles Forbes, General Briggs, F. C. Brown, William Howitt and George Thompson. Mr. Brown being Honorary Secretary, and from thenceforth the soul of the movement, though others might be its figure-heads or mouth-pieces.

I have no intention here of giving the history of the British India Society. I was not openly mixed up with it, except as a subscriber and as attending its London meetings. But my uncle Mr. F. C. Brown used to dine generally two days a week at our house; the subject was always uppermost in his thoughts; I knew from him everything that was going on, and sometimes helped him directly in the way of correcting proofs and otherwise. The first meeting I attended with him was one of the 'Society of Friends', at their Hall in Bishopsgate Street, Devonshire House, in June 1839 . . .

George Thompson's speech in Devonshire House, though very powerful, was not so fine as some I afterwards heard from his lips. He was, so to speak, scarcely yet in the saddle; he had entered upon a vast subject, but was not yet familiar with more than some of its leading features. But it was a crushing one for my Uncle, who was set down to follow him, owing to an announcement (true no doubt literally, since my Uncle was born in India) on the placards and handbills of the meeting, that 'F.C. Brown, a native of India' would address it. Now in those days a native of India in the ordinary sense of the word was a rare sight in London. When therefore instead of a dark-skinned creature in picturesque oriental costume a tall English gentleman in a frock coat stook up there was a movement of palpable disappointment of which my Uncle could not but be conscious. With military firmness he nerved himself to say what he had meant to say, but he did not know how to manage his voice (not a very powerful one) so as to fill a large building, was not heard by the greater part of the audience, spoke in appearance coldly of what he felt passionately, and on sitting down, vowed to himself that he would never speak in public again, which he never did (unless reading an address to a Chamber of Commerce can be so called) except perhaps to second a vote of thanks to a chairman. I always much regretted this determination, as I am convinced that when his nervousness had worn off, after a few more trials, he would have proved an effective speaker, if not to the largest audiences. As it was, he remained always in the background, and his immense labours on behalf of India have remained virtually unacknowledged, while many a speaker has obtained a reputation for knowledge of the subject whose whole matter was supplied by him, unstinting as he always was of his vast resources of information . . .

Although I had known Sir Charles Forbes from the first, on account of the old connexion between our families, I began to see much more of him from this time. He was a very striking-looking old man, tall, square-built, large-browed, with piercing eyes and a determined mouth, the face well set off by the abundant grey hair, – very different from the red-headed Scotchman whom my mother recollected on her first arrival in India, always ready for a 'bobbery' hunt after some jackal, or perhaps unlucky *pariah* dog. His manners too, through much association with natives, were remarkably dignified. And this grand air of his was in striking contrast with his surroundings. He lived in a large house in Fitzroy Square, and till 4 o'clock in the afternoon at least, in his bedroom, a very large front room over the front drawing room. Here he was to be found in a ragged dressing gown and slippers, and would make his visitor sit down on a big couch piled up at one end, like every table and mat of the chairs in the room, with books and pamphlets, so that there was just room for himself and his guest. His conversation was always interesting and often striking, and when he gave a dinner, you could not have a more courteous host. Although it had been a love-match, he lived apart from his wife, whom he had married as a widow, – a very clever warm-hearted woman, but alas! anything but sweet-tempered. Husband and wife had both strong wills and the time came when the clash became too severe. They parted by mutual consent. But the strange thing was that they continued to live in the very same square, and Sir Charles would always watch out of his window to see Lady Forbes (his house being on the south side of the square and hers on the west, so that it was within his view) go out for and come in from her drive. His third son George (the two eldest being married) lived with him, his youngest son and daughter with their mother, besides a daughter of hers by her first husband, and all the children were constantly in and out of both houses. On one occasion, when we had gone (with tickets from him) to the flower-show of the Horticultural Society at Chiswick, my mother and Lady Forbes were sitting on a bench in the gardens, and Sir Charles happened to pass by. He made most gallant bows to both ladies, and Lady Forbes turned round to my mother and observed quite with pride what a fine-looking man her husband was and what a courtly bow he made . . .

A more interesting meeting than the Devonshire Hall one took place in Freemason's Hall, July 6, 1839, when the British India Society was publicly inaugurated. Here there were on the platform real coloured natives of India, Princes of Oude and of Mysore, vakeels or envoys of the Rajah of Saltara, etc. Lord Brougham was in the chair. The speaking was very good, and it was the first occasion on which I heard Daniel O'Connell, already an old man, but to my mind the greatest speaker there, though both Lord Brougham and Thompson spoke their best. Except Lord Lyndhurst's, O'Connell's voice was the noblest I ever heard from a speaker's lips, and if it had not the deep bell-like resonance of that, its range was marvellous, and no less so its distinctiveness, so that even if he dropped it almost to a whisper, that whisper was heard throughout the hall, whilst his action was marvellously effective. I have seen him – quite an old man – bend literally double to express humiliation, and yet the movement was as rhythmic as that of a tree-trunk bending under a breeze. With all his faults, I believe there were generous and noble elements in O'Connell's character which might have developed into real greatness, and that if Repeal of the Union had been conceded to him not only would the experiment have been safely carried out in those days, but it would have led ere this to developments in the organisation of the empire which are possibly now beyond reach, [No, within reach by the time I now re-read these pages, March 22, 1900.] while the progressive debasement of Irish agitation and agitators had rendered repeal ever more and more dangerous to the empire. In his efforts for India he showed himself always absolutely disinterested, and in his willingness to render service in any way that might be thought fit, and his total freedom from all conceit and self-seeking, he exhibited a remarkable contrast to various much smaller personages in the movement. He was no doubt, as father Kenyon said of him, a *grand homme manqué* but I doubt if he was really 'alloyed' much 'below the standard'; I incline to think that much which seemed alloy was not more than outward tarnish, which friction of great opportunities for good would have rubbed off, had he attained to high office, as I believe he deserved to have done. From what I saw and heard of him in those days I have retained a very soft place in my memory for 'ould Dan' . . . [At the time I speak of, the term 'Esquire' had already practically ceased to have any meaning beyond one of courtesy in London. But at the meeting of July 6, 1839, sitting at the end of the platform with a wand-bearing steward from Lancashire standing beside me, I asked him the name of a speaker whose face I did not know. 'I don't know, Sir, but he's an Esquire'. Rich manufacturers were only 'Mr.'s' then in the north.]

The position of the British India Society soon grew to be a peculiar and indeed precarious one. The great philanthropic movement of the day, the anti-slavery movement, had almost spent itself through the completing of emancipation in the British Colonies. Yet there were those who clung to its principle, and would not hear of giving up the work till Slavery was exterminated throughout the world. The Sturges might be taken as the chief representatives of this feeling, and although present at the inaugural meeting of the British India Society, Joseph Sturge never joined in its work – another fraction prominently represented by the Buxtons, aimed at the extinction of the Slave-trade, their efforts resulting in the premature and disastrous (first) Niger expedition. The number of abolitionists who joined British India Committees led to both attractions and jealousies between the old movements and the new one. On the other hand a vast movement was being organised in the North for a nearer object, which itself both attracted and interfered with the British India movement, that for the repeal of the Corn-Law. Most of the Anti-Corn-Law leaders, in particular Cobden and Bright, were in full sympathy with the British India Society, but they felt, and felt truly, that their own object was one of more immediate consequence to this country. Internal differences, moreover, sprang up on the subject of the opium question, some of the members of the Committee holding that it was, and others that it was not, within the sphere of the British India Society's exertions; hence several secessions. And below all, the Chartist agitation stood in the way of all others, the practice being, whatever the object of any meeting might be, for the Chartists to attend in a mass, and pass amendments in support of the People's Charter. Without questioning the earnestness of several of the Chartist leaders, I must say that the present generation has no idea of the terrorism which was at that time exercised by the Chartists. For whatever good purpose a meeting might be proposed, the first question always was, 'shall we be upset by the Chartists?' – To revert however to the British India Society: in the midst of all these older, or more powerful, or more energetic forms of agitation, it was too new, too weak, its objects were too remote, to conquer for it a secure foothold. It had one valuable asset, so to speak, in the services of George Thompson, its lecturer. He was in full sympathy with the anti-corn-law reformers, and eventually, in the summer of 1841, a formal treaty was concluded between Messrs. Joseph Pease and F. C. Brown on the one hand, and Richard Cobden and John Bright on the other hand, of which Mr. Bell has done well to publish the text (see page 192 of his work). The British India Society agreed that Mr. Thompson should render his gratuitous services to the Council of the National Anti-Corn-Law League in its efforts to promote the establishment and secure the recognition of the principles of free trade, and the Council of the League recognising the legitimacy and importance of the objects sought by the British India Society, approving of its principles, and regarding its objects as kindred to its own, and as inseparably connected with the establishment of free trade and the promotion of the best interests of the British Empire, pledged itself as far as it was competent, upon the settlement of the question of the Corn Laws, to co-operate with the British India Society for the attainment of justice to India. By the time the Corn Laws were abolished, however, the British India Society was virtually extinct. [The minute books of the British India Society were left by my uncle and have remained in my possession. The very last entry in my uncle's handwriting of a Committee meeting held 7th October 1842 runs as follows: 'Mr. Brown was requested to invite his nephew, Mr. John Ludlow, to join the Committee'. My own memory is a complete blank on the subject. It will be observed that I had just then returned from the West Indies and was now of age.] But the treaty was in its spirit always faithfully carried out by Richard Cobden and John Bright.

My uncle soon won the friendship of both the great Anti-Corn-Law Leaders, and I naturally came into contact with the movement. I knew of the negotiations for the above-mentioned treaty, and saw the original. My first actual experience of the League agitation was acquired in the autumn of 1841, when my mother and I, having determined on a trip to Ireland, stopped on our way amongst other places at Manchester. It happened to be just the time when the League had called together a conference of ministers of all denominations to consider the subject of the Corn-Laws, though it had reached its last day. George Thompson was there, and the Joseph Peases, father and daughter. Calling on the last (who had already left for the conference) we met her father and the ex-boroughman[8] of Manchester, John Brooks, a most worthy and very popular man, bought from him two tickets for a tea of the 'Operative Anti-Corn-Law Association' which was to take place in the evening, and were taken by him to the Conference, which was being held at the Town Hall.

This conference was a very curious sight. The Hall which could hold 1000 persons was about two-thirds full. The assembled ministers might number from three to four hundred black coats, men mostly stout, dirty, low-browed, unintelligent in face. The conference was almost exclusively composed of dissenting ministers, with one or two Roman Catholic priests, and two or three clergymen of the Church of England. One of the last, the Rev. Thomas Spencer, was in the chair, and I must say that no one who judges from the dissenting minister of to-day, generally a well educated, often a highly educated man, can have any idea of the inferiority of the class fifty years ago. Mr. Spencer was simply a head and shoulders above his dissenting colleagues, and I feel satisfied that the whole thing must have foundered had he not been in the chair. Many of the speakers were fluent, some of them had a certain amount of rough eloquence, but the poverty of thought and narrowness of mind which were generally manifested, were all but incredible. The only exception was an Irish Priest, Peter Gunn, full of native eloquence, combining both pathos and humour, a humbler specimen of the type of which O'Connell was the supreme exemplar. He began by saying how happy he had been on the previous day to hear the applause which had been lavished on another Roman Catholic priest (a Rev. Mr. Hearne), how bent was Ireland on showing her gratitude to England, how she loved her as a twin sister and was ready to clasp her to her breast inviting the action to the words with a most vehement and resounding embrace of an imaginary sister. Then, I know not how, he went off upon Ireland's wrongs, and launched in a moment into a violent tirade against the aristocracy, against England generally. Twenty times he was recalled to the question, and twenty times he strayed away from it, always beginning again with the greatest good humour: 'Well, about these Corn-Laws', and in a moment forgetting completely but the wrongs of Ireland. There was a moment of tumult caused by a minister from Chichester with the face of a bulldog, who declaimed furiously against 'Poor Law Bastilles', but he ended by apologising, and all went off well. A layman, Thomas Gisborne, ex-member for Carlow, made, though a clever man, a very second-rate speech, but much better than those of the ministers.

We dined (an early dinner) at the Peases, meeting the Irish barrister Robert R. R. Moore, whom I had heard declaim on the opium trade, and returned at half past four to the Conference, when a deputation from the Anti-Corn-Law League was introduced. Cobden spoke shortly, but was listened to with rapt attention, and often with warm applause. But his popularity at this time was nothing to that of George Thompson – it seemed as if the hall would crumble down beneath the thunder of applause that greeted him. I found his speech rather too long, but very firm, – less rhetorical than what I had heard from him before. Some of his intonations were 'models' of oratorical art, – there was an 'abject slaves' for instance, the 'abject' ringing out sharp as a lark, with a fall upon 'slaves' into a grand baritone, which was really splendid.

From thence we went to the working men's tea at the Cornmarket. I had never been better pleased than with the sight of this great hall, seating 900 persons, and filled mainly with working men, all clean, civil, quiet, listening with attention even to the bad speakers. Mr. Brooks was in the chair, and a most comical speech he made, both as to words and gestures. He began by disclaiming all pretension to oratory, – 'You all know me,' he said, 'you know that when I speak the words just come tumbling out of my mouth like potatoes out of a sack.' Still there was so much good sense and good humour in the man that he evidently deserved the friendly regard which he certainly enjoyed. There were thirteen speakers in all, beginning with the Rev. Thomas Spencer, who spoke admirably, though without other eloquence than that of a spirit at once brave, pure and moderate . . . Here I heard for the first time John Bright, then a black-haired young man, and it is a remarkable fact that I did not even name him in my notebook, so little had he as yet given signs of rising above the ordinary level of platform oratory. Indeed, he was evidently far less popular than Robert Moore, for whom there were deafening cries. The latter spoke last, preaching peace with most warlike fury. The proceedings were not over till past midnight. Altogether I came away with a very high opinion of the decorum, the order, the intelligence of the Lancashire working class. Used only to London audiences, I now saw the very different demeanour of Lancashire men, their energy and enthusiasm on behalf of a cause which they have at heart. The tea-party as a means of agitation was also a novelty to me. [We were also much struck with the *ensemble* and correctness of the singing of 'God save the Queen'. The Lancashire working class, it is well known, supplies largely the best oratorio choruses.]

At the period I speak of Cobden, as an educated, travelled, thoughtful man, stood completely out from among the Lancashire manufacturers of the day, who on the liberal side at all events (I cannot speak from personal knowledge as to the Conservative) were apparently to a man unlettered and untravelled. I do not indeed think that Cobden ever rose beyond his Lancashire level of those days, but it was already that of a statesman. Bright on the other hand, though remarkable already for his command of language, was at that time far behind Cobden in his power of dealing with facts, and too much carried away by his own passionate rhetoric. It must be recollected that he was far less of an educated man than Cobden; he told a friend of mine in after years that till he was 21 he had scarcely read anything but Shakespeare; and even after he had entered Parliament and had begun his crusade against the Game Laws, the late Mr. Empson told me that Bright had come to him to ask for some information, and that he had been surprised at his ignorance of English history. I do not say this to disparage him; on the contrary, I have all the greater admiration for one who had the courage to educate himself as he did, not only when he had already grown to be a man, but a leader of men.

I knew Bright in after years, in the first instance through my uncle Mr. F. C. Brown, latterly through sitting with him on the Jamaica Committee, and still later I used sometimes to come across him at the Athenaeum. Widely as we differed in many of our views, I had the greatest personal liking for the man. What struck me especially on the Jamaica Committee was his thorough candour. More than once I have known him to say: 'Well, I came into this room thinking such and such a course was the right one, but what I have heard has greatly ((or 'altogether')) changed my views'. W. E. Forster had a very high opinion of him. 'If I were in a great strait for money, and no security to give, John Bright is the man I would go to', he once said to me. As to his powers as an orator, as our English Demosthenes, no one I presume now disputes them. His speeches have taken their place among English classics.

On my return to London in the autumn of 1841, talking politics with Mr. Ker, and the repeal of the Corn-Laws being mentioned, I said that the question had become much more serious, now that the Corn-Law reformers had a man like Mr. Cobden to lead them. 'Who is Mr. Cobden?' was his question. A man hand-in-glove with half the political notabilities of the day, and thoroughly *au courant* of London politics, had not even noticed Mr. Cobden's name. 'I met your Mr. Cobden at —'s,' Ker said to me a year or two later, 'and he is certainly a very sharp fellow'.

I became a member of the League, and in that capacity after my return from my second voyage to the West Indies, attended frequently the Drury Lane and Covent Garden meetings which were certainly sights to be seen, on a night when some popular orator was to speak, the whole house crammed from top to bottom, mostly by men, but to the extent of perhaps one quarter by women. I heard repeatedly Cobden, Bright – who had developed enormously in power and quality of eloquence – and a third who was considered by many the equal to the other two, W. J. Fox, then Minister of Finsbury Chapel, afterwards M.P. for Oldham. He had a great choice of language, and was really eloquent, but I never liked him so well as either of the other two; there was always a sense about his eloquence of effort and elaboration. What Sir Robert Peel called 'the simple unadorned eloquence of Richard Cobden' derived its value essentially from this – that the man was evidently full of matter and, entirely forgetting himself, was solely intent on putting his matter in the clearest and most convincing form for his audience. Words he cared nothing for, provided he could outsay his thoughts. Bright's mind was much less supple and was narrower than Cobden's, but higher, and with him the right word was necessary to the thought. He spoke far more slowly than Cobden, but every word told. In his highest moments he was the grandest political speaker I ever heard. I never knew a man so dominate his audience. I heard him deliver himself of that famous passage in his Covent Garden speech of December 1845 (see page 278 of Mr. Rogers's Collection, vol. II) –

> Since the time when we first came to London to ask the attention of Parliament to the question of the Corn Law, two millions of human beings have been added to the population of the United Kingdom. The table is here as before, the food is spread in about the same quantity as before, – but two millions of fresh guests have arrived, and that circumstance makes the question a serious one, both for the Government and for us. Those two millions are so many arguments for the anti-Corn-Law

League – so many emphatic condemnations of the policy of this iniquitous law. *I see them now in my mind's eye ranged before me, old men and young children, all looking to the Government for lead; some endeavouring to the stroke of famine, clamorous and turbulent, but still arguing with us; some dying mute and uncomplaining. Multitudes have died of hunger in the United Kingdom since we first asked the government to repeal the Corn-Law, and although the great and powerful may not regard those who suffer mutely and die in silence, yet the recording angel will note down their patient endurance, and the heavy guilt of those by whom they have been sacrificed.*

When he came to the words which I have italicised, the common phrase 'you might have heard a pin drop' would have been literally true. The whole huge audience was hushed as by a spell, while slowly and with solemn pathos the words went forth from his lips, the clear tones of that grand voice carrying them to the furthest ends of the theatre.

Now that by common admission Bright's name stands high on our English glory-roll, no one who is not old enough to have witnessed it can have any conception of the wild passionate hatred borne to him – in those days, by the squirearchy and all its belongings, to say nothing of the bulk of the farmers. I have known young fellows, already liberal enough in other ways, and who afterwards grew to admire and love the man, become red in the face and almost ready to foam at the mouth at the mere mention of Bright's name. No doubt Bright used very strong language, and not unfrequently against persons who did not really deserve it; how strong, I do not think he himself felt. And the odd thing was that when you heard him say those things, you hardly knew how strong they were. For Bright was essentially, both in appearance and in fact, in scriptural phrase, a 'lionlike man', but the lion he was like was a good tempered lion, who must roar, but might roar without being savage; it was only in print that the savageness appeared.

I must now observe that though a member of the Anti-Corn-Law League, I was only a shilling member, and never sought to be anything more. For whilst I sympathised with the League as to the abolition of the food duties, just as I sympathised with the Anti-Slavery Society as to the abolition of slavery, my stay in the West Indies had made me feel the claims of the colonies, of which both organisations appeared to me equally oblivious, whilst as respect the Anti-Corn-Law League in particular, I felt it to be essentially a middle class movement, of which a large and representative portion of the working class were distrustful.

These reminiscences having no claim to being in strict chronological order, I may here mention that during the General election of 1841, (I think it was), I went about a good deal to the London elections. The generation is already dying out which has witnessed the 'nomination', that curious exercise of literally universal suffrage, unchecked and unregu-lated, by which, in default of a poll being demanded, a candidate could be sent to Parliament. The most characteris-tic specimen of the kind was the nomination for Westminster, the hustings being held in front of St. Paul's, Covent Garden. On the occasion I speak of Captain, afterwards Admiral, Rous stood on the Conservative side against Sir de Lacy Evans and Mr. Leader. The choice of the site was a peculiar one, as it furnished the most abundant supply of missiles in the shape of cabbage stumps or vegetable refuse of all kinds, to say nothing of eggs, and how they did fly at the luckless candidates! The noise was deafening. I was not thirty yards off the hustings, yet I declare that except now and then half a sentence from Captain Rous, whose voice I suppose had been trained amid Atlantic gales, to a proper shouting diapason, I heard not a syllable from the platform. Col. Evans, a tall man with a good deal of action in his speaking, was simply ludicrous, looked so like a huge puppet pulled by wires and bobbing up and down in dumb show. Leader, on the contrary, well dressed, a flower in his button-hole, took the whole thing with smiling composure, even when an egg came to decorate his frock coat, and was evidently the favourite. Rous had the largest share of missiles, and though obviously the unpopular candidate, his pluck evidently told. I am bound to say that some years later when I sat on the hustings behind George Thompson at his nomination, I found the Whitechapel crowd infinitely less noisy and ill-behaved than that of Covent Garden and Charing Cross.

I may mention that although I have had a vote for the greater part of my life, I have never voted but twice, – once at

Finsbury, to keep out a money-lender from Parliament, – once at Wimbledon, after the establishment of the ballot, mainly to see the process. Had I been member of a smaller constituency, where every vote tells, I should probably have acted otherwise. But living as I have always done in large ones, I have thought it useful to show of how little worth is the exercise once in every three to six years of a right shared with thousands of others, and that a man may have his fair share of influence in proportion to his capacity, as I believe I have had, without caring to exercise such a right. Moreover, not being a party man, and having unfortunately never been in a position to vote for a candidate whom I personally cared for, it has almost invariably appeared to me that the candidates on both sides were just as good the one as the other, so that I was practically indifferent as to the result. For the same reason, though I have been on the committee of particular candidates, I have never belonged to a party organisation. If I subscribed to one registration association I should feel bound to subscribe to all, for I am just as anxious that the Conservative should have a vote if he pleases as the Radical, the Home-Ruler, as the Liberal Unionist.

As I am on the subject of politics, I may mention that a chance was offered me, when I had been, I think, two years at Ker's, which had I accepted it could probably have altered the whole course of my life. The Chancellor of the Exchequer in the Second Melbourne ministry, F.T. Baring, (afterwards first Lord Northbrook) wanted a private Law Secretary, and applied to Ker for one. Ker offered it me and pressed me to take it. The salary was only '£100 a year', but 'in ten years you will be in the House' he said, and the prediction, so far as the holder of the post was concerned, was fulfilled in the person of W. Dougal Christie, who took it. But I had not sufficient sympathy with the Whigs to care to identify myself with them, and the Melbourne ministry was evidently on its last legs (it went out in 1841); I declined the offer. Ker renewed it in a perfunctory manner to Greenside, who was on the other side in politics, and who declined it on that ground.

From a worldly point of view this was of course great folly on my part. Had the offer come to me some years later when my mother and I had become straitened for money, I am afraid I must have accepted it. We were not so then, and I think I was justified in principle in the refusal. But at any rate, acceptance would have prevented our going to the West Indies in 1841, which I feel certain we were called to do. Even therefore if I was wrong in my decision, my blunder subserved God's will.

*   *   *   *   *

I had now completed three years' pupilship with Mr. Ker, but as I could only be called after five years, we determined to go out again to the West Indies, again intending to stay about a twelvemonth, and to return by the United States . . .

We reached Fort Royal under very painful circumstances. My sister Maria had very recently been confined of a third child (a girl having been born since our first visit), a very fine boy, who had died in convulsions on the very morning of our arrival. It was my second death, but that of my sister Julia, eight years before, had been purely pathetic. The mother's agony of grief over her beautiful boy was now heart-rending, and it was long before she could in any way be comforted. The first nephew I thus saw was a dead one. On the other hand my elder niece was now a bright quick little girl, and by her side was a delightful little sister . . .

I soon found unexpected occupation, which lasted me during five months of my second stay on the island. One of my brother-in-law's principal clerks had died just before our arrival, and so backward was then the colony – he was at his wits' end to find another. Living as I did on his hospitality, and expecting to do so for nearly a twelvemonth, I was very anxious to render myself serviceable to him, and with the bumptiousness of youth asked him to try me. He did so after a little demur, and I found myself in a whitey-brown linen jacket sitting at a desk in his office before a large folio. He had two old principal clerks, who had been in the office for years . . . He explained to me himself what I had to do – the books were kept by double entry, on the pure Italian system, – and as soon as I had fully mastered the fundamental principle of all book-keeping, that every debit implies a credit . . . I found substantially no difficulty in the work and really liked it . . .

I have thus had the experience – unique, I should fancy, for an Englishman – of having kept for five months the ledger of a French colony. And I must have done it satisfactorily, for copies of all colonial accounts are sent to the *Cours des Comptes* in Paris (answering more or less to our Audit Office) and overhauled there, to be sent back if found wrong in any particular, and I was told years after by my brother-in-law that no fault had ever been found with my handiwork. As for the two old clerks, when they found that I really could do the work, they looked upon me as an absolute prodigy . . .

CHAPTER VII

A Crisis in My Own Life*

(Knowing the feelings of my nieces on the subject, I have purposely omitted the matter of the present chapter from that relating to my second stay in the West Indies, although to me it appears to cast not the slightest slur on the memory of their father, whose only fault was that of over-chivalrous trustfulness, whilst it puts in the strongest light their mother's high qualities. My own life is virtually unintelligible without the story).

When in 1841 my mother and I went out for the second time to Martinique, we could not but observe a settled gloom resting upon both my brother-in-law and my sister. We attributed this at first to the death of their baby, which we had almost witnessed, but soon became convinced that there was some other cause for it, of a more abiding nature. We succeeded, not without difficulty, in getting at the cause of this, and it proved to be a very serious, I may say an appalling, one.

My brother-in-law's first years of office had been exceedingly prosperous. He was very active, very able, very energetic. His father had probably been negligent latterly, and at the time of his arrival there were very heavy arrears of

---

* Not to be published till after the death of my niece, Constance Desgraz.

76

revenue outstanding. He was paid (beyond his salary) by a percentage of 10 per cent on receipts, and in 1849 he told his wife and me that he had received from the *taxations* or percentages no less than 100,000 francs (about £4000) upon a million francs of *revenus,* largely composed of arrears got in. (Let me observe at once that this was the result of strictness, not of severity, not one single defaulter had been prosecuted. But my brother-in-law had such a happy manner that I might say he coaxed half the public debtors into payment, and made the other feel that it would be the worse for them if they did not pay.) Then came the earthquake; the utter loss of the 60,000 francs which he had spent on his first house, – 40,000 more to be spent on his second one; perhaps half that sum on his country house at the Pitons, a real necessity of life in the hot weather. Meanwhile the island was well-nigh ruined, taxes had to be remitted, or the payment of them deferred; the treasurer's earnings fell off, while the demands on his income increased with his growing family. He had moreover lent a sum of, I think, 16,000 francs to the English contractor, Thorp.

Then two blows, one an overwhelming one, fell on him at once. Thorp had gone beyond his means in his various undertakings, even though M. Guillet, the *ordonnateur*, had sought to prop him up by signing (illegally) drafts for work not yet done. He was known to be in difficulties – a few months later he was bankrupt. Moreover news had come that owing to irregularities having been discovered in the books of the Colonial Treasurer at Bourbon, the French Minister of Marine and the Colonies had determined to send out a Commission of two French officials as Inspectors, to examine the books of all colonial treasurers. One of these, M. Armand Béssic, in later days Minister himself, was an official of noted strictness and severity.

Liot had just begun to make ready for the inspection. I must here observe that while the cash for current purposes was kept in a strong iron safe with two keys, on the ground floor of the *Trésor*, one key being held by the Treasurer and the other by his cashier, so that two persons had to be present every time it was opened, the reserve was kept in a vault underneath, in sealed boxes. These seals were never broken unless a box was required upstairs, which occurred very seldom, as the two-keyed safe contained more than enough for ordinary purposes, and in balancing accounts, the boxes in the vault were taken at the value marked upon them. In preparing for the Inspectors' visit however Liot felt it right to verify the contents of the boxes in the vaults (the same being done also at St. Pierre). He found, to his horror, a deficit of about £6000 (150,000 francs). And he had only 10,000 francs (£400) of his own to meet it, besides the 16,000 owing by Thorp the insolvent contractor.

How the deficiency could have been created Liot could only surmise. He felt certain of the honesty of his sub-treasurers; two out of the four indeed dealt only with comparatively small sums, and their balances had always been strictly verified. There were indeed two possible explan-ations, which might be concurrent ones. His father had had a dishonest cashier who had levanted to a foreign island (there were no 'Extradition' treaties then) some years before my brother-in-law took office. Again, when the earthquake had tumbled down the Treasury and rent open the vaults below, there might have been abstractions from it for which my brother-in-law had quixotically rendered himself responsible by acknowledging to the balance of the evening before.

My brother-in-law was utterly crushed by this discovery. He had absolutely no means of meeting the deficiency. If ascertained by the inspectors, not only would he lose his place at once, but the £3000 of his caution-money, forming the bulk of my sister Maria's fortune, would be applied towards filling up the gap, with the further probability of a prosecution for embezzlement for himself. To crown all, it was found that calumnious reports were going round about his relations with Thorp. The whole of the Creole community had been jealous of the Englishman's enterprise and success, and several of Liot's friends had found fault with him for being in friendly relations with Thorp. It was now said, here, there and everywhere, that Liot was Thorp's secret partner.

From what happened subsequently to myself, I am morally convinced that these reports originated with Guillet the *ordonnateur*, who wanted to divert to the Treasurer the responsibility of his own signatures to drafts upon the

Treasury illegally given – advances of public money without security to a contractor. But not the slightest suspicion of him was entertained by my brother-in-law, who indeed reckoned on Guillet's own responsibility in the matter as a great security for himself.

Talking over the matter with my mother, it was clear that the only help to Liot could come through us. Our position was this: my father had left £12,000 to be divided among his four children after their mother's death, but a Col. Smith, my sister's godfather, had also left her £2000 to be accumulated till she was 21, and this by my father's will had to be thrown into hotch pot with his £12,000. By the time she was of age, it amounted to something like £3000, and this was applied to for her husband's *cautionnement* or caution money, before he could take his treasurership.

Including this, the whole sum of money to be divided amounted to nearly £15,000. Moreover by the death of my youngest sister Julia in 1833, her share became divisible into fourths amongst her mother, two sisters and brother, so that the total share of each of us children came to about £1000, and my mother's to about £9000. We determined, my mother and I, that my sister Eliza's share should not be touched (though she was quite willing it should be); that if possible the balance of my sister Maria's, above £1500, should be paid over to my brother-in-law, and that I should hand him about £4000, my mother contributing a small balance, so as to make up 140,000 francs. Liot was to pay 5 per cent interest on my £4000 . . .

But this was as yet but a scheme, although we succeeded in persuading my sister Maria and her husband that there would be no difficulty in carrying it out; and he then told me it had been his fixed resolve to shoot himself before the arrival of the Inspectors, as this would have been the only means of securing to my sister the small allowance of 6000 francs a year payable to her as a treasurer's widow, which would be forfeited if he were dismissed from office. Her 75,000 francs advanced for his caution money must go at any rate if he were dismissed, together with all his property in the islands in payment of his defalcation. This was the only occasion on which an angry word ever passed from me to my dear brother-in-law, and yet I am not sure that I should not have come to a similar conclusion.

We left Martinique on our return to Europe in July 1842. It had again been our intention in going out to return by New Orleans and the United States, but there could be no question of this now. We reached Europe in August.

The difficulty for us was to get at the money, and to get at it in time. For (except my mother's trifle, which was in consoles) it was invested in the names of no less than four trustees, my uncles Edmund Ludlow, F. C. Brown, Gordon Forbes and Thomas Allport . . .

I now went over to Paris to confer with my brother-in-law's legal advisor and agent, M. Grémion, holding a joint power of attorney to act with him. My brother had also impressed upon me to take counsel with the *ordonnateur*, who had returned to France on leave of absence, and whom, as I have said, he looked upon as an entirely trustworthy friend. I found that Liot's family were informed of the matter, but his brother-in-law M. Trigaut de Latour was the only member of the family with whom I held any communication. He called upon me, and wrote to me. He had seen M. Guillet, and I was surprised to find that he considered Mr. Thorp was the person who most needed help. I called on M. Grémion and found him of the same opinion – 'Charles and Mr. Thorp,' he said, 'are the same thing, and if he is made a bankrupt it will be Charles's ruin.' I met M. Guillet at his office and he expressed the same opinion still more forcibly. I was entirely taken aback. I said to them, 'Mr. Thorp owes Liot money, not Liot to Mr. Thorp, why should I advance my money to Mr. Thorp?' They shook their heads most gravely – M. Grémion took me apart and impressed upon me the weight of M. Guillet's advice. I was in a terrible dilemma. Each of the two men was twice my age, one was Liot's old school-fellow, friend and confidential advisor of many years' standing, a man of high reputation for prudence and good counsel as a lawyer, jointly named with me in Liot's power of attorney; the other was also a friend of my brother-in-law's whom I had been specially enjoined to consult with, whose ability I knew. After a long conference with the two, I told them that I would give them an answer the next morning.

That night, I lay awake for hours, and I believe it aged my mind by ten years. I felt it terribly bumptious of me, a young fellow only a few months past 21, to go against two men so greatly my seniors. Strange as it afterwards seemed to me, I did not as yet suspect M. Guillet's sincerity. Cowardliness naturally tempted me to leave the responsibility on these two grey-haired men who were ready to take it. Still I always came back to these two points: (1) In all my talks with Liot and my sister, there had never been any question of lending money to anyone but him; (2) Thorp was already his debtor; why should I lend more money to him?

At last, after earnest prayer, I made up my mind that, come what might, I would not advance a penny to Mr. Thorp without my brother-in-law's express sanction, and having made up my mind, fell asleep. The next morning I went to Grémion's, where I found Guillet already awaiting me, and I told them that I could not take their advice without hearing from Liot. Grémion gently shrugged his shoulders, Guillet lifted up his eyes to heaven, 'I was ruining my brother-in-law!' they both warned me. I returned at once to London, and wrote to my brother-in-law. And on the journey back, I saw that Guillet was a false friend; that he was simply seeking to save himself at Liot's expense by making him appear responsible for Thorp's proceedings, including the obtainment of those illegal advances of public money which he had himself sanctioned.

It took then about a month to receive a reply from Martinique. When I got mine from my brother-in-law, I found to my infinite joy that I had done right. It would have been a fatal step for his interests, he said, if I had made any advance to Mr. Thorp. And he and my sister took the same view as I now did of M. Guillet's conduct, and all relations between them were at an end from thenceforth. God had enabled me to save my dear brother-in-law from moral ruin. 50,000 francs (and something more) raised on a promissory note signed by myself and my Uncle Frank were thereupon at once sent out in drafts on the Colonial Bank's Barbados branch. On the return of Mr. Bellenden Ker to chambers after the Long Vacation, a case was laid before him. Of course nothing was said about my brother-in-law's position; the question was simply treated as that of a loan. The opinion was favourable, and means were devised for carrying out the scheme. There remained now the trustees to deal with. My Uncle Allport, a man of mediocre capacity, was perfectly willing to follow my Uncle Frank's lead. My Uncle Ludlow thought the course imprudent, but saw that my mother and myself were determined on it, and that the trustees could not legally oppose us, and acquiesced. But Mr. Forbes – no doubt with the very best intentions – at first entirely refused his consent and resisted all my Uncle Frank's persuasions (it would have been very awkward for the latter, having signed the promissory note for £2000, if the transaction had fallen through). At last I took upon myself to write to him in somewhat sharp terms, threatening legal proceedings if he did not give his consent. He now at last gave in, but though our mutual relations remained outwardly friendly I found many years after (as I shall mention further on) that the remembrance of the proceeding rankled still in his heart.

So now the stocks were sold out, Messrs. Forbes, Forbes and Co.'s advance was repaid, and the balance – about £3600 – remitted to the West Indies, again in drafts to Barbados. I may mention that my brother-in-law and sister, on going to Barbados to receive it, were kept waiting for a fortnight, lodging indeed at the house of the agent for the bank, and being most hospitably entertained. It turned out that the head office had drawn for a sum quite in excess of the balance at its Barbados branch and the agent had been obliged to make up the amount by borrowing from other West Indian branches – an experience which led to the establishment of a rule limiting the amount of drafts presentable at branch offices without previous notice.

The whole deficiency was thus covered up before the arrival of the inspectors. But the strain had been too great for my brother-in-law; he fell seriously ill soon after their first few visits. What would have been the result but for his wife, no one knows. But she was a woman of the rarest mental energy. She mastered all the details of his acounts, and spent days and weeks (the inspection lasted, if I recollect aright, between two and three months), in going over them item by item with M. Béssic, who had come to this task with strong prepossessions against Liot. He could not of course travel beyond the date of Liot's installation as treasurer, and had to admit at last that every item of receipt and expenditure since then was covered. Liot's position was safe, and hard as he was M. Béssic at last relaxed into absolute kindness. 'Quelle femme, mon Dieu, quelle femme!' he said afterwards in speaking of my sister. His colleague was always most kind.

The doctors were anxious to send my brother-in-law to Europe, but he shrank from going there on account of the expense. His symptoms puzzled them; one of them said once that it looked like a case of slow poisoning. Now it so happened that he had in his service an excellent black cook named Rodney, but who belonged to a family of poisoners. This man, who I believe was really attached to my brother-in-law, volunteered to nurse him in his sickness and was singularly attentive. His own dream all his life had been to go to France and see Paris. At last my brother-in-law was obliged to give in to his doctors' advice, and the family returned to France, Rodney being taken with them as a reward for his careful nursing. My brother-in-law rapidly grew well after leaving Martinique, and never had any recurrence of the same symptoms. Rodney remained with him while in France, and returned with him to Martinique, my brother-in-law continuing to treat him with the greatest kindness. But the kinder his master was, the more impatient Rodney seemed to become, and at last he told him that he must leave them. My brother-in-law remonstrated with him, and asked him what fault he had to find with his place. Rodney could allege nothing, but blurted some strange words: 'People are astonished that I remain in your service, as everybody says that I poisoned you.' Liot himself saw nothing but a morbid feeling in the words, but the suspicion occurred to my sister, which I own amounts for me to a conviction, that during Liot's illness Rodney had been administering small doses of poison to him, in order to ensure his master's being sent to France, and in the hope of accompanying him. At any rate he left his place and the island, and died shortly after.

Our income, (i.e. my mother's, as I was not yet earning money) never more than very little above our strictest needs, was now diminished by some amount of interest foregone.

Moreover, a large portion of it, instead of being paid to the day, had now to be remitted from the West Indies, with every probablility that for some time my brother-in-law's payments would not always be punctual (which they were not for the first year or two). My mother and I determined at once to forego all amusements. We had till now been in the habit of going every few weeks to one or other opera, taking what was termed box-pit tickets, i.e. tickets admitting to the pit by the box entrance, (box-holders being entitled if they chose to turn their seats into pit seats at pit prices), now and then to concerts, and very rarely to other theatres than the opera. Moreover we had been accustomed, as has been seen, to travel in the holidays. All this we determined to give up at once, (without of course saying a word to my sister or my brother-in-law) and as a matter of fact, I have never been to a theatre since then, except once to take my nephews Malcolm and Edwin to a Christmas pantomime. Picture Galleries are the only form of amusement (except a very rare concert) for which I have paid a few shillings yearly, and the Crystal Palace I have perhaps visited – never alone – half a dozen times. And our Long Vacation rambles were to be confined to a trip to Paris, where we knew how to live very cheaply. The only pleasure we reserved to ourselves was that of continuing to send out a bag of books to my sister every year. In spite of all economy, we were more than once greatly straitened, and had at the year's end to borrow money, generally of my cousin John Ludlow, to tide us over.

Moreover, I found myself in a very false position. Having no security for the £4000 I had lent to my brother-in-law, my whole capital was restricted to under £1500, whilst at the same time the real facts had had to be kept secret, though something of them had oozed out, and I was supposed to have money invested in the West Indies at a high rate of interest, or even to be a large West Indian land-holder! Thus, whilst utterly crippled for want of resources, and anxious for every penny I could earn, I saw myself treated as having no need of earning.

But a far more poignant trial – though at the same time the most precious of God's blessings – was at hand. In the spring of 1843 I had parted with the great bulk of my future fortune.

On the 29th December of that year I fell in love with the daughter of the very trustee and executor of my father whom I was conscious of having offended, my cousin Maria Forbes. Her father was then and for many years afterwards in embarrassed circumstances, constantly exceeding his income. She was – I could feel it at the time – on the eve of a severe illness, which came on in the following year, brought her near death, and left her more or less of an invalid for life, except during a few too short years after our marriage. I am thankful that this love only came to me when I had done all I could for my brother-in-law, saving him from dishonour and ruin, if not death. I believe indeed that if it had been otherwise, I should have still found strength to make the sacrifice, but instead of making it joyfully, I should have made it, as the French say, *la mort dans l'âme*. God had graciously hidden from me at the time the dread trial that was awaiting me. For I could say nothing of my love. I had practically but a few hundred pounds that I could call my own, since all the rest depended on Liot's life . . .

CHAPTER VIII

My Love for My Cousin Maria Forbes
Amor Vatem

... I fell in love – suddenly, at a particular time and place, within the space of half-an-hour, with a girl whom I had known all my life ... Till now I had never cared for any woman outside my own family. My sister Maria had been for me the ideal of womanhood. I felt that, up to a certain point, I owed myself to her. I did not even fully appreciate the angelic simplicity of my mother's character, the best woman, *haplos*[9], that I have ever known, not even excepting my dear wife. Outside my own family, the only woman that had at all attracted me was Mme. André of Tours, essentially good, and notwithstanding her stoutness graceful and charming, but I was as far from being in love with her as with one of my own canaries. I had known my cousin from childhood. As mites of two years old or so (she is only by a few months my junior) we had been fond of each other, and there is a tradition of my having fought her elder brother Gordon on her behalf. Some years later when we met as children on the occasional visits to England which my mother used to make with me, I played with her elder sister Margaret (now Mrs. Unwin) and herself, and rather preferred the former. Nay, as a confession of stupid conceit, I may say that I have a distinct recollection of both sisters being brought in to see me when supposed to be asleep, and of a sense of offended dignity that a little girl my junior should be up later than myself. Moreover, when on the same occasion (which my wife has wholly forgotten) my mother told me laughingly that Maria Forbes had announced her intention of marrying cousin John, my whole manly pride of, I suppose, 8 years old had been aroused, and I had resolved that this should never be. When we came over to England to reside, there was no real intimacy between the families. My mother had never liked her much younger sister (who was now dead) marrying so old a man comparatively as Mr. Forbes (though he long outlived his wife). We differed on almost all points; he was a high Tory, we were Liberals; he was a narrow churchman, we instinctively broad, besides other sources of friction. Maria had met with a severe accident at Malta, from the effects of which she has suffered all her life; she had moreover had a love affair, or something very near it, with a man who had played with her. I found her at first narrow and conventional – the only thing which wore upon me, during a two or three days' visit she paid us in Cadogan's Place, being her evident gift for music. We seldom called upon the Forbeses at Ham Common; I found no pleasure in such visits, and on this particular occasion, 29th Dec., 1843, I was very loth to go. My mother comforted me by saying that we need not call again till the spring. In those days, I used never to eat lunch; Maria did not do so either, and the consequence was that we two remained in the drawing-room alone for half-an-hour. It was a fairly low room opening into the garden, but blocked up along nearly the whole of the garden front by a conservatory. To my mother and myself – to all of mine, and I think I may say generally to all but dwellers at Ham – the air of this room in particular, and more or less that of the whole house, felt heavy and vaulty. I have never been able to believe but that Maria's long years of invalidism before her marriage were protracted by her stay in this house. But it is treason to say so to a Forbes. That half hour changed the whole course of my inner life – she was looking dreadfully ill, with almost black marks under her eyes. I cannot now remember of what we spoke, but scales seemed to fall from my eyes. I looked into a soul so pure and humble that I felt the deepest shame of myself. The feeling suddenly came over me that this woman or none must be my wife. I loved her then and there with my whole heart at two and twenty, as I love her at the date of writing this, in my 76th year. [I write this in my 80th year, and find cause to love her, if possible, better than ever.] The house which I had been so reluctant to enter had become the very centre of my life. Singular to say – notwithstanding my foreboding of evil for my cousin, and the crushing sense of being as it were bound hand and foot from helping her in any way – the joy of loving was at first greater than the pain. I loved dearly my mother, my sister Maria, her husband and children, but as it were only as continuations of myself. But now I was taken out of myself. I emptied myself, so to speak, into my cousin. I felt that God had as it were struck me down before her. She was, I might say, the last person whom I could have deliberately wished to love. Poor as a rat, indeed, I had no business to fall in love at all. But I had no sense of any sympathy in her with me, nor indeed was there any, for years to come. And yet there was a strange joy in thus giving myself away without return – I might almost say without hope of return.

From this time I became more anxious than ever to begin to earn money, though it was years before I was able really to do so, as for a long time the expenses of my chambers exceeded my fees, and notwithstanding various attempts to wriggle into literature, my first paid literary work was an *Edinburgh Review* article in 1848. Great as the wrench would have been of tearing myself away from my cousin, I should have been ready to go out to India had I received satisfactory opinions of the chances of a conveyancer in Calcutta, where I would have gone either to die or make my fortune.

One curious result of my love was, that it made an Englishman of me. England, not France, was now my heart's home. It was in English that, as will be presently related, my verse came to me (though I might once in a way write a piece in French), I began to think habitually in English. Till now my notebooks, whether relating to life in England or travel, were all in French.

If I have been worth anything morally – as men reckon worth – it is mainly to my love that I owe it. That in the first place brought out for me the immeasurable superiority of the moral over the intellectual. It was not as my intellectual but as my moral superior that I looked up to my cousin. I could not but feel that she was better, unspeakably better, than myself. I set myself to struggle up to her level. From her constant, habitual self-sacrifice I strove to learn, and to some extent learnt, how to sacrifice myself. The time came when I was able, in obedience to what seemed to be for a while the call of God, to contemplate the giving up even of all hope of her love in order to do His work in France. It was distinctly to my love that I owed the power to do this, knowing that, she probably would not have approved or perhaps even understood the act. My love has been to me through life a beacon and a consecration. Knowing this by experience, I cannot say how utterly I abhor that false and prurient asceticism which would put the very slightest shade of contempt or depreciation on the holy love of man to woman, the only true earthly reflexion of Christ's love for His church.

I will add here one other observation. I have said that my apprehensions as to my cousin's health on that 29th December 1843 were fully justified; that she was shortly after very seriously ill, even to danger of life. I cannot describe my anxieties, my agonies about her at that time. Now, looking back, I cannot but think that my own break-down in the summer and autumn of 1844, leading to my stay in Madeira during the autumn of that year, was mainly attributable to those anxieties and agonies ...

*   *   *   *   *

On my return home from that fateful visit to Ham of December 29, 1843, I felt that I must conceal my love from my mother, from my sisters, as it could but pain them (my mother indeed ultimately discovered it by her own observation) – and yet I could not remain altogether silent, passive. Involuntarily, I may say, I bubbled new *English* verse. At the age of 14 to 15 I had written some amount of French verse (besides a few English stanzas in the metre of *Childe Harold*, Byron being then my favourite English Poet), but after a time I had very rationally come to the conclusion that all I had written was trash, and had burnt it; not however before my sister Maria, when with us in 1837, had somehow found the French verse and had copied out 2 or 3 pieces, which may possibly still remain among my papers in her handwriting. Still, I could not get rid of the itch to versify, as may have been seen by my unfinished stanzas on Mme. André, and I had also tried my hand on English verse, chiefly in translations from the Spanish, but also, at the time of the second Afghan war, in some furious stanzas, 1842, and an unfinished piece dated in this very December, 1843. And once, a few months before the time I am speaking of, I had been so touched by Miss Martineau's *Life in a Sick Room* that I had written off from the Reform Club and sent to her (anonymously) a bit of blank verse thanking her for it. But now, day by day, verse poured forth from me irresistibly, and continued to do so for the better part of a quarter of a century, until that dread time, that seven months of hell upon earth, when I had ceased at last to hope. Occasionally, after I had proposed to my cousin, I had sent her a bit of my verse; but when after our marriage I put the whole collection into her hands, she told me that had I done so before, she could not have continued to refuse me.

Now let it be clearly understood that if this mass of verse should be found undestroyed at my death, it must never be published as a whole. Much of it is of too private a character; most of it is far too poor for publication. A few pieces of a less personal type have appeared in print already... I must observe that I never published any verse out of mere literary vanity, but at first in the hope of earning money by it, or, when I found that it never brought me anything more than a copy of the magazine in which it appeared, because it seemed to me that it might do some good. I tried – in 1847, I think – to publish a small volume of verse, but it was returned by Moxon with thanks. I showed the collection to Mr. Maurice when we became intimate, and when *Politics for the People* was started he asked me if some of it might not be published in the journal, but I did not think it up to the mark. A good many years later I sent a different collection to Charles Kingsley, but though some of the pieces have eulogistic remarks by him his endorsed opinion on the whole was that it was 'second-rate' – a judgement in which I entirely acquiesced, and so gave up finally all thoughts of publication . . .

Quite accepting Kingsley's judgement as to my second-rateness, I am inclined to think that a small volume of really good verse might be winnowed out of the mass which I have written. The rest, I hope, will be conscientiously destroyed (*midiscribus esse poetis & c.*).

I should perhaps add that I had when young a real – perhaps a fatal – facility for verse-making. I remember, Kingsley having strongly recommended to me Clough's *Bothie*. When I found it at the Reform Club, to which I then belonged, after reading it through at a sitting, writing off to him in about twenty minutes, *currente calamo*, at the rate of rather more than a line a minute, a letter in English hexameters, which I literally signed and closed and stamped while the Club page-boy was waiting. I, of course, kept no copy, but the three first and the last line remained in my memory –

> Sweet is the poem as mountain breezes over the heather,
> Sweet as the mountain tarns it singeth of – fresh [I am afraid I wrote 'sweet', which 'sea-spray' certainly is not.] as the sea-spray
> Tossed by the plunging bow in the face of the noon-day steerer.
> . . . .
> Comes for the letters the page. John Townsend's words are ended.

N.B. Disagreeing with Kingsley on the point, I consider that trochees are fully admissible in the English hexameter.

CHAPTER IX

A House in Cadogan Place 1843 and Following Years –
Beginnings of Work – My Health Breaks Up –
Madeira, Spain and Portugal

Till our second return from the West Indies we had always lived in furnished houses or at the boarding house which I have before spoken of. In the spring of 1843 (after three months' stay in a furnished house in Sloane Street) we furnished a house on the south side of Cadogan Place and occupied it for 10 years. Incredible as it may seem now, Sloane Street formed then to a great extent a boundary line for London. Close out of it, to the West, in Cadogan Terrace there were actual fields, a wheat field (or was it rye?) in 1843, succeeded by oats, some years later on by a nursery ground for the commoner flowers – the purple stocks, I remember, were quite a blaze of colour – afterwards for the more expensive plants. But by 1853 the ground was already marked out to be built over . . . [My after pseudonym 'John Townsend' was suggested by the then position of our town.]

I was called to the Bar in Michaelmas Term 1843, and took chambers in Chancery Lane, practising only as a conveyancer. It was seven months before I had my first business. I tried, meanwhile, to do something in the way of literature, – offered articles to a Review and a Magazine, a translation of a foreign work to publishers. Nothing saw the the light, except two letters to the *Spectator* . . .

The weariness of life lay upon me at this time – say just before my falling in love with my cousin – like lead. Coming back from chambers after dusk, I used to pass through St. James's Park. I was now assailed with a real desire to kill myself, and be rid of a life in which it seemed to me that I had no work to do, and might perhaps do more good to others by dying than by living. This was quite different from the singular impulse which has repeatedly come over me when I have been looking from the end of a pier or from a ship's side, on a fine day, into clear deep water, to throw myself over. This, I believe to be purely physical. I was not tired of life on such occasions; there was only a sort of craving to melt my own physical life into a larger one. Whether this points to an aquatic origin of the creature man, and represents only the inmost desire to get back to the primordial conditions of his existence, I cannot say . . .

But at the time I speak of there was a longing of actual death. I remember once, in the Park, opening my penknife and feeling for the exact spot where the heart beat, in order to plunge it in. I reckoned that in the misty greyness [By the way, although suicides are most frequent in summer, it was on the dark misty autumn evenings – say between 6 & 7 – that I was at this period specially haunted by the suicidal impulse.] no one would see me fall, and that I should be dead long before the keepers came round to shut the gates. On this occasion the thought of my mother preserved my life. I felt that I could not leave her all alone, that it would be cowardice to do so, and so I wearily restored my penknife to my pocket. This must have been, so far as I can judge, in the autumn of 1843. It was not till about 11 or 12 years afterwards that the craving came back to me with almost irresistible force.

Throughout 1843-4 I had been put to a very severe stress both of mind and body. A new and absorbing interest had entered into my life. The need of earning money had become urgent. And I as yet failed to do so. In the summer of 1844, though having had measles pretty badly as a boy, I caught them a second time severely, and the measles were followed by haemoptysis, abundant on several occasions! I remember one night, when the blood was gushing up from my throat, wondering whether it would choke me and whether death was indeed at hand . . . If death had come, without fault of mine, through the onrush of blood from within, I must own that it would have been welcome.

Dr. (afterwards Sir Thomas) Watson was called in. He was the first great (English) doctor I had come across, a tall, thoughtful, serious, yet benevolent-looking man, reputed admirable in his diagnosis, but not always reliable in following up a case. He found consolidation of quinsy and prescribed a winter in Madeira, for which we started in the beginning of October . . .

I may mention that whilst the haemoptysis for which I had been sent to Madeira never stopped completely whilst I was there, it disappeared altogether with me in Spain, and never returned afterwards. Unjust towards Madeira I know I shall be considered by many, but when I got away from it, I felt as if released from some wearisome prison, and consider the time almost altogether wasted which I spent there.

CHAPTER X

Return to Chambers – Louis Meyer –
Beginnings of Literary and Social Work (1845-7)

I went back to Chambers, and renewed my work at the Bar. Business began slowly to come in. I 'checked' a good deal for Ker, and sometimes earned money thereby. I tried my hand still at literature, and got a bit of verse 'Fairies and Railroads' inserted in *Honest Tait* (January 1846). But as I received no pay for it, I did not trouble myself to go further in the poetical line for publication, beyond some scraps published after 1848 . . .

I was in practice for over 31 years in all, 1843-74, barring the time (1844-5) that I spent in Madeira, Spain and Portugal through the break-up of my health. In only eleven of those years were my takings over £100, exceeding £200 only in three, viz. £218 in 1865-6, £224 in 1851-2 and £263 in 1874-5. Pecuniarily thus I cannot say that I was successful. Yet I have no hesitation in saying that I did my work well, had staunchly attached clients, and was employed on important business. The main fact was, that in confining my practice to conveyancing I had as it were embarked on a sinking ship. The bases of conveyancing as a profession were (1) the complexity of the law in relation to property, real more especially, (2) the low range of education of solicitors. Now within my lifetime, on the one hand the law has been constantly simplified (I have taken my part in this simplification), on the other the education of solicitors has constantly gone on improving. The days when a 'pure conveyancer' like Duval could make his £3000 a year have long since gone for ever. Ker, I think, was the last successful one of the tribe, but then his income, besides that derived from his business and his pupil room, latterly in particular was largely made up by his salary as Commissioner under one or other of the Royal Commissions to which he was very successful in getting appointed. From a business point of view therefore, I ought to have combined equity-drafting with conveyancing, sought pupils, and accepted briefs. I had intended to do the first, and had fixed upon my tutor, but two things prevented me from carrying out my intention. On the one hand, a very pressing need of an old school-fellow demanded, as it seemed to me, the money which I should have had to give. On the other hand, I went to hear my intended equity-tutor plead before the Vice-Chancellor of the Duchy of Lancaster, and was so disgusted at the pains he took, in plain English, to humbug the judge, that I felt in myself, 'Well, if this is what I am to be taught in an equity counsel's chambers, I don't care to learn it.' So I never got any equity training . . .

In my visits to Paris I saw more and more of M. Vermeil and his deaconesses, and came to be so far connected with the work that on one occasion I had, in the absence of both M. Vermeil and Mlle. Malvesin, the singularly able and high-minded female head of the institution, to receive in the first instance a German lady of title, who was wishing to establish an institution of the same kind (I need hardly say that Pastor Stredner's institution at Kaiserswerth was the original pattern of all such houses). I greatly enjoyed the hearty, cheerful self-devotion which reigned throughout the establishment. One *Soeur I—*, who was over the infant school, particularly attracted me, she was so devoted to her charges, and they so devoted to her. As soon as class was over, they clustered round her, they hung upon her, like bees about a honey-comb. Still, I was told, she could never be a deaconess, and at last I got at the reason. This bright, sunny creature, beaming with love and goodness, had by a worthless mother been sent out into the streets at 13 or 14 to earn money by her shame. She was entirely uneducated – she had seen no good at home. Yet by instinct she hated the life. Somehow one of the sisters, – if I remember aright *Soeur* Malvesin herself – had come across her and rescued her, though her parents made a fierce fight to retain her. She made such progress in her schooling that she became first an assistant in the infant school, then was entrusted with the sole charge of it, and gave the greatest satisfaction. I could not help feeling reminded of the passage in I Cor. VI, 9-11, in which after enumerating some of the vilest sins of which mankind is capable as among those of which some of the disciples had been guilty, St. Paul bursts out with the triumphant joy-cry, 'But ye are washed, but ye are sanctified, but ye are glorified'. [There is a pleasant sequel to the story. Some years later, a Protestant peasant, i.e. small landowner, from the South of France came to be treated at the little hospital attached to the *Maison des Diaconesses*. During his convalescence he used to go about the different branches of the establishment. He was particularly attracted by the infant school and by *Soeur I—*, fell in love with her, and asked her in marriage of *Soeur* Malvesin. On hearing this, the poor girl burst into tears and said – 'He must know all – he must know all'. He knew all, and married her. They were very happy, and she afterwards came up to Paris every few years to see her dear *Maison des Diaconesses*, always with fruit for the infant school.]

To dispose at once of what I have to say of the 'Maison des Diaconesses', I spoke of it to Mr. Ker, and gave him a letter to Mlle. Malvesin, when I think in 1847 he and Mrs. Ker went over to the Continent during the Easter vacation. They went to the institution and were very much interested in it. In the same year, when my *Letters on the Criminal Code* came out, I laid a copy of it on a table in the reading room of the Athenaeum Club – Empson, Ker's friend, then editor of the *Edinburgh Review*, took it up and was interested in it, and spoke about it to Ker – the doings of his Criminal Law Commission forming, as I have said before, the subject of the work. Ker said (as he told me) that it was a pupil of his who had been blowing the Commission up. He must have gone on to speak to Empson about the Paris Deaconesses, for the result was an invitation through him from Empson to write an article on the subject for the *Edinburgh*. This was in proof for the number of January 1848, but only came out in April, as I shall relate further on.

Mr. James Guillemard and Mr. Lane, of whom I have spoken in a previous chapter, with M. Vermeil, had been my first clerical friends, but I cannot say I was intimate with any of them. Through dear old Mlle. Dumas I won – God knows by no merit on my part – my first really intimate clerical friend. A Lutheran herself, she was very fond of one of the two Paris ministers of the *Confession d'Augsbourg*, as the Lutherans are officially designated, Louis Meyer – she spoke to him of me. He asked me to call upon him. I had never seen him. I called, and that visit was a turning-point in my life. He addressed me at once as a friend (he must have been at least 10 or 12 years my senior), and with that open directness which is the surest key to another's heart. I had wished – wanted to work for God and my fellow-men. Arnold had given me the wish, Meyer gave me the impulse which made me do so. Arnold, Meyer, Maurice – to those three men I owe under God my better self. Dear Meyer! With Charles Mansfield he was the most lovable man I ever knew. He simply overflowed with love and goodness. Of all men I ever met with, he is the one who has made me most realise St. John. I have heard him preach, alas! – not more than two or three times, and have never heard anything to equal the affectionate fervour of his appeals. He seemed quite carried away by the passion of his love for his fellow-creatures. And what joyous altar of the newest pattern could ever equal the eloquent symbolism of his communion table at the Lutheran church of Rue des Billettes – a plain cross standing upon it, and at the foot of the cross an open bible. Two or three volumes of his have been published since his death, sermons and fragments of sermons and letters, but though often entrancingly vivid (take for instance the sermon *La Foi* in the sermons' volume) they give only a poor idea of the power that lay in the man himself. I avow myself no great partisan of Sunday schools; I have never taught in one myself. But Meyer's Sunday school, which he always taught himself, and which was preceded by a very short service, the only one which the children attended, was simply a delight to them, all the more so that it was always followed, in fine weather, by a two hour walk out with him into the country (his church was in the then outskirts of Paris), in bad weather by a two hour talk with him, when he would tell them stories and sing with them. A more beautiful sight than Meyer among his school children I have never seen. The love he inspired was the 'perfect love' which 'casteth out fear', and yet it could be seen to be instinct with reverence.

But even nearer to Meyer's heart than his school was a society formed amongst young men of both the Protestant churches, over which he presided with a worthy clergyman of the Reformed Church (Meyer however being the soul of the society), called the Friends of the Poor (*Amis des Pauvres*). This was a visiting society, consisting only of young men, who under the direction of the two *pasteurs* occupied themselves with the relief of distress. I have since then seen societies of young men in London, but for a simple earnestness I have never seen any to equal the *Amis des Pauvres*[10]. They met once a month, if I recollect aright, one of the two *pasteurs* dropping in to give a very short address or prayer. The only time I ever spoke French otherwise than in mere conversation (till the first International Co-operative Congress in 1895) was at one of these meetings, when I remember telling my fellow-members (I was a subscribing member) that according to my view '*Nous sommes trop Protestants, et pas assez Chrétiens*'. But I am anticipating.

On the occasion when I first called upon Meyer, he spoke of the need there was for the well-to-do and cultivated classes to help the poor – of the *Société des Amis des Pauvres* (to which I did not then subscribe), of the good it was doing, of the interest the young men took in it; said that from what he heard, there was nothing of the sort in London, and in the most delicate manner pressed me as to whether I had been doing all that I could for my poorer brethren. Although the appeal only gave voice to what had been stirring for some time already in my own heart, I am afraid I answered him very self-righteously, only admitting in his own words that I had not done 'all' I could. In point of fact I had done nothing.

I told Meyer that I would see what could be done in London towards interesting young educated men in the condition of their poorer neighbours, but that the circumstances of the two capitals were so different that the work could not be done in the same way.

That interview laid the foundation of a friendship between us which lasted throughout Meyer's earthly life – may I renew it in another! . . . On my return to London the first step I took was to address myself to the Strangers' Friend Society, and to obtain permission to accompany one of their visitors on his rounds. It was very interesting and I was much struck by the unaffected piety of one sick man, a working printer. But it seemed to me a haphazard sort of charity. There was no telling by what other agencies relief might already be supplied to the persons visited. Though the method of the society came nearest to that of the *Amis des Pauvres*, it seemed to me unsatisfactory in a city like London where so many different agencies were at work (and there are many more now) from the Poor Law upwards. Thinking over what I had undertaken to do, it appeared to me that my special field of duty lay within my own profession, and in the neighbourhood where it was exercised. I had felt already what potencies of good work lay in the number of young men, whether students or young barristers not yet in full practice, who came day by day to the Inns of Court to earn or to learn how to earn their livelihood. And just outside was on nearly all sides a poor and in some parts a foully vicious neighbourhood. Shire Lane, next to Bell Yard, now pulled down, was a place which no decent man could pass through in broad daylight. Few people however could have had less facilities for such undertaking than myself. I don't think I knew a single young man of about my own age in England beyond those who had been my fellow pupils at Mr. Ker's. Owing to my having discontinued dining in Lincoln's Inn Hall as soon as I found that it was enough to put one's name down on the dining list, I had none of that casual acquaintance with the young men of my Inn which diners in Hall acquire. Still, I talked over the matter with one of my former fellow-pupils, then like myself in practice on his own account, W. C. Belt. He entered into my views and together, I think, we called towards the end of 1846 upon the Preacher of Lincoln's Inn, the Rev. J. S. M. Anderson, to say something of what I wished to see done, and

wished to do ourselves, in helping the poor of the neighbourhood. He was very kind, but referred us to the chaplain of the
Inn, the Rev. F. D. Maurice. I called upon the latter. He
explained to me that the Inn as such, being extra-parochial,
had nothing whatever to do directly with helping the
poor, but that the produce of the offertories at the chapel
was divided amongst the three neighbouring parishes. To the
incumbent of the smallest and poorest of these, that of the
Liberty of the Rolls, Mr. Z. K. Redwar, he advised me to
apply. And he asked me to bring my friend Belt with me to
tea at his then house in Queen Square, Bloomsbury . . .

This was my first contact with one who was to fill so large a
space in my life, and whom in a couple of years more I was to
look up to as my spiritual leader. I knew nothing of him at
the time beyond having read and been rather struck by a
review of one of his works in the *Spectator*. I had no suspicion
of his greatness, beforehand, nor did it reveal itself to me at
first sight. Nor did I know at the time that he was suffering
under a very heavy sorrow, that of the death of his first wife. I
felt his kindliness, but had hoped for direction and found
none. I recollect no details of that evening at his house, as
stated in Mr. Maurice's *Life*. In relating the matter to my
mother, I described him as 'a good man, but very imprac-
tical.'

The parish of the Liberty of the Rolls, in which I now
worked for part of two years, is at the present day, I believe,
blotted out of existence . . .

My district consisted mainly of one Court, and was one of
the most respectably inhabited of the parish. I shall never
forget the feelings of moral terror with which I commenced
my visits. I was shy and reserved, and I don't know if I should
have felt more depressed if I had been going to execution – in
fact, if I had had the choice offered to me between a halter
and my district, I believe I should have chosen the halter.
[The same feeling has, by one of the most admirable of the
lady-workers at St. Margaret's House, Bethnal Green, been
admitted to me to have been her own on first visiting among
the poor.] Still, I persevered, and I believe did my duty fairly
well, and was liked by my people. I liked some of them, and
there was in particular one dear old invalid, a Mrs. New,

whose husband was a free-thinking compositor, she herself being a woman of the deepest and most unaffected piety, whom I visited till her death, and whom I felt it a privilege to have known. There were a few other decent and pleasant people – a few doors remained closed to me, but I do not recollect to have met with much ill-behaviour. [Let me here enter a protest against a view and practice not infrequent among the clergy, and tacitly approved of by Bishop Winnington Ingram in his admirable little book on *Work in Great Cities*, that the visitor has a right to force his presence into people's houses. I am sure it causes the bitterest feelings of opposition in the hearts of the poor. And it seems to me inconsistent with the spirit of our Lord's injunctions to His apostles and disciples, as well as to His own practice, which clearly imply the departure from a place where you are not received. The trick which Bishop Ingram implicitly recommends, of putting your foot in if a door be ever so little opened, had better at all events be first tried in Belgravia.]

I forget now how I was brought into contact with a society then in its infancy, which has expanded since to enormous proportions, the Young Men's Christian Association. I saw a good deal at the time of its very able Secretary, Mr. Tarlton, who afterwards became a beneficed clergyman of the Church of England. I attended some of their meetings, which were then purely devotional, or perhaps I should rather say, emotional. I pressed upon Mr. Tarlton the importance of visiting the poor, as a check upon religious excitement. At last he used to me the expression: 'What is all religion but excitement?' That was a view I could not take, and we parted company from thenceforth.

After seven months' visiting I formed a plan which since then has been more than realised, of a legal visiting society for the benefit of the neighbourhood of the Inns of Court, the visitors being chiefly the junior members of the profession, the seniors being only subscribing members. I reckoned that there must be at least a thousand young men in and about the Inns of Court possessing sufficient leisure for the purpose; that if only one in ten could be persuaded to devote one hour a week to visiting, this would be nearly equal to three visitors employed six hours a day (the average of a Scripture-reader's labours) during the whole year, an amount of power far exceeding anything that could be required. Considering however the nature of the agency employed, I did not think the visiting could be indiscriminate, but should be confined generally to that of the good poor and to the really distressed. The work would thus not be a substitution for District visiting, but supplementary to it, and above all preparatory for it. In anticipation of the Charity Organisation Society, I considered that a society so formed might place itself in connection with schools, almshouses, penitentiaries etc., so as to give completeness to the work – I proposed that the society should be composed first of a nucleus of tried moral men, all if possible district visitors, and next of as many men as would choose to take any share whatever in the society's work, the link between the two being some (purely voluntary) act of religious worship, not to extend – except from the mouth of a clergyman – beyond the reading of a portion of scripture, and to follow, not precede, the regular weekly meetings of the society.

I opened myself on the subject to a remarkable man, also a Lincoln's Inn barrister, Edward Thornton (since dead)... He did not enter into my views, being wedded to the larger scheme of the Metropolitan District Visiting Society, of which he was the originator. But we remained friends. He was always visiting amongst the poor, and I accompanied him on some of his rounds... He was a thoroughly well-meaning and self-devoted man, with a perfect genius for very small details. I remember being much amused by the practicalness of his objection to the ticket system of the Mendicity Society. 'You give a ticket to a man in Bayswater, and then he begs all the way to Red Lion Square; or if he is familiar with London, stops halfway at some house where he knows that he can sell his ticket'. His cautions to me as to dealing with the poor were singularly shrewd, and often made me laugh, which he took in very good part...

I must not conceal the fact – I myself remained unsatisfied with my visiting work. It seemed to me that no serious effort was made in any single instance to help a poor person out of his or her misery, but only to help him or her in it. (The parish indeed was so poor, that anything beyond ticket relief was impracticable, unless out of the visitor's own pocket). I found that in general I was avoided by the men, seeing only the women. I found that petty relief was expected from me by the less respectable – I came to doubt whether I was not pauperising instead of elevating the people. It sickened me to see that as soon as I made my appearance in the Court, certain well-known faces looked out of the doors, to make sure that I did not go away without their getting a ticket. I was constantly provoked by the unsocial boast with which even the undeserving sought to ingratiate themselves with me, that they 'kept themselves to themselves' and had 'nothing to say to their neighbours'. I attended one district visitors' meeting, and found myself the only man present besides the incumbent. One of the 'lady' visitors annoyed me by what I may call the crudeness of her notions of charity. Some years after, I saw that she was convicted of embezzlement.

Still, unconsciously to myself, I was learning social facts which would be valuable to me thereafter. In my district was a young cabman, and I heard from him the hard terms on which cabs were let out by the proprietor, and how often he had not been able to make up the daily toll of shillings. [I have in one way or the other seen so many poor homes in the neighbourhood of the Inns of Court that I have quite forgotten when precisely or how I did some visiting in the neighbourhood of Clare Market. In one room in the top floor of a house in Wych Street there was a family in which two of the sons, lads of from 14 to 16, made money by selling sandwiches at the doors of theatres. I unthinkingly said that their trade was at least one that ought to save them from starving. 'Oh no! sir,' said one, 'we never eat our own sandwiches – if we have any over we put them in a cloth and damp it' – showing me some thus kept over for next day – 'This is what we eat'. And he opened a cupboard and showed some hard crusts of bread, evidently the leavings from richer people's tables. The mother added, like the cabman in the Liberty of the Rolls, 'But we often go supperless to bed.'] I saw a woman working on a soldier's scarlet jacket, and complaining of the way in which it burnt her eyes to look long at it. I came across a painter on strike, and realised the power of a trade union, and talked to him, God forgive me!

rubbish about half a loaf being better than no bread, and how strikes never succeeded in the long run.

Above all, I met and became friends with one of the best men I ever met with on God's earth, John Self, scripture reader for the parishes both of St. Dunstan's and of the Liberty of the Rolls. A man not of great mental capacity, but the earnestness of whose faith gradually expanded his intellect – aye, visibly broadening his brow – so that I always found myself more and more in sympathy with him. He had been originally a grocer, but had found it impossible to carry on the business in a small way, and be honest. He had therefore given up his shop, and accepted the poor pittance of a scripture reader. When I first knew him, he had, I think, only two rooms, in which he lived with his wife and two sons, and yet the most perfect neatness and order prevailed . . . Eventually he became parish clerk of St. George's, Bloomsbury, and was able to add to his income by several other means. In various ways, he was to help me efficiently in after days.

Dissatisfied as I was with my experience of district visiting, I felt doubtful whether that experience might not be peculiar to the parish. I therefore as early as February 1847 called upon the incumbent of one of the district parishes of St. Giles's, who pleased me much better than Mr. Redwar, the Rev. Robert Morris . . . Mr. Morris accepted me, and I had some interesting cases under my charge, chiefly in and about Great Queen Street. One was a terrible one. In a large house on the north side of the street, on the ground floor, I found a married couple, the wife dying of consumption, in a simple closet off the hall, formerly no doubt, when the house was better inhabited, a housemaid's closet, with no light but from a pane in the door, and absolutely no communication with the outer air, no ventilation except through the door leading into the hall. When I asked the man how he could possibly have taken such a room, his plea was that it was cheap – only 1/6d a week, giving permission to cook in the kitchen below. How the surgeon attending the poor woman could have allowed her to remain in this dark closet, where she never could get a single breath of fresh air, I could not understand. But the poor creature was too far gone to make it worthwhile or even possible to move her. She died between two of my visits. I cannot recall the place to my mind without horror . . . Another case placed under my charge was that of a slop-tailor in Duke Street, working with his family for Messrs. Moses. He was a very decent man, and strange to say, whilst telling me of the low prices he received, and how he could only get on by putting his children to work on his garments as soon as they possibly could, and before they ought to do so, he bore no grudge to his employers, and according to the most orthodox doctrines of competitive plutonomy said he knew they could not pay more, or they would be cut out by their rivals. Although already a socialist at heart, and by this time in full intellectual and moral revolt against plutocracy, I could not find it in my heart to disturb the poor fellow's faith in the Mammonite gospel, so long as I had nothing else to offer him. But as I went down those Duke Street stairs, I felt somewhat like the Scots Laird, about Charles the First's execution, and wished that kings of slop work too might be taught that they 'had a thirl in their necks' . . .

Before the end of 1848 however I had so much other work on hand that I gave up all district visiting, and I am bound to say that I have remained dissatisfied with its results, so that I have never resumed the practice. One link however with my old parish of the Liberty of the Rolls I did not give up for years, so long as the thing itself remained, a little Christmas-tide supper instituted by Mr. Self for a few of the real spiritual *élite* of both the parishes in which he served, given in the room of one of the members, which was always very delightful. Here there was no standing aloof of neighbour from neighbour, but the most kindly and brotherly feeling amongst all. And let me say that the goodness of the good poor is unspeakably beautiful and Christ like. An assured prospect of work for a week, nay for a few days, is sufficient to take away all care for the morrow, and is accepted as a blessing direct from the hand of God. And their helpfulness to each other puts the rich to shame. They are constantly giving out of their necessity, if not in money, in time, the worth of which mounts up so terribly for the underpaid. And along with the ready, childlike enjoyment stands often the calm, fearless readiness for death. I don't think our little suppers had ever all the same guests two years running; death was sure to have taken away one of them at least during the twelvemonth. I remember one man, an old soldier, who came, I think, twice. On the latter occasion he had a racking cough; he knew that he had but a few months to live, and said so openly. Yet he spoke with beaming face of his happiness, and joined as heartily as a child in the 'Praise God from whom all blessings flow' until stopped by a fit of coughing. [I was once in an omnibus in which there was a very poorly dressed man, very far gone in consumption, whose coughing went to one's heart. 'Beg pardon, gentleman,' he said during a brief respite, 'I know it's very disagreeable but I can't help it – I'm going to the Hospital.' Surely the self-forgetfulness of one who in his own deadly disease sees only its disagreeableness to others and apologises for it, is sublime.]

So things went on with me until 1848, I groping after some means of uniting class with class, especially amongst young men of my own profession, with Arnold still for an ideal, yet dissatisfied with existing methods, so far as I had tried them. All this time I was receiving brief occasional visits from Mr. Maurice, when he came, (once a quarter I think, since February 1847), to hand over to me the share of Lincoln's Inn alms which was set apart for the Liberty of the Rolls parish. We never got further in intimacy during this period; he was shy, and I was shy; but I seem to have been deriving unconsciously from our intercourse an impression of his perfect sincerity, which was to draw us close together on the appointed day.

CHAPTER XI

The French Revolution of 1848 – Beginning of Intimacy with Mr. Maurice

In an amusing book, but one of which the popularity has been far beyond its merits, *The Englishman in Paris*, it is stated of the Revolution of 1848 by the anonymous author that neither his friends nor himself 'have ever been able to establish a sufficiently valid political cause for that upheaval'. For myself, who lived in intimate relations with many French families, I can only say the exact contrary. I will not here repeat what I said in far more vigorous language than I am now capable of, on the 'Reign of Louis Philippe' in the first number of *Politics for the People*. But I, a man of 79 (1900), declare that what I wrote on the subject at 27 is not in the slightest degree exaggerated; that a baser, more corrupting rule than that of Louis Philippe never weighed on a great people, except that of the Second Empire, for which it was in reality the preparation.

Hence, though not on the spot, I was not altogether surprised when the news came of a revolution in Paris. I had seen lately from year to year growing up a feeling of absolute hatred towards the Orleans dynasty, very different from the half-humorous contempt in which *La Poire* [Louis Philippe's face, narrow above, broadening below and further broadened by his whiskers, had been caricatured so exactly as a pear by Philipon, the great *chargeur* of the day, that the pear had become his well known emblem, which had come to be deemed seditious. There was a story that the Duc de Montpellier when a boy, seeing pears drawn all about on the walls, and not knowing what they meant, had amused himself by chalking them on the walls of an inner courtyard at Neuilly, to the consternation of the household, until it was discovered who the unwitting caricaturist of royalty was.] and its *juste-milieu* supporters were viewed ten years before. Charles Lesseps used to declare that he was ready to strangle Louis Philippe with his own hands, or hang him on the first lamp-post. But the tidings were very alarming to my Mother and myself, as my two sisters, with the family of the married one, who had returned from the West Indies, were all there. Nor was there any certainty as to what had taken place. I went to the Foreign Office, saw an Under-Secretary of State, and found that they were there as much in the dark as the papers, and could not tell me whether or not I could get into Paris, all railway communication having for the time been cut off. I provided myself with a Foreign Office passport and started, taking my chance, and as it happened, entered Paris by the very first train that was allowed to do so on the line I took.

One or two things struck me on the way. The first was that on landing, although it was literally not known who was uppermost in Paris, passports continued to be asked for and luggage to be examined (though, I must say, somewhat perfunctorily). The second was to observe a peasant ploughing in his fields by the railway side, as gracefully and soberly as if everything were in its normal state, and not even stopping to look at the train. [I was myself so preoccupied that in an article I afterwards wrote for *Fraser's Magazine* on the subject, I found eventually that I had mentioned the wrong railway as the one by which I had gone to Paris. (N.B. I have been eased, I am afraid, by one now dead, of the volume of *Fraser* for 1848 containing the article above referred to, as well as others on which I set a value.)]

I was soon relieved of all personal anxiety by finding all my dear ones safe and sound. For a few days I lived almost in the streets. My recollections of July 1830 were then fairly fresh, and I could compare the two events. There had been much less resistance this time than before, notwithstanding the much larger army, the wall of Paris, the detached forts. Yet there had been more destruction. In an article published in *Politics for the People* I described my visit to the wrecked galleries of the Palais-Royal, under a pass from General Courtais, *Commandant* of the National Guard, when

> scarcely a pane remained unbroken in the lofty windows, not a mirror, not a chandelier, not a piece of furniture. The fragments of glass and rubbish on the floors lay in two or three inches deep. Here and there a statue or a picture had been preserved scathless, thanks to the words 'National Property' (*propriété nationale*) written in charcoal upon the pedestal or upon the frame, or to similar inscriptions running in all directions upon the walls, e.g. 'Works of art are National property', 'Respect the works of art', etc. Yet some of the most beautiful marble statues had lost a leg or arm, whilst sword cuts gashed most of the pictures. In their blind fury, the mob had not even spared pictures commemorating the birth of the former Republic, as that of Horace Vernet, which represented the assuming of the green cockade by Camille Desmoulins, in the Palais Royal gardens, one of the earliest scenes in the first Revolution.

At that very first moment, the victory of the people appeared to me too sudden, too complete, to be lasting. Lesseps took me to two ministerial receptions, the first held by either Minister, those of Ledru-Rollin, Minister of the Interior, and Crémieux, Minister of Justice. It seemed to me impossible not to measure Ledru-Rollin's mediocrity at a glance. Portly and pompous, wearing his tri-coloured scarf round his waist with anything but dignity, you could see at once that no one was more astonished than himself, to find himself where he was. Little old Crémieux, though not an imposing personage either, had a good deal more *sang-froid*. But the pitiable, revolting thing was to see Louis Philippe's trusted officials, Prefects noted for their servility, *Procureurs Généraux* hated for their partisan violence, crowding in to pay their respects to the new ministers. As Lesseps pointed them out to me one after another, one repeatedly heard a low murmur of amazement run round the room when some notorious Louis Philippist pushed forward to make his bow.

People seemed in the greatest hurry to efface all traces of

the Revolution. Scarcely a barricade remained on the Monday after it. Workmen were busy repairing as far as practicable all damage done. The streets were crowded. I twitted Lesseps with his former deadly hostility to him only the year before. He only shrugged his shoulders. One remarkable feature of the Revolution was the complete absence of all demonstrations of hostility towards religion. Priests were seen everywhere, walking about in their *soutanes*, with or without tri-colour cockades. But the most curious thing was the transformation of Paris into one vast *Agora*. Everywhere you met men standing on a bench, on a chair, on anything that might raise them a little above the crowd, and speechifying on almost any subject under the sun, and anyone who pleased might answer. You could hardly call the thing a meeting, or even the pretence of a meeting. Anyone who chose spoke; anyone who chose listened.

Barring some childish vandalism as above mentioned, I believe that never was there a more harmless – if I may use the word, a more innocent – revolution than the French one of 1848. The 'Democratic and Social Republic', the worshipped ideal of the French working class, was now to be a fact. There might be trying times at first, and if so, as the workmen said, 'We place three months of suffering at the service of the Republic'. The popular song of the 1848 Revolution (though it belongs to a somewhat later date) was no longer the flashy *Parisienne* of 1830, but the really noble *Chanson du Travail* with its burthen:

> *Travaillons, travaillons, mes frères,*
> *Le Travail, c'est la liberté.*

All *Chauvinisme* was at a discount. The hereditary hatred between French and Teuton which the Germans had invented was nowhere. Republicans – I might say Liberals – of all nations 'fraternised' freely with the French. No doubt there were dark and evil elements in the vast fermentation of the day. But the conscience of the nation was not deadened as it is now. Attempts against public morality were promptly suppressed.

I believe now, as I believed then, that if the French Revolution of 1848 had been, if I may use the term, taken in hand by earnest Christian men, able to understand and grapple with social questions, it might have regenerated France and Europe. Atheistic socialism did not then exist. We may smile at Fourier's ranking God as a primordial principle with nature, humanity and mathematics; but he, of all contemporary French thinkers, the most untrammelled by tradition, could not construct his new world without a God. The St. Simonians had aimed at forming a church as well as a polity. Cabet sought to make out that Communism was 'True Christianity'. Proudhon's first published work, never afterwards disavowed by him, was on the necessity of Sunday observance. Buchez, practically the founder of the early French working men's associations, was a convinced Roman Catholic.

For myself, the sight of all I saw around me in France impressed on me the conviction, on the one hand, that this was an essentially socialistic revolution, the principles of which would spread from France throughout the world; and on the other hand, that Socialism must be made Christian to be a blessing for France and for the world. I felt that what was needed was not so much professional teachers of Christianity as practical witnesses of it. I urged, as strongly as I could, upon three French Protestant ministers of my acquaintance, one Lutheran (Meyer), and two Calvinist, the necessity of having Protestant ministers in the Constituent Assembly. But besides ministers, there should be men restrained by no conventionalities from mixing up in the whole social life of the nation, accustomed to look upon society as it is, and as a whole, and not from any peculiar professional point of view; free to use any means in their power, speech or action, whilst feeling themselves thoroughly responsible to God at any step. And beyond Agénor de Gasparin and one other, I could see but one French Protestant layman capable of acting such a part. I did not see how any Roman Catholic could do so with full freedom, notwithstanding the existence of Buchez and his school, one distinguished member of which, the late excellent Dr. Hubert Valleroux, I had the pleasure of knowing in later years. Lamennais – when I had been at College eleven years before the idol of the younger Roman Catholics – had been driven out of the Roman Catholic Church and, as it eventually appeared, out of Christianity, for daring to face openly the social and political problems of the day, and the author of the *Paroles d'un croyant*, one of the noblest works of the 19th century, was now what continental public opinion of almost all shades absolutely looks down upon, an unfrocked priest.

I returned to England penetrated, overwhelmed by the sense of the gravity of the crisis. To whom should I address myself? I did a strange thing, considering the complete absence of anything like intimacy till now between myself and Mr. Maurice. I unbosomed myself to him, in a letter referred to in his volume of lectures on *Learning and Working*, as having had a very powerful effect 'upon his thoughts at the time and having given a direction to them ever since'. He answered me in terms which showed that I had taken the right course in applying to him. Thinking over the matter meanwhile, the idea had grown up in my mind of founding a newspaper in Paris. It was not indeed the first time I had had the idea. In 1840, when the anti-English spirit raged highest in France, I had already felt a strong impulse to go over to Paris and bring out a newspaper on the English side, but the expense would then have been enormous. In 1842, about the time of the Afghan and China Wars, I had also strongly felt the need of setting up a Christian newspaper in this country, and had written on the subject to Joseph Staye. One of the things that had delighted me in Arnold's *Life* was to find that he had had the same feeling. Believing as I did (and do) that 'the great daily teacher of mankind in modern times is not the pulpit, but the press', and that 'if that teaching be not thoroughly Christian, mankind will never be thoroughly Christianised'. (I quote from a memorandum on the subject for Mr. Maurice, dated 16-17 March, 1848, from which several of the above details are also taken). I thought now that France was the country where the attempt most needed to be made . . . I proposed to call my paper *La Fraternité Chrétienne*, and believe I have still somewhere three 'Leaders' which I wrote in anticipation of its existence. The sacrifice which it would have been for me at this time to repatriate myself from England is indeed not to be told. But I was ready to face it, and my Mother was ready to do so with me. But I was very poor, and the question of expense was an essential one. I wrote to friends in Paris on the subject; one did not reply; the answers of others showed the cost to be far beyond my means. I wrote to one rich English friend[11], many years older than myself, asking him to lend me the money. He declined. Baffled, I was obliged to give up the plan.

Mr. Maurice wrote to me that probably God had some work for me to do in England. I began now to go pretty frequently to his house of an evening. I saw more of his friends . . . I also became acquainted with several members of Mr. Maurice's family, his father and mother and sisters. His father (a Unitarian minister, as is well known) had nothing of Mr. Maurice's power, but otherwise I felt was essentially akin to him in that keen generosity of feeling, that noble indignation against falsehood, meanness and cruelty which were so characteristic of the son. Mrs. Maurice, his mother, was a most venerable old lady, obviously of much stronger intellect than her husband, but more self-contained and reticent, so that on the rare occasions on which I met her, I was not able to enter into her intimacy. Nothing, I may observe, could be more beautiful than Mr. Maurice's conduct towards his father. Although he had broken loose from the teaching of the Unitarian minister, who had refused a fortune rather than forego his tenets, there was a tender filial reverence in his bearing towards the old man which was inexpressibly touching . . .

I feel as if I had been coasting round and round what must be the central subject of these *Reminiscences*, shrinking from landing upon it. The lesser nature can never fully take in the greater, and Mr. Maurice was immeasurably greater than myself. Not only was he by far the greatest man I have ever known, but I look upon him as one of the greatest leaders of religious thought who have ever been raised up in the Christian Church. The succession of such leaders, starting from St. Paul and St. John, appears to me to be as follows: Athanasius, Augustine, Anselm, Luther and Calvin, Maurice. On this side of the 16th century, I see no one to compare with him. His main characteristic (something of the same may also be found in Anselm) is the inseparableness of his philosophy from his theology. In what I may call his intellectual and moral dramatism – in his power of placing himself at the point of view of characters the most opposite in themselves, and following out hence the sequency of their thoughts and lives, I do not think he has ever been equalled. His *Moral and Metaphysical Philosophy* is thus a work unique so far of its kind, which will remain among the great masterpieces of the world's literature, and in the 30th century will very likely be considered the best title to remembrance of the 19th. [The one weak part of the book is its treatment of philosophy among the Romans. Mr. Maurice had overlooked the fact (and admitted to me that he had overlooked it) that the philosophy of the Romans is characteristically to be found in their jurisprudence.] As a writer, he suffers from his extreme facility of expression. You might say that words did not exist for him, but only sentences. Intent only on expressing his thoughts, he would cancel whole pages at a time without a moment's hesitation, simply because another development of his idea seemed to him preferable, and has destroyed more good writing than would have made two or three other men's reputations. [I speak from abundant experience, more especially from having prepared for the press the six volumes of his *Lincoln's Inn Sermons*, and more recently that on the Acts of the Apostles.] Yet there are abundance of pages in all his works which rise to the highest beauty. His range of style is moreover remarkable from the highest philosophic culture to the simplicity of the sermons preached in country churches; from brightness of wit, sharp-edged at times, to deep pathos. One of his powers, that of sarcasm, he of set purpose restrained. What a satirist he might have been may be seen by a review of his (which unfortunately I do not possess to refer to) of one of poor Robert Montgomery's poems in the *Athenaeum*.

But it is in passing from the writer to the man that I feel the main difficulty of saying all that ought to be said. I think I have met as good men as Mr. Maurice, but the combination of greatness and goodness in him was absolutely unique. As far as I could ever see, he was entirely without selfishness. Hence, though intensely sensitive, and almost morbidly conscientious, I never could detect in him the slightest, I will not say rancour, but tendency to rancour against those who misrepresented and calumniated him. Again, he was absolutely fearless; there was nothing he would not have faced in behalf of what he thought right. His tenderness of heart was unspeakable. One feature of his character which is not perhaps sufficiently brought out in the admirable biography of him by his son John Maurice, was his keen sense of humour. No one enjoyed a good bit of honest fun more thoroughly, or told a good story better. I had the pleasure, on one occasion when he came down to me at Wimbledon, for the purpose of meeting Dr. Norman Macleod (who unfortunately went to the wrong station and never turned up) of introducing him to Mr. Lear's delightful *Nonsense*, and to have heard his deep voice rolling out the 'jumblies' in full enjoyment of their mock-heroics, or his incessant laughter over the 'Voyage of the four children' is one of the pleasantest recollections of my life. [It is remarkable how many-sided is really good fun. I have heard the same piece read out in the same room with the same fullness of enjoyment by another friend, Lloyd Jones. Not one intonation was the same in the two cases, and yet the reading was equally enjoyed and enjoyable in both.]

Were there no shadows to this portrait? Morally, I know of none. I never knew Mr. Maurice say or do one untrue, unkind, unjust, one selfish word or act. As to vanity, the thing was as foreign to his nature as drought to the ocean. Now and then I may have seen him a little hasty, but how anxious, how very over-anxious always to make amends! His very intellectual defects, if they may be called such, sprang from his good qualities. He suffered himself, through excess of humility, to be too much swayed occasionally by men greatly his inferiors, as Archdeacon Hare, Bunsen, and others. In his tenderness for others' liberty, he did not always exercise sufficient authority, and gave some colour to a clever saying of Jules Lechevalier: 'Mr. Maurice's system is a very good one for bringing men in, but it is all door'. But that which kept him down in the world's eyes so as to prevent his ever showing himself in his full greatness to his contemporaries generally was, I am persuaded, a constitutional weakness. There would come upon him occasionally fits of morbid depression in which it seemed to him that all he had ever done was a complete failure, and that he must tear all his work of the day to pieces, as I have seen him tear away sheet after sheet of his MS. It was, I feel persuaded, the instinctive feeling of this weakness which more than any other consideration, prevented him from seeking or taking any position of prominence in the world's eye, as the Mastership of the Temple, or the Preachership of Lincoln's Inn. I did not myself fully realise the strength of this morbid tendency till the later years of his life, when he was so unhinged by the fall from the faith of Bishop Colenso as to deem it his duty to throw up the pastoral charge of the Vere Street church in order the better to fight for the faith. I then first understood his turning away years before from the leadership of the Co-operative move-ment to the founding of the Working Men's College, an act which had caused me the most crushing unspeakable pain at the time. I saw that just such a fit of depression must have prompted the act. I saw how terrible must have been the result, had a similar fit come over him – as I believe it must certainly have done – in the far more prominent position to which he would have been pushed on, had he continued to lead us in our Co-operative campaigns. And I felt, therefore, for the first time that he had instinctively done right in his timely withdrawal from a perilous leadership, and that it was I who had deluded myself through want of sufficiently realising all the ins and outs of a most peculiar character. I then understood fully the words I had more than once heard from him: 'I am not a builder, but one who uncovers foundations for building on'.

I think there was deep instinctive truth in the name by which we his intimates used to call him among ourselves – of course, not to his face: 'The Prophet'. He was essentially a prophetic man, using the word not in its narrow vulgar sense of foretelling, but of forthtelling, which may often include foretelling, but is not limited to it. When the spirit was on him – for one can use no other term – as he stood erect on a platform, we will say, his hands clasped behind his back, his small form actually towered before you. [It is a curious fact, arising probably often from a first impression thus received, that some people could not think of Mr. Maurice as being short. Speaking on the subject once with Mrs. Maurice, she instanced in this respect particularly her brother, Gustavus Hare, who was actually several inches taller than her husband, and yet was unable to understand how people could call him short, declaring that he appeared to him as tall as himself.] His face was radiant with a solemn beauty, the tones of his voice pealed upon your ear as those of a magnificent organ.

In strange contrast to this really statuesque grandeur, when he had to speak out the truth that was in him, was a trick he had of what Kingsley called 'prophesying into his waistcoat pocket'. This I have chiefly seen take place at his Bible Class, when some question was unexpectedly put to him which thoroughly 'fetched' him, stirring up the very depths of his thought. He would, by a sudden movement, plant his two elbows upon the table, put his two hands before his eyes, plunging them into the waves of his dark hair, and after a few moments' silence, with bent head, deliver himself in low tones – so low sometimes as to be difficult to follow, drawn as it were out of the very depths of his soul – of his reply, which always addressed itself not to the words of his questioner, but to the thought beneath them, often to the questioner's own bewilderment. On such occasions, I have often been reminded of the dialogue of our Lord with the woman of Samaria, and have felt vividly its reality.

That deep insight, reaching far below mere speech, was what many persons must have felt as a reality when they have gone to seek advice of Mr. Maurice. Direct advice, in the strict sense of the word, he scarcely ever gave, but he seemed to cast a light into the seeker's own mind and heart, enabling him or her to disentangle that which was puzzling it, to see a way where there had seemed to be none. It was the fashion to call him obscure; I have never known a man so full of God's light.

And although I am fully conscious of having on more than one occasion behaved wrongly to him, still – never having known what it was to have a father (my own, as I have told before, died when I was two years old) – I had towards Mr. Maurice a feeling of reverence which I have never had towards any other man. I have no doubt known many a greater man than myself, and yet none who did not, so to speak, stand on the same level with myself, and whom I could not thus measure with myself. Mr. Maurice stood on an altogether higher level; him I could not measure.

It was of course only by slow degrees that I realised the strength, the beauty, the greatness of Mr. Maurice's character, and I have here anticipated the experience of many years. At the beginning of our intimacy, in March-April, 1848, I do not think I appreciated more of him than his kindliness, his sincerity and his earnestness. I had not yet learnt to love him. But it was new life to me to find sympathy in my highest aspirations, to be able to pour them out to another. The friendships I had made hitherto either belonged, so to speak, only to the surface of me, or else lacked in breadth what they possessed in depth. This new friendship both reached to the depths of my being and spanned its whole breadth. But I had no idea as yet what other friendships would grow out of it, how it would fertilise my whole life. The famous 10th April, 1848, was to initiate these other friendships.

CHAPTER XII

*The 10th April, 1848 – Charles Kingsley – Charles Mansfield –
Politics for the People –
New Friendships*

The wave of revolution which had swelled and burst in France was now sweeping over the rest of Europe. In our own country, it gave a great impetus to the Chartist movement – a very harmless one on the whole, now that we can look back on it. Two or three of the 'points' of the Charter are the law of the land, and the suffrage has been so widely extended as to be not far from universal. Several of the Chartist leaders were worthy and honest men, William Lovett more especially, whose acquaintance I made in after days. Before long, I was to have more than one friend who had been a Chartist. But Chartism had in the main surrendered itself to the leadership of a tall, red-haired, brazen-tongued Irishman, who ended his life in a madhouse – Feargus O'Connor. For myself, advanced Radical as I was for those days, I had long ceased to look for any substantial results from merely political reform. Even had universal suffrage and annual Parliaments been granted the right, exercised once a year, to (say) one 50,000th part of an M.P., it did not seem to me worth a struggle. Social reform alone was worth living for, and if need were, dying for.

When, therefore, the Chartists announced their intention of holding a monster meeting on what was then Kennington Common (now Kennington Park), and proceeding from thence to Westminster to present the People's Charter to

Parliament, I felt no sympathy with them. But neither did I sympathise with their opponents, and never cared to apply for the truncheon of the special constable. I fully recognise and admire the wonderful demonstration in favour of law and order by the people of London, through the enrolment of special constables by the ten thousand. But I had no wish to break any Chartist's head, nor for the means of doing so. It seemed to me that the best way of preserving public tranquillity was to go about one's ordinary avocations, in full reliance that the government and police would do their duty, if need arose. [I may observe that the occurrence in the Psalms for the morning of the 9th, of that noble 46th Psalm, 'God is our hope and strength, a very present help in trouble' came to many persons, myself among the number, as a pledge of deliverance from threatening evil.] Moreover, having, I may say, seen two revolutions, I felt perfectly satisfied that I was not going to see another one now. Revolutions do not come off at fixed dates, like dramatic events. The solemn preparations for this one in the streets, at Somerset House, and all the principal government offices, were to me almost ludicrous. So when the 10th of April came, I just started at my usual hour for chambers, having first had a good laugh with my Mother over the special constable who had been told off for the protection of the south side of Cadogan Place; a small, slight, lame man, whom a stout Chartist coal-heaver might have taken up with one hand by his trousers and swung in air, but who walked up and down our foot pavement with an air of intense patriotic self-importance. Had I been in his place, as very likely I might have been if enrolled, the whole current of my life might have been changed. I should not have met Charles Kingsley under circumstances which made us intimate from the first. Looking back dispassionately from a distance of over 40 years, I believe now as I did at the time, that I was not called of God to be a special constable on the 10th April, 1848.

At my chambers (I was then at 69 Chancery Lane, now renumbered and a shop) I, of course, found no business waiting for me. But towards 12 o'clock, if I recollect aright, my clerk told me that a gentleman wished to see me. There entered a tall young clergyman with strongly marked features

[Some of the details given in this and the following chapters have been published in a somewhat different form in the papers and in the *Economic Review* already referred to.] who, stuttering something about M-M-Maurice, presented me with a note from the latter. The note told me that a young friend of his, rector of Eversley, had felt so strongly about the Chartist demonstration that he had come up to London to see what could be done to prevent a collision, and that I should find that he had great sympathy with my own views. I told Mr. Kingsley that as far as the Kennington Common meeting was concerned, my own feeling was that no danger was to be apprehended from it; that the preparations against any rising appeared to me overwhelming. Yes, but would the poor fellows themselves believe that? Would they not be goaded on by their leaders till a collision took place? Ought not one to do all one could to prevent this? He for one meant to go to Kennington Common and see what he could do. If he went, I said I would go with him, though I did not think anything would happen, or if it did, was afraid we should be too late to stop it. So I told my clerk I should not return that day and off we sallied. While we were in Wellington Street, before we had actually set foot on Waterloo Bridge, [In Kingsley's *Life* it is stated that we walked to Kennington Common. This is a mistake.] we found crowds of people coming from the south side. Kingsley stopped one of them, who told him that the meeting had dispersed. So there was an end of his anxieties. We decided to go at once to Mr. Maurice and tell him the good news. We had been talking all the way to the Bridge, and we talked all the way to Queen Square. By the time we reached it, we found we were quite at one as to the necessity for some organ in the press as a broad outspoken Christianity, ready to meet all social and political questions. We were intimate from that day.

When we reached Mr. Maurice's house in Queen Square, the subject of a Christian journal was again discussed. Mr. Maurice inclined to a series of tracts. The further discussion of the question was however postponed. In the meantime, the first blow was struck by Kingsley's noble 'poster' to the 'Workmen of England', which forms part of the prospectus of *Politics for the People*, and is also printed in Kingsley's *Life*.

Archdeacon Hare was then in London; his influence over his former pupil and now brother-in-law was always great. He favoured the idea of a journal. The idea was taken up by John William Parkers, the publishers, and as related in Kingsley's *Life*, by the afternoon of April 12th, 1848, the issue of *Politics for the People* was decided on – a penny journal, of which Mr. Maurice and I were to be joint editors, and of which the first number was to appear on May 11th . . .

Kingsley in those days was very thin and gaunt, lanthorn-jawed, I may say; the large mouth indicative of great resolution, but at the same time singularly mobile. A single glance at him showed that you had no ordinary man before you. His conversation was fascinating by its originality – keen observation, strong sense, imaginative power, deep feeling, righteous indignation, broad humour, succeeding each other without giving the least sense of incongruity or jar to one's feelings. His stutter, which he felt most painfully himself as 'a thorn in the flesh' in fact only added a raciness to his talk, as one waited, for what quaint saying was going to pour out, as it always did at full speed, the stutter once conquered. For instance, on the Sunday evening in 1851 when Mr. Drew made a fool of himself by protesting against Kingsley's sermon on the 'Message of the Church to Labouring Men', and many of us were met together at Mr. Maurice's, all in a state of hot indignation, the conversation falling on F.W. Newman, and reference being made to his work, *The Soul, her sorrows and aspirations*; Kingsley, who from first to last had preserved an admirable *sangfroid*, burst in with the words, 'Oh yes! the s-s-s-soul and her stomach-aches', making us all roar at the shrewdness of the wit. On another occasion at Eversley, I had been descanting to him on one of my pet terrors, that of a half-million or more of Chinamen taking into their heads to march westwards. He listened in unusual silence, and then lifting his head, 'W-well, J-John T-Townsend, if there is to be a great Ch-Chinese invasion, it will do one good thing – it will c-c-confound all the interpreters of prophecy!'

Kingsley now told me how it was he had got the parish – a story which cannot here be fully set forth, and the whole details of which his biographer probably never knew. The rector of his parish, when he came first as curate, was a well-to-do man, of cultivated tastes, who had travelled and lived much abroad, a great lover of art. But beyond preaching two or three times a year a well-written scholarly sermon, and sending a cup of broth now and then from his kitchen to a sick parishioner, he did absolutely nothing in the parish, which was frightfully neglected. He was always very courteous and kind to Kingsley, but only met with a smile his suggestions for any improvement in the parish. On the other hand, the squire, Sir John Cope, an old boon companion of George the 4th when Prince Regent, was the very type of Kingsley's bad squires, and the big house was a centre of demoralisation for the parish. Between the do-nothing rector and his bad squire, the ardent young curate was eating his heart out in impotency. He had accepted, as told in the *Life*, a better curacy in Dorsetshire, when suddenly the rector of Eversley disappeared. And as an instance of the reticence of the English peasant, Kingsley told me that although he had already made friends with every labouring man in the parish, and none of them had ever breathed a word on the subject, yet when the rector took to flight, he found that every one had known for years of his disgraceful conduct. Nor are the circumstances of Kingsley's presentation to the living quite correctly stated in the *Life*. No doubt the feeling was universal in the parish, for his appointment, but, as he expressly told me himself, this would never have swayed a man of Sir John Cope's character. But he hated his nephew, and next heir, who was a clergyman, and he was determined not to give him the living, as would have been the natural thing to do. On the other hand, Kingsley had attracted his notice by following the hounds on foot (he was a splendid runner); he knew that the young curate had brought with him from Cambridge the reputation of a first-rate horseman, and he expected to find in him a man of his own kidney, ready to hunt and drink with him. So he presented him, and great was his amazement when the new rector not only refused to get drunk at his table, but began pestering him with demands for this, that and the other parish improvement, or complaining of the conduct of his gamekeepers and other underlings. Of course, Sir John swore at Kingsley and drove him from the house, and so long as he lived there were no permanent friendly relations between the Hall and the Rectory, although when he was more than usually ill and afraid of death, the squire would send for the rector and make maudlin promises of amendment, to be broken as soon as he got better again. [When Sir John Cope died, and was succeeded by his clergyman nephew, Sir William, Kingsley was for a time in Elysium. Here was a squire who himself taught in the schools, and would help in the services. A year or two later I asked him how things got on at Eversley. 'Well,' he replied, 'I sometimes regret old Sir John. This man is thoroughly well-meaning, but he is fussy, and worries the people about trifles. The old man at all events left them alone.']

I need hardly say that Kingsley made an admirable parish priest. The population of Eversley was very scattered, and one excellent plan of his was to make the cottages of a few bed-ridden or infirm old people in outlying hamlets a kind of chapel of ease, fixing days for administering the sacrament to their occupants and inviting all the neighbours to join, as often as possible taking the opportunity of delivering very brief addresses. Whilst always willing to be perfectly friendly with dissenters, he virtually exterminated dissent from the parish. His fearless manliness, his keen eye for the point of a horse or a steer, even to his skill with the rod, were so many openings for him to the heart of English peasants. They might come to hear him in the first instance not because he was a parson, but because he was a man; but once in church, the plain, direct language of his sermons came home to them. I certainly never saw a village church so well filled with the very class that ought to fill it as Eversley church. It is remarkable that in the reading desk or the pulpit Kingsley's stutter completely disappeared, arising as it seemingly did from nervous hurry to express his thought. His sermons he always read. Like all men who have a rich store of their own thoughts, he could afford to borrow from others. 'N-now, J-John Townsend,' he said to me one Sunday, 'I am going to commit p-petty larceny; I am going to take a sermon of M-Maurice's, and t-turn it into l-language understanded of the p-people.' (I am bound to say that when I heard it, I did not discern the theft, so completely had he made the idea his own by his treatment of it). Of the interest, the charm, the preciousness of his society in those the most earnest years of his life, I cannot give an idea. You could not but be at your best with him. His fiery sincerity forced you to be sincere; his upwardness, if I may use the word – the strain of his being toward the highest – lifted you with him.

At this time, Kingsley had only published his *Saint's Tragedy*, with a Preface by Mr. Maurice. I strongly suspect that this, with some of the Poems, will survive most of Kingsley's prose works, brilliant though he was as a novelist. It seems to me to contain the promise of a Christian Shakespeare . . . The purpose is thoroughly noble. For Kingsley was above all things, especially a poet, and might, I am convinced, had he chosen, have stood forth in his age as Tennyson's true successor. Unfortunately, he took up a crotchet which he sought to inculcate upon his girl pupils at Queen's College, that prose was a higher thing than poetry – for no other reason, I believe, than that he found prose more difficult to write, it not being his native language – and so neglected and kept down his special gift instead of cultivating it. That he wrote many and many a page of splendid prose is, of course, not to be denied, but I think the finest pages will always be found to be poetical prose, not prose pure and simple . . .

With Charles Mansfield, Charles Kingsley was undoubtedly the man of most original genius whom I have ever known. I need not dwell upon his character, which is set out under its best aspects (though not in the least untruly so far as these are concerned) in his wife's admirable biography of him. I cannot, however, but protest against a fashion which some persons have of putting him on the same level with Maurice, or even before him, so that I have heard the names joined together more than once as those of 'Kingsley and Maurice', whilst the true sequence is essentially 'Maurice and Kingsley'. Kingsley was emphatically Maurice's disciple – his greatest, no doubt – and within the realms of ethics and theology, I do not think there is a true thought in Kingsley which has not its root in Maurice's teaching. But besides his knowledge of art and natural science (which latter was wanting in Maurice – he speaks somewhere of the 'hoof' of a bear) he had the gift of expression, and the poet's instinctive craving after the very truest and most telling expression. Mr. Maurice's affection for him was unspeakable; in fact, with all his kindness and friendly benevolence, I have often doubted whether he ever really loved anyone except Kingsley of all the young men who from this period began to gather round him. Mrs. Maurice, I may observe, was herself of the opinion that there was 'something quite peculiar' in Mr. Maurice's affection for Kingsley.

I need hardly say that Charles Kingsley was a delightful host, nor was his wife less delightful as a hostess. I have said that he and I were intimate from the first; but if we had not been, this visit must have made us so. Kingsley, I must own, was very prompt in flinging himself into intimacy with anyone whom he found congenial to any side of his many-sided nature. He was essentially an artist, and was fascinated for the time by any new type that struck him. He was thus apt to overvalue a newly made friend, as I am afraid for a time he overvalued me. I learnt perforce in course of time not to take his ardent commendations of this man or that altogether *au pied de la lettre*; in one case I and others had excuse bitterly to rue having done so; in another, he was certainly greatly mistaken as to the character of the man whom he recommended. On many points he also changed his opinions, sometimes for the better, sometimes for the worse. When I first knew him he swore by Professor Owen; he lived to make excellent fun of some of Owen's teaching in the *Water Babies*.

One change in him which eventually put an end, not to our friendship, but to all fellowship in work between us, was his passing from ardent abolitionism in the days of Fremont's candidature for the American Presidency, to actual partisanship for the South in the Civil War – a change of view the possibility of which, I am convinced, would have been as incredible to himself as it would have been to me in the early years of our intimacy. He once indeed laughingly defended to me change of opinions. 'A fool,' he said, 'when once he gets an opinion, sticks to it like grim death, because he does not know when he shall ever get another. But a clever man knows that if he parts with one opinion he can always get another just as good, and so he is always ready to change on cause shown.'

I am bound to say that he and Mr. Maurice – Mr. Kingsley more especially – made me change my views on one important ethical topic, the lawfulness of capital punishment, which I used to think morally unjustifiable, but which I now believe to be perfectly lawful and often necessary for the sake of the living. I had a good deal of correspondence with Kingsley on the subject.

Of my two fellow-guests, ((on this first visit)) one – Corrington – I never met again, though I saw plenty of his handwriting. With Charles Mansfield, on the other hand, an outline of whose story so far as he knew it then Kingsley gave me, requesting me to ask him to my house and exercise what influence I could upon him for his good, I was to grow ere long into a friendship deeper and more intimate than with Kingsley himself. The time is not yet come, probably it will never come, when the whole sad story of Charles Mansfield's life can be profitably told. But I cannot help thinking that by degrees his memory will emerge as that of one whose character was as unique as it was attractive, though long misunderstood and misjudged; whose faults, whose sins, were so strangely mixed up with the misunderstandings and mistakes of others, and with his own best and highest qualities, that they are beyond human judgment, and must be left to the judgment of the All-seeing and All-merciful One.

Whilst Kingsley's mind was bold and curious, that of Charles Mansfield was daring, far-reaching, subtle to a degree . . . He had studied medicine and walked the hospitals, and though he had given up the study because he said he found there was no certainty in it, he was deeply versed on the one hand in physiology and natural history, on the other in chemistry and physics, and I could see that medical men fully recognised his knowledge even of their own profession. He had gone deeply into mesmerism in the early days of its practice in this country, and his range of philosophical reading had extended to neo-platonism and the occult sciences. As an observer of natural phenomena, he could not be surpassed. No one could be more delightful as a companion, especially in a country walk, when he saw everything, understood everything. He knew every bird by its flight, let alone its call. Physically, though slightly built, he was a model of agile and supple strength. Fear he did not know. But the childlike simplicity, the utter unselfishness, the tenderness of his nature, were what most endeared him to his friends ... [Some justice is for the first time done to Charles Mansfield's genius in Mr. Leslie Stephen's *Dictionary of English Biography.*]

I asked Charles Mansfield to dinner soon after my return from Eversley. My mother took to him at once, and it was soon settled that he should dine with us on a fixed day every week – Tuesday, if I recollect aright. Since the beginning of March, I had begun, with my mother, attending Lincoln's Inn Chapel on Sunday afternoons, when Mr. Maurice preached. Charles Mansfield, who had ceased going to church, on my recommendation came too. That little upstairs chapel, with its horrible window flaring with Boucher's arms, the most discordant jumble of colour I think I have ever seen (there were just one or two patches of dark blue on which I used to fix my eyes for rest) became ere long the spiritual centre of our beginning movement. 'Bring him to hear Maurice,' came to be the usual counsel when a possible recruit was spoken of.

I need hardly say that Charles Mansfield, though he did not contribute much, took a lively interest in *Politics for the People* and helped me much by his encouragement and counsel. Of course, I now threw myself heartily into the work, contributing sometimes under the signature 'John Townsend', or the initials 'J.T.', or without ear-mark, about one-quarter of the matter, besides supplying all the extracts. Take it all in all, I don't think many unsuccessful journalistic ventures brought together more distinguished contributors. Besides two or three whom I have forgotten, the list includes Mr. Maurice, Kingsley, Charles Mansfield, Arthur Stanley, Archbishop Whately, the future Archbishop Trench his successor, James Spedding, Corrington, Arthur (afterwards Sir Arthur) Helps, S. Godolphin (afterwards Lord S. Godolphin), Osborne, the well known 'S.G.O.' of the *Times*, Edward (afterwards Sir Edward) Strachey, a still surviving veteran, H. Bellenden Ker, M.J. Brickdale, Dr. Guy, (but not Archdeacon Hare, as stated in Kingsley's *Life*) ...

The greatest anomaly about the paper was that, springing as it did out of the conviction that Socialism must be Christianised, it never dealt with the subject of Socialism otherwise than by giving the report of a lecture by Mr. A. J. Scott, afterwards Principal of Owens College. The fact is, looking up to Mr. Maurice as my leader, I was always waiting for him to strike the key-note in this direction, intending to follow him up to the best of my power; whilst he, considering me better acquainted with the subject, was waiting for me to open it. So we went round and round it, so to speak, instead of facing it. In his 'More last Words' which closed the volume, Mr. Maurice admitted that this had 'Perhaps been our greatest mistake,' that we had 'not fairly entered upon the subject which we hoped would have been most prominent in our pages, the relation between the capitalist and the labourer.' Still, I believe that *Politics* contains much that is of sterling, permanent worth.

I need hardly say that we were greatly disappointed when we were informed by the publishers that *Politics* had to be dropped. Yet its short three months' life was anything but wasted. Its small public of two thousand subscribers were most of them enthusiastic in their support of it. It was not read in England alone. Copies for instance used to reach a young engineer then in charge of the works on a lighthouse in Jamaica – afterwards Sir George Grove. More important still, *Politics* was read by a few working men, who distrusted it so long as it lived, and only realised its sincerity when it died, so that, as it turned out, its death was of far more worth than its life.

But it must not be supposed that it was ever viewed by us as an end in itself. It was only held a means for social action. At an early period I had introduced Charles Mansfield to Mr. Maurice, who, of course, appreciated him at once, and asked him to tea at his house with me, and the regular process at first was that either by introduction from Kingsley or Mansfield or directly, a man should come to me, and if found apparently fit, be passed on to Mr. Maurice, whose teas soon became a weekly institution. It was thus that Charles Mansfield brought his cousin, Archibald Mansfield (commonly called Archie) Campbell, a young architect, who was found a valuable recruit – bright, lovable, unselfish ... Like Charles Mansfield he was a vegetarian, and married a charming vegetarian wife, but they were sore tried by the death of child after child. Archie Campbell in turn brought – but if I recollect aright, only some months later – another architect, his fellow-lodger, F.C. Penrose, late successor to Sir Christopher Wren as 'Surveyor' of St. Paul's Cathedral and President of the Society of Architects, our greatest contemporary architect of the classical school, a most dear friend of my own, and as sterling an Englishman as ever breathed.

Another acquaintance I made at this time – for some years I deemed and called it a friendship – was that of the present Dr. F.J. Furnivall; for the Furnivall of those days was a very pleasant companion, and seemed to be a most desirable recruit. A pleasant looking young man called on me one day at my chambers, to express his sympathy with what we were doing, and his admiration for Mr. Maurice. I offered to introduce him to the latter, but so great were then his diffidence and humility that he shrank from accepting the offer at first. He was persuaded, however, was invited to tea, and became zealous in beating up recruits and generally promoting what was already beginning to be felt as a social movement. He is thoroughly kindly and *serviable*; full of zeal and pluck in everything he undertakes; above all mercenary considerations. But he is utterly wrong-headed, and has developed an overweening vanity which, joined to that wrong-headedness, has for many years rendered him first a most troublesome then a most mischievous, latterly an impossible element in any undertaking involving a purpose of moral good. During all the latter years of the Working Men's College, he was a perfect thorn in Mr. Maurice's side, and of late years he seems scarcely to miss an opportunity of discrediting Mr. Maurice's intentions and work. When I first knew him he was devout; he seems now to have lost all faith. Now and then, when he speaks, he may still say something worth the hearing, but as a rule his utterances are wholly mischievous. I may seem personally ungrateful in saying this, as one of his habitual misrepresentations has been to exalt my share in the Christian Socialist movement as against that of

Mr. Maurice, and whilst I have never spared him, I do not recollect that he has ever attacked me. But I must distinctly say that owing to his wrong-headedness he has for many years been incapable of giving a correct account of anything he has been connected with, and latterly a failing memory has rendered his utterances still more untrustworthy. But I am anticipating.

Through my new friends again I came in several instances into relation with their families . . . for example, Mr. Kingsley's. His father I think I only saw once; his mother, an active, clever old dame, from whom her son seemed to me to have inherited a good deal of his quickness, several times. She did virtually all the work of Chelsea parish, her husband nothing virtually beyond the preaching. I remember one quaint axiom of hers, that the only outward test of honesty was the nose; that a man or woman with a good big nose was pretty sure to be honest, but that a small nose was characteristic of a thief. Another of her fixed ideas, which was of a nature to horrify many philanthropists of her day, and still more of ours, was that it was always best to punish severely a first fault. And she told me how a girl of 14, who had been a great favourite of hers at the schools, had been taken up for stealing some cakes, and on finding out who she was the Inspector of police had sent to say that if Mrs. Kingsley wished it he thought he could get her off. 'Don't do that by any means,' had been her answer, 'let her be punished, and I'll see that she will never do it again.' So she was sentenced to a fortnight or three weeks' imprisonment, I forget which, during which time old Mrs. Kingsley found a good place for her, in which she would be kindly but strictly treated, and at the expiration of the term, Mrs. Kingsley sent her to it, after giving her a tremendous jobation. The girl never went wrong again, married, and made an excellent wife and mother . . .

Then again, through *Politics*, I made great friends for two or three years with little John William Parker, a very kindly man, and the most honest publisher I have ever known . . . He used to have a Tuesday evening at home, in a big room over the shop in the Strand, where smoking was allowed, which I frequently attended. Kingsley was generally there when in town. Once I met Mr. George Meredith, then simply 'an interesting young man' who had published a volume of poems. I cannot say I took to him, there was such a striving for effect in all he said. There I also first met two excellent men who both became intimate friends with me, Samuel Clark and Charles Robert Walsh ... I am not sure that I ever met Archdeacon Hare at Parker's after the first conference when the starting of *Politics* was decided on, but I used to meet him at Mr. Maurice's. As Arnold's friend, I should have been strongly prepossessed in his favour even had he not been Mr. Maurice's brother-in-law. He was still a delightful talker, and was always to me all that is kind and friendly. But it must be said that he was now past his prime. I heard him in Lincoln's Inn, and it would have been difficult to recognise in him the preacher of the *Victory of Faith* ... It was always my impression that he acted rather as a drag on Mr. Maurice, and I suspect that even *Politics for the People*, which he had so powerfully helped to found, though falling far short of its original purpose, was too advanced for him ...

I think that with Erskine of Linlathen and Mr. John Hamilton of St. Simon's, the Rev. A.J. Scott was Mr. Maurice's most valued friend of his own or about his own standing. He was certainly among the ablest and noblest men I have ever known, and one whose reputation has fallen infinitely short of his deserts. As a lecturer, I never heard him surpassed. His thoughts seemed to flow naturally into the most telling language. He had a most wide range of sympathy. No branch of learning seemed foreign to him; he was an admirable critic both of art and music. You were sure to meet remarkable men at this house. I have met there Carlyle; I have met Chopin. Scott was, I consider, a victim of our civilisation, of the divorce which it creates between physical and intellectual labour. Had he been bred to the plough, his greatness would have been recognised. Strongly built, and of a very sanguine habit of body, the effort of writing and to some extent even of thinking sent the blood to his head and tortured him with head-ache; in fact, the little he ever wrote was written at the risk of apoplexy. One of our number, Archie Campbell, rated him above even Mr. Maurice (the pride of Scot in Scot had perhaps something to do with the judgment). But I am inclined to think there was a deep truth in Charles Mansfield's estimate of the two. 'Scott,' he said, 'charms you by giving you back your Highest thoughts in the most beautiful language; Mr. Maurice gives you new thoughts altogether'.

Four or five months had made a vast change in my life. At the beginning of March 1848 I had not one intimate friend in England; at the close of July I was one of a group bound together by their common veneration for one of their number, and most of them by their ardent yearnings for social development. My ideal was to form, what in Nelson's life it will be found he sought to make of his captains before Trafalgar, a 'band of brothers'. We were anything but daunted by the failure of *Politics*. To a certain extent, our connection with it opened to several of us the pages of *Fraser's Magazine*, published by the same firm, and the same month of July 1848 which saw the issue of the last number of *Politics* saw also that of the first chapters of *Yeast*, which certainly have no ring of discouragement about them.

But for one abiding source of pain, these days of 1848-1849 were among the happiest of my life . . .

CHAPTER XIII

Little Ormond Yard School –
Mr. Maurice's Bible Evenings –
Ruskin – Froude – Artisan Allies –
Conferences with Working Men

I have shown how during the brief life of *Politics for the People* there had come together round Mr. Maurice a group of men, most of them young, and I may say, penetrated with the 'enthusiasm of humanity', as it has been called . . . The question was, what practical work could be done by us in common for our fellow-men.

The first that suggested itself was the taking in hand some black spot in London, and try to moralise and christianise it. There happened to be such a spot within a stone's throw, so to speak, of Mr. Maurice's house in Queen Square, a yard leading out of Great Ormond Street, where, as mentioned in Mr. Maurice's *Life*, a policeman durst not venture alone at night. As a beginning, it was determined to set up there a night school for men. It was during our deliberations on this subject that, as related in Mr. Maurice's *Life*, T. Hughes called upon Mr. Maurice to offer his services in any good work, and Furnivall was the man who pooh-poohed the idea of getting any good from him, though I think someone else expressed something of the same opinion, but it was soon found that his accession to our number was the most important that had taken place since that of Charles Mansfield, whilst I in particular met in him the man who next to Charles Mansfield was to become my most intimate English friend. What shall I say of him but that he was as genuine and sterling a character as the world can show? incapable of a base or unkind thought, always ready to fight single-hearted and if need be single-handed for what he deemed true and right; withal one of the cheeriest of companions, full of humour, thoroughly enjoying a good laugh. In those early days he was indeed imbued with a good deal of the country squire feeling (his father lived at Donington Manor, Berks) and would get red in the face in speaking of the anti-corn-law leaders – in later years when in the House, he struck up a cordial friendship with John Bright. He came to us, if I may use the term, a full-fledged Maurician, having been an attendant at Lincoln's Inn Chapel since 1846. He was then newly married, and his house became a new and most valuable recruiting ground for our little band.

A house was taken towards the entrance of Little Ormond Yard, and a housekeeper engaged (through my friend Mr. Self the Scripture-reader) in the person of an old national school mistress named Mrs. Troubridge by this time rather behind the requirements of the age, but an excellent creature, who undertook at the same time a class for girls, and did very well. Before setting up the school, visits were paid by several of the future teachers, two of them together, to most of the houses in the yard. [In the neck of the yard were some houses inhabited by working people of a more decent class than those at the back. Calling at one of these, we found an energetic little woman, the mother of a family, who complained bitterly of being spiritually neglected; 'One might as well live in a heathen land – haven't had a tract for the last three months!' Alas! I do not know what she must have afterwards thought of *our* heathenism, who never gave out one single tract during the whole course of our Ormond Yard work!] Bills were then put up on our own house announcing the opening of the class. The adult school was very wisely carried on not on the ground but on the first floor, so that we had the command of a staircase down which unruly pupils might be pushed, but I do not think this extreme measure had to be resorted to more than three or four times during the existence of the school. [A piece of verse headed 'Death, Rest, Labour' published in Fraser, October 1848, was bred for me out of one such 'bad night' in Ormond Yard.] Some of the nights were no doubt noisy enough, but there were various good fellows also among our men pupils, particularly a former sailor named Regan, who helped us to keep order. Unfortunately the existence of the class for girls soon gave occasion for loud complaints that the boys of the yard were neglected; we had only one room available, and were persuaded to admit boys with the men, the admission of the boys, as usual, drove away the men. Some of the boys were very nice young chaps also. There was one beautiful little fellow, a McCarthy, younger son of an Irish bricklayer's labourer (his elder brother was also a pupil) who in particular won my heart. One thing that struck me when, as winter came on, we had a fire lighted in the room, was the extreme chilliness of the boys, arising no doubt from insufficiency of nourishment. At times when the room was quite too hot for the teachers the boys would cluster round the fire as if they could not get enough heat out of it. Again I observed the important part which hop-picking plays in the life of the poor of London. Every year in the hopping season some of our boys would disappear for a fortnight, to come back so altered by the change of air and the wholesome out-of-doors life that on one occasion I literally did not recognise a boy – a favourite pupil, too – when he returned to school after his two weeks' absence . . . Not the least result of our Little Ormond Yard School was that acting as a leaven, it led to the establishment of another free school by volunteer agency. This was set up in a narrow street parallel to the Strand near Charing Cross . . .

An excursion which we made to Epping Forest with our boys was then in its way a good deal of a novelty. The children were simply amazed at the trees. It was on this occasion that an expression was used by one of them which Kingsley has somewhere quoted; the boy supposed that 'the country was the yard where the gentlefolk lived'. One incident was the appearance after dinner of a number of gipsies. Our boys were warned of their thievish propensities, and it was amusing to see them at once put forth all the alertness and vigilance learnt in London slums in defence of law, order and property. And the Londoners had the best of it. Not a thing was lost.

I may say that I was much better satisfied with the results

of our free schooling than I had been with those of my district visiting. Partly, perhaps, because the work was evening work, we got into contact with men of the labouring class, and not merely with the women. And the relation itself was a much more friendly one than mine had been with my poor people, since we gave no money (unless occasionally where there might be illness affecting a pupil or his family). We came across strikes, as I had done already when a district visitor, and learnt more or less consciously, much that would be of use to us. Still, we had only reached the lower ranks of the labour army, – the labourers properly so called, not the artisans. We were anxious to reach the latter, but though we had already come into some kind of intellectual touch with them through *Politics* – Millbank and Shorter having written letters to it which we printed – it was not till the Spring of 1849 that we entered into personal relations with them.

To dispose of Little Ormond Yard School at once, I may say that the usual difficulties of securing regularity in attendance by voluntary teachers soon made themselves felt, and already in the Spring of 1849 I find that I was in treaty with Mr. Self for his taking the school as a (poorly) paid teacher one night a week. The arrangement ended by his taking the teaching altogether off our hands. We used however still to meet the children at a Christmas tea, and on one occasion there was an excursion to the seaside which I believe was very successful, but which I did not attend. Eventually the clergyman of the Parish took over the school.

Meanwhile my house in 86 Cadogan Place continued to be a recruiting ground, at all events for bachelor fellow-workers. Precisely because my house was small, with only a maid to wait, so that we were rarely even six at table, it enabled one all the better to test the likeliness of each new guest for our purposes . . . Hughes himself, being married, came to us, I think only once with his wife, but through him I made another lifelong friend in his brother Rugbeian and brother Oxonian – called away alas! whilst these Reminiscences were being penned (January 1895), Septimus Hansard. Short, sturdy, plain-featured, loud-voiced, with no pretensions to elegance, Hansard was a splendid specimen morally of the best John Bull type. He had been head of Rugby at the time of Dr. Arnold's death, and formed with Hughes the main link between the Arnold school and the pure Maurician one . . . He afterwards received the living of Bethnal Green, which he retained till his death. Without any *ad capitandum* features, Hansard's sermons had a weight and directness which often rose to real eloquence, nor have I ever heard a nobler sermon than one I heard him preach at St. George's in the East, at the time of the Bryan King riots, while yet the parish was seething with disturbance, his strong voice rising above all clamours from without and from within. The working men of the East End know well that they have had no better friend in the church than the late Rector of Bethnal Green . . . One achievement of which he was very proud was when a fire took place near the church, on the Thursday in Holy Week, 1891. Twenty-five people were burned out of house and home, and lost all. He always had on the premises some bales of stout coarse clothing stuffs. He made an appeal to the women of his parish, and two or three of them were set to work at the Rectory. They worked – barring church hours – all Good Friday – all Easter Eve – and with the help of an appeal inserted in the *Times*, by Easter Day all the homeless ones (whom he had managed also to get sheltered) had Sunday clothes to wear. Of course he fed the workers, and his daughters read to them or otherwise amused them, and though they worked hard for those two days, they said they were the pleasantest they had had in their lives.

Hansard took a great interest in the London school board and in Board schools. He was very proud of his own Board Schools and took me over them the last time but one that I saw him. He was evidently well known to the children, many of whom grinned from ear to ear on seeing him. He said the effect of the schools had been wonderful. The children were at first dirty, unhealthy, using (girls as well as boys) the filthiest language. They were now, as I could see, for the most part clean-faced, cleanly dressed, healthy-looking, some of them quite rosy, and punishments for foul language were almost unknown. The last time that I saw him was at his daughter's marriage. Not only church and church yard, but the adjoining streets, were perfectly crammed with people . . .

By this time most of our little band were attendants on

Sunday afternoons at Lincoln's Inn Chapel. Many of them, as I found out, had doubts on various points connected with the Christian faith, and more than one such point was debated amongst two or three of us in a friendly way at my house. All, I think, wished for a closer approach to Mr. Maurice in matters spiritual. I suggested the idea of asking him to have Bible readings at his house, and it was warmly accepted. I then broached the subject to him, and he gladly entered into the plan. The year 1848 thus closed with the establishment of those Bible evenings at his house, of which so deeply interesting an account by Charles Mansfield is given in Chapter 24 of Mr. Maurice's *Life* . . .

In a No. of Mr. Ruskin's *Praeterita* the following passage occurs:

I loved Frederick Maurice as everyone did who came near him, and have no doubt he did all that was in him to do of good in his day . . . But he was by nature puzzle-headed, and though in a beautiful manner wrong-headed, while his clear conscience and keen affections made him egotistic, and in his bible-reading as insolent as any infidel of them all. I only went once to a Bible lesson of his, and this meeting was significant and conclusive.

The subject of lesson, Jael's slaying of Sisera – concerning which, Maurice, taking an enlightened modern view of what was fit and not, discoursed in passionate indignation, and warned his class in the most positive and solemn manner, that such dreadful deeds could only have been done in cold blood in the dark Biblical ages; and that no religious and patriotic Englishwoman ought ever to think of imitating Jael by nailing a Russian or Prussian's head to the ground, especially after giving him butter in a lordly dish. At the close of the instruction through which I sat silent, I ventured to enquire, why then had Deborah the prophetess declared of Jael 'Blessed above women shalt the wife of Heber the Kenite be'? – on which Maurice, with startled and flashing eyes, burst into partly scornful, partly alarmed denunciations of Deborah the prophetess as a blazing Amazon, and of the song as a merely rhythmic storm of battle rage, no more to be listened to with edification or faith than the Norman's sword-song at the battle of Hastings. Whereupon there remained nothing for me, to whom the song of Deborah was as sacred as the Magnificat, but total collapse in sorrow and astonishment, the eyes of all the class being also bent on me in amazed reprobation of my benighted views and unchristian sentiments. And I got away as I could, and never went back.

I had long ceased, out of pity to one who had written some noble things in his day, to read Mr. Ruskin's later publications. But the above passage was brought to my notice by T. Hughes. We wrote the following letter to Mr. Ruskin:

Dear Sir – In a late no. of your *Praeterita* we find an account of a Bible-reading (not lesson) which took place many years ago at Mr. Maurice's house. Having both been present on the occasion as well as yourself, our recollection of the circumstances differs in some respects so materially from your own that you must allow us to express our dissent from your account.

We would observe in the first instance that Mr. Maurice's views on the subject of Deborah and Sisera are fully set forth in the 18th discourse of his book on *The Old Testament,* a work which is substantially a reflex of the Bible-readings in question. You will find in this, as according to our distinct recollection there was not at the Bible reading you refer to, no contemptuous reference to the 'dark Biblical ages', still less any 'partly scornful and partly alarmed denunciation' of Deborah, but simply the assertion that whilst 'a brave, noble woman', she is not to be installed as 'a teacher of ethics'.

Mr. Maurice seldom began the discussion unless by a few remarks. He certainly did not do otherwise on the occasion referred to, and the terms 'discoursed in passionate indignation ... at the close of the instruction' by no means answer to the facts as we recollect them.

Your own part in the discussion, we also distinctly recollect, was not confined to a mere question, but was a vehement and somewhat lengthy outpouring in praise of Jael. The 'startled and flashing eyes' were not those of Mr. Maurice, whose self-possessed demeanour on the occasion is still before our eyes, but your own, and struck forcibly another of our number, now with God. [Charles Mansfield, who as we were coming away said to me, 'Did you notice Ruskin's face whilst Mr. Maurice was speaking? It was that of a man possessed'.]

You consider Mr. Maurice to have been puzzle-headed. We, who knew him a good deal more intimately than yourself, used to find him whilst he lived the greatest solver of puzzles, and that not by direct explanation, but by the true Socratic method of enabling others to see clearly what was in their own mind.

We are,
   Dear Sir,
      Yours faithfully,
         Thos. Hughes
         J.M. Ludlow.

Of this letter no acknowledgement was vouchsafed to us, nor is it referred to in the second edition of *Praeterita.*

Had Mr. Ruskin been a younger man, I should consider the account he gives of the occurrence a malignant travesty of what took place. At the age when it was written however, I am willing to believe that it was the result of a still active imagination coupled with a failing memory. His inability to understand Mr. Maurice's character, is simply ludicrous. Mr. Maurice 'Puzzle-headed'!

Although in after days I acquiesced in the giving up of the Bible-readings, for which there was no more time to be found, I deeply regretted the necessity, and I have always thought that the spirit went out of our movement with them.

But I have been anticipating for years, and I must not forget that I am still on the years 1848-9. Here let me record one or two of my failures in the endeavour to extend the 'band of brothers'.

Kingsley had read Arthur Clough's first poem, *The Bothie of Toper-na-Fuosich,* and had been immensely taken with it. I read it on Xmas Day 1848 at the Reform Club, and felt as enthusiastic about it as himself. So did Charles Mansfield, when I had bought it and lent it to him. We both agreed that Clough was a man to enlist if possible in our brotherhood. So I wrote and asked him to dinner. Besides my mother, Charles Mansfield was the only other guest. We both liked him much personally. I suppose he did not so much like us, for he remained singularly reserved – *boutonné,* as the French would say – on the defensive as it were against the friendship which we were seeking to form with him, never allowing the slightest opening towards any intimate talk. I never saw him but once again, at Mrs. Hensleigh Wedgwood's. I did not think him improved. We left together. 'Which way do you go?' I asked, hoping that perhaps I might walk some way with him. 'I go north.' – 'And I south,' I replied. Those were our last words together, as we shook hands. [I was able to be of some use to Clough in getting him the principalship of Gordon Hall, having at the time a good deal of credit with one of the chief promoters, E.W. Field, to whom I strongly recommended Clough.]

In the spring of 1849 the publication of Archdeacon Hare's

Life of Sterling raised, it may be said, the first storm around
Mr. Maurice. In a letter to Kingsley of the 20th March I see I
speak of 'articles in the *English Review, Record, Herald, Globe*
and now finally *Morning Chronicle* which have made our dear
Prophet's piety (!!!) a subject of controversy in the leaders of
the Daily press'. Just about this time came out Froude's
Nemesis of Faith, which for some days literally crushed Mr.
Maurice. He thought to have reviewed it, but feeling that it
would be most imprudent for him to do so, I reviewed it
myself in *Fraser,* with the result however of seeing the article
attributed by the *Record* to Archdeacon Hare, and made the
ground of a fresh attack upon him as well as upon Mr.
Maurice. Of course I wrote to the editor claiming the
authorship, and a note was inserted to the effect that they
were informed that the article was written by a gentleman
whose name was 'totally unknown to them'. [I believe my
style, as well as my personal appearance, to be singularly
indistinctive. There is a paragraph by me in Kingsley's *Cheap
Clothes and Nasty,* which formed the subject of one of Mr.
Greg's strictures upon Kingsley in the *Edinburgh Review,* and
which nobody has ever detected as being by a different hand,
on the other hand Kingsley was unhesitatingly identified with
the 'John Townsend' of *Politics* by newspaper critics. I wrote
half of a Preface by T. Hughes to an edition of Lowell's
Biglow Papers, and Sir Fitzjames Stephen in the *Saturday*
animadverted upon what I had said as T. Hughes', without
the least suspicion to the contrary till I told him. An article of
mine in the *North British* was attributed to Carlyle. On the
other hand Kingsley attributed to me till I disabused him an
article by Sir J. Herschel on *English Hexameters,* and I found
difficulty in convincing another friend that I had not written
an article in the *Edinburgh* which I had never even seen. On
the other hand I have been repeatedly accosted by strangers
as an acquaintance, and as late as the autumn of 1891 a
young girl whom I had never seen came up to me with
outstretched hand in the street and addressed me as 'Pa'. (As
she was very plain I failed to be touched by her appeal).]

Froude knew Kingsley, and in consequence of a letter of mine
which the latter sent to me, opened a correspondence with me,
which led to my asking him to my house. He stayed a few days,

and I suppose our disappointment was mutual, for though always on friendly terms neither of us afterwards ever sought out the other, and we scarcely ever met. The *Nemesis of Faith* had, I must say, led me to expect in him a man deeply in earnest as to spiritual questions. It seemed to me that there was no depth at all in him. A very charming man, no doubt, but merely charming men have never been much in my line . . .

To express my whole feeling as to Froude – he appears to me like Ruskin – to have had a woman's soul in a man's body. That is, I take it, the main explanation of the extraordinary fascination for both of that essentially male genius, Carlyle. Historic accuracy is not to be looked for in him. He sees with his imagination, and without the least dishonesty tells what he sees. Of course you must never rely upon him *a priori;* that 'Froude says so' is absolutely no guarantee of the truth of any historical assertion, but his brilliant imagination may enable him to discern what others have not discovered, and often leaves vivid pictures on the mind, even if not strictly accurate ones, of much that in duller pages would have passed altogether unnoticed. Kingsley was greatly taken with him, but his influence upon Kingsley was not, I think, a good one.

Of much more importance however than these occasional failures to gain recruits for our band amongst men of our own class were the friendly relations which we now succeeded in opening up with men of the artisan class . . .

*   *   *   *   *

It was in the latter part of 1848 that we first succeeded in establishing friendly relations with working men. *Politics for the People* had, without our yet knowing it, paved the way for so doing, but the actual achievement came to pass through my friendship with John Self. I had sent him *Politics,* in which he had taken a great interest, and he in turn had lent his numbers to others. I was speaking to him of our wish to get at the artisans, and he told me that in Fetter Lane there was a remarkable man, a tailor named Walter Cooper, a professed chartist and infidel, and Lecturer on Strauss[12], but a man of great independence of character, who when out of work and really starving had actually refused relief. He was one of those to whom he had lent *Politics,* whose distrust of the journal had been disarmed by its failure. I said I should like to know him, and the man's consent having been obtained, I called on him at his shop in Fetter Lane, and we had a good talk, I seated on the board at which he was working. He was then, I am bound to say, a perfectly genuine man, whatever he may have become eventually. A Scotchman, he had been brought up a strict Calvinist, and had lain awake whole nights praying to be delivered from hell-fires. Then his faith had by degrees dropped away, and he had become a Chartist, Socialist and unbeliever. He knew Thomas Cooper, the poet of the *Purgatory of Suicides,* and they had agreed to call each other cousins. I think that what impressed him most was my pointing out that the Lord's Prayer contained in it the promise of God's will being done on earth as well as in heaven. I spoke of Mr. Maurice, and he promised to come and hear him. He came accordingly to the afternoon service at Lincoln's Inn (February 1849). I saw him in the course of the week, and asked him what he thought of Mr. Maurice's sermon. 'Well,' he replied, 'I didn't quite understand him. But there were some very fine things that I did understand. And I think the man is in earnest'. I suggested that he should go again. 'Yes,' he replied, 'I told you that I did not quite understand him; I mean to go till I do'. He became a regular afternoon attendant and I am sure the best and most earnest period of his life was whilst he was such. By the 20th April 1849 I could write to Kingsley that he, 'was rapidly becoming a devoted Maurician'.

Through Walter Cooper in turn I made the acquaintance of some of his friends, two of whom were to become closely connected with us, Joseph Millbank and Thomas Shorter. Both were men several years younger than Walter Cooper, watch-case-joint finishers by trade, but Millbank worked chiefly in gold and Shorter in silver; they worked at the same shop, were intimate friends, and were perhaps the most remarkable young working men I have ever met with. Millbank had the better presence, and the more aplomb, so that he made the best of his abilities; Shorter was more retiring, but was the deeper thinker, and was looked upon by the working men as 'very long-headed'. Two or three years later, when Millbank had given evidence in favour of co-operation before the House of Commons Committee on the Savings and Investments of the Middle and Working Classes, Mr. John Abel Smith, our ablest opponent in the Committee, took him apart afterwards and said to him: 'Mr. Millbank, surely you are too clever a man really to believe what you have been telling the Committee?' And Millbank had to reiterate his profession of Co-operative faith to the banker M.P. personally. On Sully's retirement, as will be hereafter mentioned, Millbank and Shorter became joint secretaries to the Society for Promoting Working Men's Associations, but Millbank, with a family of five young children, went out to Melbourne, where he had a brother-in-law settled, to better himself, and Shorter then remained sole secretary to the Society, becoming afterwards secretary to the Working Men's College ... Of Shorter I will say, that a more thorough gentleman in the truest sense of the word I have not known ... He published *A book of English Poetry, Echoes of by-gone days,* a volume of original verse, etc. He was editor for some time of the *Quarterly Journal of Education,* also for a short time of a provincial newspaper, and he was an admirable speaker ...

In March 1849 came from Walter Cooper the suggestion that Mr. Maurice should hold meetings with the working men, and the appointment of a joint committee of gentlemen and working men for the purpose was, I think, the first thing that brought me into contact with Walter Cooper's friends, I being on it with T. Hughes, and I think Charles Mansfield. We took ... a room at the Cranbourne Tavern near Leicester Square, where all our subsequent meetings of the time were held.

I never recollect any meetings so interesting as were these. They were, it must be recollected, fortnightly conferences between a clergyman of the Church of England and working men, the first of the kind that had ever been held by one in Mr. Maurice''s position, although something similar had already been done by a very manly and worthy congregationa-list minister of South London, the Rev. John Burnett of Camberwell. Mr. Maurice generally began the proceedings by a brief statement of the objects of the meeting, inviting a free expression of opinion, and the great feature of the evening was his final deliverance . . . We, the young men who gathered round Mr. Maurice, seldom or never spoke, with the exception of clergymen. As a rule, the tone of the working men who spoke was courteous as well as manly, though now and then there was an insolent fellow, who tried his best to wound. Sometimes we found unexpected allies, or opponents who seemed as if they might become such, and in such cases Mr. Maurice would generally ask them to come back with us to his house for a cup of tea. Amongst the last were Lloyd Jones, of whom more anon . . .

Amongst others who did not come to Mr. Maurice's house were . . . Mr. Holyoake, who if I remember aright attended only the first meeting, sneered at it in the *Reasoner* and did not impress me favourably; Mr. Bronterre O'Brien, a chartist leader and former school-master, who tried afterwards very hard to be taken on in the Christian Socialist movement, a man whom I disliked and distrusted, the very type, I should say, of a misleader of the people, and a much more sincere man, Hetherington, the chartist and atheist printer and newspaper editor. But with the exception of those men with whom we were already in relation, Walter Cooper, Millbank and Shorter, the most remarkable man by far whom the conferences brought out on the working men's side was Lloyd Jones, an ex-Owenite lecturer, then a tailor in Oxford Street, and at the same time an editor of a weekly Socialist newspaper, who was to become later on an efficient ally of ours, and to me for many years a very dear friend.

Lloyd Jones was descended – as stated in the short memoir of him by his eldest son prefixed to his work on *The Life, Times and Labours of Robert Owen* – from a Glamorganshire family, and . . . learned his father's trade, that of a fustian-cutter . . . I do not now remember under what circumstances he first came across Robert Owen or his works. But in 1831 he was a member of a store set up in Salford, the members of which were for the most part young unmarried men, which did not however prosper and had to be wound up; that he and his friends thereupon set up a free night school for boys, girls and adults, which in six months had 170 scholars in regular attendance, between the ages of 12 and 40, most of them being factory boys or girls, and which was carried on for six years during which time they held also Sunday meetings chiefly on social subjects. By 1837 he was sufficiently prominent to be appointed by the Manchester Socialist Congress of that year one of six 'provincial directors' of the movement . . .

Lloyd Jones was one of the best speakers of the Parliamentary type whom I have ever known and had he entered the House of Commons I feel certain that he would have reached a very high place there. Always addressing himself to the reason and not to the passions, he was eminently clear and convincing, but he could be very pathetic, and very powerful. Walter Cooper was much more of the popular orator, yet Lloyd Jones obtained a much firmer hold of his audience, and knew that he could do so. I remember once at a Crystal Palace meeting in after years his beginning a speech with abstract considerations so far off, it seemed, from his subject that for a time he left his hearers cold, and I was beginning to wonder whether for once his speech was not going to be a failure. Yet with the most consummate art he went on weaving the tissue of his argument till what had seemed to be detached threads took one by one their place as essential elements of the design, and by degrees he obtained the most absolute mastery of his delighted audience, and was greeted at its close with cheers such as no other speaker of the day obtained, though several were eloquent and popular men . . . He was again an admirable journalist, though he virtually wrote only for the provincial press. He founded the *Leeds Express,* now a valuable property, and French journalists – and they are good judges – have expressed to me their admiration of his journalistic talent. As a companion he was delightful, as full of knowledge of books as of men. But his books were mainly old books. He had a philosophical indifference to new publications, unless by really favourite authors, but was a determined haunter of book-stalls, and got together a very valuable library . . . His was a character that remained young beneath the abundant soft white hair that framed a greatly broadened brow and made him a most striking looking man, his heavy moustache giving him something of a military look. To the last he would kindle into sudden fire over some act of oppression or wrong, or burst into genial laughter over a good joke. He caught his last illness in the course of an unsuccessful canvas for a seat in the north, undertaken at an age when if he had entered the House at all, it should only have been oñ an uncontested election. I must add that, though singularly reticent as to spiritual matters, Lloyd Jones once told a friend of mine that since he had come to know Mr. Maurice and his friends he believed that Jesus Christ was the Son of God, and I can vouch from personal experience that his death-bed was that of a christian.

The result of the conferences was to satisfy us that social and not political questions were what lay nearest to the heart of the thinkers among the working class. The path was opening more and more in the direction which had revealed itself to me as the true one in March 1848. And our leader, Mr. Maurice, was growing greater every day in our eyes. I never knew him speak more impressively than at these conferences; it would in fact be impossible to exaggerate the height of spiritual power to which he rose. I deeply regret now that we did not provide a reporter for his utterances. The only scraps of report that came out were in hostile newspapers now and then, and were always poor, and generally distorting. [I have said that Hughes' *Prefatory Memoir* is hopelessly incorrect as to date ... Moreover, my recollection of the scene when Kingsley first came forward at the conferences does not exactly tally with Hughes' ... I sat next to Archie Campbell; by him was a then young Scotch friend of his, David Masson. Two or three seats off sat Kingsley, who, I think, had come in a little late. When he rose up and began his speech with 'My f-friends, – I am a p-p-parson and a ch-ch-chartist' (not a Church of England parson, as Mr. Hughes had it; the words 'ch-church of England, I mean' were added almost *sotto voce*) I was horrified to see Campbell by my side stuffiing his pocket handkerchief into his mouth, and evidently writhing with all but uncontrollable laughter. I felt desperately in earnest myself, and looked daggers – I might say cutlasses and broadswords – at him, which of course only excited his risible muscles the more. It was only after the close of the meeting that he explained to me how his neighbour, who had never come across Kingsley before, and still spoke with a strong Scotch accent, when Kingsley began to deliver himself after the fashion above-mentioned, had leant over and whispered to Campbell in a tone of deep concern – 'The man is drunk!']

On one occasion after one of the conferences, going with Mr. Maurice to Mr. Scott's, I met at the latter's house for the first time Mrs. Gaskell, who had just glided into well-deserved fame as the authoress of *Mary Barton*. I found her, as I described her to Kingsley, an 'entirely lovable creature', an intense admirer of the *Saint's Tragedy,* and ardent for the revival of *Politics for the People,* of which she had been a regular reader, and I may say we were friends from that hour.

One observation has to be made. We who joined together in such work as I have mentioned or have yet to mention were none of us idlers. Lawyers, Architects, Chemists etc., we were almost all trying to do our best in our several professions. I do not believe any one of us neglected any professional work he had to do for the sake of our social efforts. Except on the 10th April 1848, which was practically throughout the Inns of Court a *dies non* for professional purposes, I don't think I was ever away from my chambers till after business hours for social work. Mansfield was certainly working unremittingly at his chemistry; Penrose and Campbell at their architecture; Walsh at his medical practice or sanitary reform.

In point of fact, 1848 and the following years were those of greatest business activity till the time of the Friendly Society's Commission. In 1848 I was devilling on the Joint Stock Companies' winding-up act 1848, of which I published an edition early in January 1849, with a long introduction (pp.58), urging many reforms in legal procedure which have since been carried out... In 1849 I drew up an amending act, the Joint Stock Companies' Winding-up Act 1849. Of this I also published an edition... [The publication of this second book had for a sequel one of the only two law suits in which I was ever concerned on my own account. The publishers, Messrs. Stevens and Norton, had kept it back longer than I liked, on the ground that the publication of another work on both acts, by a barrister whom I will not name, was being itself delayed, as they suspected, for purposes of piracy. It came out at last, and mine followed immediately. My publishers sent me a copy of the rival work, and I was amazed – and to some extent amused – to find that whilst fulsomely praising me, it had appropriated the whole of my first work, including the entire index. A suit for an injunction was immediately instituted by Stevens and Norton, and with entire success, the whole of our costs as between solicitor and client being paid by the defendants . . .]

CHAPTER XIV

Stay in France – Sanitary Reform –
The Working Tailors' Association –
1849 – 50

In the long vacation of 1849 my mother and I spent a pleasant month with my sisters at a house which my married sister had taken at Bellevue, running often up to Paris... During the sixteen months which had elapsed since my last visit to Paris, things had not yet settled down – owing no doubt in great measure to the terrible insurrection of June 1848. Partly from caution, partly of set design to damage the republic, the wealthy classes, who mostly belonged to the two defeated monarchical parties, the Legitimists and the Orleanists, did not spend their money freely. There was little demand for luxuries, material or intellectual. Artists could scarcely sell a picture. From the lower ranks of the art world, starving men were driven into other callings. The man who used to bring round the newspaper to my sister's house at Bellevue turned out to be an artist by profession, chiefly in water colours, or in black and white as a designer for illustrated papers, M. St. Ange Chasselat. He was a pleasant-looking man, and attracted the attention of one of my nieces, and of a cousin of hers, rather older, who for a few months was staying with my sister. They first found out his story. He told them that after the revolution he and a friend in the same line, finding themselves absolutely starving, had taken to selling newspapers in the streets. Yet even this was not very profitable, and he confessed that on one occasion, when evening came and they had not made a sou to break their fast by calling out true news, they had laid their heads together and determined to try the effect of false. So they began shouting out: 'Deputation of three hundred hunchbacks to the President of the Republic . . . ' They sold off their whole stock in ten minutes, and had a good supper. 'But,' he added, 'we never dared afterwards show ourselves in that neighbourhood'. He was now in receipt of regular wages, and was comparatively well off, but hankered of course after his old calling. He brought some of his tinted sketches, which showed him to have real talent. Two or three were bought; his name was mentioned to friends; he got back to art, and I saw many years after in the *Illustrated London News* a whole series of illustrations by him of Normandy Churches.

On my visits to Paris, I went to see several of the *Associations ouvrières* which, as I wrote to Kingsley on my return, were 'endeavouring to beat down and moralise competition by competition itself'. I visited those of the tailors, arm-chair makers, cabinet-makers, upholsterers and cooks. The main results of my observations are set forth in No. 4 of the *Tracts on Christian Socialism,* 'The Working Associations of Paris' . . .

At that time I was more of a Fourierist than anything else; subscribed to the *Démocratie Pacifique,* the able Fourierist organ of the day, bought while I was in Paris, a hat of a zealous Fourierist hatter, who with every hat that he sold gave a cheque for a proportionate amount to be expended in Fourierist publications. Now I need hardly say that Fourier's Socialism was all-embracing; that he contemplated a new industrial and social world (*Le nouveau monde industriel et sociétaire* is the title of one of his works), and that therefore the working associations of the day in Paris could not be the satisfaction of my social aspirations. I mention this, because the formation of co-operative associations of producers in the various trades has been treated as if it had been the be-all and end-all of our socialism (I shall revert to this hereafter). In the same letter of 11-12 October 1849 in which I described my visits to the Paris associations, I used these words: 'Mind, I don't think these men have got hold of the right handle to Socialism, which our men have in the "Home Colonization" movement, but such handle as they have they are working with heartily'.

Home Colonies, I may observe, had formed one of the subjects of Debate at our conferences, and in an earlier letter to Mrs. Kingsley of 20 July 1849 I had written: 'The Master' (i.e. Mr. Maurice) 'is all but professing Communism (literally, Communism) with a boldness and a vigour quite amazing: telling our working men friends that what has chiefly engaged his mind during their debates on Home Colonies has been "far less what he should say, than what he should do in the matter." Fancy our all squatting in Kerry, with the Prophet at our head! I for one have openly pledged myself to join his colony, if he should found one'.

But during the autumn of 1849 sanitary reform was the subject uppermost in the minds of most of us. Kingsley, Walsh, Mansfield, were all ardent sanitary reformers. The cholera was upon us. In the course of a series – or rather several series – of letters on *Labour and the Poor,* which were being published in the *Morning Chronicle,*[13] an account was given of a place called Jacob's Island in Bermondsey which was simply terrible. A tidal ditch forming the island had been blocked up at the entrance, and its henceforth stagnant water had become literally an open sewer, whilst remaining the only source of water-supply for the inhabitants. Walsh was an inspector during cholera time for the General Board of Health; Jacob's Island was in his district; he got as many of us as he could to go down and see it for themselves; there was no exaggeration in the story. We obtained permission to reprint the *Morning Chronicle* article, which (with a few words of Preface by me) was circulated to the extent of a thousand copies (October 1849). The authorities were stirred to action, and in the meanwhile we put up three (I think) water-butts, supplied with water from the nearest water company, in different places of the island for the convenience of the inhabitants – though I am sorry to say they were so little appreciated by many that the brass cocks were being constantly stolen to sell, so that all the water ran out. [The Jacob's Islanders had been utterly neglected, and were a bad lot generally, with a few exceptions; indeed the hardest drinkers were the best lives, as the sewage dignified by the name of water was a worse poison than gin. The motives of Walsh and his friends in coming among them were incomprehensible to them with the exception of one old man, fairly well-to-do and respectable, who afterwards, cross-questioning Walsh as to the water company or other body in whose interest he supposed him to work, at last came to the conclusion that 'he supposed it was pure patriotism', adding after a pause, 'Well, I also joined the Volunteers for six months when I was a young man!']

At the present day, when the preservation of the public health is almost universally recognised as a matter of general concernment, it is almost impossible to realise what a novelty were our operations in Jacob's Island. I saw for instance (as I wrote to Mrs. Kingsley, 31st October 1849) a 'stirring attorney of the name of Slee', who promised to circulate our hand-bills, and was marvellously struck with the prodigy of 'certain benevolent people – not in any way connected with property in the parish – contributing their time and money to remove other people's nuisances'. We projected a series of Sanitary Tracts; Walsh had plans of Lectures and Evening Talks; Mansfield was analysing water and studying the Public Health Act. My own notion was that of a Sanitary League, with shilling members, a regular show-room and local committees. The idea was ventilated at a meeting at Mr. Maurice's; without at once accepting it, he it was that suggested the title 'Health-League'. At a meeting at my house with Walsh and Charles Mansfield the plan of such a League was settled as I give it below.[14]

My first draft of it was – what shall I say? – somewhat bellicose, but the *mitis sapientia* of my dear friend Walsh judiciously toned it down, and I can still hear him chuckling quietly over the words '*stimulating* and assisting all public bodies and private persons'.

There was at this time, if I recollect aright, no voluntary body in existence for sanitary objects. There have been many since, and sanitary reform has no doubt progressed immensely since 1849. But I suspect it would have advanced still more rapidly on the democratic basis of the shilling subscription and the recognition of the 'right to live'.

To our dismay however, Mr. Maurice finally declared

himself against the 'Health-League', as will be seen from a letter of his to me of November 24, 1849, printed at p.23 of Vol.II of his *Life*. At the same time he fully approved of our practical work at Bermondsey, and at a meeting at his house one Friday evening a memorial to the Commissioners of Sewers on the subject of Jacob's Island was read, approved and signed, and a sort of nameless society constituted to meet weekly on Friday nights alternately at Mr. Maurice's and at Walsh's, when the work was to be partitioned out, the whole being cemented by a shilling a month subscription. Kingsley indeed, was inclined to rebel, but I warned him that I was not prepared to do so. 'It is a very great principle', I wrote to him through Mrs. Kingsley, 'to establish this of the right of a number of persons not connected with a parish to go in and help to remove a certain number of parochial nuisances because they are also national crimes, and if we begin with too coarse a wedge, we shall never get the small end in, without which the block cannot be split'. Still, I was myself greatly disappointed, and expressed my disappointment to Mr. Maurice.

Looking back, however – although I am far from agreeing in Mr. Maurice's denunciation of Societies and Leagues[15] – I feel that he was rightly inspired in checking our sanitary crusade. It would have taken us away from our appointed work, that of co-operation – would have tended to render barren the progress we had already made in opening friendly relations with the working class. For at that time at all events the working men generally were not – I doubt if they are now – alive to the importance of sanitary questions: the question of work and earnings was and is still for them the main one of all. A Health League with shilling subscriptions may possibly yet some day be established; but it will never have in their eyes the importance of a single sitting of the Trades Union Congress.

The conferences with the working men, interrupted during the summer, had been resumed. The *Morning Chronicle* letters on 'Labour and the Poor' were revealing from week to week the horrors of the lower strata of the labour-world. A strong feeling was being shown at the conferences in favour of co-operative labour, which, from my visits to the Paris

Associations, I could not but share. Moreover, since June 1848 when the bourgeois reaction represented by Cavaignac had not only put down insurrection as it was its right and duty to do (though not with the savage violence which it exhibited) but had slowly discountenanced all the strivings of the working class, fettered their associations and at last crushed all the more advanced forms of Socialism such as Proudhon's *Banque du Peuple,* there had come to England a first batch of French refugees, who were more or less leavening, for good or evil, the English working class. Amongst these was my old acquaintance Jules Lechevalier, the West Indian abolitionist, the former St. Simonian and Fourierist, latterly under Louis Philippe the projector of a plan of emancipation and colonisation combined on co-operative principles to be tried in French Guiana, and who finally had been associated with Proudhon in the *Banque du Peuple.* He was also the Paris correspondent of the *New York Tribune,* and so highly were his services appreciated that when he came to London in 1849 the *Tribune* retained him as an extra London correspondent. I had not been much inclined to him in former days, but he was always a pleasant companion and whenever I went to Paris or he came to London, he always sought me out. When he came to England in 1849 I had found him greatly changed for the better. I have always thought it was the companionship of Proudhon, a man of terrible sardonic insight, who would bear with no shams, which had brought out all that was most sincere and earnest in him. He was full of wit, of knowledge, of resources, a charming companion. I introduced him to my friends, telling them frankly that I had not cared for him formerly, but that he appeared to me now really sincere. He won his way with them greatly – too rapidly, I feared. He went to hear Mr. Maurice, read his books, lent them to dissenters. A new prospect of religious socialism, I believe, opened itself to him, and I think that for a time he really gave himself up to it. Kingsley was greatly struck by his conversation; Mr. Maurice welcomed him. If, in after days, he sank to depths of hypo-crisy and intrigue which helped seriously to mar our co-oper-ative movement, there is all the greater reason for doing justice to the services which he rendered to it in earlier days.

On the 17th December I wrote to Charles Kingsley:

Have you or have you not read the letters in the *Morning Chronicle.*
Have you or have you not read the Metropolitan series, and
especially that number of Friday last, describing the slavery of
the journeymen tailors to the great slop sellers. If you have not,
read them forthwith; if you have, tell me whether I am not right
in saying that operative associations or partnerships such as they
have in Paris *must* be set up forthwith, whilst the subject is yet
fresh in people's minds; that all who have not, by deed and word,
perpetually protested against the false seeking after cheapness
which has created this slavery, are *bound* now to spend our money
in creating that machinery which more than any other must tend
to destroy it. The thing as it seems to me, is to be done in two
ways, 1st by giving money for setting up workrooms and shops,
2nd by promising custom. Mansfield, Hughes, Campbell, are
quite at one with me on the subject, and on its urgency. I have
not spoken with the Master since I have got thoroughly roused
myself upon it. I mean to go to Walter Cooper as soon as he
returns to town, i.e. tomorrow, and press him upon it. Could you
come up in the holidays to whip the thing through? Tailors and
shirtmakers seem the trades for which association is most needed,
but a good many associations might be started at once; they
might help one another immensely by a due carrying out of the
'mutuality' principle, i.e. the exchange of labour and products at
cost price, so as to dispense with all capitalists' profits after the
single one on the purchase of the raw material; i.e. if a bakers'
association supplies food for coats to a tailors' association, two
profits are saved, the bakers' and the tailors'; if a millers'
association enters into the League, there is a third profit saved; if
a corn-chandlers', there is a fourth. This is Proudhon's really
great idea, which our friend Lechevalier has helped more than
any other to carry out – I seem to see work crowding in upon
us with such rapidity that one knows not whither to turn.
For instance, petitions should be got up against the very first
day of Parliament for a discontinuance of the Government
clothing contract system. And yet Jacob's Island is not to be
neglected . . .

In short, before the end of December 1849 we were determined to try the experiment of co-operative workshops, with
or without Mr. Maurice's sanction. I need not tell again the
story of the dinner which took place at my house in the last
week of that month, to which Mr. Maurice invited himself,

and at which it was resolved at the same time to start a workshop or workshops . . .[See T. Hughes' *Tract on Christian Socialism*, 'History of the Working Tailors' Association, 34 Great Castle Street'.]

Before however any organised work began, two blows were struck almost simultaneously – the publication of Kingsley's *Cheap Clothes and Nasty*, issued originally as an independent Tract by 'Parson Lot', and an article by myself on 'Labour and the Poor' in *Fraser's Magazine* for January 1, 1850, precedent in date to the former . . . *Cheap Clothes and Nasty* was a tremendous blow to the competitive fetish. Of all the publications in connection with our movement, it was unquestionably the most telling. Published at first at 4d, it was indeed at that price beyond the reach of the working class generally till its reprint by us at 1d, as one of the *Tracts by Christian Socialists.* A petition to the House of Commons against the Government contract system was drafted by me early in 1850; Hughes and Hansard worked especially hard to procure signatures for it. Of course it produced no effect at the time . . .

I think also I may say that my article on 'Labour and the Poor' in *Fraser's* for January 1850 was not without effect. Published as the first paper in a new volume, it would attract additional attention from the Editor's note prefixed, that 'altho' agreeing entirely with the spirit and object of the following communication, we beg to guard ourselves against being supposed to go along with the writer in the whole of his opinions'. I do not think I have ever spoken out more strongly than in that paper what I believe to be the truth. I denounced underpaid labour as 'robbery' and claimed that the

> Government should put a stop at once and for ever to the iniquitous system of accepting contracts at the lowest tender, without inquiring into the morality of each contract . . . If the contract system is to be persevered in let a fair price be settled by competent judges, and the contract given to some fair-dealing man, who shall pay to his work-people living wages; if it be needful, let a Government officer see that living wages are paid to them.

I anticipated the creation of the government workshops at Pimlico by going on to say that if it should seem impossible to do as I had suggested, 'then let government take the work into its own hands; deal directly with the workpeople; pay them living wages without the intermediate profit of a contractor. There is no reason why there should not be Government clothing workshops as well as Government dockyards'. Besides recommending co-operation both in consumption and production, labour marts or exchanges, which might be usefully annexed to model lodging houses, as then proposed in France, – the moralising of private custom by endeavouring to ascertain who are the fair dealing and honourable tradesmen, and dealing with such, or dealing directly with the workman, I went on to say:

> Let every sect and creed enter into friendly competition for improving the conditions of the people of England. Let the Church especially put forth all her strength to grapple with the hundred-headed evil. She has been supine far too long. As I look upon her benefices and Episcopal Sees, upon her deacons, her priests and her bishops, I seem to see the skeleton of a great army, the battlefield of a holy warfare; all the strongholds are occupied – officers to command them there are plenty, but the privates are nowhere, or fighting on their own account. What are all our religious societies but irregular partisans, guerilla bands, ill-armed and ill-disciplined, too often unscrupulous in their means of warfare? Let the parish on one hand cease to be a mere geographical expression. Let it become a centre of radiating life; let it be no more for the working classes a mere dispenser of legal relief, but the source and focus of all local action. ((The Parish Councils, first elected in 1894, are a carrying out of this idea)); let the poor be no more helped in their poverty but helped out of it, by parochial lodging houses, parochial baths and wash-houses, friendly meetings between class and class, and every other institution which can be brought within the compass of parochial effort and parochial control. On the other hand, let all those scattered seeds of Catholicity, which the huge landslip of the 16th century has buried far underground, be set free to vegetate once more . . . Let the Protestant orders of both sexes, bound by no vows, go forth in joyful self-devotion to battle with the wretchedness and the vice of England . . . We must have Orders of Nurses, Orders of prison attendants, Orders of workhouse masters, workhouse matrons, workhouse teachers, perhaps parish Surgeons – bodies of men and women that shall show forth in its purity the essential communism of the Church, and leaven the whole of society with a spirit of self-devoted industry. Let such a

spirit once go abroad, let it raise labour to the rank of a sacred duty, the holier in proportion to its very arduousness and repulsiveness, and the word 'Penitentiary' will no more be a shameful lie, nor the word 'workhouse' a promise unfulfilled; our wastes will be cultivated, our marshes drained; the soil shall no more be robbed of its fruitfulness by wasteful cultivation, nor neglected by slothful ignorance; our polluted rivers shall no more bear away to the sea the exhaustless wealth of the sewage of our towns, nor our atmosphere grow thick with the wasted carbon of our smoke; the well drained houses of our airy streets shall no more breed fever and pestilence, sweeping away children, shattering the frames of the adult, already weakened by starvation wages ... ; feelings of mutual sacrifice, confidence and love will gradually supplant those feelings of mutual encroachment, of mutual distrust and hatred, which now estrange class from class and man from man; the very word 'society' will seem too foreign and merchant-like to express the mutual relations of Englishmen, and in good old Saxon phrase used of yore for kingdoms as for republics, England will be truly one commonwealth and one church.

Do these words seem visionary and ideal? Would it not be better anyhow ((to try)) how nearly we can realise them, than to measure how far we fall short of the promise they hold out, as an excuse for folding our arms and doing nothing.

In accordance with the decision come to at my house as above mentioned, the Working Tailors' Association was started in Castle Street East, Oxford Street (before January 22, 1850), with Walter Cooper for Manager, and Gerald Massey, afterwards a well-known author in prose and verse, for Secretary. The starting of a Needlewomen's Association, which soon after was opened in Red Lion Square, was also decided on.

CHAPTER XV

The Society for Promoting Working Men's Associations –
1850 and Following Years; Promoters, Officers and
Well-Wishers – Some French Refugees of 1848

Fresh recruits were now frequently joining us, especially
... Edward Vansittart Neale, a member of the Vansittart
family, and a man of singularly acute and ingenious mind,
and of devoted benevolence. He had strong socialist leanings,
and had already lost a good deal of money on a co-operative
experiment by some Frenchman whose name I had never
heard and have completely forgotten.

Whilst our little group was in fact 'promoting' the formation of co-operative associations for production (as they might
now be designated) there was as yet no actual organisation to
correspond with the team. Mr. Maurice, in his horror of
'system', would not for some time hear of any central office or
council to advise the associations, which soon began to spring
up on all sides, and to prevent their competing with one
another. 'This,' I wrote to Kingsley (25 March 1850) 'is a
much more serious affair than the Health League, and if our
dear Master has not yet nerve to carry the thing on, some one
must have it and soon' – He gave in however, and the
selection of the Council was left absolutely to him. Most of us,
I must confess, if not all, were surprised to find Vansittart
Neale among those named, as we scarcely knew him as yet
and notwithstanding the regard and friendship which I grew
to feel towards him, and my reverence for his self-devotion to
the co-operative cause, I still think it would have been more

162

judicious not to have included him, at all events at the first, for he had taken no part in any of our previous work; he could not in any sense of the word be termed a Maurician; (this indeed was probably the very reason why Mr. Maurice named him) we could not feel sure of him as we did or seemed to do of each other . . .

A Society being thus constituted, it became necessary to provide it with a Secretary. Lechevalier recommended one, in the person of Charles Sully, a Englishman who had resided many years in France. He was a bookbinder by trade, had been a Chartist and was a Socialist and a professed Republican; had fought in Paris during the insurrection of June 1848, when, as he told me himself, he (fighting bare-armed) had been splashed with blood up to the elbows – After June 1848 he had at first gone over to the United States, examined into mormonism [He showed me the only copy I have ever seen of the '*Book of Mormonism*'. How such an obvious travesty of the Bible, and so poor a one could ever have exercised the slightest influence over anyone who had ever read the latter, I could not and cannot understand.] but had eventually returned to England. He was a man of known ability and integrity. He had however become convinced that politics must be kept out of co-operative work, and was quite ready to help our movement . . . There were details in his career which we only knew of later on, and though they compelled us to part with him, I am bound to say on the one hand that no abler or more faithful secretary could our association have had, – on the other, that he remained for me a valued personal friend. In after years, I had occasion to employ him, and found him most absolutely trustworthy . . .

Sully undertook the framing of the constitution 'to serve as a basis for the laws of the United Associations of working men' and of a Code of Laws for an Association in conformity with the constitution . . . The object of the Union was declared to be –

> To carry out and extend the principles and practise of Associated Labour
> 1st. By forming associations of working men, who shall enjoy, among themselves and their families, the whole produce of their

labour, subject to the re-payment of borrowed capital (if any), with a fair interest thereon.

2nd. By organising, both among such associations and among any others of combined capitalists and working men who may be admitted into the Union, the interchange and distribution of commodities.

3rd. By establishing among all the associations admitted into the Union institutions for the benefit of the members, as Friendly societies, model lodging houses, schools etc.

4th. And by the full development of every means of brotherly help and support, which the capital, the credit, the custom, the knowledge and the influence of the association can afford.

Members of associations were to receive a periodical allowance (representing wages) 'which shall be a fair day's remuneration for a fair day's work, according to the talent and industry of the individual receiving it and shall as far as possible be the same in all the associations of the same trade, in the same place, and for the same nature and quality of labour'. The net surplus, after deducting current expenses, sums due for repayment of borrowed capital, and interest on the capital employed in outfit and other standing expenses, was to be equally divided among all the associates, in proportion to the time they have severally worked, one third of the profit at least being left to increase the capital, which was to be equally divided among the members for the time being, and might bear interest, which alone could be paid to the members, the capital thus formed being inalienable. Each association was to contribute a sum to be fixed by the Central Board towards provident purposes. Hired workmen were to be paid the same wages as an associate would receive for similar work by way of allowance, and unless dismissed for misconduct, a certain sum in lieu of profits. No new associate could be received until after a period of probation. Rules were to be made for the admission and remuneration of apprentices. Each association was to have a manager who should have no other commercial business . . . No work was to be done on Sundays. 'The hours of daily labour must not exceed 10, excluding meal times, unless with the assent of the Central Board and of the Council of Promoters.' . . . No association whilst in debt to the Council of Promoters could borrow money without its consent, and whilst any money advanced by the Promoters remained unpaid, the Promoters reserved to themselves a right of veto on the appointment of the Manager, and on the regulations of the association relating to his powers and duties.

A clever writer, – or if I may use the term, writeress – upon social subjects, Miss Beatrice Potter, now Mrs. Webb, in her work on *The Co-operative Movement in Great Britain*, insists upon fathering all our association work upon Buchez. No doubt the primary form of our associations being modelled in great measure upon the Paris ones is derived mainly from Buchez. But anyone who has a larger acquaintance than Mrs. Webb seems to have with continental socialism will see the traces of various other influences in Sully's 'Code of Laws'. The 'organisation of labour', to begin with was essentially Louis Blanc's watchword. But beyond this, we looked to the organisation of 'Exchange' and 'Consumption' – the 'interchange and distribution of commodities', – thereby linking our objects with those of Owen on the one hand, of Proudhon on the other. In fact, considering the presence in our midst of Lloyd-Jones, an old Owenite, of Lechevalier, successively a follower of St. Simon, Fourier and co-worker with Proudhon (to say nothing of myself), it was simply impossible that we should have blindly followed Buchez, whom not one of us had ever acknowledged as his teacher, and whose socialistic works not one Englishman amongst us except Sully had ever read before our associations were started. I say this without in the slightest degree wishing to depreciate Buchez, with whom I had considerable sympathy, but simply to restore the facts.

I must now warn the reader that there may be much that is loose and uncertain in my future references to the proceedings of the Society for Promoting Working Men's Associations. Through a most untoward mischance, all the official records of its proceedings, which had been carefully preserved, have passed long ago now into unknown hands, and have no doubt been destroyed. On the break-up of the Working Tailors' Association in Castle Street where they had been kept, they were transferred to the Working Men's College, as to a perfectly safe depository, the freehold being vested in the College. Some years after, I wished to consult these records, and on applying to see them at the College I was amazed to learn that Mr. C. E. Rawlins, the Secretary, a very well-meaning, zealous and devoted little man, but fussy and without much judgment, who had never taken part in our co-operative movement and felt no interest in it, finding these papers to block up some shelves of a cupboard which he wanted to utilise for the purposes of the College, had, without consulting a single person, taken upon himself one fine day in vacation time to sell the whole as waste paper. The loss was simply irreparable.

Owing to this loss of documents, I am unable to give the sequency in which new members joined the Council, and cannot be quite sure of all who did. One of the earliest of these at all events was Vansittart Neale's cousin, Augustus A. Vansittart, with him the richest man of our body . . .

Another well-to-do member who joined the Council after a while was the son of the then Earl of Ripon, Lord Goderich. Shortly after the establishment of the *Christian Socialist*, in January 1851, I was called upon at my chambers by Henry Ellis, then a young Bombay civilian, who . . . told me that our views were taking hold upon persons in the higher ranks of society, and instanced his cousin, Lord Goderich, then an attaché at the Brussels Legation, whom he offered to bring to me, and did so, I think on the following day. Before long Lord Goderich was elected on the Council, of which he always remained an efficient hardworking member.

I may here make a clean breast of it; I rather shunned and discouraged for my part any actual intimacy with Lord Goderich, as I always have shunned through life such intimacy with persons much above me in position or for-tune . . . But – although at present widely separated from Lord Ripon both as to politics and religion, – I have a very great regard for him. He is a first-rate worker, a man, for the most part, of very good judgment; (though I cannot agree with or thoroughly understand his conversion to Romanism); his aims are high; he did noble work in India. I believe him indeed to have had in him the possibilities of much more than he has ever risen to; that at one time in his life he had the makings of a great and really noble popular leader. And if I might mark the turning-point from which he fell away into the subordinate, though honourable politician, I should place it at the suppression of his pamphlet on *The Duty of the Age* . . . I have never ceased to think that Mr. Maurice's conduct in that matter was a grievous and lamentable error of judgement . . .

Another new Promoter, though certainly not of the more pecunious type, was Cuthbert Ellison. Introduced by Hughes, he soon won popularity on his own account, though not with all. He was very good-looking with black curling hair, slow of speech, an unmistakeable man about town – the original, I believe unquestionably, of Thackeray's *Arthur Pendennis* and the only man I ever knew who took a positive pleasure in dress. I remember now a wonderful plum-coloured cravat of his, and the childlike delight with which he looked at it in the glass . . . Associating with richer men than himself, Ellison had contracted extravagant habits, and was a good deal in debt. Moreover he was in love with a young lady, – an orphan under the charge of an uncle, – with whom he had been thrown together, amongst other occasions, at the Canterbury Theatricals (Ellison's affairs generally, including his love affairs, I may observe, were the common property of all his friends). And the cruel uncle would not think of handing over his niece to a young man, however gentlemanly and popular, who had nothing but debts to keep her upon. Ellison was at the bar – I am not aware that he had any clients, – and when I first knew him formed part of an odd little community, 'Three Fig Tree Court' . . . It was here I first knew Tom Taylor . . . I never knew of any other man who gave up the bar not because he failed at it, but because he succeeded, – 'it was making too great a rascal of me' he said. So he threw it up, and became for many years Secretary (and a very good secretary) to the then General Board of Health, yet vexing the soul of its President, Lord Shaftesbury, by continuing to write farces, in his spare time. He leaving, and Thackeray having withdrawn, the community broke up and Ellison migrated to Old Square, Lincoln's Inn, within a stone's throw of the set occupied by Neale, Hughes and myself. It was here that I became really intimate with him, and was initiated into his love-troubles. A singular instance of the influence of the mind over the body was this – that whenever Ellison's love affairs had taken a turn particularly to the bad, he was sure to be laid up with rheumatism – real *bona-fide* rheumatism. This I have seen more than once.

The great question was how he might make money to marry upon. Writing a law-book was one of the means seriously pondered over. Finding a place under government was of course always looked to, though Ellison being a Conservative and the Liberals having about this time an awkward trick of being in power, the hope was but a flickering one. Hughes hit upon a plan for helping him which at first greatly tickled my fancy. We were setting up, as will be related later on, an 'East End Needlewomen's Association', which was to be co-operative in intent, but carried on practically on the principle of benevolent mastership. Hughes proposed Ellison for treasurer. I was surprised but said nothing at the time. After the meeting was over, I asked Hughes why he had done so. 'My dear fellow,' he replied, 'he is over head and ears in debt, though he is a perfectly honest fellow' (which no doubt he was) 'and it may help him with the old uncle of the girl he is in love with if he sees him treasurer of an institution with Lord Shaftesbury for President and Chevalier Bunsen and others on the Committee'. The joke of making a man Treasurer because he was over head and ears in debt was so good a one that I had not the heart to remonstrate. But although no harm came of it, I must own that my conscience often smote me afterwards. For it turned out that practically the Treasurer was not Ellison himself, but 'the Buster', Ellison leaving all his cash affairs to his boys. 'Boy', I have heard him call out when laid up with one of his rheumatic attacks, 'where is Lady so-and-so's cheque?' Fortunately 'the Buster' himself was strictly honest, [It used to be related of Ellison that one day leaving his club with a friend, he hailed a cab and invited the friend to get up with him. At the end of the ride he searched his pockets to pay the fare; then, striking his forehead 'That boy! He has put no money in my purse.'] but I felt a cold shudder run through me on such occasions.

One day there was a great crisis in Ellison's love affairs. His engagement had been going on for years without his prospects improving. Hughes at last remonstrated with him, telling him it was fair neither to the young lady nor to himself that things should so drift on, and extorted from him consent to let the engagement be broken off, and heroically offered to do the thing himself. *Illi rebur et aes triplex circa pectus erat* – assuredly for he had never even seen the young lady; but he got invited to a ball where she was to be, was introduced to her, danced with her, and in the course of a quadrille broke off the engagement. There was to be a mutual restoration of presents. But when Ellison received back amongst other things such trifles as a rosette which had fallen from his theatrical costume some years before when he had been acting at Canterbury, his heart melted altogether; he rushed off to the lady's house, undid all the undoing, and behold! they were more firmly engaged than ever!

I never saw Ellison really angry, but for some time he was more nearly so with Hughes than I have ever known him. I reasoned with him, reminded him that he had himself authorised Hughes to do what he did. 'Ah, but not that way, my dear fellow', he almost shrieked out, 'not that way'. However, very soon after the Conservatives came in, and his friend Lord John Manners got him a magistrateship at Newcastle. There was now no more obstacle to his marrying, and so void of resentment was Ellison, that the first visit the new married couple paid was to Hughes's house at Wimbledon, which he had built at the same time as mine and they always remained fast friends. In course of time he was promoted to Manchester, and eventually reached London, and having if not much law plenty of good sense and much right feeling, proved one of the best of the Metropolitan police magistrates. A kinder-hearted more lovable man never lived.

Another Promoter whom I must mention was George Hughes, T. Hughes's elder brother . . . but he never took a prominent part in our work . . . [The only list of the Council of Promoters I can find is a late one prefixed to the first report of the Council, 1852. The ordinary members of the Council were then – Maurice, Neale, Hughes, Furnivall, Crease, Ellison, Lord Goderich, Grove, Hansard, H. J. Hope, G. Hughes, Lloyd Jones, Lechevalier, Louis, Alexander MacMillan, Mansfield, D. Masson, Millbank, A. Nicholson, Penrose,

Vansittart, Walsh, Woodin – 22 in all. Extraordinary ones (9 in all) were G. Bradley, Rugby (the present Dean of Westminster), A.M. Campbell, Rev. J. Ellison, W. Johnson of Eton, Kingsley, Rev. T. G. Lee of Pendleton, W. Lees, Capt. Lawrence Shadwell, Strettell. My cousin General Ludlow's name is struck out in ink, and I have a faint recollection of his having withdrawn.] Moreover, besides actual promoters, a good many of the so-called higher classes began to take interest in our proceedings and seeds were sown which no doubt bore fruit afterwards in a better understanding of social questions. Lord Mandeville, afterwards Duke of Manchester, came to one of the early meetings of the Council. The Hon. I. Cadogan at an early period called upon Mr. Maurice and showed great interest in the subject of association, which he had been talking about with the then Duke of Cambridge, Uncle to the Queen. Chevalier Bunsen, then Prussian Minister in England, was an early friend to our movement, and was on the Committee of the East End Needlewoman's Home and Workshop. The Roman Catholic Bishop (as he was then) Wiseman, gave his custom (for liveries) to the Working Tailors' Association; in the Church of England the Bishop of Oxford (Wilberforce) did the same, and I think another Bishop also.

In literature too, if some channels of publicity became closed to us, others were opened. Professor A. C. Fraser of Edinburgh, when appointed Editor of the *North British Review* . . . sought contributors among us, and later on I availed myself of his invitation on several occasions during his editorship. Mr. Cook the then editor of *The Morning Chronicle* (afterwards of the *Saturday Review*) without at all espousing our views on the subject of association, was generally willing to open his pages to our communications. So, to a certain extent, was Mr. Wilson of the *Globe*, an old acquaintance of Mr. Maurice . . . More gratifying still was the sympathy we met with from two eminent female novelists, Mrs. Gaskell and Miss Bremer . . .

Before parting with the subject of the Council, I should say a few words as to the Secretaries to the Society . . . An untoward count had compelled us to part with him ((Sully)). As a former revolutionist who had rallied to peaceful socialism, he had incurred great obloquy amongst his former associates. Mr. George Julian Harvey, a physical force chartist, now dead, in order to damage our movement, began to cry it down as pretending to be religious whilst having a bigamist for its secretary . . . On enquiry the following proved to have been the facts. Sully as I said before, was a bookbinder by trade. A skilful worker and a clever man, he had married the daughter of a rich leather-seller, but had set his father-in-law against him by his extreme opinions and lost his support in trade, and eventually failed. He then applied to his father-in-law to help at least his wife and their only child, a daughter, and the old man had offered to take the whole charge of them on one condition, – that they should break off all relations with Sully himself. And to this Mrs. Sully, a woman it would seem of weak character, had consented. This cut Sully to the heart; it was then that he went over to France, and threw in his lot with the extreme reds. In course of time however he fell in with and became attached to a decent young English woman, housekeeper, if I recollect aright, at the British Embassy, and being under the impression that he was legally at liberty to marry again if he had simply heard nothing of his wife for seven years, after waiting till the expiration of that time, he married the housekeeper. Being an excellent workman, he had got on well in Paris, and had paid off all his creditors (the sum for which he failed had not been a large one), so that when he returned to England he was in no fear of any pecuniary claim against him. The second Mrs. Sully, a very respectable, good woman . . . declared that she considered herself Sully's wife before God, and that he had always been the best of husbands to her.

On deliberating upon the matter, we all felt that there was really no moral guilt attaching to the second marriage, but at the same time we could not keep as our Secretary a man in such a position, nor could he safely stay in England at the risk of a prosecution for bigamy. They went to America, and he got on very well there from a pecuniary point of view, becoming manager of a large establishment and continuing to interest himself in social matters (several communications from him will be found in the *Christian Socialist*). But he returned to England after some years, his second wife, if I remember aright, having died . . . Millbank and Shorter were appointed in Sully's place joint secretaries to the Society and did their work very well; and on Millbank's emigrating to Australia, Shorter remained sole secretary.

*   *   *   *   *

Our movement of 1848-53 was so intimately connected with the contemporary French one that I may as well group together here at once my recollections of several (not by any means all) of the French refugees in this country with whom I was thrown in contact, although my more intimate relations with some of them belong chiefly to my stay in Wimbledon . . .

Of the Republican refugees I have already spoken of Jules Lechevalier. I call him a Republican, but in point of fact he never cared anything about political forms, but only about socialistic organisations, of which he himself should be the pivot – not necessarily the visible one . . .

Another far better known Frenchman who took refuge in England about the same time, though I did not know him till afterwards, was Louis Blanc. I was rather prepossessed against him personally by what I had heard of his vanity and pushing himself forward from Charles Lesseps, and his reputation was that of a wild revolutionist. Carlyle's description of him as a 'harmless little man' was much nearer to the truth. My first letter from him is of December 1850, thanking me for having sent him the numbers of the *Christian Socialist*. He spoke at one at least of our meetings or conferences, and I met him a few times besides, once at dinner I think at Lechevalier's. He lived in this country a quiet, honourable life, maintaining himself by his pen, took pains (as Ledru-Rollin did not) to understand England, learnt English thoroughly, so as to speak it in public with facility, and always kept himself from the schemes of the wilder revolutionist refugees. I never read his *Lettres d'Angleterre*, though I believe they speak somewhere of me in friendly terms.

But indeed though friendly, I never grew intimate with him, as he seemed to me shut up, if not in himself, yet in a certain narrow circle of ideas which had become himself. I remember for instance at the dinner of which I have spoken above, on the occasion of a dish of potatoes, his launching forth, *more Gallico*, into an eloquent eulogy on Parmentier, delivered in those slow modulated tones in which you could see that the little man *s'écoutait parler*; and his utter amazement afterwards when I told him, what is almost a truism amongst ourselves, that it was at least doubtful whether the potato was not more of a curse than a blessing to Europe, and that as respects Ireland at all events, it appeared to have been one of the main factors in the debasement of the population, since civilisation is practically impossible where cookery, the earliest of the arts, is reduced to the boiling of a root. Still, though it was easy to make fun of the little man, and I for one did not see that I had anything to learn from him, he really deserved respect . . .

Another French working-class leader whom I saw, but I think only once, was Cabet. Cabet was a man of second-rate abilities, taking rank far below the St. Simons and Fouriers, or even the Buchez's and Louis Blancs, but thoroughly sincere and well-meaning. He had sacrificed his position in the French *magistrature* to his principles, and whatever fortune he had to the carrying out of those principles. He had taken up Communism, probably because the limitations of his mind prevented him from understanding any less simple method of carrying out social views. He had seen its connections with religious faith, as shown by his *Vrai Christianisme* however limited might be his comprehension of Christianity.

Beloved by his family, by all working men who came across him, he was anything but a vulgar demagogue. He never flattered the working man as such. Nadaud (of whom I shall speak presently) who had seen a great deal of him, told me that Cabet always corrected among his working class friends any roughness of speech or manner, and took pains to teach them correct and even polite behaviour, giving them lessons how to enter or leave a room; or in pronunciation or in spelling, when their speech or their letters showed them to require it.

I saw him on the occasion of his return through England

to go to face trial and conviction after a sentence by default (what the French call *purger sa contumace*) after settling his Communist Colony in America – an act which was enthusiastically celebrated by Hughes in his article on 'Cabet the Icarian', *Christian Socialist* I, June 28, 1851. He was then 65, and died not very long after.

In strong contrast to Cabet stood Pierre Leroux, whom indeed I only knew some years later. Leroux was essentially a thinker, and the journeyman printer who produced such works as his *Du Christianisme et de son origine Démocratique* was no ordinary man. Of all the leading French socialists, he was the one with by far the most cultivated mind. He had been at one time a St. Simonian, and was to the last a religious – I do not say really a Christian – Socialist . . .

The largest shoal of refugees which was driven on our shores (and to this indeed I think Pierre Leroux belonged) was that of the 10th December 1851. This brought us amongst others the French friend with whom I have been most intimately connected in later life, Martin Nadaud. Nadaud came from the department of the Creuse, which chiefly supplies Paris with plasterers, and was a plasterer . . . He was fairly educated generally for a French working man in those days, spoke French correctly, and in public with great power, and wrote it also with correctness. He was the most prominent French working man of those days, having been not only, with several others, a member of the Legislative Assembly, but also personally a working man candidate for the Presidency, for which he received some thousands of votes. He had the honour of being among those first arrested on the occasion of the *Coup d'Etat*, and refusing to swear allegiance to Napoleon, was shovelled out of the country, to which he did not return till the fall of the empire.

Nadaud joined one of our builders' associations as a plasterer, and maintained himself for some years by his work, although labouring under the disadvantage that the French method of plastering is different from our own, so that, as I was told by the manager of the Association in which he worked, he never could have earned the highest rate of English wages. He was, however, thoroughly steady and dependable, and was always liked by his English comrades.

(In those days indeed there was nothing of that jealousy of foreign workers which has grown up amongst us of late years). But he was from the first struck with the contrast between English workmen and French, and wanted time for investigation and study. Louis Blanc advised him to try and get a situation as French master in an English school, and another refugee friend gave him some help in improving his knowledge of his own language; after some experience in three smaller schools, he obtained a place at an important Wimbledon school . . . He proved an excellent teacher by sheer force of hard work and painstaking perseverance, and his pupils won many successes at the Sandhurst and Woolwich examinations . . . He was always anxious to improve himself, and after a time spending all his holidays at the British Museum library, collecting materials for his *Histoire des classes ouvrières en Angleterre*, a work which in spite of various mistakes of fact, shows remarkable insight into the meaning and effects of the past as regards the working class . . .

Later on, when the fall of the Empire had set all French exiles free to return to their country, Nadaud threw up his English place and returned to France. He was made Prefect of his own Department, and earned some local obloquy by the strictness with which he enforced the conscription . . . He dismissed all the servants of the *Préfecture*, except one man whom he retained to receive and introduce business callers, and one old woman as housemaid and housekeeper. The savings thus effected, amounting to I forget now how many thousand francs, he paid into the Patriotic fund for carrying on the war. After the war, he threw up his Prefectship, and in spite of his past rigid rule as respects the conscription, was elected Deputy by a large majority for his Department, and sat many years for it . . .

It was only in one particular transaction that I came across another refugee having a name well known in France, who had been in the French Parliament and returned to it after the fall of the Empire, M. Madier de Montjau. There was some arbitration in which refugees were concerned and M. de M. and I were selected as arbitrators. He was an *avocat* and, of course, treated me as a *cher collègue*. I found him clever and agreeable, and we were in no difficulty as to coming to an agreement for our award. I was, however, rather surprised when, on the third occasion of our meeting, he asked me to lend him, I think, seven pounds, and offered to sign a *bon* for the amount. I would rather not have lent the money, but I felt at once that my only security would be the man's honour, and that if I got any document from him I could never sue him for it. So I told him that I would trust him, and declined to take any security. He insisted with extraordinary pertinacity, and for a moment I hoped that he would not borrow the money at all. However, I stood firm against accepting any acknowledgement, and at last he took the money. I rather thought at the time I should see no more of it, but some time after he called and repaid it (in two instalments). I told the story to one of my refugee friends. 'What,' said he, 'M. has repaid you money borrowed! You are the only man in England to whom such a thing has happened. His habit is invariably to do as he offered to do with you, give a written acknowledgement, and then when asked to repay shrug his shoulders and say, "You have my acknowledgement – do your worst against me." And I really believe he considers himself virtually to have repaid a debt when once he has acknowledged it'.

The man, however, was liked, and had many good qualities. He had had a fairly good position at the French bar, which he had sacrificed, together with his seat in the Assembly, for conscience sake. He had done many acts of personal kindness to his brother refugees. He always behaved to me like a gentleman. But I never saw him again after the day when his sense of honour had compelled him to repay my unacknowledged loan. The sacrifice, I suppose, was a too cruel one . . .

One remarkable thing I feel bound to add as to the Republican refugees. Putting aside the case of Lechevalier, who substantially was a man of no politics, and that of Madier de Montjau, I never knew any one of those of my acquaintance to do a dishonest act. Whether workmen or peasants, advocates or professors, they were honourable men. And yet, instead of being received with honour in a free country as the opponents of despotism, as men who preferred exile to dishonour, they were abused and calumniated by more than half the press, and by the great bulk of 'society' so-called. Because there might be among them some one pompous fool like Ledru-Rollin, who from the depths of his ignorance could write upon the 'decadence of England', men who were struggling to maintain themselves by their labour, often undergoing the greatest privations, were too often treated as a set of lazy good-for-nothing revolutionists. When I think of the different treatment that was meted out to the *émigrés* of the last century, few of whom were worthy to lick Nadaud's shoes, I feel as indignant at the recollections as I often was in past days at the reality.

A different and far muddier wave of French emigration came indeed to us after the *Commune*. I am far from saying – indeed what I shall have to say will show that I was far from feeling at the time – that all who had to do with the *Commune* were rascals or fools. But the actors in that awful tragedy, never exceeded in its horror since the days when rival factions fought each other to the death within beleaguered and starving Jerusalem – when peals of German laughter rang to the crash of every French shell on the houses of the capital of France, may deserve our deepest pity, when only the sincere victims of a dread mistake, but cannot claim that sympathy to which the opponents of Napoleon's tyranny had a right . . .

With the Bonapartist refugees I have never had, and never wished to have, any relations. The second Napoleon I can only look upon as God's scourge sent upon France, to punish the selfish and narrow-minded *bourgeoisie* for the crushing out of all that was noble and hopeful on the part of the working classes in the Revolution of 1848. I never heard of one honest and sincere Bonapartist of the Second Empire (I am of course not speaking of the poor ignorant peasantry who, Nadaud assured me, really believed in some places that Napoleon III was Napoleon I) and cannot conceive that any should have existed. So far, they are morally far below the honest *Communard*. [I have spoken here only of French refugees. I have known and do know, some German refugees now settled in this country, of whom I shall have occasion to speak in reference to the Working Men's College. Karl Marx was good enough to present me with the first volume of *Das Kapital*; I own I did not avail myself of the opening to claim his acquaintance. I was once invited to dine with Mazzini, but I had already at that time given up dining out, and declined. I need hardly say that I had a great regard and admiration in many respects for Mazzini, notwithstanding his cruel (which in another man would have been cowardly) attack upon the Socialists, at the time when the name carried most of obloquy. But I had been warned that Mazzini was so surrounded by spies, who completely befooled him as to their patriotism, that confidential friendship with him was impossible, and as he was a man whom I should not have cared to know as a mere acquaintance, I thought it best to let him go his way whilst I went mine . . .]

CHAPTER XVI

The Christian Socialist Movement, 1850-4;
Associations and Institutions – Literary
Work and Lectures – Parliamentary Work

The first Association in connection with the movement, the Working Tailors' Association, was started in Castle Street East, Oxford Street, as Hughes has related in the second 'Tract on Christian Socialism' on the 11th February 1850, with 12 associates under Walter Cooper as Manager. When the Tract itself was published there were already four other associations whose addresses were advertised on its cover – the North London Needlewomen's Association (afterwards the Working Needlewomen's Association) a body of a less democratic character, with a superintending committee of ladies, – two associations of Working Shoemakers, and one of Working Printers. In June of that year Lechevalier was delivering a course of lectures in the small room at Exeter Hall, Neale projecting an agricultural colony, and there were eight associations at work, comprising besides those already named a Working Bakers' Association, a Working Builders' Association (succeeded by a North London Working Builders' Association), an additional Association of Shoemakers, and the London Co-operative Stores afterwards developed into the Central Co-operative Agency. Two of the three Working Shoemakers' associations however soon came to grief, and had to be amalgamated into one, which now became simply the Working Shoemakers' Association ...

As I have mentioned the shoemakers, I must dwell for a

moment on ((Christmas)) one of the most remarkable and sterling characters connected with our movement. He was the one wholly uneducated man amongst our managers; could neither read nor, except so far as signing his name, write; was so ignorant that at the time he first joined us he imagined that the movement was set on foot by the Government; for many years at least, though strictly sober during the whole of the working week, he used to get drunk every Saturday night. On the other hand, the man was, to begin with, an admirable workman, taking a pride in sending nothing out that was not thoroughly good of its kind (he told me in latter years that he was the only remaining bootmaker in London, – so his leather-seller informed him – who required leather to remain two years in tan). Again, the man's natural ability was so great that without knowing how to read or write, he so thoroughly mastered his book-keeping that he was able to turn at will to any particular entry, carrying the amount in his head with unerring certainty, so that no one not aware of the facts could have imagined that he did not read it off. Moreover, he was a ruler of men; kept his association together when two others were foundering; took in new branches of work; mastered the conditions belonging to each; made himself, a mere strong shoemaker, (those who know the curious gradations of rank in the labour-world will appreciate the achievement) obeyed by gentlemen's bootmakers and ladies' shoemakers. And the man himself went on always mellowing. At first rough, suspicious, hard-mouthed, the very expression of his face gradually changed, till in old age it became kindly and gracious. His associates one by one fell away, so that for many years he carried on business on his own account, but his face always lighted up with pleasure when he saw anyone who had been connected with our movement, and he was always ready to render any service in his power. Dear old Christmas! Co-operation certainly brought out the best that was in you, and that best was in its way very good.

Later on other associations were formed; the Pimlico Working Builders; the Working Piano-forte Makers; the City Working Tailors, the Broad Silk Weavers of Bethnal Green; the Working Smiths (a very unsatisfactory body); in the provinces the Southampton and the Liverpool Working Tailors, the Salford Hatters and Banbury Plush Weavers. [Besides these, there were various almost still-born bodies of Working Lithographers, Basket-makers, Brick makers, Wood-cutters, Metropolitan Working Builders, etc. To the associations of engineers which arose out of the Engineers' strike I shall advert hereafter. In the provinces other bodies were springing up in large measure through Walter Cooper's and Lloyd Jones's lecturing tours.] Unconnected directly with the Society, but having amongst its trustees several of the promoters, was the East London Engineers' Association, arising out of the Engineers' strike or rather lock-out, of which more will be said presently. Then again besides the Working Needlewomen's Association, of Red Lion Square, there was the East End Needlewomen's Institution of Wellclose Square, Whitechapel, and also a Ladies' Guild. Lastly, there was an association founded quite apart from and in opposition to our Society, by malcontent tailors of the Castle Street body, the differences in which were the first really untoward event in the career of the Society, (there were also serious divisions later on among the Pimlico Builders).

Of the associations proper, a few words should be bestowed on the Broad Silk Weavers. The Spitalfields Weaver is, I believe, nowadays an extinct species in the Labour family. He was not so then, although his days were numbered through the competition of steam machinery. So far as my experience of the Labour world goes, he was an altogether peculiar creature. It was one of the family employments, in which children of two or three could be made useful, and in fact it was only by the united efforts of the whole family that a living could be obtained. Hence a distinct race, so to speak, had been developed generation by generation, since the revocation of the Edict of Nantes had flung the silk weaver and his trade into England. They were, with the rarest exceptions, altogether undersized, the men very little if at all over 5 ft., often with delicate features, and a quite peculiar quietness of manner. Sober, for drunkenness would spoil their work altogether; patient and self-contained, for violence and even emotion would do the same. In the old time they had been famed for their high standard of education, their acquirements, their scientific tastes; and even at the time I speak of it was rare to find one who could not read and write. The physical conditions of labour moreover were not altogether unfavourable, for silk-weaving demands large, well-lighted rooms. In the palmy days of the trade every weaver's family had what the French call its *pied-à-terre*, a little strip of ground in the country two or three miles off, where it cultivated vegetables and flowers, and a few of these remained still in weavers' families. And the earnings of the whole family might at the utmost, for 16 hours work a day, come to 16 or 17 shillings a week, a day or two in the fortnight being lost in setting the loom and unsetting it! But nothing among them surprised me more than this. We went some of us one evening by invitation (7 Jan. 1852) to address a meeting in a chapel, nothing doubting but what the minister of it was virtually our host. There was no minister of the chapel! It belonged to the weavers themselves, and they invited whom they liked to come and preach in it of a Sunday. So that these extraordinary little creatures, indulged in the luxury of a proprietary chapel of their own, which they used as a meeting place for any collective purpose. Tea was served before the meeting. So quiet an audience I never addressed.

The Broad Silk Weavers' Association began with very good prospects, and the Manager, Rice, had borne a very good character as a workman. But alas! in this as in several other cases the temptation of handling money in sums which he probably never dreamed of having at his disposal before was too strong for him. He levanted, and the weavers had no courage to make a second attempt . . .

I must own that personally I never had at the time any hope of successful co-operation except in the better paid trades. Still, manful attempts were made to encourage it in the lowest paid ones. There were great efforts for the formation of a slop-workers' association, which eventually resolved themselves into the adding to the Working Tailors' Association, under Walter Cooper's management, of a branch for ready-made clothing in the Westminster Bridge Road, which foundered with the association itself.

Something must now be said of the bodies established for

the improvement of the condition of women. I have mentioned already the Needlewomen's Association in Red Lion Square. It was under the superintendence of a committee of ladies, including Mrs. Maurice, Lady Inglis, Mrs. Bellenden Ker... It was different with the East End Needlewomen's Workshop, established to benefit the slop-workers of the East End. It was considered that self-government in any shape was probably beyond the reach of these poor creatures, and the plan adopted was to vest the whole business management in a superintendent, to whom the capital was to be advanced at £5 per cent interest, and who would receive, besides a salary, a percentage on profits, subject to which, and to the payment of out-goings (including interest on capital) the whole profits were to be laid out for the benefit of the workers, forming a fund for emigration (a West End sales shop was also contemplated, but never established).

As this institution was to give no rights of self-government to the workers, but to be purely benevolent, it was thought that we might fairly address ourselves for its support to those for whom the association of self-government and working men was still a bugbear, or who at all events were not prepared publicly to countenance it; Lord Shaftesbury was again applied to, and this time with success. A private meeting was in the first instance called at his house, to which were invited all who appeared to have taken any interest in the subject, including Mr. Carlyle, who had recently written a letter to the papers on the difficulty of finding a decent needlewoman for love or money. He came, – Chevalier Bunsen amongst others being also present – and the plan which we had drawn up was read. Mr. Carlyle did not pay the slightest attention to it, or to anything that was suggested, but simply delivered himself of an outpouring to the effect that he heard that women were greatly wanted in California, and the best thing that could be done would be to send off there some shiploads of them – 'Ship 'em to Californy, that's all, ship' em to Californy, they'll all get married there!' Of course the advice was cynically one-sided, as it took no account of the old, nor yet of the wives and mothers, and I never liked Lord Shaftesbury so well as on witnessing his genuine indignation at it, as well as his efforts to master that indignation. A very promising committee was formed . . .

We secured very good premises in Wellclose Square . . . The first thing to be done was to put the premises in order. Jackson[16] stated that he had among his reformed *protégés* a sufficient staff to do all the work required, if materials and tools could be provided for them, and that they would do the work much cheaper than any regular builder, he looking after them. His suggestion was accepted at once by Lord Shaftesbury. But alas! a few days later Jackson made his appearance at my chambers, pulling a very long face. His converted thieves, after I think two days' work, had levanted one evening with the whole of the portable stock with which they had been supplied, to the amount of, if I recollect aright, some £20 or £30! I could not help laughing, though it was a bad beginning. If I recollect aright, the work was now put into the hands of our Pimlico Builders' Association, who did it very well. The next question was that of the choice of a superintendent. Calling on Lord Shaftesbury to speak to him on the subject, he handed over to me a letter which he had received from a former schoolmistress and governess, thrown out of work by illness, asking if he could help her to a situation, and requested me to inquire about her. With another member of the managing committee I saw her, a tall, strongly built woman, accompanied by her brother, a sergeant in the guards, his chest plastered with medals. She produced good written testimonials, having had, she said, a severe illness since her last place – and her brother backed her up in the highest terms. We reported to Lord Shaftesbury and it was decided to engage her. She displayed great energy in getting up the establishment, obtained the custom of one or two good firms, and Jackson in turn reported very highly as to her efficiency. As many private orders as possible were obtained, and the work proved stoutly, if coarsely, done.

The thing was now fairly launched, and seemed only to require maintaining. It was visited, at least once a week, by some member of the managing committee. But the £500 starting capital which had been subscribed before long required to be supplemented. We thought it was now fairly time to apply to the public, as had from the first been contemplated. But Lord Shaftesbury never would consent to this, always promising (and always in vain) to obtain what was wanted from his friends, and so the thing struggled on till at last Louis brought us the startling intelligence that Miss Dennington was a drunkard. I was for dismissing her at once, and so I think, were all the managing committee except Jackson, who, I am afraid, took rather the missionary than the business point of view. He declared that he had spoken with her and prayed with her, and that he believed that she would now go straight. Of course she broke out again, and her accounts began to be unsatisfactory. We reported to Lord Shaftesbury that she must be dismissed and asked for leave to engage a new superintendent, again however insisting on the need of raising fresh capital, without which it was hopeless to go on. But as he still set his face against any appeals to the public, and no more even held out any hopes of finding for us what was wanted, there was nothing to do but to close the establishment. This Miss Dennington at first resisted, and attempted to carry on the business on her own account, and I had the pleasure (the only time in my life) of being county-courted on a claim thus raised for goods supplied. The claimant was indeed non-suited without my having to open my mouth, having failed to bring forward any evidence of my liability to his claim beyond the fact of my being on the managing committee, although Miss Dennington came forward as a witness on his behalf.

The result of this experiment as to carrying on our work with the help of great people, at the lapse of eighteen months during which that first experiment was carried on, was that no single member of the general committee except, I think, Chevalier Bunsen, had ever gone to the place, not even the Chairman, and that we poor devils and our personal friends supplied all the capital expended. I resolved then and there never to seek help in any social undertaking from a class above my own. I cannot help suspecting that Lord Shaftesbury's narrow theology led him from the first to distrust Mr. Maurice and any fellow-worker of his, and that the fear of being too openly connected with us had been the main reason why he had throughout set his face against any public appeal for funds.[17]

There was also brought into connection with our society another institution for women, the 'Ladies Guild', founded by a very energetic lady, Mrs. Southwood Hill, to work out a patent taken out by a Miss Wallace for decorations in glass . . . I can say nothing of it at first hand, as I had no faith in it, as I never set foot on its premises, and declined to be connected with it. But the 'Ladies Guild' had at least one important result. It brought into our movement Mrs. Hill's daughters, one of whom was the now well-known Miss Octavia Hill, one of the ablest, most self-devoted and most admirable of the distinguished group of female workers in the latter half of the 19th century. When it is recollected that Mrs. Nassau Senior, another member of that group, was T. Hughes's sister, I think it may be claimed for Christian Socialism that it has been a powerful leaven in the work of both sexes.

Its secondary workings can indeed hardly be traced except, here and there. Thus the Pimlico Working Builders' Association became the main agent in the establishment of a Westminster and Pimlico People's Institute (August 1851), and also took interest in the establishment of a School of Art for Artist Workmen (November 1851). A member of the City of London Working Builders' Association set up a Co-operative and Temperance Coffee House in Camden Street, Islington Green, (April 1852). The North London Builders' Asssociation enrolled a North London Co-operative Builders' Friendly Society (May 1852). Both the Pimlico and North London Builders had Co-operative stores of their own.

But the mention of the two Builders' stores brings me to the direct connexion of the movement with consumption. Lloyd Jones in particular amongst the promoters was well-acquainted with the co-operative stores of the North, knew with what comparative ease they could be set up, and was always urging their establishment. Lechevalier was, I think, the next who took up the idea, and pressed it upon me. I was lukewarm in the matter – I felt then, as I feel now, that consumption, however important an element in human life, should be kept subordinate to production (and, to tell the truth, the *Christian Socialist* was taxing my already slender pecuniary resources to the uttermost). Lechevalier then turned to Neale and Hughes, who turned to him a more willing ear, and these two, Neale more particularly, advancing the necessary capital, the London Co-operative Stores were opened at the same house where the meetings of the Society were then held, 76, Charlotte Street, Fitzroy Square, Lloyd Jones being manager, Lechevalier the supervisor. I do not recollect whether any part of the capital was contributed by any other of the promoters besides Neale and Hughes, but I believe we all loyally supported the stores by our consumption, purchasing any articles supplied by them. The Central Co-operative Agency, into which the London Co-operative Stores developed, exercised so strong an influence on the movement, that it will require a separate chapter hereafter.

The work which we did during those few years was largely literary[18]; of course the two books which mainly fix the place of the Christian Socialism of those days in English literature are Kingsley's *Yeast* and his *Alton Locke* . . . His sermon, *The Message of the Church to Labouring Men* (J. W. Parker & Son, 1851), was absolutely forced into notice by the inconceivable folly of Mr. Drew, the incumbent of St. John the Evangelist, Charlotte Street, Fitzroy Square, who after settling a course of sermons, selecting the preachers, and causing the course to be advertised in the *Christian Socialist*, took upon himself to rebuke Kingsley from the reading-desk.

The most important group of publications for which the Society for Promoting Working Men's Associations was directly responsible was that of its *Tracts*. Of course we had to publish these at our own cost. J. W. Parkers had taken fright at us, and *Cheap Clothes and Nasty* had, on their refusal to have anything to say to it, been published by Pickering, virtually through his Cambridge correspondents, Macmillan & Co. The *Tracts on Christian Socialism*, 1 to 8, were published by George Bell who then in his turn took fright. Of these four were by Mr. Maurice, one by T. Hughes, two by myself, the remaining one, already referred to, giving the constitution of the Society for Promoting Working Men's Associations and the model laws for the associations. The *Tracts by Christian Socialists* were four, one by Mr. Maurice, another the reprint of Kingsley's *Cheap Clothes and Nasty*, the remaining two a reprint of my article on 'Labour and the Poor' in *Fraser's*

Magazine. The best by far of the series is, I think, the first, Mr. Maurice's *Christian Socialism, a dialogue between Somebody (a person of respectability) and Nobody (the writer),* of later years reprinted by the Oxford branch of the Christian Social Union, a body which in several respects must be considered a continuator of our work. In these days, when the term 'Socialism' is sought to be narrowed in the using to this or that particular system, and the patent meaning of the word, and its history in this country as well as elsewhere, are so grossly overlooked that 'Co-operation' and 'Socialism' are actually treated as antagonistic, both by men who call themselves Socialists and by men who call themselves Co-operators, one cannot too strenuously insist upon the cardinal value of Mr. Maurice's declarations in the Tract in question:

> I seriously believe that Christianity is the only foundation of Socialism, and that a true Socialism is the necessary result of a sound Christianity.
>
> The watchword of the Socialist is Co-operation; the watchword of the anti-socialist is competition. Anyone who recognises the principle of Co-operation as a stronger and truer principle than that of competition has a right to the honour or the disgrace of being called a Socialist.[19]

. . . But the publication which, I venture to think, did most to keep the movement together while it lasted was the *Christian Socialist*, with its continuator, the *Journal of Association*. It will be recollected that my starting point in Christian Socialist work had been the idea of a newspaper. *Politics for the People* very inadequately fulfilled my ideal. It will be seen from Kingsley's *Life* that by February 1850 we were both of us desirous of starting something in the way of a journal. On July 6, 1850, I wrote to him:-

> You are informed that the undersigned, having £10 in his pocket, has projected a twopenny paper, stamped, to be called *Brotherhood, a journal of association,* to be edited by J. T. until the return of the 'Prophet' (Mr. Maurice was then in Germany), and sub-edited by Furnivall; to tread avowedly in the steps of defunct *Politics*, yet to be differently organised, and to be mainly a channel of communication for the associated workmen with the Promoters and the public, the *Tracts* being considered the chief organ of the Promoters themselves. To consist of two parts, like

the *Leader* [An able weekly of the time, edited by Mr. Thornton Hurst, and professing Communist principles.] articles and correspondence; association being the only common ground, and everything else as free as possible. No remuneration except contingent upon profit, and then shared amongst all contributing in proportion to length of contribution, with a certain share for Sub (Editor), Editor pledging himself to devote his share to the general fund of the associations, so as to make sure that he will not make money out of the rejection of others' labour. No news, except association news, although Furnivall suggests that a first page of weekly summary like the *Spectator* would greatly aid; which is to be considered. Political articles or any others, however, not excluded.

The reason for the establishment of such a paper was, as I wrote to Kingsley, that we had a large amount of literary talent amongst our working men which was either lying idle, or forcing its way through wrong channels, – instancing Gerald Massey, then Secretary to the Working Tailors' Association, whom I had had regularly to blow up for having publicly connected himself with a thing called the *Red Republican*, patently treasonable. He was bullied out of it by my offering him the choice between Association and the *Red* and in the note in which he consented to withdraw, he had told me 'that if I for instance had set up an organ of Christian Socialism he should have been quite willing to write in it' . . . In course of time my idea expanded into that of the combination of a newspaper at the then received price of 5d with a penny periodical extracted from it, so as to speak to all classes, and to make the richer pay for the good quality of the matter supplied to the poorer . . . However, Mr. Maurice objected *in toto* to a newspaper, and only accepted the idea of a penny journal, and he carried with him Charles Mansfield, Furnivall, and Walsh, Lloyd Jones and Lechevalier alone supporting with me the idea of the newspaper. So the penny journal was decided on, and became the *Christian Socialist* . . .

The first number of the Christian Socialist appeared on November 2, 1850. I eventually claimed the sole responsibility of the editorship, in order to disengage that of Mr. Maurice, who I felt had had too much to bear as joint editor of *Politics for the People* and round whom the clouds were already gathering of that storm which ultimately cost him his King's College Professorship. Furnivall was an excellent sub-editor. The paper was of course printed by the Working Printers' Association . . . and we set up a publisher of our own in the person of an ex-chartist and ex-political prisoner, friend of Walter Cooper's, a costermonger (!) by trade, of the name of Beyer, a queer little one-eyed man, very clever, who for a time, I verily believe, was sincere and honest, and whose *Autobiography of a Chartist Rebel*, contributed to Vol. 2 of the *Christian Socialist*, is a really remarkable thing. Eventually, I am sorry to say, he levanted to Australia with another man's wife (having left his accounts in a very unsatisfactory condition), and there, I am afraid, went from bad to worse. [I fear that the ex-costermonger was quite unaware of the duty of forwarding copies of publications to certain institutions, – hence the fact that even the British Museum Library possessed no complete lot of our publications till Dr. Furnivall presented one.]

I now sought to rally round me as far as possible all the contributors to *Politics*. Of course I did not succeed in doing so. No Archbishop, Bishop or Dean ever wrote for the *Christian Socialist*; no Arthur Helps, no James Spedding, no Bellenden Ker; Bridedale, as 'Tory Bill', wrote only to say why he could not work with us. Still, I think we had no reason to be ashamed of our contributors. Mr. Maurice wrote a series of eight papers on education, a touching and beautiful tale, which unfortunately remained unfinished, 'The Experiences of Thomas Bradfoot, schoolmaster', and four other articles . . . Kingsley, besides the articles mentioned on p. 263 of his *Life*, Vol. I, contributed several others . . . – Gerald Massey gave us some good bits of verse, and amongst other papers some noticeable studies on 'Tennyson and his Poetry'. We had a number of valuable sanitary articles from no fewer than three medical contributors . . . Furnivall wrote a number of excellent papers; Hughes – though he wrote chiefly for the *Journal of Association* which he edited – contributed several papers to the *Christian Socialist* . . . Two distinguished 'G's' were contributors, Geo. Grove to the *Christian Socialist*, Lord Goderich to the *Journal of Association*. Neale contributed pretty frequently, A.M. Campbell wrote a valuable series on 'The Evils of Competition', Charles Mansfield gave us several articles, Alexander Macmillan also . . . Our working men contributors did their part very well and amongst our correspondents were old Robert Owen and his son-in-law Dr. Travis . . .

Of course we were attacked on all sides. The *Reasoner*, Mr. Holyoakes' atheistic organ, was liberal in its sneers, and appealed to our good sense not to 'Mix up Christianity with Socialism'. The dissenting *Eclectic Review* charged us with trying to 'repeal a law of nature'. The *Guardian*[20] accused us of 'pious fraud', 'temporary abandonment of the expression of great and holy truths', 'making converts on false pretences' and a 'fraternisation wholly indefensible' with atheists. The *Daily News* spoke of 'the revolutionary nonsense' which we termed 'Christian Socialism', of our 'vague, misty, absurd and mischievous . . . communism'. Mr. Ernest Jones on the other hand, the Chartist leader of the day, attacked the movement as anti-social and reactionary. The *Weekly Dispatch* and *Morning Advertiser* volleyed abuse into us. On the whole, our only fair opponents were Mr. W. H. Greg in the *Edinburgh Review* and two writers in the *Inquirer* and *Prospective Review*, both Unitarian organs. We were treated in a friendly way, by the *Spectator, Robert Owen's Journal,* the *Leader* (occasionally), the *Northern Star* and a few now forgotten reviews or other periodicals, and more or less so likewise by the *Globe* and *Morning Chronicle*. In many cases we simply noticed in a short paragraph the attacks made upon us. In other cases, where the matter seemed to deserve it, an article or even several articles were devoted to the reply . . . We showed ourselves at all times *respondere parati* whenever we met an opponent worth being stood up to, whilst showing disdain for mere abuse.

But if we feared nothing from attacks, we were helpless against what the French call *la conspiration du silence*. Publicity was in many quarters absolutely denied to us. The *Times* was only one out of several newspapers which refused to admit an ordinary business advertisement of the *Christian Socialist*. Newsvendors refused to supply it, and insisted it was dead when it was vigorously struggling to live. Literary institutions refused its admission to their reading rooms as a gift. If we ever succeeded in getting it taken in at a railway bookstall it was hidden out of sight. I believe it is impossible for this generation to conceive of the uphill work it was in the fifties to speak out truth unwelcome to Mrs. Grundy ...

But can the literature inspired by the Christian Socialist Movement be confined to the publications I have noticed? No, it leavens, it really comprises, the whole work of Kingsley and T. Hughes, and *Tom Brown* belongs as truly to it as *Alton Locke*. As to Mr. Maurice, though, I believe from a morbid self-distrust, he latterly turned away from practical connection with the Co-operative Movement, the principles of Christian Socialism pervade all his subsequent teaching. Take for instance his noble series of Cambridge lectures on 'The Conscience'. Not a word occurs in the volume about co-operation as such. But passages like the following enunciate the bases of all true co-operation, give its full *raison d'etre*:

> The Conscience takes no account of Power except as it is joined to Right, except as it has its ground in Right (Lecture VII). You cannot contemplate the individual man out of society. But you must indicate his position in order that you may show what society is, of what it consists (Lecture IX) ...

I claim therefore that the Christian Socialist Movement begun in 1848 has profoundly influenced the literature, the thought of the 19th Century in England so that, as I have ventured to say in the *Economic Review*, 'boys fresh from school, girls who have just left their governess' now freely express views which in us were howled at as revolutionary in 1852.

I think I may say that outside of our practical work in the setting up of associations, the two reforms which we specially set before us were that of the Government contract system, and that of the law relating to associations. In the latter object we succeeded; in the former we failed for the time, to be how ever abundantly justified in our endeavours forty years later.

I have shown that we were already getting up a petition for the reform of the Government contract system early in 1850 ... A second petition, that Parliament might be pleased 'forthwith to take such steps' as might in its wisdom 'seem needful and expedient towards procuring a thorough reform of the Government clothing contract system, so as to secure "living wages" to all workpeople employed under it' was sanctioned by the Council of Promoters, and lay for signature at the Society's rooms and at its publishing office. When Lord Derby's Ministry came into power in the following year, Hughes was able to announce in the *Journal of Association* (March 8, 1852), that a cabinet minister was considering the question of the contract clothing system, with a view to having a hit at the slop trade. Eventually the establishment of the Army Clothing factory in Pimlico at least cleared the Government from any direct responsibility for the horrors of the slop clothing system. But it was to the legalisation of co-operative associations that we found ourselves compelled to devote our chief energies.

It is hard at the present day to realise the hindrances to almost every kind of commercial association that existed in 1850. The principle of the law of partnership was the unlimited liability of all the partners for all contracts, and the bringing all the partners before the Court in all legal proceedings relating to the partnership – very excellent principles both of them, so long as the partners are few, and can all take part in the management, but utterly inapplicable to large associations. Exemption from unlimited liability, and even the power of suing and being sued by officers, were privileges obtainable only by Act of Parliament or Royal Charter . . .

All we lawyer members of the Council of Promoters were fully aware of this; none probably so much as myself. In advising working men to associate under such legal conditions, we were therefore morally bound to pledge our utmost energies to the reform of the law. In the meanwhile, the real safety of the members of our associations lay in this, that very few of them had anything to lose. At the same time, I must note as a remarkable fact that, whilst the legal knowledge of English working men is generally very slender, we found amongst them a very general fear of the unlimited responsibilities of partners . . .

At an early period of the movement, with the assistance of my legal colleagues on the Council, I drew up a bill to legalise our associations by bringing them under the Friendly

Societies Acts. We had most of us a great admiration for Lord Ashley[21], the most prominent among public men who had done practical work on behalf of the working class. I wrote to him, asking him to bring in the bill. Some correspondence ensued (May-June 1850) but he declined to do so. Zealous as he had been for the protection of the working class by legislative enactment from oppression under the factory system or the mining system, to enable them to better themselves by association of their own was apparently too revolutionary for his philanthropy. Again, we tried without success to obtain the insertion in the Friendly Societies Bill of that year of a clause legalising our associations. But an unlooked for opening had meanwhile offered itself to us through Mr. Slaney's Select Committee on the 'Savings and Investments of the Middle and Working Classes', which was destined to lead us to unexpected success in our undertaking . . . I saw a good deal of Mr. Slaney then and afterwards; a very worthy, well-meaning man, but hazy-minded, so that while always fumbling after some good end or other he was seldom able either to see it clearly or to grasp the means for carrying it out. As it was however, through coming in contact with us he was able to achieve a momentous good in the legalising of co-operation . . . H. R. Vaughan Johnson, employed by Mr. Slaney to hunt up evidence for his Committee, and being a friend of T. Hughes's, and knowing that the latter was engaged in starting co-operative associations, called on him in chambers and asked if he would wish to bring their case before the Committee. Hughes saw at once the value of the suggestion, and brought the matter before the Council of Promoters, and I was appointed to open the case of the associations before the Committee, none of us being in the least aware that it had as yet found no evidence to take, and that we were conferring a real favour on its Chairman by coming forward.

Although I had before been present at sittings of select committees, and had contributed papers to the appendices of several of their reports, I had never been examined as a witness before one. The experience is not a pleasant one, owing to the difficulty of constantly readjusting so to speak, one's mental focus, backwards and forwards, to meet the varying lines of thought of the different members of the Committee. And having had to renew the experience many years later, it appeared to me that the Select Committees of the 80's compared very unfavourably with those of the 50's. A witness going before a Select Committee of the House of Commons nowadays must expect no longer to be examined only, but badgered and bullied, the chairman perhaps taking a leading part in the sport. In the Select Committee of 1850 to which I have referred, although the views which I and my friends presented to the Committee found a very able and strenuous opponent in Mr. John Abel Smith, a London banker, whose opinions were shared by several other members, I cannot call to mind one discourteous word from him or any of them.

Not that I wish to represent the Committee in question as a model one. Mr. Slaney's special hobby was *commandite*, or the unlimited liability of directors or managers only, with limited liability for the other partners – which the experience of the Continent shows to be only beneficial to small concerns, the result in large ones being almost invariably to put men of straw at the head of affairs, whose unlimited liability becomes thus a mere trap to outsiders. As chairman he addressed none but leading questions to witnesses, so that, as I wrote to Kingsley, the only chance a witness had of speaking his mind lay in some searching question from Mr. John Abel Smith, put for the express purpose of knocking him down. Still, all the members seemed to take a great interest in the subject of our working men's associations.

I was in some respects pretty well qualified to give evidence. I have mentioned in one of the earlier chapters of these Reminiscences that in Mr. Ker's chambers I had developed a certain aptitude in matters relating to joint stock companies as well as in the drawing of Acts of Parliament ... Above all, I had formed a strong opinion on what was then a moot question, though now long since settled in the sense which I advocated, viz. that of limited liability in association trading. I had thus really a good deal to say, and I said it, though I did not always hold my own as well as I ought to have done against Mr. John Abel Smith. Other witnesses were Vansittart Neale, Hughes, Millbank, Walter Cooper, Lechevalier, Clarkson, Manager of the Working

Bakers' Association, Lloyd Jones . . . Altogether, before the competitive world generally was aware of what we were doing, we were able to pour in a really overwhelming amount of evidence in support of our objects . . . Two other witnesses gave indirectly favourable evidence. The only actual official examined, Sir Denis Le Marchant, Secretary to the Board of Trade, was not adverse. And, to our great surprise and delight, the one really eminent witness, John Stuart Mill, who had refused somewhat ungraciously to attend a meeting called by the Council of Promoters, on the ground of the divergence of views between himself and our society, being applied to by a deputation from our body, not only consented to be examined, but gave evidence of the most decided character in favour of our objects. 'I think', he said, 'there is no way in which the working classes can make so beneficial a use of their savings both to themselves and to society, as by the formation of associations to carry on the business with which they are acquainted and in which they are themselves engaged as workpeople.' He dwelt on the small proportion of the price paid at a shop for an article which goes to the real labourer; insisted that 'an increased number of mere distributors has no tendency to increase the quantity of wealth to be distributed, but only quarters an additional number of persons upon it', held it decidedly 'desirable' to give 'all possible facilities but no premium' to associations of a co-operative character; thought that 'associations by the working class to carry on, as their own capitalist, their own employment' had 'a very great advantage over any other investments for the working classes.'

The Report of the Select Committee was however much less explicit than we could have wished, ((and a further year of petitioning and negotiation were necessary before success became a possibility)) . . . On 27 February 1852 Lord Derby came into power, and in place of an effete Whig ministry which had shilly-shallied with us through two Sessions, we had a dashing Tory one, anxious to curry favour with the working class. By March 8 Mr. Hughes could tell the readers of the *Journal of Association* that a cabinet minister had 'taken up the question of giving working men's associations fair play in good earnest.'

On March 18, Mr. Slaney moved for leave to bring in a bill to legalise the formation of industrial and provident partnerships, and Mr. Walpole, as Home Secretary, offered no objection to his doing so. A trick was however played upon us by the enemy, through the insertion in the bill of a clause which applied the principle of unlimited liability to all associations under it, and Hughes and Vansittart Neale were deputed to wait on Messrs. Slaney and Sotheron with respect to this. I must however say at once that all we could eventually obtain was a limitation of the liability for the collective debts and engagements to two years after a member's ceasing to belong to the society, being one year less than in joint stock companies. Through the time of the House being talked out by the Irish Members, the second reading of the bill had to be postponed till after Easter. The second reading was carried without opposition on April 21st, and the bill referred to a Select Committee. We were greatly helped this time in our efforts by the appearance in the *Edinburgh Review* for April of an article on 'Investments of the working classes', from the pen of Mr. Greg, which whilst protesting against being suspected of an approaching conversion to Christian Socialist views, zealously recommended the legalisation of working associations and of co-operative stores dealing with the public. Another great help was that the Select Committee on the bill invited the presence of some of its promoters, not as witnesses for examination, but for enabling them to discuss the clauses of the bill with the Committee. Mr. Neale, Mr. Hughes and myself were appointed for the purpose, and I have the most agreeable recollections of the friendly way in which we were met by the members of the Committee and of the efficient work which we were enabled to do . . . The Committee held its last sitting on May 18, and the bill was reported under the title of the Industrial and Provident *Societies* Bill. It passed through the House of Commons, the House of Lords, and received the Royal assent on 30th June . . .

A Royal Commission was appointed towards the end of 1853 to report 'whether it would be expedient that any and what alteration should be made in the law of Partnership so far as relates to the limited or unlimited liability of partners.'

... I was asked to inform the Commission whether my views on the subject had undergone any change ... Three years later, in 1856, the principle of limited liability for companies was admitted by law (the Joint Stock Companies Act 1856) ... Four years later the same principle was extended to Industrial and Provident Societies, and instead of its adoption being left optional as with companies and banks, it was essentially bound up with the Act of Registration, the society becoming thereby incorporated (Industrial and Provident Societies Act 1862). In 1871, under instructions from the Central Co-operative Board, I drew up a bill to amend and consolidate the law (the Industrial and Provident Societies Act 1871). Finally by an Act based upon the Friendly Societies Act 1875 which I had drawn, but drafted by Mr. Neale (Industrial and Provident Societies Act 1876), the law as to co-operative societies was placed substantially on its present footing, and it will thus be seen that from 1852 to 1876 we old Christian Socialists practically had everything to say to it ...

Again, I believe I may claim to have been the first who openly claimed the legalisation of Trade Unions. In the first of my lectures on *The Master Engineers and their workmen*, delivered at the request of our society 13 February, 1852, I said: 'I must take this the earliest opportunity of expressing my conviction that Trade Societies ought henceforth to be legalised in the same manner as Friendly Societies' and went so far as to say that their existence unlegalised appeared to me 'a great evil'. Later on, with Hughes and other friends, I urged similar views on the Committee on Trades Societies and Strikes, appointed in 1859 by the Social Science Association, and which reported in 1860. I also, at a deputation to Mr. Walpole, suggested the principle of a temporary act which for the first time legally recognised Trade Unions by name, the Act to protect the funds of Trade Unions from embezzlement and misappropriation (Trades Unions Funds Protection Act) which became Law 9th August 1869. And if I had nothing to do with the passing of the permanent act on the subject, the Trade Union Act 1871, I was able, on its amendment in 1876 by the Trade Union Act Amendment Act 1871, to obtain the enactment of a series of clauses which have placed trade unions almost on the same footing as Friendly Societies. Of the many useful measures which Mr. Hughes was able to promote whilst a member of the House of Commons I have unfortunately kept no account. But what has been said will, I think, show that we Christian Socialists made our mark on the legislation of our day.

CHAPTER XVII

The Christian Socialist Movement 1850-4 –
My Co-operative Tour in Lancashire &
Yorkshire – My Work at the Bar

I think I must devote a chapter to that 'Co-operative Tour' in Lancashire and Yorkshire in the autumn of 1851, 'notes' of the business part of which were published in the second volume of the *Christian Socialist*, and to the friendships which grew out of it. It was intended to be, and was, in great measure, a walking tour, in company with T. Hughes, but he had to leave me before long, and in the latter part of it I was alone. I was determined to have as little luggage as possible, but did not care to carry a knapsack, so I instructed W. Cooper to make for me a garment with the largest number of the largest pockets possible. The result was a masterpiece of the sartorial art, with no less than fourteen pockets, big and small, holding a nightshirt and slippers in one, a spare shirt collar and a shirt in another, brush, comb and toothbrush in another, and altogether sufficient to provide for two or three days' tramp, without recurring to any other luggage; dust-coloured moreover, [In anticipation of *khaki* (1900).] so that it did not take the dirt. I had never travelled on foot before, and twelve miles walk between Gloucester and Newnham, done within 7 minutes of the three hours, had been my greatest pedestrian achievement. Hughes, on the other hand, was a splendid stepper, and had once walked up from Brighton to Charlotte Street, to attend one of our lectures. The consequence was that I was at first dead-beat, but he prescribed for me a restorative till then unknown to me, a pint of half and half made hot with a small jorum of gin in it. I am not either a beer or gin-drinker, but I am bound to say his prescription acted on me like a charm, so that after food and rest I was able to walk several miles further that evening.

The Inn where we put up at Manchester was one where Lloyd Jones always stayed on his visits, and to which we were recommended by him, kept by an old Owenite Socialist named Ingham, both he and his wife very worthy people. Nothing like drunkenness or impropriety was ever allowed on the premises, and our accommodation, though not stylish, was very comfortable. One scene of which I was a witness in the big room, as we were smoking there of an evening, amused me not a little. Two old gentlemen, one certainly over 60, the other but a little below, sat down one after the other at the same table. They evidently knew each other, and began talking. After a time their voices rose higher and higher, and the talk became first a loud discussion, and at last an angry dispute. And the subject of it – of all subjects to be passionately discussed in a Manchester tap-room – was the virtue of the deceased Empress Josephine. Each had read some book or books on the subject, but neither of them the same as the other, so that there was absolutely no common ground between them. Ludicrous as the scene was, they both got so angry that I was really afraid of their coming to blows. At last one rose, and speaking loud enough for any one of the 15 or 20 people who were in the room to hear, took leave of his opponent with these words: 'Well Sir, all I can say of you is that you are a dahmed fool!' Now if ever I looked for fisticuffs and said so to the waiter when his customer had gone out, 'Why sir,' was the reply, 'they go on at it like that every night – they are old friends!'.

At Manchester I had the pleasure of calling on Mrs. Gaskell . . . ((who)) gave me an introduction to a mill-owner at, I think, Harpers Key, Mr. David Morris . . . We had a long and friendly discussion, though not approximating much in our views, he resolutely limiting his ideal to benevolent mastership . . .

At Bury, where I eventually took up my headquarters while in Lancashire, I made the personal acquaintance of one of my most valued working men friends, John Bates. I arrived there, as I mentioned, after nightfall, and at first saw only his wife, then a very pleasant-looking Lancashire lass, and learning that he was in bed, did not wish to disturb him, but she so insisted on the disappointment he would feel at not seeing me, that I allowed her to do so. And so he came down, a well knit, strong-built young fellow, the very ideal of an English working man, with an open, handsome face, and a look of strong will about his firm mouth and blue eyes. They insisted on giving me some kind of meal, including the standing buttered toast of a Lancashire workman's hospitality. John Bates was a very noble specimen of an English workman. He had great natural abilities, and I have piles of his letters containing much that is still worth reading. I think he was at this time working at the shop of the Lancashire and Yorkshire railway; in after days he wished to come to London, and I got J. P. Wilson to take him on as foreman engineer (he was then under-foreman) at Price's Candle Factory, where he remained till his death, giving always the greatest satisfaction . . . Of the powerlessnesss of the most upright of employers to check outside rascality he gave me one striking piece of evidence. It may not now be generally known that Prices were the first to manufacture glycerine on a large scale. The Wilsons were intent upon selling it pure to the public, and for this purpose confined the sale of it, under stringent conditions, to two agents, both of whom had been foremen to the company. In after years I wanted some for my mother, and speaking to John Bates, who used pretty frequently to come and see me of a Sunday at Wimbledon, and to join us at our Sunday early dinner, said I proposed to order some of one of these agents. 'I'll bring you some,' he said, 'we are allowed to take away with us' (I forget now what quantity). 'It is the only way you can get it pure. Before it reaches the agents, within half-an-hour of its leaving the factory, it is adulterated'. How and through whom, I could not get from him . . .

The latter years of John Bates's life were embittered by a sad family catastrophe. His wife drank. He had always had a special horror of drunken women, and knowing this, I suppose, the poor creature succeeded for years in concealing the fact from him. But he discovered it at last, and I don't think I ever witnessed such a change as that made in him by the discovery. His face was a singularly frank and open one, and though capable of great sternness, combined habitually something sunny with its thoughtfulness. But there was now a cloud upon the brow, a sadness about the corners of the mouth, which never left him from henceforth, except for a passing moment ...

At the time of my tour Mr. Maurice was staying on the shores of Ullswater, and there were staying with him two of the Miss Sterlings – bright charming girls who grew into noble women ... I could not resist the temptation of spending a day or two with him, but only reached his house very late, the autumn day having already closed when my train stopped at Carlisle. I walked the eight miles to Ullswater, much to the astonishment of a worthy turnpike-keeper's wife, of whom at one place I asked my way – 'But it's so dark!' she exclaimed. However, I arrived without the shadow of an adventure, spent two very pleasant days, and brought Mr. Maurice back with me to Bury to a conference.

In Yorkshire again I made several valuable acquaintances, and one lifelong friend. Leeds had then a remarkable group of co-operators. There was David Green the Communist bookseller, the leading spirit of the 'Leeds Redemption Society'; James Hole, author of the excellent volume of *Lectures on Social Science and the Organisation of Labour;* and John Holmes the linen-draper, in his prime an indefatigable and enthusiastic co-operator, besides other men of less mark ...

Whilst staying at Bradford I made the acquaintance, which ripened at once into friendship, of one whose comparatively early death has always seemed to me one of the greatest moral calamities which have fallen upon England in my time, W. E. Forster. 'Be sure you go and see Forster of Rousdon', had been Lloyd Jones's injunction before I had left London; 'I have spoken to him about you'. In his *Life of Forster* (p. 156) Mr. Comyns Reid has printed a letter which I wrote to him on the subject of this visit telling amongst other things how I recognised in him the long lanky lad who used to attend the meetings of the British India Society, and how he opened his religious doubts to me ... From this time we Christian

Socialists looked upon Forster as a sort of outside ally. He will be found thus taking the chair at a Christmas meeting of co-operators at Bradford, 1851, though expressing great doubts as to the success of the co-operative production . . . I saw a good deal of Forster from that time for some years. He came to dine with me in Cadogan Place, in company with Mrs. Forster and her widowed sister. I think I was able to help him a little at one of his elections for Leeds, when there was some talk of working class opposition, through Allan of the Amalgamated Society of Engineers. The last time he called upon me he used the words: 'I have not seen as much of you of late years as I ought to have done.' Very soon after . . . he sickened of the disease which cost him his life.

I cannot measure the loss which [W.E. Forster was to his country. I have known no statesman of such high aims, of such rigid honesty, of such dauntless courage. Mr. Gladstone no doubt aimed high, and always meant to be honest, but his matchless powers of self-delusion joined with his extreme intellectual subtlety were perpetually driving him into self-contradiction and seeming equivocation . . . I was greatly disappointed when after Gladstone-Achilles' temporary retirement from the fray, Forster failed to obtain the leadership of the Liberal opposition, and I still think that under him it might have sooner returned to power. His one great mistake, I consider (though it was one which casts anything but discredit upon his memory), lay in his acceptance of the Irish Secretaryship, a hopeless task. I warned him at the time that as long as Irish Home rule was not combined with the admission of all the self-governing colonies to a share in the government of the empire, it would be so. He afterwards admitted that I was right, and took up that idea of so-called 'Imperial Federation' – in which I firmly believe lies the sole key to the solution of the Irish problem. Death however came too soon to enable him to carry that question to a practical issue . . .

My co-operative tour of 1851 thus brought me useful acquaintanceships and more than one friendship – that with W. E. Forster an invaluable one. As an experience moreover it was of the utmost worth to me – bringing me into contact with the pick of our provincial working class. I must repeat here that I found them as a rule thoroughly frank and friendly. Although for the most part I bore with me no credentials, I was seldom received with any distrust – in one place only did I receive any rough play, as I have related in my *Notes*. I doubt if there is any country in Europe, save Switzerland and the Scandinavian kingdoms, in which I should have fared as well; certainly neither in France nor Germany. [My friend Nadaud has told me that there were places in his department (the Creuse) where any stranger to the place not accompanied by a resident would have stones thrown at him. Furnivall tells an amusing story of how he narrowly escaped being marched to the *corps de garde* for having gone in search of an out-of-the-way place called Fourniville in Normandy, which he supposes his ancestors to have come from, to not being able to prove that he was a commercial traveller, the only class of stranger ever known to go there.]

CHAPTER XVIII

*The Christian Socialist Movement 1850-4 – The Establishment
of the Central Co-operative Agency – Relations with Trade
Unions – The Central Co-operative Agency and the Council
of Promoters – My Retirement from the Latter –
The Educational Trend of the Movement*

I come now to the less pleasant part of these Reminiscences.
It could not have been told in Mr. Maurice's lifetime, and
scarcely in that of Vansittart Neale. I have lived to see that
neither of them was really to blame, and that it was
altogether for the best that things happened when they did. I
have to trace the causes which led to the abandonment of
organised effort on Christian principles directly to promote
co-operative production, and to the concentration of such
effort on the work of education only, whatever individual
promoters might continue to do in the sphere of co-operation.
Very various influences combined to produce this result, and
it is not altogether easy to disentangle them.

One important factor in the matter was a pecuniary one. It
was no easy job for us in the first instance to find funds for the
various associations. The only one of the first promoters who
was really well off was Vansittart Neale. I believe we every
one of us did our best. Charles Mansfield found the means of
liberally contributing through such rigid economy in food
and drink as bordered on starvation. Furnivall, I believe,
joined Mansfield and A. Campbell in their vegetarianism in
great measure for economy's sake, that he might be able to
give the more, Mr. Maurice was always nobly liberal in his
contributions to any good purpose. Hughes's hand was never
closed whilst there was anything in it to give. For myself, I

206

was at that time, I may say, beginning to be decidedly successful as a conveyancer and a very extraordinary wind-fall[23] gave me unhoped for means of helping the work. Careful economy alone however enabled me to do what I was able to do, and I felt the benefit of having given up all amusements since the year 1842 . . .

I will not conceal my opinion that the presence among the early promoters of one really rich man, and he even more liberal than his fortune allowed, was less a help than a hindrance. Neale's purse gave a premature and factitious expansion to the movement. It soon became known – and all such knowledge tends inevitably to exaggeration – that there was money to be had at the office of the Society for working men who chose to set up associations, and applications poured down upon us for help. Neale offered indeed to make all his advances through the Council of Promoters, provided it made itself responsible for them, but this we declined to do, and I still think, were wise in so declining. The consequence was, however, that Neale's advances became individual ones, and this paralysed the moral authority of the Council. If the Council lent £50 from its funds, and Neale lent £500, it is obvious that he became in the eyes of the Association his debtor, ten times as important as the Council at large. And in those days – he mellowed immensely afterwards – if things went a little wrong with an association his debtor – he was apt – if I may use the vulgarism – to 'come the capitalist over it' by threats from his position as a creditor, which further impaired the moral authority of the Council. [T. Hughes, in an 'In Memoriam' article published in the *Co-operative news* for Oct. 1. 1892, speaks of Neale as the 'naturally impatient, self-relying, rather scornful worker of 1849 (1850), fond of short cuts and subtle ways of baffling difficulties and defeating opponents.'] In short, I believe that Neale's liberal purse was one great means of flooding the field of association with mushroom bodies devoid of all self-reliance. The Association first started – and started before he joined us, that of the Working Tailors – was indeed the last or nearly the last of the group which remained on the field.

Let it be observed that the raw material of our association was not a promising one. The London working men, I consider, do not possess the energy, the enthusiasm, the spirit of self-sacrifice which in 1848-9 at all events distinguished those of Paris. Co-operative Associations had not been preached among them by popular and eloquent voices, as it had been in France. Moreover, the special working class organisation, the trades unions, as a rule looked askance upon co-operation – the only noteworthy exception being the Amalgamated Engineers. The consequence was that our associations, instead of containing the pick of the trades, as they should have done, tended often at least to be made up of what the French call the *déclassés*. And I, for one, instead of wondering at their failure, am rather surprised that several of them should have continued so long, and have turned out such good and solid work as many of them did.

Still, the breaking up of several of them under discreditable circumstances – and in two cases levanting of the manager with the cash – in two or three others, hopeless internal discords – in another, the first association of builders, the mere intoxication of success (it went to pieces after receiving money on the completion of its first contract) – was very disheartening to those of our promoters who were more enthusiastic than myself. One cause of failure, then, in the movement was the absence of sufficient perseverance, discipline, a self-sacrifice and self-reliance among the associations, Neale's too great liberality adding greatly to the mischief; another, the discouragement thereby produced among the promoters. Another and more potent cause was the influence of organisations more or less rival to the Society and in particular the Central Co-operative Agency.

The subject of co-operative consumption had not been overlooked by us. But in Sully's 'Constitution' it was just glanced at, under the 2nd head, as a means of carrying out and extending the practice of associated labour, 'by organising, both among such associations and among any others of combined capitalists and working men who may be admitted into the Union, the interchange and distribution of commodities.' So in the 'model laws' for association, the second object was 'to work with other associations for a complete organisation of distribution, exchange and interchange.' The principle of making consumption subordinate to production is, I consider, essentially the right one. But I freely confess that we quite under-valued at first the importance of co-operation in consumption. I have told already of the rise of the London Co-operative Stores, with Lloyd Jones for manager ... But there was no actual room for the stores in Sully's constitution, so that although I believe all or nearly all of the London Promoters dealt with the establishment in Charlotte Street, it was not subordinate to the Council, and remained outside the society.

The longer the stores went on, the more alive did Lechevalier become to the possibilities of development involved in them, and when dining with me he would press me to take up the subject. In the course I pursued I was half-right and half-wrong. I saw from the first rightly that the principle of the co-operative store was not itself auxiliary, but opposed to that of the productive association. But I misjudged the possibilities of its development, and the dangers involved in them. Finding that I would not take the lead in the matter, he betook himself for good to Neale and Hughes.

I have said before that I believe Lechevalier, when he first came over from France, in 1848, to have been in earnest ... But this did not last. In the first place his mind was essentially constructive, not seminal. He had no patience to plant the good seed and let it grow; he must be building a visible fabric. And he had not sufficient depth to converse long with principles; forms made up his habitual world. Hence, after once recognising, both spiritually and intellectually, the necessity of combining Socialism and Christianity, he soon turned away from the more spiritual faith of Mr. Maurice to the formality of the High Church. Bred up originally a Roman Catholic, it is possible that the similarity of High Church forms to those of the church of his childhood increased for him their attraction. At any rate, it was in Mr. Richard's church, in Margaret Street, that he first became a communicant, I think in May 1851.

One day, after he had been a long time without coming to see me, Lechevalier came and gave me mysterious hints of a grand undertaking which was being prepared, to bring together all the co-operative stores throughout the country under the direction of a great establishment in London for supplying them with all goods sold or used by them in which Neale and Hughes as well as Lloyd Jones and a man named Woodin were to take part. On hearing Hughes's name I pressed him as much as I could, and at last extracted from him the facts as to the original scheme of the Central Co-operative Agency, according to which there were to have been five partners – Neale, Hughes, Lechevalier, Lloyd Jones and Woodin.

My old distrust of Lechevalier had for some time been coming back to me. The man, disciplined into straightforwardness by Proudhon, startled into sincerity by Maurice, had seemed to be giving place to the well-meaning but deviously working schemer of old. Those, be it remembered, were the days of unlimited liability, where every partner in a commercial undertaking must be responsible to creditors to the last farthing of his means. Now Lechevalier and Lloyd Jones had nothing to lose; Woodin, I think, I had never yet seen; Neale, I considered, must look after himself. But I could not bear the thought that Hughes, with his young wife and young family, should risk all he had. I went to him the next day, entreated him not to become a partner, and at last prevailed upon him to reconsider the matter. The upshot was that Neale and he were no longer to be partners, but trustees, the firm consisting only of Lechevalier, Woodin and Jones. Whether or not, in the events that happened, I was the means of saving Hughes from ruin, God only knows. In the only conversation we ever had afterwards on the subject, he said he thought that if Neale and he had been partners, they would have looked more closely into matters, and perhaps kept straight at the agency. For my own part, I have never regretted what I did on that occasion.

And so the Central Co-operative agency was launched 7 months after the establishment of the London Stores. Lechevalier was to be the first manager; Lloyd Jones was in the first instance to give general attention to a Manchester branch which was to be at once started, besides another branch in Great Marylebone Street, Portland Place. Joseph Woodin, a friend of Lloyd Jones, more or less tinctured with Owenism – a man as it turned out of quite exceptional business abilities, and of great experience in grocery matters, was to have charge of the shop. Be it observed that the scheme of the Agency had from the first to the last been got up – I attribute this to Lechevalier – without in any wise consulting the council of Promoters, and without its knowledge.

Now let me say at once that the principle of the Central Co-operative Agency was an admirable one. It prefigured, on a higher moral level, the great Wholesale centres of the present day. Its professed objects were:

> First, to remove the opposition of interest which now exists between the buyer and seller. Second, to destroy the system of fraud and adulteration, which taints so deeply and extensively the retail dealing of this country. Thirdly, to save the labour and capital now wasted and the time now lost under the present mode of distributing articles of daily consumption . . . and fourthly, to facilitate the formation of associations by which the labourer may secure to himself the profits of his own labour – after defraying the necessary expenses of capital, promotion and management – by providing the arrangements requisite for disposing of the results of that labour.

And among the means by which it was proposed to effect these objects were 'Thirdly, to supply the means for establishing Working Men's Associations; and Fourthly, to promote the formation of local co-operative stores throughout the country, to which London stores might serve as a central agency.'

But it appeared to me that however good might be the object this was to a greater extent setting up a rival organisation to that of the Society for Promoting Working Men's Associations. Who could understand the nice distinction between 'promoting' such associations, and 'facilitating their formation', or 'supplying the means for establishing them'? And at the same time, the fact that several members of the Council of the society were also publicly connected with the Agency must tend to produce an impression that the two bodies were virtually the same, and render the society responsible in the eyes of the public for the agency. Therefore, although a member of the stores and thus entitled to attend the meeting for the establishment of the Agency, 30th May 1851, the report of which was afterwards published as a pamphlet, I purposely abstained from attending, and except

Strettell, Neale's brother-in-law, no single member of the Council not connected with the Agency was present. [A really valuable publication of the Agency, I may say here at once, was Mr. Woodin's *Catalogue of Teas, Coffees, Colonial and Italian produce and wines etc.*, with the retail prices affixed, sold by the Central Co-operative Agency. Not only did it anticipate the great catalogues of the 'Army and Navy Co-operative', 'Civil Service Supply' and other stores of the present day, but it gave details as to the nature and origin of the principal articles of grocery, and all the more important parts of the revelations then (1852) recently given by the *Lancet* as to the adulteration of such articles.]

Lechevalier made a very able speech, but the speech of the day was Woodin's. Woodin – in appearance the very type of the bland, smiling tradesman, was, at bottom, a man of indomitable resolution. He denounced by means of scathing facts the tricks and frauds of competition trade. 'The tradesman,' he said, 'pretends to do one thing, and means to do and does something very different. To ensure success, it is necessary to be a good story-teller.' And I fear that every word he said is true to this very day.

In the *Christian Socialist* for the next day, May 31, 1851, there appeared a leading article by me on 'Working Associations and Co-operative Stores', to every word of which I would subscribe to this day, pointing out, together with the moral superiority of associations for production over associations for consumption, the dangers involved in either form; exclusiveness in the working association, over-inclusiveness in the co-operative store; pointing out also the possibility that co-operative stores might engage in production, and out-Moses Moses. I concluded as follows:

> The great security for the present is, that the co-operative store movement is now in the hands of men who have been tried in the cause, and who distinctly avow the determination of using its machinery for the purpose of promoting associations for purposes of production, and for organising exchanges. That machinery is, I believe, one so dangerous when left to itself, that it requires to be wielded with something more of despotic power than is needed for the success of a working association, if that despotic power can be trusted to fit hands. It is, therefore, I believe, essential that the

associations for purposes of consumption, having by this time so far outstripped those for purposes of production, should find a fixed centre, and the partners of the Central Co-operative Agency have stepped forward to meet this want. Their task is a great one, and I trust they will prove equal to it ...

Neale replied to the article in a much longer letter under the same title in the number for June 14. He avowed an open preference for the co-operative store:

I think, indeed, that the co-operative store better deserves to be considered as an expression of association feeling than the working association ... I regard the co-operative store as the more important institution at the present time, just because it is that one by which the poorer classes can most easily help themselves, and pave the way for advancing to higher forms of association life. Secondly, because it makes a nearer approach than the working association to the solution of one of the great social problems, the union of interest between different classes ... It is, at present, not only the most important institution, but the one most truly expressive of the true principle of association.

I rejoined in two articles on 'Co-operative Stores and their management', in the numbers for June 21 and June 28. In the first I pointed out the confusion into which Neale had fallen, and which is still of constant occurrence, between the elements of distribution and of consumption in the Co-operative store, consumption being the purely selfish element, distribution a labour of which the fruit should be justice and fair-dealing. I expressed my dissent from his claim that the store was the most important institution, and the most truly expressive of the true principle of association ...

It is not for me to say which of us two had the best of the argument. [In after days indeed, Neale, himself, in his *Manual for Co-operators*, pointed out 'the impotence of distribution taken alone, for effecting the ends of co-operation', and that 'the idea embodied in the distribution store ... may even become an obstacle' to 'social progress', and that 'what the distribution society cannot do ... it is the function of the productive society to do, or prepare the way for its being done ... Thus co-operation will be raised from the ignoble office of a union merely in order to get things cheap and good, to the noble fruition of an institution by which men may be gradually made better, and therefore both happier and better men.'] Nor does it, I think, matter much. It seems to me sufficient to point out that a prominent Member of the Council of Promoters, and the wealthiest, averred his belief that a different form of association from that mainly advocated by the Society was the more important, and 'the one most truly expressive of the true principle of association'. This was surely setting up banner against banner. The two institutions could not co-exist in juxtaposition on such a footing. One must make the other subordinate to itself.

*   *   *   *   *

I now come to a side of our work which I consider to have been of the greatest importance, our connexion with Trade Unions. I have already told how I had come across strikes in the first instant as a district visitor, in the second as a teacher in Little Ormond Yard. The *Morning Chronicle* letters on 'Labour and the Poor' on the one hand, and our Conferences and our more intimate acquaintance with the working class generally, helped much to clear our ideas – I may speak, at all events from myself – on the subject of trade unions, and strikes. In my third 'Letter to the Working Men's Associations of London', published in the *Christian Socialist*, December 7, 1850, I already wrote on the subject in much the same terms as I might use today, except that I too hastily considered Trade Unions to have done their work, and evidently did not deem them capable of being extended to the agricultural class. By degrees we began to find relations actually opening up to us with Trade Unions. In April 1851 I find that 'a deputation from a trade society of Plasterers waited on the Central Board on Monday, April 14, with reference to a projected association of that trade union, the parties supplying their own capital'. Three weeks later some cabinet makers desirous of forming an association are stated to have been referred to the 'East End Cabinet Makers' Trade Society, who are considering the question of co-operation'. By July a plan had been set on foot in the amalgamated iron trades for organising some large iron works on co-operative principles, and 'The Windsor Ironworks Company' ((was)) the first result of the plan . . . The Woodcutters' Trade Society and Journeyman Bakers ((were also)) considering the question of association . . .

But it was the breaking out of the great conflict between the Master engineers and their men which brought our society for the first time in direct connexion with a great trade union. Before that conflict began, a deputation from the Council of the Amalgamated Society of Engineers, consisting of W. Allan, the permanent secretary, and W. Newton, the leader of the movement in the trade (with, I think, Joseph Musto, the then President of the Society) attended, on the Council of Promoters, in order to discuss the question of Association (i.e. co-operation) both under its legal and commercial aspects. This was my first acquaintance with two remarkable men, of whom afterwards I saw a good deal.

William Allan was essentially a Scotchman, canny, cautious, somewhat narrow, whose confidence it was difficult to win, and who perhaps never trusted himself entirely to anyone; an ungrammatical speaker with a strong Scotch accent, and with no pretensions to eloquence; but strong-willed, clear-sighted, self-possessed – a man more of deeds than of words, but whose words were never idle ones. William Newton contrasted with him strongly. He was at once a thoroughly shrewd man of business, and a born orator; in his earlier days too prone to play upon the passions of his audience, but a masterly reasoner whenever he chose. At the time I speak of he had already ceased to work at his trade, and was Secretary to an insurance company (the London & Counties Fire & Life). When he became a marked man after the close of the great fight he set up a weekly political and industrial journal, the *Englishman*, the first number of which appeared in January 1854. Later on he became proprietor and editor of a local paper, the *East End Observer*, which during his time was an excellent one of its kind. He was elected to the Metropolitan Board of Works on its formation, and became its Vice-Chairman, but was carried off by death at a time when he was practically certain of entering the House of Commons at an early date, and would assuredly have won aprominent place there.

William Allan, on the other hand, died churchwarden of his

parish . . . He, a Presbyterian originally, succeeded in getting a church-rate voted, the first time for years, his one argument perpetually repeated being – 'We have a fine church, we mustn't let it tumble down'. At his death I believe it was agreed on all sides that the parish had never had so good a church-warden. [A most painful scene occurred at the cemetery where he was buried – one of the South London ones. There was a vast crowd, both of parishioners and engineers, to say nothing of private friends. We were kept a very long time waiting, and when the cemetery priest appeared, he was so drunk that he could hardly get through the service! There were there, besides many dissenters, many who never attended public worship at all, and the opportunity would have been a rarely precious one to any true minister of Christ. As it was, believer and unbeliever alike were disgusted, the latter scornfully so. I wrote on the subject to the Bishop, who replied – while deploring the occurrence – that he had no jurisdiction over cemetery ministers. In those days indeed cemetery chapels were used only for the solemnising of the burial service. This I believe was found so conducive to madness in the chaplain that it is now made a condition of the license that full Sunday services should be performed.]

The opening up of relations between our society and the greatest of the trade unions was of momentous consequence to at least several of us. It must be borne in mind that the Amalgamated Society of Engineers occupied at that time a unique position, as a combination not of workmen in one trade, but of a whole group of such combinations in the iron trades throughout the country, the principle of the combination being that of an 'equalisation of funds' every year between all the branches, so that in whatever quarter a conflict might arise, the society should be prepared both for defence and attack – a plan since frequently adopted in other trades, but of late rather cried down. Although only established in 1850, by the 31st December 1851 the Society had 11,829 members, a balance of nearly £22,000, an income of nearly £23,000, the expenditure being not far from £13,500. Such figures were, till then, unheard of in the Trade Union world, though now largely surpassed. Thus for the year 1897 the same Amalgamated Society of Engineers returned to the Registrar 87,313 members, £305,882 funds and £347,867 income – five other unions returning over 50,000 members, five over £100,000 funds and seven over £50,000 income.

The Central Co-operative Agency were not slow to avail themselves of the opening made by the leaders of the Amalgamated Engineers and to them belongs the credit of having set on foot the first attempt formally to connect the trade unions with the co-operative movement of which one has heard so much of late years. In the number of the *Christian Socialist,* Vol. II, for November 15, 1851, will be found an 'Address and proposals of the Central Co-operative Agency to the trade societies in London and the United Kingdom'. The proposals were that the members of the various trades should be 'invited to consider the expediency of forming a model association in each trade, of organising co-operative stores, either in each trade or by a combination of working men of various trades' . . . The stores . . . and the model association would 'be found also by the funds of the trade, or by special subscriptions amongst the members', and the latter would 'undertake to execute orders for the articles of their production, and employ for the execution of those orders such of their members as are out of work.' The model associations and stores were to 'be organised by and under the control of the Committees of existing trade societies'. The Central Co-operative Agency undertook '1st to supply to the model stores and association, at wholesale prices, all goods' they might require, 'either as articles of consumption or materials, 2nd, to warehouse, show and sell their produce, on their account and on commission. 3rd to advertise and collect orders on their behalf, 4th to provide for any feasible and profitable operations of credit or exchange among the associa-tion, 5th to put the associations on the one hand in communication with capitalists, and on the other hand with traders or customers'. It was to be 'for the trade societies to consider what kind of connexion they would establish with the already existing Society for Promoting Working Men's Associations'. Copies were forwarded to the Committees of the various trades of the prospectus of the Central Co-operative Agency, the report of the meeting for its establishment, a list of the wholesale prices of their grocery stores, a catalogue of the articles sold by the shops in connexion with the agency; rules framed for establishing co-operative stores under the Friendly Societies Act, and a prospectus of the Windsor Iron Works Company 'as showing the best framed scheme of a legal association of working men'. A consulting Committee, it was stated, had been called together by the agency, comprising with Shorter eleven Managers or Members of existing working associations . . . , Mr. Thornton Hunt of the *Leader* newspaper, a professed Communist, Mr. Fleming, sub-editor of the *Morning Advertiser,* an old Owenite, Mr. Richard Hart, and also William Allan and Newton. The address was signed by the trustees and partners of the Central Co-operative Agency and seven members of the 'Acting Board' of the Consulting Committee, including Newton and Allan. The 'Proposals' were of course eulogised by the *Leader,* and also by the Chartist *Northern Star,* and relations were opened up, I believe, with some trade societies.

Little however was done before the struggle broke out, for which the Amalgamated Society of Engineers thought itself to be so well, and was so little, prepared.

It is characteristic of young trade unions – and this the Amalgamated Society of Engineers was, though composed of older, and to some extent old ones – to be ready to rush into a conflict. This readiness arises partly from that desire to try one's strength, which is common to all healthy young creatures, partly from the absence of the vested interests if I may so call them, of aged or invalid members, and of the widows and children of deceased members, which act in a well established trade combination as a check of always increasing weight on any rash exercise of its power. Now there were two customs in the iron trades which had long been objected to by the men – systematic overtime and piecework. One of the very objects of the amalgamation had been the hope of obtaining more efficient means of concentration of power for getting rid of these. In July 1851 the Council of the association issued a circular to all the branches which requested information as to the number of men (distinguishing between unionists and non-unionists) working piecework and systematic overtime, and the votes of members on the abolition of both, the schedule of replies to be returned by August 14. Out of 11,800 men over 9,000 voted, and of these only 16 in favour of systematic overtime and piecework. The Council now addressed (24 November 1851) two circulars, one to the men, the other to the employers. The latter, drawn up by W. Newton (to what extent altered from his draft I cannot say) is in most respects an admirable document, but contained one unfortunate expression, certain to set up an Englishman's back, that the Council had 'come to a resolution to abolish' piecework and overtime. Yet, oddly enough, the actual conflict broke out in reference to a local dispute between Messrs. Hibbert & Platt of Oldham, and their men, the story of which and of the consequent lock-out is given by T. Hughes in an admirable paper in the volume on *Trades Societies and Strikes* (1860)...I may here observe that this great struggle affords a striking illustration of the fact that in almost all trade-conflicts the interest of the men coincides with at least the immediate interest of the public at large, whilst the interest of the employer is always opposed to at least the immediate interest of the general public. For the interest of the men (I do not say that they are *always* wise enough to see this) is to confine a trade dispute within the narrowest possible limits, both in order to minimise the strain upon union funds for the support of those thrown out of work, and to maximise the amount of financial support received from members still at work, and this restriction of the area of the conflict causing the least possible disturbance of general business, is obviously also in the interest of the community at large. But the interest of the employer in dispute is always that the area of the dispute should be extended, if possible, so as to embrace the whole trade. For however bitter an employer may be against his men, he knows very well that his most dangerous enemy is the nearest employer in the same business, carrying it on while his own works are stopped. His private interest is thus essentially antagonistic to that of the community at large; he would maximise the mischiefs of the dispute while the men, if left to themselves, would minimise them. (It is true that the large modern Federations of Trade Unions, especially in the less skilled trades, too often fall in with the employers' interests by extending instead of limiting the field of contest – 'calling out' men who have no grievance).

This was naturally the course followed by Hibbert and Platt, and the issue of the Amalgamated Society's circular as to piecework and overtime gave them the required leverage for working upon their fellow-employers. They succeeded in getting the Lancashire employers to pledge themselves to a lock-out. Then of course it became in turn the interest of the Lancashire employers to extend still further the area of the conflict, and they prevailed on the London employers, by what I must consider false representation, to enter into the lock-out, the avowed object of which was now to put down the Amalgamated Society. A lying writer taking the name of Amicus whose first letter was conspicuously printed in *The Times*, and formed the basis of all its subsequent treatment of the subject, added further venom to the dispute. Virtually the whole press followed in the wake of *The Times*.

It is almost incredible at this time of day to realise what I may call the universal conspiracy of the press at that time to stifle the voices of the working class and their friends. Personally, as I stated in my lectures to be presently noticed, I 'endeavoured in vain by means of letters duly signed in my own name at full length, to obtain a hearing through the columns of the daily press ... addressed in vain three newspapers in succession, belonging to different shades of opinion.' A much more important man than myself, though not then in Parliament, W. E. Forster, wrote a letter to *The Times* simply to put questions on the facts as alleged by the men. His letter was not inserted. The paper which had started from the lies of Amicus was determined to brazen them out. Collective letters signed by Lord Goderich, Hughes, Vansittart Neale and A. A. Vansittart were, if I recollect aright, besides an important letter from Mr. Fielden, almost the only attempts to rectify misrepresentation, which obtained publicity in the press at large. The Amalgamated Society had indeed a weekly organ of its own, the *Operative*, very ably edited by W. Newton, but it was habitually ignored by the ordinary press.

We Christian Socialists at all events did our best to secure a fair hearing for the men ... in ((many numbers of the

Journal of Association and in pamphlets, lectures and meetings.)) . . . On February 5th 1852 it was decided by the
Council of Promoters that a course of Lectures should be
delivered on behalf of the Society on the relations between
capital and labour, with special reference to the pending
contest between the operative Engineers and their workmen. I
was requested to deliver the first, but found that I could not
treat the subject adequately in less than three lectures, which
were afterwards published under the title of *The Master
Engineers and their workmen* . . .

We had of course pressed upon the Amalgamated Engineers the advisability of forming co-operative associations of
their own, and there was a strong feeling among them in this
direction. Long before the lock-out a plan was on foot on
behalf of the Bury branch of their Society, for the purchase of
the Windsor Foundry near Liverpool, one of the owners of
which, Mr. Finch, was a well-known Owenite Socialist . . . The scheme however we never carried out. Attempts
on a smaller scale were made after the lock-out by the
engineers of Greenwich, at Deptford, on land belonging to
Mr. Evelyn, under the name of the b'deptford Engineers'
Company', who eventually set up their works at Cambridge
Road, Mile End, as the 'East London Iron Works' . . . and
again by the Atlas Company, Emerson Street, Southwark
Bridge Road. The Amalgamated Society itself took up the
question. In April 1852 a circular was issued on behalf of the
Executive Council, 'How shall we set about the work of
preparation for a coming time'. It began: 'There is but one
way, we must co-operate for production. We have learned
that it is not sufficient to accumulate funds, that it is
necessary also to use them reproductively, and if this lesson
does not fail in its effects, a few years will see the land studded
with workshops belonging to the workers.'

But nothing substantial came of this. The East London
Ironworks Association indeed promised for a time to be a
complete success, and I believe, might have been. It had at its
head one of the Musto's, a man of splendid energy, and great
self-devotion was shown at the starting of it. But what I may
call the fascination of business seized upon John Musto. In
the excitement of fighting for contracts, prudence first and

then honesty were thrown to the winds. And though two or three other trades – Clock-makers, Builders, Moulders – also took up the subject of co-operation, the movement among the trade societies proved an abortive one. I cannot help thinking however, that much of the future of productive co-operation lies still in that direction.

Let me indeed observe that from the first the iron trades had been warned not to expect too much in the way of immediate results from co-operation. In an excellent article, 'A word of advice to the Iron trades' (*Journal of Association* Jan. 24, 1852) Hughes had told them: 'You are not likely to get funds or work enough to employ all, or even a majority of your members, in Association. Neither are a majority of your members fit for association yet, had you work and funds enough.' Two or three carefully founded associations, he had pointed out, 'will do more towards winning in the end, than a thousand ill-manned, ill-furnished shops, set up all over the country. The long game is the sure one in the long run'. And the event no doubt showed that even the pick of the 'aristocracy of the trades' were less fitted for co-operation than we had hoped.

But I consider nevertheless that the moral effect of our efforts in bringing the trade unions into friendly contact with the educated and professional classes was of inappreciable value. The barrier then broken down was never raised up again, and never can be . . .

Although we did not hesitate to side with the men in the Engineers' lock-out, we from the first had urged the employment of other means for the settlement of trade disputes. As early as in *Politics for the People* I had taken occasion of the then recent decree of the French National Assembly for the reorganisation of the *Conseils de Prudhommes* to write an article on the subject of these Courts, which, acting primarily by way of conciliation, have power, when conciliation fails, to settle trade disputes judicially. I urged that 'the convenience of such an institution is far less even than its social worth, as a means of associating the working classes to the dispensation of justice, the execution of the laws.' (June 17, 1848). When the dispute in the Engineering trade broke out, we backed the men to the best of our power in their offer to submit the dispute to arbitration, which offer the employers (let it never be forgotten) scornfully rejected: 'with every respect,' they wrote, 'for noble and distinguished referees, whose arbitration has been tendered to us and with no reason to doubt that their award would be honest, intelligent, and satisfactory, we must take leave to say that we alone are the competent judges of our own business' . . .

Let it moreover be remembered that if we urged upon the men the practice of co-operative labour, it was never as a nostrum. In the *Journal of Association* five articles are devoted to 'a righteous joint stock company', i.e., Price's Candle Factory, where the results of the benevolent labours of the two managing Directors, Mr. G.F. and Mr. J.P. Wilson, but especially of the latter, were then in their first bloom. Neale, in his *May I not do what I like with my own* quoted the successful instance of profit sharing of Messrs. Laurent & Deberny, a firm of Paris founders. In my lectures on the *Master Engineers and their workmen* I took care to refer to a whole series of cases of benevolent employers . . .

*   *   *   *   *

All my fears about the Central Agency were speedily realised. It did put itself forward practically as a rival to the Society for Promoting Working Men's Associations, and seek to supplant it. It organised a series of lectures, the first of which by Mr. Thomas Ramsay (a lecturer obtained by Lechevalier from his High Church friends) *Is Christian Socialism a Church matter?* delivered August 8th, 1851, was separately printed. In the whole of its 43 pages, the Society for Promoting Working Men's Associations is not once referred to, but the Central Co-operative Agency is virtually put in its place . . . It is but just to say that the next lecturer on behalf of the Central Agency, George Dawson, a then celebrated non-conformist preacher at Birmingham . . . fully recognised not only Christian Socialism, but the existence of Christian Socialists. But a still more decided step was taken by the Agency in issuing their *Address and proposals to the Trade Societies in London and the United Kingdom* spoken of in the last chapter, in which, after referring to the establishment of

Co-operative Stores and Working Men's Associations, it went on to say, 'These efforts being destitute of a centre of business, the Central Co-operative Agency has been established as a legal and financial institution for aiding the formation of stores and associations, for buying and selling on their behalf, and ultimately for organising credit and interchange between them.' The Society for Promoting Working Men's Associations being almost contemptuously referred to in the words, '*it is for the Trade Societies to consider what kind of connection they will establish with the already existing Society for Promoting Working Men's Associations.*'

Admirable as may have been in many respects the ideas contained in those proposals, they carry still further the mistake originally committed in selecting London as a main centre of co-operative distribution. And they promise what must have been seen very soon, where not at once, by the hard-headed men for whom they were intended, to be impossible. Granting even that the Agency might supply all goods required by the model stores and associations – and experience has shown that even the gigantic wholesale societies of the present day, with capitals enormous in comparison with that of the agency, cannot supply all goods to the existing stores only – how could it 'warehouse, store and sell' the produce of the Associations and 'collect orders on their behalf' in any general way? The implication seems to be that all the trade of the country is carried on from London, the fact being ignored that the nearest market is always the most convenient and the most sought after; unless indeed the Agency expected to be ubiquitous, which it obviously could not be.

On the other hand, the reference to the Society for Promoting Working Men's Associations must, I fear, have been understood by many as meaning, 'Deal with us; we know how to help you; as for that other preacher-preacher affair, you may please yourselves.'

I think I recognise Lechevalier's hand throughout the above proposals. Fertile as was Neale's brain, Lechevalier's was even more so, and prolific of larger conceptions. The idea of virtually centralising all the trade and business operations of associations and stores alike seems to me to belong essentially to a French intellect, accustomed to the centralisation of all rule. To Lechevalier at all events alone do I attribute the set purpose of supplanting the Society for Promoting Working Men's Associations. Still, I am afraid that Neale himself would at this period have willingly seen it out of the way. [In the last year of his life Neale, in two letters to T. Hughes, which I have been privileged to see, seeking to counter-balance the 'overwhelming' influence of the consumers, was projecting a 'Productive Co-operative Union' which he expressly spoke of as a 'resuscitation' in an enlarged form of the old Society, in the hope of eventually 'merging the whole movement in one healthy Co-operative action.' If he could only have anticipated such conclusions by forty years!] This, I think, is shown on the one hand by the *Laws for the government of the Society for the formation of Co-operative Stores* (1851), a printed scheme by him which wholly ignores our Society; on the other hand and still more evidently by another printed *Scheme for the formation of the working associations into a general union* . . .

For now, Lechevalier was like the man out of whom the evil spirit departed for a time, to return with seven other spirits more wicked than himself. All his old intriguing habits returned with multiplied intensity. He saw that whilst Mr. Maurice and his immediate friends were a mere group of earnest men, devoid of all self-seeking, the High Church party were numerous, organised, powerful and aggressive. He attached himself thoroughly to them, and borrowed Mr. Ramsay from them, his first lecturer for the Agency. He succeeded in carrying Walter Cooper away with him to Margaret Street Church (though to me he spoke always with contempt of Cooper), got him made churchwarden. It seemed at last as if he could do nothing in a straightforward manner. To quote one instance of his duplicity – after borrowing money of T. Hughes (which I believe he never repaid) he went straight to Mrs. Hughes, and dilated upon her husband's imprudent generosity. When he wrote to me his letters, which at first had always been clever and interesting, became censorious and arrogant, besides growing to enormous prolixity. I think he wrote to me once 16 or 17 sheets! My reply was that when he wrote to me at such length, while

I would carefully read what he said, I could not undertake to answer it . . .

A few of us, Mansfield, Louis and myself, felt before long that the Agency and the Society as independent authorities could not safely go on together; that the Agency must recognise the supremacy of the Council. A choice must be made between the two. A ground for the necessary conflict was supplied by complaints which reached us that the *employés* of the Agency received no share in the profits, and were arbitrarily dismissed. The scheme of the Agency provided indeed that one quarter of the profits was 'to form a fund for the additional remuneration of such of the parties employed in the business as may seem to the partners to deserve such remuneration in addition to their salaries.' I brought forward the subject of the relations between the Society and the Agency at the sitting of the council on October 30th, and I dwelt at length on what I considered the encroachments of the Agency on the Society's province. I called upon those Members of the Agency who were on the Council to carry out in spirit that part of Article 3 of the Constitution, requiring that all hired workmen, unless dismissed for misconduct, were to receive a certain sum in lieu of profits, to be fixed by the Association; or in case of dispute by the Central Board. On the other hand, having heard in the interval between the two meetings that Neale was very much afraid that if I left the Council (which I made no concealment of my intention to do if its vote went against me) I might use the *Christian Socialist* to attack both Council and Agency, I offered to resign the editorship of the paper to any member of the Council designated by the President.

The members of the Agency refused to submit in any way to the Society. They had contributed the capital for setting up the Agency; the Society had nothing to do with them. I thereupon moved the exclusion from the Council of such of its members as were trustees or managers of the Central Agency (i.e. Neale, Hughes, Lechevalier and Lloyd Jones). Only Mansfield and Louis voted with me. The same or the next day, as I had resolved with myself beforehand, in the event of defeat (which I had expected), I resigned my own membership, on the ground that there could be no unity of action between one who had sought to exclude certain of his colleagues and the colleagues he had sought to exclude.[24]

I remain as firmly convinced as ever that the Agency as it was then constituted could not have worked with the Society. It would indeed have been more logical to have moved the exclusion of its members, if they refused to bring it into organised relations with the Society, when the Agency was first started, than after the lapse of some months. But the result would have been the same. Mr. Maurice had a rooted dislike to exclusion as such. Lechevalier once said of him that 'his system was all door', and there was a certain amount of truth in the saying. I can only say that I acted from a sense of duty, and that the pain of proposing the exclusion of so dear a friend as Hughes, of a colleague I was so fond of as Lloyd Jones, to say nothing of Neale after all he had done for the Associations, was very great. I was right moreover to resign my place on the Council when defeated. I was wrong ever to have resumed it afterwards.

Mr. Maurice read my letter of resignation to the Council at the meeting of the 13th November, and the next day I received a letter from Shorter, stating that resolutions had been unanimously adopted expressing deep regret at my resignation, and a hope that I might still be reckoned an 'extraordinary member', and that I might 'occasionally attend the deliberations of the Central Board', and also requesting me 'for the present' to continue to edit the *Christian Socialist*. I replied thanking the Council for their kind intentions towards myself, but stating that I could not accept the invitation conveyed by their resolutions as to my remaining an extraordinary member of the Council, or attending the deliberations of the Central Board. As to the latter, I pointed out that 'viewing as I did the *Proposals to Trade Societies* as containing a still more deadly attack upon the Central Board than upon the Council, it would be even more impossible for me to promote any purpose of peace and harmony by attending the Central Board than by retaining my seat at the Council.' But if the Council allowed, I should be glad to remain a 'Promoter'. As respected the third resolution, 'I must beg leave to observe that I never offered to surrender the editorship of the *Christian Socialist* into the hands of the

Council generally, but simply into those of such member or members of it as the President should name. Not, therefore, recognising the jurisdiction of the Council in the matter, I left my former offer "open for his acceptance".'

It must be observed that Mr. Maurice himself felt fully the difficulties of the situation. The two bodies, the Society and the Agency, working from the same address in London, were being constantly confounded together and the Society made morally responsible for all the proceedings of the Agency. A paper drawn up by him and adopted by the Council on the subject was printed in the *Christian Socialist* for December 5. I extract the following portions of it.

> The function of the Society, as defined by its laws, is to apply the principle of Christianity to trade and industry. Any persons who try systematically to sell good wares at a fair price are applying one part of the principle of Christianity to trade and industry . . . They affirm that the smallest transactions of life are under moral law. If they are not merely honest for their own sakes, because it is wrong for them to be otherwise, but have a direct purpose of helping their poorer brethren, and of enabling them to help themselves, they still further apply a Christian principle to practice. This the Agency, by its very name, by its constitution, by its acts, undertakes to do. In this sense, therefore, it is fulfilling a portion of that task which the Society has imposed on itself. But there is another part of the task which it cannot fulfil, or can very imperfectly fulfil. It is not, and cannot be, the main business of those who are establishing and promoting an honourable trade, to teach men their relations to each other. Now this is the main, the characteristic work of the Society. It desires . . . that working men should understand that they are brothers, and can work together as brothers . . . The Agency cannot be held responsible for such teaching. On the other hand, the Society cannot be held responsible for the plans which the Agency may see fit to adopt for promoting the success of its business. It cannot be held responsible for any particular method of co-operation which the Agency may recommend . . . In these proceedings, the Society has taken, and can take, no part. To do so would be to sacrifice its own position, and to involve itself in words and acts which it may disapprove. The Society, therefore, with the full concurrence of those of its members who belong to the Agency, has resolved as soon as possible to leave the house which it has hitherto occupied, and to seek another place for its

meetings, rather than to seem to take part in a work over which it has no control . . . The continuance of members of the Agency in the Council of Promoters is a witness that objects and methods which cannot be confounded need not therefore contradict each other. If they should hereafter prove to be inconsistent, the members of the Council must be content to separate . . .

The statement goes on to set forth a note by the Trustees and partners of the Agency, to be sent to their friends with the statement. Part of it runs as follows:

> The Members of the Agency are anxious, by a positive declaration on their part accompanying that of the Society, to prevent the friendly relations which subsist between the two bodies being misunderstood. On the other hand, they are desirous that the independence of the two bodies should not be lost sight of. The Agency is ready and willing to act with and for all associated bodies whatever their internal constitutions or external relations, if they are desirous of carrying on their business in the genuine co-operative spirit of fairness, and the desire to render to each other mutual assistance. The partners and trustees feel confident that from the central position of the Agency in London, and the resources at its command, it possesses the means of facilitating to a very great extent the formation of those common institutions by which the collison of separate associations can be prevented, while their individual freedom is preserved, and stability be given to the whole association movement, by making use of the important business of distribution as a means of calling forth and securing a market for productive industry.

It is clear from Mr. Maurice's statement that he was alive to the dangers arising from the proceedings of the Agency. But how those dangers could be sufficiently averted by merely shifting elsewhere the headquarters of the Society (the Agency being left in possession of the old quarters) I could not see then and cannot see now. To apply a homely proverb – there would be no less two Kings of Brentford, though one should live in East Brentford and one in West.

It was a great wrench for me to sever myself from the Council, but the wrench of giving up the editorship of the *Christian Socialist* was greater still. The *Christian Socialist* was peculiarly mine. It had not cost a penny to the Society; I had raised all the funds for its establishment. It had won for itself by this time its own public – small, but devoted and enthusiastic. Neale's fear that I should run the paper in opposition to him and to the Council was wholly unjustified. Such a thought never entered my head. My highest wish, as I have I think said before, was to be Mr. Maurice's chief lieutenant. To have set up my flag against his would have been abhorrent to me, as an anti-social act. But even in the sacrifice of giving up the paper there was to be a peculiar sting. Not indeed that it should pass into the hands of Hughes, whom Mr. Maurice designated for the purpose, the man whom I knew to be fit for it and whom I had myself proposed to him. What pained me was the giving up of the title *Christian Socialist*, which appeared to me virtually equivalent to the hauling down of a ship's colours in the very hottest of the battle. But there was no alternative after the following expression of Mr. Maurice's views in a letter to me. (It was the time, I should observe, of his first difficulties with King's College Council):

> I hear in so many quarters, especially from Bunsen and Hare, complaints of the political articles in the *Christian Socialist* as damaging the cause of English associations, that I feel convinced the only course for Hughes is to publish a mere Journal of Association *and call it by that name*. I think the well-being and progress of the movement is very much involved in this change. I have not liked to speak of it because it holds out a remote – a very remote but still a possible chance of my retaining my position at King's College . . . [Compare with this the last paragraph in Mr. Maurice's letter to Dr. Jelf of 30 November 1851: 'It may be right to mention that the paper called the *Christian Socialist*, to which I said in a former letter that I had contributed, will, after the beginning of the year, cease to contain any political or general articles, and will be entitled the *Journal of Association*'. (Life, Vol. II, p. 96).]

No one who knew Mr. Maurice could ever for a moment have suspected him of yielding to a selfish motive. But he was, I believe, yielding to the influence of men greatly inferior to himself, on whose opinions he set an exaggerated value. [Since Mrs. Maurice's death, I may say that observing her total want of interest in Co-operation as such, I have suspected that her influence had largely worked on Mr. Maurice to dissever him from active connection with the movement.] It will hardly indeed be believed that throughout the whole of the second volume of the *Christian Socialist* the only three political articles, properly so called, were one on the Admission of Jews to Parliament, one on Kossuth and one on the Napoleonic *coup d'état*. But had there been ten times as many, I never should have expected that Mr. Maurice, who had started with me *Politics for the People*, would have found fault with me for writing on politics in a Journal intended for working men, and would have sought to tie it down to a mere record of Association progress. [Only a few months later, as I find stated in a letter from Mr. Maurice, Lord Goderich and Hughes went to him and told him 'the *Journal of Association* could not go on upon its present footing, but must take some general line or be given up.' He himself suggested 'reviving our old *Politics for the People*', thinking from what I had said to him, that I might enter into the plan. But I declined, considering that it would imply retrogression. Now that we had professed our Socialism, that must be the avowed ground for our politics. The scheme came to nothing. But I need hardly say that on taking up the editorship of the *Journal of Association* during the last few weeks of its existence, I did not hesitate to deal with politics, from the Socialist point of view, so far as space allowed.] However, there was no discussing the matter, when the question of his continuance at King's College was once raised. So in an address 'To our Readers', signed in my own name, I announced (Dec. 27, 1851) that the editorship would pass into Hughes's hands, 'who by his position at once as a Member of the Council of the Society for Promoting Working Men's Associations and as a trustee for the Central Co-operative Agency, embodies most aptly the idea of the harmony which should reign between these two bodies.' And that as the limited space at his disposal would preclude him from travelling beyond the immediate subject of Association – 'which we have always considered to form only a part, and not the whole of Christian Socialism' – it seemed but natural that the former half of our title should be dropped, and that our paper should remain simply *The Journal of Association*. I offered at the same time to refund the subscription of any contributor to the fund then open for the maintenance of the paper, who should consider the change 'a breach of faith towards himself.' No claim for reimbursement ever came in. [Pecuniarily, the change of name was disastrous. The expenses of the *Christian Socialist* had amounted to nearly £200 a year. By halving the size, and some other trifling economies, the cost should have been brought within £100. I handed £50 over to Hughes to cover six months' publication (January-July 1852). It was all spent by mid-April, and the death of the *Journal* had been announced April 5, after the next number. I felt greatly vexed, as it seemed to me a breach of faith towards our subscribers, and undertook to carry it to the end of the volume at my own cost resuming the editorship . . .

In my calculation of cost, be it observed, nothing had been allowed for receipts. These fell off greatly on the change of name and size . . .]

On being informed of my retirement, the Central Board also passed a resolution expressive of their deep regret, and stating that they would feel grateful if I would occasionally attend their deliberations. They moreover expressed a wish to hear my reasons for my retirement. With the Council's permission, I met them, and made the longest speech of my life, lasting over an hour. I sought to be as dispassionate as possible, and George Hughes, who had been deputed by the Council (as it were with a watching brief) to be present on the occasion, told me afterwards, that I had said nothing to which he could object. I was, however, perfectly frank, and avowed to the Central Board my conviction that the Society and the Central Agency could not co-exist on their present footing, and that the present relation must lead to the ruin of either or both. They listened to me attentively, but said nothing. I expected nothing else. Half of the members perhaps were on the 'Consulting Committee' formed by the Agency; the associations to which the others belonged were mostly in Neale's debt.

No mention either of my retirement or of the meeting last referred to was published in either the *Christian Socialist* or the *Journal of Association*. Our dirty linen, such as it was, was effectually washed at home. I continued to do what I could for the cause and contributed to the *Journal*. I continued on the Committee of the East London Needlewomen's Workshop, which was not formally under the Council; took an active part, with Hughes, Neale and Lord Goderich especially, in trying to obtain fair play for the men in the Engineers' Lock-out. Remained, be it observed, a member of the Society whilst ceasing to be on its council, and at its request delivered my three lectures on the *Master Engineers and their Workmen* . . . Be it observed, moreover, that I remained throughout on perfectly friendly terms with the members of the Agency (except Lechevalier, of whom I felt always increasing distrust), and continued to deal with it as far as possible. From Lloyd Jones in particular, I received a most friendly letter (March 3, 1852) in which he wrote: 'If you suppose that anything, either concerning the Agency, or anything else, has caused me to have any other feelings towards you than those of admiration and respect, and something more even than either of them, you were never more mistaken in your life.' Woodin too, though I cannot say we ever came to be intimate, I appreciated more and more . . .

Lechevalier, on the other hand, had by this time become, in plain terms, a thorough humbug and hypocrite. He had taken into his head to find an English wife and had pretended to fall in love with Miss Craigie, a young lady who attended Mr. Richard's church. In order, apparently, to look better in her eyes, he began in 1851 to tag 'St. André' to his name. At the Agency, to use Woodin's words of him, he could not go straight. They had at last to get rid of him, some time before 27th October, 1852, as I find from a note of Walsh's of that date. He then threw himself more than ever into the hands of the High Church party, and for a time got completely round the Rev. Charles Marriott of Oriel and some of his friends, and having in the meanwhile paid a visit to France, returned with a wife (Miss Craigie had become a Roman Catholic and entered a convent) and succeeded at great cost to his new friends in setting up and conducting to its ruin a 'Board of Supply and Demand', i.e. a 'Universal Purveyor' establishment, professing to guarantee the purity and genuineness of all articles . . . He published a work in 1854, entitled *Five Years in the Land of Refuge*, full of mis-statements as to his old colleagues in our movement, and which I only mention to guard any of my readers who might come across it against giving any credence to it. Finally, being no longer able to live upon anybody in England, he returned to France, made his submission to the Emperor, and was taken into the Secret police . . .

I may refer here to a rather curious experience which belonged to this period, but which remained almost without consequences either in the history of the movement, or in my own after-life. In May 1852 Mr. Maurice received a letter from a young barrister, and poet, Arthur Munby, telling him that Sunday after Sunday crowds gathered to hear the preaching of all sorts of doctrines religious, political, social, and suggesting that some of ((us)) might be sent to try and give some wholesome direction to this spiritual fermentation on Kennington Common . . . There we attended for some time every Sunday; the scene was very much what it is now every Sunday in Hyde Park . . . I felt myself very cold and tongue-tied, and spoke but seldom, but A. H. Louis spoke frequently, and often with great power and eloquence. I remember one really shocking sight, – a red-haired Calvinist of the narrowest predestinarian type, who preached to a throng of donkey-boys and others, all jeering at him, the anti-gospel of eternal damnation. The man was, so to speak, madly in earnest. We tried to talk to him, asked him if he did not see that he was only irritating his audience, making them worse instead of better. He said he was perfectly aware of it. Did he think he had ever done good to anybody by such preaching? No he did not think he had, but he was commanded to preach, and he must preach. A very curious pair were a father and son who had thoroughly persuaded themselves that they themselves would never die, nor any who followed their doctrine. Alas! before we ceased frequenting the Common we heard that one of them had confuted his own teaching by dying. Total abstinence men were mustered of course in great force and Louis . . . was the first, I believe, to give the retort to a once favourite teetotal axiom: 'As soon as a man begins to drink, he begins to get drunk' by saying, 'Then as soon as I take off my coat I begin to go naked?' . . . Later the whole thing was swept away by the progress of Metropolitan improvement, Kennington Common being enclosed and turned into the present Kennington Park . . .

*　*　*　*　*

The connection of Co-operation with Education, as well as with other provident purposes, had never been overlooked. Sully's Constitution provided that the object of the Union was 'to carry out and extend the principles and practice of Associated labour . . . 3rd, by establishing among all the associations admitted into the Union institutions for the common benefit of the members, as Friendly Societies, model lodging houses, schools, etc.' In the 'Code of laws for an Association', Ch. II, the object stated was '4th, To establish conjointly with other Associations such institutions as may be beneficial to all of them,' with a footnote: 'Such as general store, benefit club, *schools, library, museum,* building society etc.' Towards the latter end of 1851 the Central Board were discussing the question of a benefit society, library and reading room in connection with the society. The rules for the library were passed on October 13 and in . . . 1853 the Library contained 308 works . . . It must be remembered that the magnificent free library movement of later times had not then begun. If I recollect aright, the old foundation of Great Smith Street, Westminster, was the only public library as yet which could be used without an introduction.

But the educational work proper of the society arose out of the building of its Hall. Once having this, we naturally sought to utilise it as much as possible. It was determined to deliver lectures in it, and hold evening classes. I find from the first hand-bill relating to the subject that Mr. Maurice inaugurated the lectures by me on Nov. 23, 1852, on the Historical Plays of Shakespeare. Walter Cooper followed, ((as did many distinguished names, not all of our group)) . . . The Evening Classes were to be on Grammar, directed by T. Hughes and Vansittart; English History and Literature directed by Mr. Maurice; Book-keeping; French, by myself; singing, by Professor Hullah, classes in Drawing and Political Economy being also projected . . . In addition to the lectures and classes, free conferences were held, at first every alternate

Wednesday, but afterwards on the first Wednesday of the month only . . . [I find also in March mention of a day school we were about to set up in the Hall of Association, in conjunction with the clergy of the parish, with one-third of the admission free for associates' children, but can find no further details on the subject.]

It being clearly understood that the delivery of lectures or the teaching of classes or the attendance at either did not imply the acceptance of our views . . . the opening of our educational department drew us into contact with what may be termed an outer circle of partial sympathisers and well-wishers, bringing us also as pupils many working men in good employment, who might have no occasion for going into co-operation . . .

CHAPTER XIX

The Christian Socialist Movement 1850-4 –
Mr. Maurice's King's College Difficulties –
The Later Conferences – The Hall of Association –
The Co-operative Conference – The Association for
Promoting Industrial and Provident Societies

It is impossible to understand the latter part at all events of the Christian Socialist movement without taking into account the persecution to which its leader was made subject, and to which his participation in the movement lent, in the then temper of the public mind, the most obvious and telling pretext . . . But the largeness of his theology was at bottom the head and front of his offending.

The attack began from a long-deceased periodical, the *English Review,* and was based upon one of the most pregnant and beautiful volumes of Mr. Maurice's sermons, the series on the Lord's Prayer, published 1848. In the number of the review for the following October a short notice of the volume charged him (to use Mr. Maurice's words) 'with aversion to the name and idea of a priesthood, with counting it a misfortune that he had been ordained a priest, with being in opposition to the whole Prayer book,' and on being remonstrated with, the editor, writing to an acquaintance much interested in King's College, urged him to recommend Mr. Maurice to be prudent and threatened to make an attack upon the divinity teaching of the College generally . . .

Archdeacon Hare's *Life of Sterling* gave the occasion to the next and most venomous attack from the same review, in an article by Mr. Parker 'On tendencies towards the subversion of faith'. It was this which gave occasion to Archdeacon

Hare's pamphlet, *Thou shalt not bear false witness against thy neighbour;* but the attack was no less directed against Mr. Maurice than against himself. The reply was written with much of Hare's old force, and was a crushing one, though too lengthy; and good as it is, Mr. Maurice's letter inserted in it is by far the best part of the pamphlet. And as dogs that would fight if they met will howl in concert when out of sight of each other, the High church *English Review* was followed in its attack by the Evangelical *Record* and high-and-dry *Morning Herald.* And Dr. Jelf began to take fright, and wrote straight off to his Professor of Theology a string of questions (March 1849) . . . [see Mr. Maurice's *Life,* vol. I, p. 521.]

The next shot came from ordnance of heavier calibre than the *Record*, and scattered pretty widely. A volume of *Introductory Lectures* delivered at Queen's College . . . had been published early in 1849. The lectures – very unequal in merit – included two by Mr. Maurice and two by Kingsley. *The Quarterly Review* (March 1850) now came out with an article headed 'Queen's College, London', virulently attacking the writers generally. 'No critic,' the reviewer wrote, 'can fail to detect in this volume traces of a school of so-called *Theology*, which seems to be gaining ground among us – a sort of modified Pantheism and Latitudinarianism'. And he directed special attacks upon Mr. Maurice and Kingsley.

I tried to throw myself into the breach, publishing in *Fraser* an article headed 'Queen's College and the *Quarterly Review*'. But this time Mr. Maurice was not to be stopped. He replied in the most dignified manner not to the slanderer himself but in a letter addressed to the Bishop of London, taking the responsibility for his colleagues, and fully claiming his own double responsibility not only as Principal of Queen's College but as Professor at King's. A nobler specimen of Mr. Maurice's controversial powers is, I think, not to be found.

Till now, it will be observed, all attacks had turned upon theological, or at least religious questions. But our social work developed to Mr. Maurice's assailants quite new openings for attack. The publication (1850) of *Alton Locke* provoked a fresh onslaught beginning from the *Record* . . . From this time the attacks on Mr. Maurice merge to a great extent into those general attacks upon the movement of which I have spoken.[25]

... Nor can I have any doubt that in Dr. Jelf's mind – though perhaps unconsciously – it was at least as much the Christian Socialist leader as the author of the *Theological Essays* whom it had become expedient to get rid of as a Professor at King's, when he entered into that singular controversy represented on the one hand by his *Grounds for laying before the Council of King's College, London, certain statements contained in a recent publication entitled 'Theological Essays' by the Rev. F. D. Maurice, M.A., professor of Divinity in King's College* (1853). I hardly indeed know anything in literature more ludicrous than poor Dr. Jelf's genuine and utter incapacity to enter into Mr. Maurice's thoughts, but side by side with that incapacity it is easy to see the predetermination, or rather I should perhaps say, the irresistible instinct to condemn them.

I took a small part in the controversy, and intended to have taken a somewhat larger one, publishing a pamphlet entitled *King's College No. 1, the Facts, by a barrister of Lincoln's Inn*, March 1854. In compiling it, I addressed letters to all the members of the minorities at the two meetings of the Council of King's College at which Mr. Maurice was first censured and then dismissed, asking for details of the proceedings ... From Mr. Gladstone, from the Bishop of Lichfield, from the Dean of St. Paul's and from the Rev. J.S.M. Anderson, I received most courteous and friendly replies, which enable me to say that the account I gave in the pamphlet is a perfectly accurate one ... Archdeacon Hare afterwards said that the circulation of the pamphlet had really stopped the attacks which had at first been pouring in upon Mr. Maurice from the *English Churchman, John Bull, Church & State Gazette, Morning Advertiser* etc.

It will be observed that the pamphlet is entitled 'No. 1'. I had intended to follow it up by two others for which I had collected some considerable amount of material; 'No. 2, The Doctrine' and 'No. 3, The Man'. But on hearing my intention, Mr. Maurice showed himself so absolutely adverse to No. 3, that I was obliged to give up No. 2 also. In both I intended to have followed exactly the same plan as in No. 1, i.e., a strictly historical one, in the former setting out the course of thought in the Christian Church on the subject of eternal punishment, and indicating in what Mr. Maurice's views differed from or resembled those of others; in the latter relating simply the events of Mr. Maurice's life, and giving statements of the effect which he had produced upon others. I had collected a very valuable and interesting series of such statements . . .

I need hardly say that, with Hughes and others, I had taken an active part in getting up that address from members of Lincoln's Inn which is printed with Mr. Maurice's answer in his *Life*. On the occasion of it we got indeed a snub from the Benchers, for distributing copies of the address to the members of his afternoon congregation as they came out of Lincoln's Inn Chapel. The Benchers were probably right. But in (1891) I have known to my disgust papers similarly handed out after morning service to the congregation as they came out, from a London church, impressing on them the necessity for voting for 'Church' candidates at the London School Board election. I cannot but think that the suggestion to a congregation to sign an address to their own clergyman was a very venial breach of decorum compared to the latter . . .

It is obvious that whilst Mr. Maurice's very position as an orthodox teacher of the Church of England was at stake – to say nothing of the pecuniary interest connected with his Professorships – he could not devote to the Christian Socialist movement the time, the thought, the unremitting energy which were required of a Leader. To this cause – which must have been subtly working almost from the beginning – must perhaps beyond all others be traced the collapse of that movement. The vindication of God's love – the setting forth of the fullness of Christ's sacrifice – which were Mr. Maurice's main objects in the controversy, although it may be said that they never entered into Dr. Jelf's narrow field of vision – were of far more engrossing importance to him than the attempted realising of human brotherhood in a few commercial associations. If the latter object in any way conflicted with the former, it must by him be thrown aside. And although never was any-one more absolutely indifferent to money except as a means of good than Mr. Maurice, still there can be no doubt that many of those around him, whom he respected and held dear, looked upon his connexion with the movement, if not upon the movement itself, as quixotic, and whether openly or tacitly sought to withdraw him from it. The breaking out of the King's College storm would necessarily have quickened and given weight to all efforts and arguments to this end. I look therefore upon the King's College controversy as essentially interwoven with the story of Christian Socialism. I revert now to the general course of the movement.

An article by Lord Goderich in the *Journal of Association* (March 1852) explains the renewing of the old Conferences with working men. After the establishment of the Society for Promoting Working Men's Associations, the Conferences 'were confined to its own members, and discussed almost exclusively subjects which related to the management and details of the Associations'. But these subjects were in time exhausted, and the attendance greatly fell off. The Council on the other hand felt that they had 'somewhat to say on other matters besides Association', and determined to resume their open conferences. These took place in the first instance in the cutting room of the Working Tailors' Association, afterwards at the Hall of Association which we built in the rear of that Association's premises under Penrose's directions. The new series of conferences began 10 March 1852, and I cannot say when they ceased . . .

The subject of Trade Unions was discussed . . . and the most remarkable feature of these conferences was this, that Mr. Maurice showed himself broader-minded towards Trade Unions than even Hughes. Hughes had condemned the exclusiveness of trade unions; Mr. Maurice could not agree that such exclusiveness was essentially bad.

> He admitted the equality in which men should stand to one another; but then they must be truly *men*, something more than mere animals, following blindly the animal instincts of their nature. You could not put these latter on an equality with thoughtful, self-denying men . . . But you must bear in mind that all these men had the same capacities with yourselves, and you should endeavour all you could to make them better men, and to

bring them into your fellowship. (*Journal of Association*, April 26, 1852).

The great event in the movement for the year 1852 was, however, the calling by the Society for Promoting Working Men's Associations of a Conference by delegates from bodies engaged in co-operation, whether productive or distributive, 26-7 July, at the Hall of Association, the object being to consider the best means of co-operating under the new law. At this conference delegates were present from 28 societies, (representing about 1500 members), of which thirteen London ones and fifteen provincial, eleven others of the latter sending letters expressing sympathy, but declining to be represented on the ground of expense or distance. Among provincial co-operative societies actually represented were those of Bradford, Edinburgh, Galashiels, Southampton and Ullesthorpe . . .

To this Conference, which though not largely attended, was earnest and businesslike, and undoubtedly a success, was presented the first (and only) report of the Society for Promoting Working Men's Associations, drawn up mainly by Mr. Maurice, and the report of the Society and that of the Conference itself from a pamphlet of 105 pages . . .

((The activities of the movement are described;)) the balance sheets of the Society and of ten associations connected with it are given. The capital of the society is stated to be some £1,500, its income under £200. The high aim of the movement is put in noble words: 'The part of every honest man just now is to throw himself heart and soul into the movement, and to teach by words and deeds that men do not come together in Associations to divide profits individually, and heap up capital, but to learn to live and work together like brothers, to see justice done to the weak, and to preach the trade gospel of *the Duty to labour and the right to live thereby.*' [It will be seen that these words anticipate the whole economic gospel of Mr. Ruskin, so far as it has any value.]

Thirteen resolutions were passed by the conference ((including recommendations for the formation of a co-operative Friendly (benefit) Union and an Investment Society; an executive committee to deal with business during the year was also appointed)). On the evening of the first day's conference the festival of the society took place, Mr. Maurice presiding. The attendance was much larger than the Hall, built to seat 300 persons, could accommodate at once, and the tables had to be made up a second time . . . Lloyd Jones spoke to the first sentiment: 'Association in trade and industry' . . . Charles Kingsley spoke to the third: 'The friends and promoters of Working Men's Associations' . . . I spoke to the fourth: 'To the Promoters of Associations on the Continent of Europe, and in the United States of America' . . .

It had not been intended that any but the selected speakers should address the meeting. But Louis Blanc was in the Hall, and in answer to repeated calls appeared on the platform, and made a neat little English speech. Vansittart Neale moved the thanks of the meeting to 'the Delegates of the Provincial Co-operative bodies, and other friends who have honoured us with their presence tonight', and T. Hughes 'the Working Tailors' Association and its manager, Mr. Walter Cooper, by whom this festival has been arranged', Walter Cooper responding to that at some length.

This Conference and festival represent, I consider, the apogee of the Christian Socialist movement. [There was, however, an occurrence at the festival which very nearly made a fiasco of it. The festival took place during one of those gusts of regal unpopularity which occasionally swept through the early years of the Queen's reign, and when at the end, Hansard being in the chair, Hughes proposed 'God save the Queen' three voices hissed including —, who had made himself indescribably troublesome during the whole evening – and to make the matter worse, he tried to argue the point. However, the opposition was put down with a high hand by means of a strong Queen, Lords and Commons speech by Hughes, and a threat of personal violence by myself to a young ne'er-do-well (I'm afraid) and 'God save the Queen' was given with tremendous loyalty by at least 5/6ths of the persons present, the passages 'Frustrate their politics, Confound their knavish tricks' being given out with peculiar gusto.]

In accordance with one of the resolutions of the Conference, a small pamphlet was issued containing the text of the

Industrial and Provident Societies Act, a statement of its advantages, and model rules for a society with suggestions for the use of the model rules. I cannot at this distance of time distinguish the shares respectively taken in this publication by the several legal members of the Council, but I believe I was mainly responsible for the statement of the advantages of the Act . . .

In spite of the success of the anniversary, things were not going on satisfactorily. I wrote to Charles Mansfield . . . (Oct. 6) on the subject:

> I have been drawing up for perusal of those whom it may specially concern certain thoughts on the present condition of the Society and suggestions for the re-constitution in the first instance of the Council[26] . . . My draft is simply that if we are not to be utter shams, we must practice the spiritual part at least of Association whilst preaching it, be united together by careful selection and exclusion and postponement of self to the common interest, under common discipline, and a true governing head, which head I still point out as the Prophet summing up my notions in the three requirements of a thoroughly democratic spirit, a thoroughly aristocratic constitution, and a thoroughly monarchical government. I don't know what will come of it, but in the meanwhile everything seems falling to pieces at a great rate . . .

There had also been a terrible scandal in connection with the movement, Beyers' bolting from wife and children with a young girl . . .

I had my *Thoughts*, according to a process then in vogue, 'anastatised', sent a copy to Mr. Maurice, another to Charles Mansfield, and showed the thing also to Louis and Hughes – I think to no others. Mr. Maurice however asked me not to circulate it amongst all the members of the Council (each of whom I had intended to have had a copy) intimating that he would take the matter in hand himself. It produced however even so some effect. The Council and Central Board requested me to resume my functions in the Society; Neale represented to me that Lechevalier who had been after all the person at the Agency against whom my opposition had been directed had been got rid of; that the Agency had now given up all lecturing, and propagandism,

and was strictly a business concern. Finally, Mr. Maurice
directed me to return to the Council, and in accordance with
the discipline which I claimed to see enforced, I felt bound to
obey him. I got a Finance Committee appointed, consisting of
Hughes, Vansittart, and myself, with very wide powers, with
a view to making 'a desperate effort' to pay off a debt of £350
still due on the Hall, and to equalise receipts and expenditure. Neale and Lloyd Jones were anxious to have 'a basis of
justice defined as the only foundation of the Council's work,
having the christian principle to those who feel it more.' I was
at first much taken with this proposal, as likely to clear up a
good deal of misunderstanding, though on further consideration I took an entirely different view of it. I had also in my
head 'some further moves in the way of organisation', which
with the rest of what was going on might, I thought, 'set the
whole thing straight again', if Mr. Maurice kept up the
temper in which he was then. Mr. Maurice, I may observe,
was then anxious to quicken and organise the Lincoln's Inn
congregation especially with a view to what I termed
'professional Christianisation'. 'I don't know,' I wrote to
Charles Mansfield, 'how he'll do it, but I am ready to serve if
he will direct. At all events there is a rattling sound of life in
the dry bones once again, which I trust may not prove a mere
rush of wind through them' . . .

I now come to the transformation of the Society for
Promoting Working Mens' Associations into the Association
for Promoting Industrial and Provident Societies. The term
'Industrial and Provident Societies' adopted by the legislature
to designate co-operative bodies, seemed to render some
alteration necessary in the title of the Society, and various
modifications in its constitution were felt to be necessary.
Towards the end of 1852 a committee of the Council was
appointed, consisting of Neale, Hughes and myself, which
was afterwards joined by Mr. Maurice to revise the constitution on the basis of certain propositions by Mr. Maurice to
take the place of the averment that the functions of the
Council were, '2nd, To diffuse the principles of co-operation
as the practical application of Christianity and the purpose of
trade and industry.' Mr. Maurice's propositions in their
original shape were 'That Society is one body with many

members; that fellow-work, not rivalry, is the law of nature; that a principle of justice must regulate exchanges.' This, it will be seen, was the carrying out of Neale and Lloyd Jones's suggestion which I had at first inclined to. But when it came out that all affirmation of Christianity was to be left out of the new constitution, I felt on reflection that (without judging others) for me at least it would be to deny Christ. There was a most painful struggle, and before it came to its height, Neale was laid up with threatened paralysis and brain fever, whilst I on the other hand got a low fever which reduced me to a state of very great prostration. [The discussions took place in Hughes's room at Lincoln's Inn, where I soon afterwards joined them. I was so weak that I remember on one occasion the tears rising to my eyes – a want of self-control which I have never been led into on any similar occasion.] The Committee thus disagreeing had to go back to the Council and seek fresh instructions. The matter remained hung up for upwards of two months, not to the benefit of the Society, Watson of the Baking Association for instance, saying at the Conference, December 1852, that 'the whole movement had become so misty that he could not understand the meaning of it any longer'. Eventually, I resigned my place on the constitution committee, and told the Council that it would be impossible for me to remain on it if it rejected the name of Christ from its Declaration of Principles.[27] I felt indeed, as I wrote to Charles Mansfield, that I had rejoined the Council too soon, and had no business there. In deference to me eventually the words to which I clung were, on Mr. Maurice's proposal, retained somewhat modified in the preamble, by way of recital.

The Promoters of Working Men's Associations, having united together for the purpose of applying the principles of Christianity to trade and industry, and desiring to state more definitely what those principles are, as they find them set forth in Christ's Gospel, that they may serve as the basis of a society to be formed for the objects after mentioned, declare;

1stly. That human society is a body consisting of many members, not a collection of warring atoms.

2ndly. That true workmen must be fellow-workers, not rivals.

3rdly. That a principle of justice, not of selfishness, must govern exchanges.

End by article 1 of Chapter 1, 'Nature of the Association':

The Society to be formed on this basis shall be called 'The Association for Promoting Industrial and Provident Societies'. It consists of persons of all classes, who will unite to carry out the above principles.

I confess I did not at the time understand Mr. Maurice's mental attitude in the matter, so that the agony, I might truly say, which it caused me was really of my own creating. He had grown by this time to distrust – he grew more and more to dislike the word 'Christian', 'Christianity', unless where they had to be accepted as terms of reproach from others. His sense of the Union of Christ with humanity itself – Christ as the 'head of every man', – Christ as 'the light which lighteth every man that cometh into the world' – was so keen, so overpowering, that there seemed to him to be something limiting, sectarian, in the use of such words. I don't say that he was wrong in principle.

Nothing is more remarkable than the fact that the word 'Christianity' occurs nowhere in the New Testament; the word 'Christian' three times only, and never as applied among the disciples themselves . . . ((But)) I believe that in relegating the name of Christ to a recital in our new constitution Mr. Maurice was not understood, and that the change not only did not enlarge our sphere of usefulness, but tended to damp the zeal of some of the most earnest among us. [In after years, when the Christian Socialist Movement *ad nomine* had ceased to exist, I felt of course no difficulty in joining organisations for co-operative purposes which involved no profession of Christianity.] But this was not all. In the scheme of the Association for Promoting Industrial and Provident Societies the characteristic feature of Sully's Constitution, the uniting in one body of the Promoters and the Associations, was given up. Side by side with the Associations there was to be an Industrial and Provident Societies Union . . .

((According to second chapter of these rules:-))

The functions of the Association are – 1. To exhibit as far as possible in its members, united in a living and organic body, the practical realisation of the principles which it exists to declare: – 2. To promote their principles by written and oral teaching, by friendly advice to, and intercourse with, all bodies of men and persons willing to help in the work, or in any portion of it; – 3. To afford legal and other assistance to all bodies constituted or seeking to constitute themselves under the Industrial and Provident Societies Act . . .

The President of the Association was elected for life, the first being Mr. Maurice; he chose the original members of the standing committees, and, with two assessors, (chosen by him out of seven members elected annually by the Council and forming with him the executive committee) exercised supreme executive authority in all matters not provided for by the constitution of laws of the association, and subject to the constitution and laws, in all matters whatsoever relative to the association. Members of the Association were elected by the Council for life . . . It was 'the duty of every member to help in the realisation of the principles of the Association, by all possible acts of self-denying fellowship'. The subscription being at least 1d per head per week; but members of a society or body contributing at least 1d per head per week were exempted from individual subscriptions. The Association was to meet three times a year, every meeting being 'succeeded, on the same or on the succeeding day, by a friendly tea-party or other entertainment, at which all members of the association, and of the societies and bodies connected with it, with their wives and families' were to attend as might be convenient.

The Council was to be composed first of the original members of the Standing Committees, secondly of members elected by the Council after having served on a standing committee . . . The Council was to meet monthly, the first meeting to take place in June 1853 . . . The Standing Committees were, however, the main feature of the association. These were to be small in number, – the Executive Committee, the Committee of social intercourse, the Committee of teaching and publication, the committee of arbitration, the legal committee, the finance committee, and the Loan and Banking committee . . . I do not know how or by whom ((these rules)) were drawn up, and am, therefore, entirely impartial in saying that I consider the scheme to have been an extremely able one. But in cutting asunder the Association for Promoting Industrial and Provident Societies from the

Industrial and Provident Societies Union, and leaving the connection between the two bodies a purely voluntary one, it reduced the old Council to a merely benevolent association, and thereby, I fear, took the heart out of its work . . .

The division of the Council into Standing Committees rendered it difficult henceforth to know what was being done in other committees than one's own, and deprived the Council of all real work. As far as I recollect, I was a member of the Committee of Teaching and Publication and of the Legal Committee. I continued to fulfil my duties in connexion with the new Association until its formal suspension of operations in November, 1854, feeling however more and more that I had made a mistake in returning to work within . . .

CHAPTER XX

The Christian Socialist Movement 1850-4 – The Working Men's College Founded – The Provinces – The Later Co-operative Conferences – The Dying-out of the Movement as an Organised Force

It cannot be too clearly pointed out that there is no formal 'solution of continuity' between the Christian Socialist movement and the Working Men's College, although as I have said, the latter virtually killed the former; so that Professor Brentano in his work on *The Christian Socialist Movement in England* is perfectly right in devoting his last chapter but one to the College. It was not raised upon the ruins of our association, for several of them survived its establishment and others were opened after. The feeling of a necessary connection between co-operation and education, as I have shown in a previous chapter, had always been with us. In the very first number of the *Christian Socialist* I had announced that the paper meant to deal with education. 'We shall all agree, probably,' I said, 'that our Universities must be universal in fact as well as in name; must cease to be monopolised for the benefit of one or two privileged classes; we may differ as to the means by which that monopoly is to be broken up, that universality attained, whether by lowering the benefits of university education to the reach of the many, or by drawing up to them the pre-eminent few of every class.' (I need not point out what vast strides have been taken in both directions since the above quoted words were published, from the establishment of University Extension Lectures to that of co-operative scholarships) ... Later on, the work of the

'People's College' at Sheffield had been brought to our notice by Lloyd Jones and the idea of something of the same nature, to be established for the benefit of London Working Men, had been mooted amongst us. Charles Mansfield's fertile mind in particular must have been stirred on the subject, for I find that I wrote to him in November 1852: 'I wish you were with us, to work out your own idea of a Working Men's College'. Mr. Maurice's dismissal from King's College accentuated, so to speak, the educational development of the movement, and as stated in the *Life*, as early as December 27th 1853 it was publicly suggested at a meeting of working men at the Hall of Association that instead of a Professor at King's College he might become the 'Principal of a Working Men's College' ...

By Jan. 10th 1854 he could already write to Mr. Kingsley of 'my College', so that on the next day a motion was made in the Council of Promoters by Mr. Hughes and seconded by Mr. Lloyd Jones for referring to the Committee of Teaching and Publication 'to frame and so far as they think fit, to carry out a plan for the establishment of a People's College in connection with the Metropolitan Associations', and that by February 7th Mr. Maurice had already drawn up an admirable plan in 12 pages headed 'College for Working Men', which should have been at least appended to his *Life*.

Historically, then, the Working Men's College arose out of the Christian Socialist Movement mainly through the success of the evening classes established in connection with the latter. Its formation was decided on by our Council; its scheme was drawn up by the Committee of Teaching and Publication of that Council. It was at first intended to be carried on as far as possible in connection with the associations. The following extracts from Mr. Maurice's plan will show this plainly:

> I understand the Committee (i.e. of Teaching and Publication) at its meeting of last Thursday to have agreed on certain maxims which they wished me to assume in any plan of education I might lay before them. 1 – They agreed that our position as members of a society which affirms the operations of trade and industry to be under a moral law – a law concerning the relations of men to each other, obliges us to regard social, political, or to use a more general phrase, *human* studies as the primary part of our

education. 2 – It was agreed that we were not bound to confine our education to our own associates, but that we should promote their interest better if we produced a scheme which should be available for the working classes generally. At the same time it was considered that we ought to do our utmost that the members of each association should as individuals enter into the general scheme, and so far as it was possible, that each association should be an integral part of it . . .

In the third of the lectures on *Learning and Working*, delivered in June and July 1854, mainly for the purpose of raising funds for the establishment of the College, Mr. Maurice refers in more veiled terms to the connection between it and the association movement, and claims at the same time for education a co-ordinate footing with the latter . . . In the same spirit is written a letter from him, undated, but probably of September 1854:

I have a letter from Hole (James Hole, then of Leeds) expressing his fears that if we take to education, we shall grow lazy and lukewarm about co-operation. The fact is that I feel it is the way to cure myself of laziness and lukewarmness. I cannot be interested in the mercantile part of the business till I feel that it has a moral basis; and I am satisfied that nothing but direct teaching will give it that character.

A few months later the lectures on *Learning and Working* were published (together with others on *The Religion of Rome*) and Mr. Maurice did me the honour of dedicating the volume to me. In the preface he refers to my first letter to him in 1848 as having had 'a very powerful effect' upon his thoughts at the time and given a direction to them ever since. He goes on to say:

We have neither of us ever doubted that the whole country must work for its blessing through the elevation of its working class, that we must all sink if that is not raised. We have never dreamed that that class could be benefitted by losing its working character, by acquiring habits of ease or self-indulgence. We have rather thought that all must learn the dignity of labour and the blessing of self-restraint. We could not talk to suffering men of intellectual or moral improvement, without first taking an interest in their physical condition and their ordinary occupations, but we felt that any interest of this kind would be utterly wasted, that it would do harm and not good, if it were not the means of leading them to regard themselves as human beings made in the image of God . . .

I felt at the time, and many years after I ventured to tell Mr. Maurice that I had felt the dedication of this book to me as the bitterest irony. Of this he was not in the least conscious, and I do not think he realised my feelings on the subject even when I told him of them. The fact is that, possessed as he always was with the burning desire to be doing with his might whatever his hand found to do, he was often not aware of the growing modification of his own purposes. Yet I think it is clear that this preface is written in a very different spirit even from his letter to me of September 1854 before quoted. Then, he looked upon the education of the College as the way to cure him of 'laziness and lukewarmness' about co-operation. Now, the 'interest' taken in men's 'physical condition' and 'ordinary occupations' (co-operation not being even named) is treated as a mere stepping-stone.

But it must not be supposed that the founding of the Working Men's College at once stopped the Christian Socialist Movement as the following ((section)) will show.

* * * * *

The proposed formation of the Working Men's College seemed at first only to promote association. More than ever, after the first Co-operative Conference, did it tend to spread throughout the provinces. Not that it had, as is often supposed, ever been confined to London. Even before the first conference it will be remembered we had affiliated societies in Salford, Liverpool, Southampton, Banbury. The *Christian Socialist* and *Journal of Association* had subscribers all over the country. A provincial society which for some time manifested considerable energy, the Bury General Labour Redemption Society, early adopted the *Christian Socialist* as its organ. Another provincial body, but of a professedly communistic character, the Leeds Redemption Society, figures also after a time repeatedly in the pages of our periodicals . . . Many stores throughout England and Scotland sent their reports and other communications. Then there were the lecturing and other tours of Walter Cooper and Lloyd Jones, whether for the Society or the Agency, the deputations to provincial gatherings and visits of individual promoters to provincial bodies . . .[28]

On April 18, 1851, there was an important conference at Bury of over 80 delegates from co-operative bodies, which was attended by Lloyd Jones and Woodin from the Central Agency, and by Walter Cooper, at which a resolution was passed 'that it would be advantageous and beneficial to the various co-operative societies if there were an unity of action established for the purpose of mercantile transactions', and recommending 'the establishment of a central trading dépôt' . . . The discussion betwen Lloyd Jones and Ernest Jones, the Chartist leader at Padcham, 28 and 29 Dec. 1851 should also be noticed. It terminated in an apparent victory of the Chartist leader, as a vote was called for, and only his supporters held up their hands, but it is sufficient to glance through the discussion to see how completely the weight of argument lay with Lloyd Jones.

The mention of Mr. Ernest Jones recalls however what I might perhaps call the reflex action of Christian Socialism on Chartism. I have already stated that several of the working men who became connected with our movement had belonged to the Chartist body. The programme of the Chartist Convention, April 10, 1851, became under the head 'Labour Law', distinctly socialistic, claiming registration for 'all Co-operative Associations for industrial purposes' with power 'to possess an unrestricted number of affiliated branches', and such an alteration of the law of partnership as to 'remove existing difficulties in the way of association.' It went on no doubt to lay down that 'all future attempts until the complete readjustment of the labour question, be modelled on a national basis, and connected in a national union . . . and that the profits beyond a certain amount of each local society should be paid into a general fund, for the purpose of forming additional associations of working men, and thus accelerate the development of associated and independent labour', a scheme which, however desirable its object might be, went far beyond our humble attempts; and also to desire that 'a credit fund be opened by the state, for the purpose of advancing money, on certain conditions, to bodies of working men, desirous of associating together for industrial purposes', which we should not have asked for. But I certainly think now, as I thought then, that the substance of the suggestions as to 'Labour Law' in the programme was 'borrowed from the practice or the efforts' of our Society.

The second Conference of delegates was held in Manchester in August 1853, in pursuance of the resolution adopted at the London one. It was attended by Mr. Maurice and others of the promoters. I did not attend it, partly because, having declined to sit upon the Committee by which it was convened, I should have felt in a somewhat delicate position, but mainly because, having by this time spent all my slender amount of spare cash upon the co-operative movement, I could not afford to travel. The Conference adopted the three objects of the Association for promoting Industrial and Provident Societies, without reference to the preamble. All formal connection of the association movement with Christianity was thus severed. In pursuance of a resolution of this Conference, the *Co-operative Commercial Circular* was established as the organ of the Executive Committee appointed by the Conference, a certain space being allowed to the *Gazette* of the Association for Promoting Industrial and Provident Societies, which thus practically sank into the background of the Co-operative movement. The first number of the *Circular* was published in November 1853. [This, with its successor the *Co-operative Circular* is, I believe, the rarest of all publications connected with or issuing out of the Christian Socialist movement. I seem to have ceased taking it in from the time when, as will presently be mentioned, the Association for Promoting Industrial & Provident Societies suspended its meetings, and consequently had no more *Gazette* to publish. I am very sorry now both not to have continued to subscribe to, and not to have retained a complete collection of the *Circular*, as it was the only contemporary record of a very obscure period of the co-operative movement started in 1850 . . . I am inclined to think the *Co-operative Commercial Circular* was at first edited by Lloyd Jones.]

On looking through a few numbers of the *Circular* for 1854, I find (July) that a 'co-operative company' had just then been started at Barrowfold, – that some Birmingham tailors had 'resolved on making an attempt to establish an association' – that the question was being mooted among the Dublin tailors, who had invited over the manager of the then prosperous Liverpool Tailors' Association, – that the Amalgamated Society of Engineers had voted £360 to the London Cork-cutters, and offered a prize of £15 for the best Essay 'on the best method of applying the Society's funds to co-operative purposes,' – that the Liverpool Ship-builders' Association had now a paid-up capital of £1,000, and had made their first purchase of a small vessel for £260, – that the Association for Promoting Industrial and Provident Societies had held its last monthly circular meeting (the meetings were now held, or intended to be held, at each association in turn) at the Westminster Bridge Road premises of the Working Tailors' Association. But each association was getting more and more wrapped up in itself . . .

The Co-operative Conference for 1854 took place at Leeds, 21 August, and was again attended by Mr. Maurice, [W. E. Forster, I find from a letter of Mr. Maurice's, was also present] who read there the first (and last) report of the Association for Promoting Industrial and Provident Societies. This was published in the *Co-operative Circular* for October 1854, and ((was)) virtually the last official act of the society before it performed the 'happy dispatch' upon itself . . .

On the 30th October 1854 Mr. Maurice delivered the inaugural lecture of the Working Men's College in the large Hall of St. Martin's Hall to about 1500 hearers. The last ordinary meeting of the Association for Promoting Industrial and Provident Societies was held on the 15th November. A special general meeting was held ten days later, November 25th and adopted the following report of the Executive Committee which I copy *in extenso* from the Co-operative Circular for December 1:

A question having arisen, whether the existence of the Committee[30] appointed by the Co-operative Conference does not make this Association useless, and whether we ought not to resign our function to them, – the Executive Committee of the Association has been desired to prepare a report on this subject.

We are of opinion that two bodies, one of them chosen by the

Co-operative Societies, the other self-appointed, cannot undertake the same functions without the risk of frequent clashing and of both becoming inefficient.

We are of opinion that the arrangement adopted last summer, of making the Conference Committee one of the Committees of Association [My memory is a blank as to this.] is not likely to work well, seeing that it does not help them to understand that they have any distinct functions, and that the subordination of a Committee, which is already responsible for its acts to one body, to another must be more nominal than real.

We are of opinion that the proposal which has been made that the Association should vindicate a special office to itself by undertaking an active proselytism among the Trades' Societies would lead us into a kind of work for which we are not adapted, and which would suggest the thought to working people that co-operation is a theory which we wish to advance, not a principle which they must work out for themselves, if it is to do them any good.

We are of opinion, nevertheless, that it is not desirable to dissolve the Association, seeing that by keeping up its different committees we testify our adherence to the principle, we pledge ourselves to advance it in all ways that seem to us right and reasonable, we tell the working people where they may obtain legal or other advice if they want it, and think we are competent to give it; we hold ourselves ready to promote social intercourse among them, and to teach them so far as they like to learn.

We are of opinion that though the functions of the Teaching Committee[31] so far as the Co-operative movement is concerned, ought to be suspended while the *Co-operative Circular* occupies that field, the College which the Teaching Committee has been the means of establishing may promote the moral ends we have in view, and the general education of the working men, more effectuatly than they could be promoted by any lectures, tracts or newspapers we might open.

We recommend therefore that the Association should continue to exist, but that its meetings for business should be suspended.

The above report, which I believe to have been entirely drawn up by Mr. Maurice, was adopted, as well as a resolution moved by Neale that it was 'a desirable object for the Association to give a greater development to its plans for social intercourse, and teaching in connexion therewith, – that it is referred to the Social Intercourse Committee to form a plan upon the subject', and 'that a special meeting of the Association be held on a day to be appointed by the Executive Committee before the 1st of February 1855, for the purpose of considering the proposal.' I have no recollection of any such meeting, or of any further proceeding connected with the Association. I do not indeed understand how its Committees could have continued to act when the Association itself had suspended its business meetings. Practically speaking, the adoption of the above report was the last act of its existence. Still, it is a remarkable fact that it was never dissolved, and may be said yet to subsist in a state of suspended animation.

The ground assigned, it will have been observed, for the suspension of the Association's business meetings was the co-existence of the Co-operative Conference Committee. This I had foreseen, from the first, refusing to sit upon the latter. In years to come, I felt no further reason for refusing to sit on the Central Co-operative Board, the existence of the Association for Promoting Industrial & Provident Societies being then only a dim memory. It is indeed obvious that the co-operative movement could not have been directed from two centres at once, and that the chief area of practical work being in the North, in the North must be the true centre. Still, had the original principle of the Society for Promoting Working Men's Associations, that of binding associations and council in one whole, not been departed from in the new organisation, it might have continued to be represented at the yearly Co-operative Conference and on its Committee, to the wholesome strengthening of the moral element in the work. All one can say is, – it was not to be.

Yet the very number of the *Co-operative Circular* for December 1854 which records the abdication of the functions by the Association for Promoting Industrial and Provident Societies speaks of a Working Goldsmiths and Jewellers Association of which a prospectus had been issued, which proposed 'to hold a large meeting of the trade' in the Hall of Association, and to which Mr. Vansittart Neale proposed to give a lecture on co-operation, of a new Silk Weavers Association formed in Bethnal Green, which had goods on hand already.

It contains favourable accounts by Lloyd Jones of the Manchester Co-operative Hatters and Tailors. It contains an address and remarks of the Executive of the Conference Committee breathing anything but despondency; the usual market intelligence of the Control Co-operative Agency, on behalf of which Lloyd Jones was touring in the provinces; the advertisement of a newly formed 'Co-operative Freehold Land, Building and Investment Society, for the purchase of lands, the erection of dwelling and workshops, the establishment of industrial colonies, and the investment of savings', enrolled as a building society, with 1/- entrance fee, £30 shares at 5/- a month subscription, entitling members to purchase, the President being Lord Goderich ...

The Working Men's College thus in no wise directly killed the co-operative movement which grew up in 1850. But it withdrew from it the greater part of the moral and spiritual influences which had hitherto been brought to bear upon it, and thus caused it to die out.

Was this a necessity? I will not shrink from saying that I do not think it need have been, had the College been planned on a less comprehensive scale, and connected with the movement, as an institution directly established for co-operators and to educate men into co-operation. No doubt very few of our associates joined its classes, but gradually, I think, more might have come, and if connected with a day-school for their children I think a permanent connexion between co-operation and education might have been established. But the admission into the council of the College of outsiders to the movement such as Grant Duff, Fitzjames-Stephen, C. E. Pearson, who in offering to lecture at the Hall of Association had expressly stipulated that they were not to be considered as accepting our social objects, rendered it necessary for us at the College to put our Christian Socialist principles more or less into our pockets. Had we had no such colleagues our scheme of teaching must have been less comprehensive, our list of Council members would have been shorn of many distinguished names, our attendance of scholars much more limited, our progress much slower and more difficult.

But the thing was not to be, and therefore, I suppose, ought not to have been.

By the end of 1854 I could no longer blind myself to the fact that Mr. Maurice was turning away from Co-operation. Nor could I conceal from myself that I had unwittingly been at least the main occasion of his doing so. For, looking back, I could see that those *Thoughts on the present condition of the Council etc.*, intended for all members of the latter, but which Mr. Maurice had requested me to withhold from circulation, had had the effect of driving him back upon education as the chief work for himself and his followers. It was because, on reading my paper, he then and there despaired of reviving what I may call the co-operative impulse, pure and simple, that he had sought to suppress the paper, in order that the initiative in the educational line might be taken by means of the Lectures and classes. I own I now bitterly regretted having deferred to his wish. If the Christian Socialist movement was to be deflected, or at all events narrowed to a merely education channel, I could have wished that my protest, if I may so call it, in favour of co-operation had been in the hands of any one of my fellow-workers.

Mr. Maurice himself at the time evidently did not feel, nor did he ever understand, even when many years after I told him what I had felt myself, the crushing nature of the blow he was giving me. For to me the very bond of our friendship lay in the resolve to Christianise Socialism. For that I had virtually risked if not sacrificed everything, – much more than I had told him of. I had wished for nothing, as I have said before, but to be his first lieutenant in the campaign, merging my work in his, never coming forward but to ward off from him a blow if I could do so. But I saw that I was myself in fault; that I had wilfully blinded myself; that the Maurice I had devoted myself to was a Maurice of my own imagination, not the real Maurice. He was not to blame; I was.

But now came the question – what was I to do? Was I to separate myself from him? To pursue without him the task to which I had thought to set myself? Could I take the leadership in the Christian Socialist movement? I had been ready five years before to give up everything in this country and go over to France, to do battle single-handed on behalf of the cause. Was it not my duty now to carry on the warfare in my own country where that cause had friends, supporters, an organisation? I had great searchings of heart on the matter. I decided at last that I could not do so; that apart from want of adequate means, I had not sufficient capacity, intellectual or moral, for leadership, especially in Mr. Maurice's place. Moreover, I had home duties. My mother was beginning to feel the weakness of age; my conscience smote me already for having left her too much alone for the sake of my work; indeed, in the spring of 1854 I had already, partly for her sake, left London for a house which I had built at Wimbledon and the leader of such a movement ought to live in London and have his hands free. So I decided, crushed in heart, to follow Mr. Maurice into the Working Men's College scheme.

Fortunately for me my own internal conflict was over when several members of the Council came to me, asking me to take the lead of the movement, since Mr. Maurice was clearly throwing the work of the Association overboard. I answered them that I did not feel myself fit for the task and could not undertake it; that Mr. Maurice was much wiser and greater than all of us put together, and that we had better follow him still.

Did I decide rightly? To this hour I cannot say yes or no to the question. It may be that God would have given me strength for the undertaking and opened to me ways for carrying it out. It may be that I should have only marred the work by the exhibition of my own deficiencies, moral and intellectual. But at any rate my life was in one respect marred from henceforth.

For this conviction was forced upon me. If I was not fit to lead in the one cause on which I had set my whole heart, I was fit to lead in nothing. I had hitherto been zealous in beating up recruits for that cause, deeming it the best one to which they could devote themselves. I could have none but a factitious zeal in pressing them to join a second-best cause. Whilst doing still with my might whatever my hand should find to do, I must be in the main *vox, et praeterea nihil*.

To that conclusion I have been true through life. I have never since sought to draw men round me. More than once, when I have perceived that some one was disposed to think too much of my judgment, to rely too entirely upon me, I have repelled him, thrown him forcibly back on himself, whilst pointing him to the source of all wisdom. When I have differed in opinion from others whilst engaged in a common work, I have scrupulously abstained from seeking to influence any who were disposed to agree with me . . .

I have been speaking hitherto of the Christian Socialist movement from the inside. It is well to note how friendly outsiders have judged it. The late Professor V. A. Huber in his *Reisebriefe* . . . written in 1854 noted that by this time the promoters had become dissatisfied with the result of their undertaking. Dr. Brentano goes on to say:

> And had the leaders of the Christian Socialist movement not been so deeply earnest in their conviction that a social reform was only possible through the moral regeneration of the individual on the basis of the subjection of all selfish impulses under the Christian's love for his neighbour and through the realisation of this teaching in all branches of social and economic life, had they in their movement had in any wise before their eyes the satisfaction of their own vanity or other self-seeking impulses, they might really have felt satisfied. For in the short time of their agitation a comparatively large number of productive associations and co-operative stores had grown up in London and in the provinces as the immediate consequence of the impulse given by them, and through their support; for the other associations also they had become helpers and counsellors; beloved and honoured by the whole working population they stood at the head of a daily growing movement of association, and in the higher classes more and more from day to day were passing away the prejudices which had been excited against associations.

One word more before closing this chapter. I have said ere this that I think Mr. Maurice was obeying a true instinct in withdrawing from the leadership of the Co-operative movement. But I am fully convinced – I wish to assert most strongly – that from the principles underlying that movement he never swerved a hair's breadth. Take this splendid passage from his Lincoln's Inn Sermons (vol. III p. 258) delivered in 1858:

> Either the Gospel declares what society is, and what it is not; what binds men together, what separates them or it has no significance at all . . . For the Bible does not accept that position which men have courteously assigned to it. It does not profess to

provide rules or comforts for men as individuals, and to leave their condition as portions of a community out of its calculations. It speaks from first to last of a kingdom. It promulgates a social law. It tells all men that this is the law which makes communion among men possible; there is positively no other . . .

It is just because I feel that from first to last he was essentially a Christian Socialist that I cannot help regretting his having turned aside altogether after 1854 from the co-operative movement. Had his voice continued to be heard at Co-operative Congresses, I don't think that movement could ever have fallen so much into the miry rut of mere commercial success as it has done.

Not indeed that the impulse given by Christian Socialism to the Co-operative movement has ever fully died out of it. Spiritually indeed, the Christian Socialist movement never died out. The literature which it inspired cannot be limited to a few tracts and pamphlets. But it leavens, if it does not altogether comprise, the whole work of Kingsley and Hughes, and *Tom Brown's Schooldays* belongs to it as truly as *Yeast* or *Alton Locke*.

Its influence again is alive in the co-operation of today. T. Hughes, in his Preface to the *Manual for Co-operators* has pointed out that 'the definition of the objects of the existing Co-operative Union, as stated in the first of its rules and orders, to which every member assents by joining it' (though too often he seems to forget most of it the moment after) 'is but a partial application of the principles laid down' (not indeed, as he states, by the Society for Promoting Working Men's Associations) but by 'the Association for Promoting Industrial & Provident Societies. Those objects set forth a moral basis for the practical working of co-operation instead of the purely mechanical one of the older Owenite Socialism. And however false to the principles so formulated may be the practice of any society which is a member of the Union, the principles are there to protest against the practice, to supply a standard by which that practice may be judged and to which it may be stretched to condemn any show of success which may be attained by departing from those principles, and if failure afterwards occur, to disengage the responsibility of the Union and draw fresh lessons in support of its principles from every such failure . . . '

In later days again, the designation 'Christian Socialism' was revived by an energetic High Church clergyman, Mr. Stewart Headlam, and a second *Christian Socialist* paper issued by him for some years, whilst the Christian Social Union is, within the Church of England, the representat ve morally of our old Association, with larger aims, but a less comprehensive constitution. To the latter I belong, but whether I shall live to reach it in order of date in these reminiscences I cannot tell.[32]

Appendix
The Co-operative League

Before quitting the subject of the Christian Socialist movement something should perhaps be said of the 'Co-operative League', a body substantially representing the *Leader* coterie of Communistic anythingarians or nothingarians, including Mr. Thornton Hunt and Mr. Holyoake, and which it is difficult for me to believe was not intentionally founded in opposition to the Christian Socialists. Its objects, as stated in the *Journal of Association* Feb. 2, 1852, were:

> First, to place those who entertain, or were inclined towards the principles of co-operative association in direct, frequent and permanent inter-communication, so that they may acquire a positive knowledge and a clear comprehension of each other's views and opinions, and thus be enabled to unite and form an active centre for the diffusion and propagation of co-operative principles. Second, to collect books, papers, documents and facts, and to communicate all kinds of general information which may either be interesting or useful to the advocates of co-operative association.
>
> Third, to enter into communication with any co-operative societies throughout the country, with a view to promote the objects and to extend the sphere of action of the League.

The League met quarterly, and published *Transactions*. I have those for May, July and October 1852, but nothing beyond.

There was nothing in those objects antagonistic to our society, but the ignoring of it and the general scope of the league appeared to most of the promoters to imply a purpose supplanting its efforts by those of a non-Christian Socialism. We had already, as we conceived, in our Society, 'an active centre for the diffusion and propagation of co-operative principles' such as the Co-operative League professed to be, and though invitations were, I believe, sent to all members of the Council to join it, few of us did. Neale formed a marked exception . . . and was followed, naturally enough, by the former Owenites among us, Lloyd Jones, Shorter, Walter Cooper, and also by Lechevalier. The League issued queries to Co-operative Societies throughout the country for collecting statistical information, as our Society did. The inevitable results followed, – on the one hand, a confusion in the minds of many co-operators, favoured no doubt by the occurrence of several names on the Councils of both bodies, between the League and our Society, – on the other a feeling of worry on the part of some stores at being asked for the same information from two quarters, which led to their refusing it to both. There was indeed never any actual conflict between the two bodies, and indeed our later conferences were at least proposed to be, I forget whether they were, joint conferences of both. I cannot say when the League ceased to exist, and do not think it long survived, if it did survive, our society. Substantially, I don't think it ever did much mischief, beyond contributing to the growing bewilderment of our associates and of co-operators generally since the establishment of the Central Co-operative Agency as to what we really were and meant. I cannot say whether or not it did any good.

CHAPTER XXI

The Working Men's College

The list of lectures and lecturers given at the end of the Preface to *Learning and Working* includes, it will observed, ((many)) of the old promoters, besides Mr. Maurice himself . . . Among the names of those who had more recently come about us . . . in the drawing class, are those of Ruskin, who after always keeping aloof from the Christian Socialist movement whilst it was fighting its fight and receiving knocks and blows from all quarters, was to take up some of its ideas and illustrate them with a perfection of language which would seem in the eyes of the public to make them his own, – of Rossetti whom he had patronised as against the critics, and whose devotion to him at that time was that of an almost adoring disciple, and of Lowes Dickinson, Archie Campbell's friend, who had by this time been brought altogether to Christ through Mr. Maurice's teaching, and who was then, and has always since remained, one of my most valued and dearest friends.

Let me say at once that the Law classes and Political Economy classes were dead failures, and were not again attempted. T. Hughes, who failed to draw an audience to Sanitary Legislation, achieved a little later immense success with a boxing class, which used to be held in the garden of the college till the weather drove us to construct a shelter for it. There was always a steady attendance at the Mathematical classes, and Brewer's History and Geography classes soon came to be recognised as among the most interesting and instructive of any, – the lecturer, notwithstanding his sturdy Toryism, soon winning the thorough regard and affection of audiences mostly Liberal and something more. But the drawing classes, so long as Ruskin presided over them, were the most popular of any. I joined them as a student, and worked on till one summer, when I fairly broke my heart in trying to render a Toby Philpot jug in all its detail of form and colour, and never went back again. [I may be uncharitable, but I have since suspected that this was the very purpose for which he had set me the task. I must say that I consider it an absolute waste of time for anyone to paint a Toby Philpot jug. And Ruskin may have felt it inconvenient to have constantly in his class an influential colleague on the Council, though God knows that spying upon him was an idea that never entered my head.] Ruskin in his class was charming. So long as you were towards him in the position of a disciple, – so long as you were in the looking-up attitude, nothing could be more interesting than his conversation, more winning than his whole demeanour towards you. But you must walk on his lines, do as he bid you. I remember the jobation he gave a young fellow, one of his good pupils, who in the coloured study of a mossy branch of a tree had introduced out of his own head at the joints, little chiselled figures such as Dicky Doyle would have put forth, – such as Ruskin himself praised when they were Doyle's – some of them exceedingly quaint and pretty, and yet so subordinate to the general truth of the thing that one did not notice them at first, and that they had to be pointed out to Ruskin himself by an injudicious admirer. It was the only occasion on which I ever saw Ruskin cross in class. Without giving the young fellow the slightest credit for the real delicacy of his fancy, he rated him fiercely for daring to think that he could improve upon nature. And as far as I can recollect the effect of the jobation was that the young man soon after gave up the class. Ruskin's influence was indeed so great that with a single disparaging remark he simply killed a modelling class formed under T. Woolner, the future Academician and which had reached the number of 17, I think, all, Woolner told me, working most satisfactorily, and some showing great promise. Woolner was bitterly disappointed, and ceased to work at the College. Had he remained associated with us, I think it might have tempered a certain arrogance of manner and tone which spoilt him in later years.

Rossetti's class was the advanced one in painting and I never joined it. I saw however something of him, and went to his studio to see his paintings. The only one which really struck me was the then unfinished one of the countryman meeting at early morning near London Bridge, beside his cart laden with straw, his old love, now alas! on the London streets. There is no doubt a singular richness and beauty in Rossetti's colouring, quite recalling that of some of the early Italian masters. And there is always thought, and mostly poetic thought, in his treatment of a subject. But he has never painted one single picture that I should call a satisfactory one, and his conception of female beauty is to me simply morbid. The man himself, slight, weakly, narrow-browed, had to my mind not an element of greatness about him and when I read how Sir E. Burne Jones came and sought him out for sheer admiration, at the Working Men's College, I can only deplore that he looked no higher for someone to admire, since although possessed himself of far greater power than Rossetti, that power seems to me to have been always marred by the morbidness which Burne Jones sucked in from the half-educated artist his master . . .

Three friends who had lectured for us in the Hall of Association, whilst disclaiming sympathy with our principles, ((were)) Grant-Duff, Fitzjames Stephen and Charles E. Pearson, though the last named soon left for Australia on account of his health. They were followed by another group of three, the Lushington twins, Godfrey (now Sir Godfrey) and Vernon, and Frederic Harrison . . . Through them I came into contact with the Positivist school, though indeed Godfrey, I think, never went beyond accepting the philosophical principles of the matter.

Their friend ((R.B.)) Litchfield was longer-headed than either, though he wanted the warmth of sympathy which made Vernon so lovable . . . He was an excellent mathematical teacher, and certainly made a first rate Treasurer to the

College . . . He married, as is well known, a daughter of Darwin's. He never went over with Vernon Lushington into Positivism, and although professedly an Agnostic, admitted that some of us 'had something which he had not got, and could not get'. His appreciation of Mr. Maurice was always most cordial . . .

A most touching instance of Litchfield's goodness is reported by Arthur J. Mun in the number for March 1903 of the *Working Men's College Journal*. Now, when about five and twenty, he rescued two young girls casually met from practically certain wrong and shame, giving them shelter in his own chambers in the Temple and sending them back at his own expense to their parents – an act at once of the noblest chivalry and of the most truly christian charity, on the part of one who seemed to be most matter of fact, and did not, as I have said, profess to be a christian . . .

With Frederic Harrison, as also with Godfrey Lushington, I was afterwards a good deal connected in Social science and Trade Union matters. Harrison was a hard-working and by all accounts most brilliant teacher of history at the College. I still come across him at very long intervals at the Athenaeum. [Of late years Harrison has made a most serious and inconceivable blunder in one of his works, connecting William Morris with the foundation of the College. He had no more to do with it than the Dalai-Lama of Tibet and never entered its door till 17 years after its foundation, to be present at a Saturday evening's discussion of his peculiar socialism.]

Hansard was at first on the Council, and so for many years was Llewelyn Davies (latterly Vice Principal) who indeed had joined our little band before the establishment of the College. Davies's *début* amongst us had been a curious one – by means of a letter in the *Spectator* flatly contradicting Kingsley as to the state of mind at Cambridge. He has been now for many years, if I may use the expression, the most authoritative exponent of Mauricianism. Had he been made a bishop 10 or 15 years ago, as he fully deserved to be, I suspect he would rather have astonished the House of Lords. He has of late years had to undergo the worst of all bereavements, – the sudden death of a wife who had been to him a most perfect helpmate. He can never be the same again . . .

The College had at first rented a house in Great Ormond Street. Funds were raised to buy this, and it became needful to give the College a legal constitution, to enable it to hold property. This was in the year 1857, before limited liability was legally allowed. The government of the College was at this time solely vested in the Council of teachers. A scheme was devised for creating a Company of which the Board of Directors should be the Council, without imposing on its members any liability to take shares, but as it was necessary to have shareholders, a few bold spirits (Hughes and I amongst the number) were found to take shares, subscribing just enough for legal expenses, but on the clear understanding that by so doing they placed their whole fortunes at the disposal of the Council. To my disgust, I own, four members of the Council (who did not so subscribe) resigned their place on it, for fear of the responsibility which the Directorship of the Company might cast upon them and did so without suggesting any better plan. The only one of these who could assign, it seemed to me, any valid reason for so doing was Harrison, who had made a promise to his father never to become connected with any company. I mention this matter because it is incorrectly referred to in Mr. Maurice's life. It is not the case that 'many of those who had been working heartily up to this point disapproved of this arrangement and threatened to have no more to do with it' as stated above; there were but four, [Grant Duff, Fitzjames Stephen, C. E. Pearson and F. Harrison.] of whom three actually did cease to have anything to do with the College but the fourth, Harrison, whilst receding from the Council, continued to teach regularly and most successfully. And instead of its being the case that 'ultimately a scheme was devised for establishing a company in connection with and subordinate to the College', this was the very scheme that was devised and objected to from the first. It was carried out and remained on foot till the Limited Liability Act, when a regular Limited Company took its place.

Another member of the Council who became very prominent and – as I have said before – was latterly a terrible thorn in Mr. Maurice's side, was Furnivall. I believe that there are few men who, meaning so well, have done more mischief. His character is the very one which Solomon must have had in his eyes when he wrote those marvellously deep and true words: 'Seest thou a man wise in his own conceit? There is more hope of a fool than of him.' . . . He worked very hard at the Working Men's College, and at the same time took a prominent part in all social gatherings, and made himself very popular . . . Furnivall had always been a man greatly devoid of tact and not to be taken into one's confidence in any matter that one did not wish to blazon abroad. But the 'wisdom in his own conceit' growing on him more and more, he became a person with whom intimacy was to be altogether avoided. He is the 'Q' whose articles in the *People* on the Sunday question and incitements to the students to select Sundays for geological excursions etc., made Mr. Maurice think of resigning the Principalship[33]. Then he, an unmarried man still young, 'adopted' a girl some years his junior, and used to take her with him to College and other functions. Of course after the lapse of a few years this impossible relationship ended in marriage, and this, after the birth of children, has been followed by separation. After Litchfield ceased to take an active part in the work of the College, Furnivall ((later)) took a lead in the promotion of its sports, founding for instance a 'Maurice Rowing Club' – a title at which, coming from him, I was much surprised, as his treatment of Maurice had been latterly of the most *de haut en bas* character, nor was I less so when after Mr. Maurice's death he voluntarily joined the F.D.M. club. Here, I consider, he generally acted the useful part of a 'foolometer', though on one or two occasions by the rarest of accidents[34] he actually made a sensible speech. [The opinion has however been expressed to me by a former member of the Club, that eventually he killed it. I had myself had to give up attending the meetings for some time before its extinction, owing to my wife's ill health, so that I can form no decided opinion on the point. I should not however be surprised at such a result. If I recollect, the last meeting I attended discussed Shakespeare's sonnets, a subject interesting in itself, but quite outside the scope of an F.D.M. Club. And I was a good deal disgusted with Furnivall's speech on the occasion.] And yet, after all I have said, the man remains an enigma to me, and I cannot help hoping that at the core of all his conceit, wrong-headedness, agnosticism, self-will, there is the spirit of a true man, which God will at the last, in His own way, purify unto himself. I need hardly say that for many years now he has mainly devoted himself to philology and has done a few useful things. His knowledge of the early literary and social history of our country is considerable, and much that he has written, so far as it is not disfigured by his conceit, is interesting and valuable. Among the students at the College he was certainly (and is still) very popular, and I am bound to say, through his ready kindliness, has deserved to be so.

I did not myself do much work at the College. I got no pupils in Political Economy. I carried on for some time an English class for Frenchmen which was at first well attended (by practically none but political refugees), but which I suppose I did not succeed in making interesting, as my pupils all gradually dropped off. [My mistake was, I believe, to have chosen the Bible for a text book. I had done this because I knew some of my pupils to be almost starving, and I could think of no cheaper book or more easily to be borrowed.] On the other hand I delivered some courses of lectures on special subjects which were, I think, found interesting judging by my audiences. One of these the students asked to be redelivered, but instead of this it was expanded into the work *British India, its Races and its History* (2 vols.), published by Macmillans in 1858 and dedicated to 'the Students of the Working Men's College'. A few years later, three lectures by me on the 23rd and 30th August and 9th November 1861, together with one by T. Hughes, were in like manner expanded into a *Sketch of the History of the United States from Independence to Secession* followed by *The struggle for Kansas* by T. Hughes (Macmillans, 1862). Again, when the *Idylls of the King* first came out, I read them to a class at the College, prefacing the course with some words as to poetry generally (which my friend Rossiter told me first interested him in the subject), and on the growth of the epics of chivalry. Out of this grew my *Popular Epics of the Middle Ages of the Norse German and Carlovingian Cycles* (Macmillans, 1865).

When in 1859 the *Working Men's College Magazine* was set up, [Three volumes of this only were published, 1859-60-61.

It has however been succeeded by a *Working Men's College Journal,* conducted altogether by students or ex-students.] I was asked to be the editor. It was at first the clear understanding that it was to be a mere record of facts and documents relating to Working Men's Colleges, and on this footing I edited three numbers. But as such it was found dull and the proposal was made to introduce original matter, which I should only have been glad to have done from the first, whilst it was hinted that the editor should be a person taking an active part in college work. I resigned of course the editorship, and Litchfield took my place. I contributed a few letters to the magazine under his editorship, either in my own name or as 'Jonathan Dryasdust', and one paper on volunteer matters, the title of which, altered without notice to me, was a complete misnomer, as implying that I was a member of the College corps (the 19th Middlesex) which I never was.

Besides the new friendships among the teachers which I owe to the Working Men's College I was enabled to form others among the students. It must be observed that the College was never sought to be limited to working men. It was thought to be sufficient if it were devised to suit working men's needs and made open profession by its title of doing so. In point of fact, I believe, the average of pure working men has always been one third of the total number of students. Many others would also be in what may be called the working man stage of the lower middle class, i.e. young men working in their fathers' shops; the remainder being assistants in shops, clerks, and in the art classes etc. men of a still higher social grade . . .

Among what I may call the genuine students the one who shot ahead foremost was Wm. Rossiter, the son of a portmanteau-maker, who worked in his father's shop, – a very small man, brimful of that nervous energy which seems to circulate with greater rapidity in the small-sized. He was the first to pass an examination as 'Associate' (October 1856), and would have been the first Fellow, but that he left London early in 1857 . . . He soon felt that his vocation was teaching, and above all the teaching of children and one of the members of our Council, H.P.P. Crease (afterwards Sir

Henry), wanting a master for a school at a mine in Cornwall, took him on my recommendation. From that, after passing the necessary examinations, he proceeded to a National school, and fell in love with the school mistress, an admirable woman who became his wife ... He then tried his wings for a higher flight, – the establishment of a 'South London Working Men's College', of which he was secretary and soul. But owing to the poverty of the neighbourhood and the much lower intellectual standard of its inhabitants, it was never more than a school, and when the Board school system was established ceased to have any *raison d'être*. Then its books were shifted to Battersea, where it became a free library, before the days of the Libraries Acts; and eventually was replaced, in South London again, by the South London Art Gallery, to which a library was added later on. But Rossiter's energies were never shut up within the four walls of any building. At one time he poured forth a series of cheap manuals on various scientific subjects, besides a *Dictionary of scientific terms*. He wrote in the *Inquirer*, in the *Spectator*; he and Mrs. Rossiter were, I believe, the very first to initiate the plan of giving country holidays to poor children ... He met the noted Secularist lecturer G. W. Foote in a public discussion, nor did the latter care to meet him again. At the Art Gallery he was in the habit of delivering Sunday evening lectures on various subjects and publishing 'weekly notes' ...

To revert however to the early days of the Working Men's College. Next to Rossiter, the two students who first shot ahead were two intimate friends, John Roebuck and George Tansley. Roebuck was a wood turner, Tansley an assistant. The former was married, the latter at first not, and a droll story was told, I forget whether by himself or by someone else, of Roebuck's having, when first taking up geometry as a study – worked out theorems on the floor of his room by using wife and child as A and B points. As his family increased the pinch of insufficient earnings led Roebuck to emigrate to America; Tansley – now dead alas! – always remained connected with the College, of which it would perhaps not be too much to say that for years he was the chief moving spirit ...

The most cheering feature of the Working Men's College and one which distinguishes it to this day, was the eagerness of the students to impart the knowledge which they acquired to their juniors. As soon as a student had received a certificate in most cases, he was put on to a preparatory class. The mastership of the adult school was never, I think, held but by a student. Since Shorter's retirement the same has till lately been the case with the Secretaryship. The Finance Committee has been always, I believe, mainly composed of students. I do not believe that the devotion of a Resident Fellow at one of our older Universities to his College can surpass – I doubt if it equals – that of many an ex-student to the Working Men's College. He may have found there for the first time not only the means of instruction but congenial society. And these he may enjoy throughout life. Besides the common room, all sorts of clubs and societies now cluster round it, a gymnastic society, an old students' club, cricket, boating, cycling, chess and draughts, Field clubs. The type of such a lifelong student was dear old Samuel Standring, tradesman and vestryman. He simply loved the College with all his heart.

I own indeed to a little surprise that the College students should not yet have sent more distinguished men into the world. Professor H. G. Seeley in the field of science, Mr. James Rowlands, at one time M.P., in that of politics, are I think the two best known to the public. It has certainly none who has attained the prominence of Sir Edward Clarke, a former student of the City of London College. [The City of London College at its establishment followed by a few months the Working Men's College, much as the Ladies (now Bedford) College followed Queen's College.]

Something however must also be said of the successive Secretaries of the College. Of dear Thomas Shorter I have spoken already. He held the post till his failing eye-sight rendered him incapable of fulfilling his duties, and was succeeded by C. E. Rawlins, whom I have already mentioned in connection with the co-operative disaster of the sale for waste-paper of the records of the Society for Promoting Working Men's Associations and the Association for Promoting Industrial and Provident Societies ... He was succeeded by H. R. Jennings, a valued friend of mine, originally a working gilder ... He was for several years teacher of the Adult School in connection with the College ...

(Except for changes of tense, I leave the above as it was written. There is a sad sequel to the notice. One fine day, Jennings disappeared from the College, but was seen by one of the students while going home on the top of an omnibus, in the midst of a group of disreputable-looking men. His wife instituted all possible inquiries after him, and at last tidings were obtained of him from somewhere in the North. It appeared that when young he had some love affair ending rather tragically, thus establishing an anniversary the recurrence of which always tended to affect his mind; so that seven years before he had already disappeared on it, though only for a day or two. Of course, under the circumstances, although his accounts were in order, he could not retain the Secretaryship of the College, and he had to fall back upon his calling as accountant, in which he continued to exhibit his old care and accuracy. But the sense of having disgraced himself with the College preyed upon his mind, and he died within a couple of years. I met several of his old College friends, such as Tansley and Marks, at his funeral in Kensal Green Cemetery. I never knew of a sadder end to a useful and honourable life).

I cannot now (Nov. 1900) recollect how long I remained a member of the College Council. But I believe I ceased to attend its meetings some time after I had given up teaching in consequence of the remarks of one of my colleagues (he was in ill health at the time), as to men who did not take part in the teaching and thought they knew better than those who were engaged in its actual work. It was not, I think, till on Hughes's resignation it was proposed and decided to choose a Principal from outside that I withdrew, as I considered this a departure from the very spirit of the College. [Personally, the choice of Professor Albert Dicey was an admirable one.] I may perhaps observe that on Mr. Maurice's resignation the Principalship was first offered to me, but I declined it as I considered that the Principal ought to live in London and be able at all events to know its working. Notwithstanding the eminence of the Principals who have succeeded Hughes I continue of the same opinion. To my mind, Tansley was pointed out in every respect as the true Principal to take Hughes's place.

Since my resignation, I have only been present at two or three Christmastide suppers of the Old Students Club (always admirably presided over by Tansley), and on one or two other occasions, e.g. a kind of reception given to Mr. Lowell. But since I have given up all evening engagements in consequence of my wife's ill health, I have ceased to attend those very pleasant and interesting gatherings, often graced with first rate speaking.

I need hardly say that the bitter feeling with which I at first regarded the establishment of the College has for many years completely passed away. I value it, as being with Queen's College the chief visible memorial of Mr. Maurice's work and as having been and being still a centre of culture and good fellowship among the working and middle class of London. It has done, I believe, untold good.[35] [Since the above was written, the removal of the College to a site in Crowndale Road has been decided on, and I was present at the laying of the first stone by the Prince of Wales, the Princess also being present (1904). The college no doubt wanted more space and the offer for its property from the Children's Hospital next door was such as could not be refused in view of such enlargement. Sept. 1904.]

CHAPTER XXII

The Firs, Copse Hill – My Experiences as a Volunteer

The years 1852-5 were eventful ones in my private life. They included in the first place a complete change of residence, and my becoming a landowner.

I had taken my house at 86, Cadogan Place on a ten years' agreement. Towards the end of that period I could not conceal from myself that my mother's strength was failing, and that London air was telling upon her. I resolved that my next move should be to the country, and that I would if possible give her a permanent home. I felt myself justified now in calling in a loan which I had made nine or ten years before, under circumstances of peculiar emergency, [i.e. the loan to my brother-in-law mentioned in Chapter VII.] and which comprised nearly the whole of my fortune, to carry out my purpose.

My mother and I had been for years in the habit, now and then, by way of a Sunday outing, of running down to Wimbledon, then still a small village, and spending some hours in the Park. We had done this from the time there was no railway, and the only way of getting down was by stage-coach, running three or four times a day and charging, if I recollect aright, half-crown fares, and when no houses had yet been built on the Park fronting the Common; when the Park itself was as lovely a piece of woodland as you could wish to see, and moorhens and kingfishers nested on the lake . . .

When we came to think of a country-house our wishes turned in this direction . . . One day in Chambers I had a call from one of my regular clients, and I happened somehow to mention that I was looking out for land to build on in Wimbledon and couldn't find what I wanted in the Park. 'But why don't you try Copse Hill?' he asked: 'Lord Cottenham's Park comes into the market on . . . next, and we reckon that the views are the best in all that part of Surrey'. Well, I got through him the printed particulars of the sale, put the thing before Hughes, and we went down together to inspect the place. The beauty of it fairly took our breath away, and we resolved then and there that here or nowhere must we pitch our tents . . . After exploring the property we made up our minds to bid either for the corner lot (about 3 acres) as being nearest to the station, or for one of $3\frac{3}{4}$ acres, halfway down the slope, with some fine oak trees below. As it happened, the corner lot was sold at a price exceeding our figure, but the other lot we got for something less.

In order to realise the fact of our acquisition, we ran down soon after by a very early train one morning and there, lying on a bank and smoking, we looked down on what was henceforth our very own. It was an exquisite autumn morning, sky of tenderest blue and meadows still of greenest green, the Surrey hills in all their beauty before us, the grandstand visible on Epsom Downs, and all the Park seemingly ours, the fencing of the lots being left to the purchasers to do. I don't think either Hughes or I will ever have forgotten the sensations of that morning. Beyond, indeed, the hope of bringing my mother down to a delightful last earthly home, I had also another dream in view, which many years after came to be realised, though there was to be a dreary time before.

Of course, we were not a pair of co-operators for nothing and our building plans were co-operative, not only in the sense that we meant to have our houses built by a builders' association, but radically so. The houses were to be semi-detached, with a common garden and paddock, and at least one common room, and the architect was, of course, to be our fellow-promoter, Penrose. By a stroke of luck we were able to secure the five fir-trees (together with some young chestnuts) which bordered the road . . . The contract for the houses was given to the North London Builders' Association, working under Penrose's supervision, and there is no doubt that much of the building – foundations, brick-work, in fact the whole shell of the work, was done admirably. Much of the finishing was so also, and so much of the painting as was done under the supervision of a very good man called Sly. But his health was breaking, and he died, – not before he had warned me that things were not going on as he should have liked them to do in the Association. Much of the finishing was thus not altogether satisfactory . . .

Let me say here at once that during the whole of the Hughes's stay at 'The Firs' the common garden, field, cow-house, hen-house, library, answered perfectly. I cannot call to mind any the slightest bickering or unpleasantness arising out of them between the two households, and the convenience of the arrangement was enormous. Whilst each of us could whenever he pleased entertain his own friends apart from the other, nothing could be pleasanter than the means afforded by the common library for taking our visitors from one house to the other, or joining forces for an evening.

One such occasion – it was in the early days of our stay – I particularly remember. Mrs. Gaskell had come down to us – charming, lovable Mrs. Gaskell – she had been much pleased with a review I wrote in the *North British* under the title of 'Ruth: the reign of female novelists', in which I had defended the ethics of her book . . . My mother and she suited each other at once, simplicity of heart the common tie between them. I rather think she stayed over the Sunday, and was present at what we termed (as I shall explain further on) 'the conventicle'. On the occasion I particularly refer to Tom Taylor was Hughes's guest, and laid himself out in his best manner for Mrs. Gaskell. We both accompanied her to the station the next morning, and being there rather early, and the conversation having somehow fallen upon Dante, Tom Taylor repeated to her that pearl of pearls, the sonnet *Il Saluto*, which she did not know . . .

Tom Taylor[36] was a frequent visitor at the Hugheses', and I saw a good deal of him. He was a charming fellow, and was especially nice with children. I had a good deal to say to him later on, during my three months' editorship of the *Reader*, when he was one of my best and most dependable contributors . . .

A still more important meeting in the joint library was one when the establishment of *Macmillan's Magazine* was decided on in 'tobacco-Parliament', Alexander Macmillan being my guest with, at his request, D. Masson, whom he did not then know. And here I must confess to a disappointment. I had for years – seconded by others – been hammering at Macmillan to make him launch a monthly magazine, telling him he had his literary staff for the purpose ready to hand, and I own I hoped for the editorship for myself, such being the aspiration of all the friends who had entered into the idea. It was therefore something of a pang to me when I heard that he had selected Masson, who had no editorial experience as I had, and whom he had first known through me. Not that I grudged anything to Masson himself, with whom I have always been on terms of friendship. I ought to have borne in mind more than I did the strength of the clan feeling among Scotchmen.

On this occasion Masson left, if I recollect aright, either on the Saturday night or early Sunday morning, Macmillan staying on to return with me to London, and attending the conventicle, a term which I must now explain.

At the time we settled in Wimbledon it was a parish with 2,500 inhabitants, and one church . . . The parish church was a mile and a half from our houses, and the living was held by a Mr. Adams, who had at one time been a Navy chaplain – not by any means a bad man, but a very dull preacher, and an exceedingly narrow Evangelical . . . The only other accessible place of worship was a school-room in New Malden . . . The consequence was that, more often than not, the two households used on Sunday mornings to meet in the library common to both houses, and Hughes and I to read alternately a somewhat shortened service, followed by a sermon, generally Maurice's, or Kingsley's, sometimes Arnold's, when it was Hughes's turn. This was our 'conventicle' and I think it continued till Hughes dissolved the partnership . . .

The evenings when there were no lady visitors to attend to

would conclude by a 'tobacco-Parliament' between us two in the library. It was here that, talking about boys' education, Hughes surprised me once by saying that he had long thought that there ought to be a novel written for boys, something far above Sandford & Merton: adding that he had tried his hand on one, would I give him my opinion of it? I was quite taken aback, for though I knew Hughes from *Christian Socialist* and *Journal of Association* to write brightly and easily, he had never looked to me to be a novelist. So he handed over a parcel of ms. written in his own clear hand (for which he must have been blessed ere this by many a printer), telling me that the chapters were not consecutive, but that I would soon see the drift of them. It was *Tom Brown's Schooldays*. The portion he gave me comprised the beginning, and with gaps here and there ended with the dormitory scene of Arthur's saying his prayers. I was amazed while I read, for I own, with all my appreciation of Hughes's abilities and high purpose, I had not conceived him capable of producing anything so original and so good. When I had finished I said, 'Tom, this *must* be published – it is first-rate'. Accordingly, within the next day or two he called upon Macmillan, who did not at once accept it, and I had to write to him (for in those days I was much more of a literary authority than Hughes), expressing my admiration in still less guarded terms than I had done to the author.

But a considerable interval occurred before *Tom Brown's Schooldays* was finished. One day two of Hughes's children sickened, and the sickness developed into scarlet fever. Hughes's eldest child, Eva, a strong-built, healthy child, succumbed. It was his first great loss, and he felt it deeply. *Tom Brown* remained for months in his drawer uncompleted. Then he took it up again, and finished it. But anyone now who can read between the lines may see that there is a great difference between the part that follows the prayer scene and that which precedes it. There is no more bubbling up of exuberant humour; there is a higher earnestness of purpose. It is poor little Eva's death which has to be read in between the two.

And so *Tom Brown's Schooldays* was published, and within a few weeks its author was famous. But misfortune struck him a second time. His second child, Maurice, was drowned (July

1859) in the Thames. He was a very fine promising boy, his father's favourite, and this second blow was far harder to bear even than the first. Mrs. Hughes now declared that she could no longer live at Wimbledon – it was associated with too painful memories. So they moved up to town again, and let 'The Firs' . . .

*   *   *   *   *

At the time of what is often now sneeringly designated as 'the invasion of England scare', but which I fully believe to have been one of imminent danger to this country, the volunteer movement on the one hand brought out a corps (the 19th Middlesex) from the Working Men's College,– T. Hughes being the most active originator and becoming the first commander of the corps – and on the other spread also to Wimbledon. I thought myself at first too old to take an active part in the movement, and declined to join the College corps, but I deemed it right to encourage the work by word of mouth, and spoke at the Wimbledon lecture hall at a meeting for the formation of our corps, which became the 11th Surrey. We had no great orators at Wimbledon, and without any vanity I think I may say that I made the best speech of the evening . . .

Our great military authority was Col. Oliphant, but the real promoter of the movement was T. Richardson, a man full of go and very popular . . . I was persuaded to join the corps, and served in it for seven years. Volunteer corps were at first very democratically conducted, and were governed by elected councils. I was on the Council of the 11th Surrey . . .

I am bound to say that volunteering, in its early days, did, I think, a great deal of moral good in bringing the different classes together, so that working men and gentlemen shouldered their rifles as full privates side by side. The gentlemen generally learnt their drill quicker, but the working men were generally more regular in their attendance. Of course, there being no standard of height, the contrast was often very ludicrous if a quite small army clothier, one of our number, happened to be next to a young fellow over six feet, but that could not be helped. Personally, I think it improved my health, and in particular my eyesight . . .

CHAPTER XXIII

The Friendly Societies' Commission – The Bills of 1874-5

... I believe I was really the occasion for the appointment of the Friendly Societies' Commission. The first and only single Registrar of Friendly Societies for England, Mr. Tidd Pratt, had become very old, and his mind was failing. My friends had settled among themselves that his place was the one for me. [This was so much the case that when in 1868 the Educational Department was for the first time really organised under Lord Ripon and W. E. Forster, both friends of mine, and I applied to Lord Ripon for the Secretaryship – a post for which I think I was qualified, and in the work of which I should have been much interested, he wrote me back that Forster, who had charge of the Education Bill, thought that I had better wait for the Registrarship of the Friendly Societies. I may own now that this was a cruel disappointment to me, as not only should I have preferred the place, but I was in terrible want of it, and had a very bad time financially till I got the Secretaryship to the Commission, and not an easy one till I obtained the Registrarship.] For it was generally considered that Tidd Pratt looked with no friendly eye to the two latest classes of societies added to his jurisdiction, in which we were specially interested, co-operative societies and trade unions ... As soon as he heard of Tidd Pratt's death, and without even telling me, T. Hughes (then in Parliament) went to Mr. Bruce and asked him to appoint me, and Bruce wrote him a letter (which I saw) acknowledging the 'claims' I had to the office, which for him virtually amounted to a promise of the appointment.

And then – hey presto! – a regular transformation scene took place. No one had taken the trouble to read the Act by which Tidd Pratt was appointed; I myself, I own, did not think of looking my gift horse in the mouth. Still there were doubts. The Board of Trade had put forward a claim to the appointment; the Treasury was said to hold it for one of its own. But behold! the Act plainly and unmistakably vested it in the National Debt Commissioners, i.e. in the Chancellor of the Exchequer. This was the then Robert Lowe, afterwards Lord Sherbrooke, between whom and me there was not a particle of sympathy. I sent in, however, an application, supported with the best testimonials I could get (which by the way were never returned to me). I have been told – though as proceedings in the Cabinet are secret, I cannot of course vouch for the fact, that the subject of the appointment was discussed in the Cabinet and that besides Mr. Bruce and Lord Ripon one or two other Ministers spoke in my favour . . . I have in my autograph book Mr. Lowe's answer (dated 4 Feb. 1870) the concluding paragraph of which is 'Mr. Ludlow's claims shall be carefully considered, but I rather think his office will not be filled up'. The explanation of which is, that finding it difficult to appoint any one else than myself in the face of the expressed views of several colleagues, he had hit upon the idea of suppressing the Registrarship altogether and transferring its duties partly to the Board of Trade and partly to local judges. [I was told later on at the office that the sole foundation for the measure was the report of a Treasury clerk who was sent to Abingdon Street[37] on a Saturday afternoon, when it was the custom to give a half holiday alternately to half the staff. Without disclosing the purpose of his visit he just poked his head into all the rooms, and then went back to say that there was very little business done, just about enough for half the staff. It is difficult to believe that the conclusion was other than a foregone one.] The bill for this purpose must have been already prepared when he wrote to Mr. Crompton, as it was introduced on the 10th February, but it was received with such disfavour that, on the suggestion of a Welsh member, Mr. Evan M. Richards, it was determined to issue a Royal Commission of Inquiry into the Friendly and Benefit Building Societies' Acts ... I received now a note from Albert Nutson asking me to step over to the Home Office. He told me that Mr. Bruce offered me the choice either of being named on the Commission or of its Secretaryship. Had I been a rich man I should have preferred the former position. I was over 49 years old, of 27 years' standing at the Bar, beyond the age when men usually accept a subordinate position. But I was just about to marry, and £400 a year was everything to me. Moreover, I felt that if the Registrarship was to subsist, the Secretary-ship to the Commission would give me the best possible training for it, and title to it. I elected for the Secretaryship, somewhat, I think, to Nutson's surprise.

The Commission was partly composed of men whom I knew more or less intimately. I had to make the acquaintance of its Chairman, Sir Stafford Northcote, afterwards created Lord Iddesleigh, of Sir Michael Beach, Sir Sydney Waterlow and Mr. Evan Richards, and nominally of my present friend Mr. Roundell, though he and I had several mutual friends, and knew each other through them already. Bonham Carter was an old ally of mine from the time of my being in Mr. Ker's chambers. Mr. Pattison was a colleague of mine on the Council of the Working Men's College. F. T. Bircham had been Charles Mansfield's solicitor, and I had called on him in reference to Mansfield's business.

Sir Stafford Northcote introduced himself to me by calling at my chambers at 3 Old Square, Lincoln's Inn, and before the end of the interview I felt sure that I should get on with him. And here I may say at once that a pleasanter and more considerate 'chief' it would be impossible to have. Though not a man of genius he was very able, but he was so unassuming that his abilities were under-rated. To lead without seeming to had appeared to be his constant aim, and I described in a letter to the *Spectator* shortly after his death the marvellous adroitness with which I have repeatedly seen him manage his colleagues on the Commission, settling with me beforehand the course to be pursued and then when the Commission met treating the matter as a purely open question, to put, so to speak, on the table for every one to give his opinion, yet by some question or observation, virtually an unfelt flick of the whip, gradually conducting the discussion to a predetermined end, so that when the conclusion was reached not one of his colleagues suspected that the Chairman had done otherwise than sum up the general feeling on the subject. But, as I also wrote, this light hand of his was ill-fitted to drive a stubborn team, or even to master a stubborn horse in a team. Hence, amongst other things, the separate reports which tend to weaken the conclusions put forth in the 4th Report of the Commission. Of his kindliness of heart it is almost impossible to speak too highly. The Commission had to examine some great rascals at Liverpool, – the House of Commons had refused to pass a bill enabling the Commission to examine on oath. Freed thus from all fear of the penalties of perjury, some of these witnesses lied palpably, audaciously, and it was quite pathetic to hear Sir Stafford's appeals to their long dead consciences: 'Now really, Mr.—, do you mean to say,' etc. During the whole period of our intercourse I never heard from him one unpleasant word, I never experienced any but the most considerate and kindly treatment. Not that I had any fault to find with any of his colleagues, (except on one early occasion when one of them chose to treat the Secretary as a waiter, and asked him to fetch the plate of sandwiches, but I saw afterwards Sir Stafford speaking with him, and nothing of the kind ever occurred again). The Commission had, in fact, been very well selected. [It seems at first sight anomalous that, appointed by a Liberal Ministry, its two most prominent members should have belonged to the Opposition. A little reflexion will however show that this must be the case with most Commissions involving long and protracted work. All the first rate and best second-rate men of the party in power have been skimmed off for office whilst there is a choice of such unengaged on the Opposition benches.] . . . all were honest English gentlemen, and it was very pleasant to see how entirely without effect on their mutual friendly intercourse were their differences of political opinion.

Next to Sir Stafford Northcote the ablest man was unquestionably Sir Michael Beach. Sir Stafford was very fond of him, and in travelling really looked after him in a fatherly way. Sir Michael Beach, though he had good powers of work, was lazy and self-indulgent, and seldom made his appearance on a railway platform before the very last moment, and it was touching at such moments to see Sir Stafford's anxiety. Sir Michael was the man to tackle a rascally witness, putting the most searching questions in the most unimpassioned, blandest tones; if anything, I always thought he had a sort of cynical pleasure in searching out human rascality. Possibly, had his friend Edward Denison lived, the better side of his nature would have been brought out more than it has been. He was at that time – I do not know whether he is still – decidedly ambitious. But beneath a cold manner I believe he meant thoroughly well, and his ambition was quite void of vanity. I have known him confess in the frankest way a change of opinion by reason of another's arguments. Bonham Carter I have mentioned before – a well meaning, steady working mediocrity. The enormous length of his questions must have greatly puzzled witnesses. Roundell I soon became intimate with; he is one of the best of fellows, and of excellent judgment. Pattison, when I first knew him at the Working Men's College, was an uncommonly nice fellow . . . Sir Sydney Waterlow has made himself a name in the history of English benevolence by his gift of the Fairlawn property at Highgate to the public. Bircham was an exceedingly shrewd solicitor and was, I may say, one of the most useful and thoughtful members of the Commission . . .

Taken as a whole, I think those years of the Friendly Societies' Commission were the happiest of my life. My work was congenial to me; I was thoroughly master of it; I found myself treated with unreserved confidence by the kindliest of chiefs; Commissioners and Assistant Commissioners were good fellows to a man, and several of them were from the first or became ere long my personal friends. I have heard most interesting conversations among them at luncheon time . . . I was struck by the free way in which on such occasions as I have mentioned Conservatives and Liberals would speak their minds to each other. I have heard Bonham Carter express strongly to his Conservative colleagues his disapproval of some Ministerial measure. I have heard Sir Michael Beach say that the one thing for which he envied the Liberals was their leader, for he really did lead, sticking to his place whenever there was or was likely to be a storm, so that his followers knew always what to do, whereas their Conservative one (Disraeli) on such occasions always walked out of the House, so that his followers were left without guidance, and had to fight every man for his own hand. Even the occurrence of a General Election before the Commission had finished its work, though it brought that work to a somewhat hastier end than would otherwise have been the case, did not ruffle the kindly mutual relations of the Commissioners. Sir Stafford was quite grieved at Bonham Carter's loss of his seat. 'There are several men on our side whom we could have better spared than him' were his words to me on the subject . . .

The Bills

When the Friendly Societies' Commission had sent in its report, all further payments from the Treasury ceased, and I had again to depend on law and literature only for my livelihood. Sir Stafford, however, now Chancellor of the Exchequer, entrusted me with the drawing of the Bill for carrying out the recommendations of the Report much, I was told, to the dissatisfaction of the Parliamentary draughtsman's Office . . . Thus, a quarter of a century after I had drawn for the Board of Trade the Joint Stock Companies Winding-up Act, 1849, I found myself again directly instructed on behalf of a Government Department.

I must admit that, from a Parliamentary point of view, the Bill of 1874 was a failure. The law as to Friendly Societies, Co-operative Societies, and Trade Unions (to say nothing of Savings Banks) is so closely akin for all three classes of bodies, that I recommended that all should be embraced in one Bill, and Sir Stafford approved of the proposal. But the great Collecting Societies, whom we were endeavouring to muzzle, stirred up the trade unions, who took perfectly needless fright at being meddled with at all – the fact being that the Bill gave them privileges which they had never had before, and only obtained two years later by the Trade Union Act

Amendment Act, 1876, – and Sir Stafford felt it prudent to withdraw the Bill after the second reading, with a view to confining it to Societies under the Friendly Societies Acts, and so re-introducing it in the following Session. As remodelled by me, the Bill became the Friendly Societies Act 1875.

In the course of the Parliamentary career of the two bills, I had frequently to attend the sittings of the House of Commons, and had the privilege of one of those seats under the gallery, dark and stuffy, but which have the advantage that members can come to speak with the occupants without leaving the House. I was usually accompanied by Walter Northcote, the present Lord Iddesleigh, then his father's Private Secretary, and who as such was an almost nightly occupant of an under-the-gallery seat. I have often had to sit there for hours without the Friendly Societies Bill coming on at all, and was able to realise how little ordinary newspaper reports often reproduce the real features of a debate. I remember one occasion when Disraeli was to have brought forward a motion for appropriating certain sittings to the Government, – a step which generally becomes necessary at a certain period of the session, but which is always looked upon with jealousy by private members, as curtailing their opportunities of action. On this occasion the claim was made by the Government somewhat earlier than usual, and it was all the more incumbent on the leader of the House to attend and make the best of an unwelcome proposal. Instead of this, we had noticed Disraeli quietly saunter out of the House an hour or two before, leaving, as it turned out, the present Lord Cranbrook, then Mr. Gathorne Hardy, to give notice of the postponement of the motion, in a pretty full House. When he did so – with less than his usual assurance – he assigned no other ground for postponement than that the Premier was not able to bring the measure forward that day, even when questioned from a back bench on his own side. Then complaints began to be made from this and that bench of the Opposition, and then from one or two ministerialists. I said to Walter Northcote – 'There is going to be a row. Mr. Disraeli ought to have been here.' As I was speaking, we heard a back-bench ministerialist declare that he was ready to vote against the adjournment. Another said the same. The unfortunate Ministers, deserted by their chief, did not really know what to say. Members began to crowd into the House, and it really looked as if the Ministry were going to be beaten through Mr. Disraeli's *poco-curante* insolence when W. E. Forster got up, and in the most good-natured manner came to the help of the Ministry, saying that having been in office himself, he knew how difficult it was to be always ready to time, and that probably the First Lord of the Treasury had not been able to procure the necessary information for this motion; that he could not refuse to vote for the postponement of the motion, provided it was to an early day. He carried the House with him, and if I recollect aright, the motion was postponed till the next day but one, when Mr. Disraeli did not walk out of the House before it came on. The whole scene was intensely interesting, and though the fate of the Ministry would not have depended upon it, the Cabinet must have been greatly weakened by a defeat on a question in which its chief was so palpably in the wrong. Yet on looking through the papers the next day I could literally find nothing from which the facts could have been put together again, or which indicated the peril in which the Ministry had stood, until saved by Forster.

This was the only time at which I used to see Disraeli. I never, however, heard him speak more than a few words. I must say the man looked to me simply like a second-rate actor . . .

CHAPTER XXIV

Trade Unions – The Social Service Association (1852-71)

In speaking of the early Christian Socialist Movement I have told how I was brought into contact with the leaders of the most important, and still the most wealthy, of the Trade Unions, the Amalgamated Society of Engineers, and with an actual labour contest on a large scale, that of the Engineers in 1851-52. The next such contests with which I was to some extent connected were the London Builders' strike of 1859-60, and that of the Stonemasons in 1861. The former was a movement for a nine hours' day without reduction which was met on the part of the Master Builders' Association by an attempt to enforce a declaration on the men similar to that of the Master Engineers, the result being a drawn battle. This contest was carried on during the sittings of a Committee on Trades Societies appointed by the National Association for the Promotion of Social Science (commonly called the Social Science Association but long since defunct).

In order to obtain information from the men about the contest, a meeting was arranged to be held at my chambers between members of the Committee and representatives of the men. There were present on behalf of the Committee R. H. Hutton, G. J. Shaw-Lefevre, T. R. Bennett, and I think T. Hughes; on the part of the men George Potter, who acted as Secretary to the men, Thos. Connolly, a big burly Irish mason, who eventually became a newspaper correspondent,

G. Howell, and I think one other. One little incident occurred, which impressed me in G. Howell's favour, and against his companions. A statement was made, which as I saw at once was misunderstood by Hutton and Lefevre, so as to tell unduly in favour of the men. G. Potter saw the mistake, and dishonestly tried, without positive falsehood, to confirm the 'gentlemen' in the erroneous view which they had taken. Connolly was silent. Howell had the honesty to set the matter right. Possibly he saw that I knew how matters stood, but I was not the less grateful to him for his outspoken endeavour to save Lefevre and Bennett from mis-stating matters in their report.

Frederic Harrison, I may say, took like myself a good deal of interest in the builders' strike, and I may here say that we wrote in collaboration a stave of verse in Cockney dialect on the subject which was printed in the *East London Observer*, then edited by William Newton. I have long since lost my copy of it, but I recollect that it partly dealt with the incident of 'Georgy Myers', the builder who had originally been a working man taking off his coat and offering to fight any of the strikers from his yard.

The Social Science Association, I may here observe, had been founded in 1857. I was early asked to join its Council, but declined to do so in the first instance, having long lost all respect for Lord Brougham, President of the Council, and thereby virtually of the Association . . . In 1859, however, on the recommendation of a conference held at the second meeting of the Association in Liverpool, in which several working men took part, a Committee was appointed, not to be confined to members of the Association, to enquire into the subject of Trade Unions. This Committee I was asked to join, and was afterwards elected a member of the Council of the Association.

Mr. and Mrs. Sidney Webb, in their *History of Trade Unionism*, refer in a note (p. 209) to the report of this Committee as containing 'the best collection of Trade Union material, and the most impartial account of Trade Union action that has ever been issued.' But it was much more than this. It was a most momentous event in the history of Trade Unionism. Taken in conjunction with the discussion at the

Glasgow meeting of the Association (which Mr. and Mrs. Webb do not so much as refer to) it opened the way for the first time to a serious consideration of Trade Unionism, and rendered possible the appointment of the Select Committee on Trade Unions a few years later (1867).

It had been very wisely decided that the authors of the various separate reports should alone be responsible for the accuracy of such reports, and the opinions expressed in them. The writers were most of them men who afterwards distinguished themselves in public life, – as John Ball, afterwards M.P. and Under-Secretary of State for Foreign Affairs, perhaps best known as an Alpine climber and geographical expert; G. J. Shaw-Lefevre, afterwards member of two Liberal Cabinets; the present Sir Godfrey Lushington, late Permanent Under-Secretary of State for the Home Department; T. Hughes; F. D. Longe, who bears the credit of having on one point revolutionised political economy; W. A. Jevons, the well-known political economist; F. A. Hill, the admirable Secretary of the Committee, under whom as editor the *Daily News* was to reach its highest reputation; to say nothing of Mr. Louis Blanc. To myself was assigned as a subject a West Yorkshire coal strike and lockout which had taken place in 1858. Through Philip A. Rathbone, one of the honorary Secretaries of the Committee, who had married the daughter of a West Yorkshire coal-owner, I enjoyed the singular advantage of being able to compare together every statement of employers and employed, so that I was able to say that my report had been communicated both to men, and masters, and that 'every fact, figure and explanation' had 'been inserted, which any of the persons consulted deemed important'. None of the other reports, however impartial, are so well authenticated. In reporting on the lock-out of engineers in 1851-52, T. Hughes in vain sought to obtain corrections from the Secretary of the Employers' Association; it was not 'agreeable to the masters generally that any authorised statement of their course of action should be made.'

In the course of my investigations, I was brought into somewhat close relations with the leading men among the Miners' Unions of the North. They were all miners themselves, except that the Staffordshire men were represented, to my surprise, by a herbalist, Richard Mitchell, the Secretary of the West Yorkshire Miners' Association, ((whom)) I particularly liked, – an upright, earnest fellow, who however died before long of consumption brought on by the hardships of his calling. But the most valuable result of my work was the necessity for weighing against one another the statements of employers and employed. One of the men's grievances in the West Yorkshire Miners' lock-out was the deductions made from wages for this, that and the other, reducing sometimes very considerably the nominal amount. I was supplied by the men with a quantity of pay-sheets (always in those days, I am sorry to say, fortnightly ones) in which these deductions were set down, sometimes in writing, more often in print. In talking over the subject with the Messrs. Briggs, I was met by a flat denial of the existence of certain of such deductions – 'such things may have taken place twenty years ago, but they are unheard of now.' For all answer I took the pay-sheets – all recent ones – out of my drawer, and laid them before them. In the course of my life I don't think I ever saw men so astonished. They had been repeating everywhere, in public and in private, that the men's statements on the subject were mere lies, and they admitted that but for this positive proof they should have continued so to consider them. The result of my investigations in this matter was to confirm me fully in the feeling which I had already as to the exceedingly narrow limits within which the testimony of the most honest employer can be trusted as to matters concerning his trade at large.

In drawing up their Report, the Committee became divided into a majority and minority. The general report, as signed by the Chairman, Sir James Kay Shuttleworth, contains itself important observations having regard to its date. The Committee was 'disposed to believe that leaders of a strike where there is no regularly organised society are likely to prove more unreasonable and insolent than where there is . . . the facts collected seem to show that the executive of a large society are more likely to take a cool and moderate view of a question in dispute than the men engaged in the heat of conflict.' They had 'not found that the constant assertion that strikes are scarcely ever successful, is at all borne out by facts.'

They were 'further disposed to believe that in some cases the existence of a regularly organised trade society has prevented the frequent occurrence of strikes.' So far as the Committee had been 'brought into personal connexion with society's officers, their experience' was 'that the leaders are for the most part quite superior to the majority of their fellow-workmen in intelligence and moderation'. 'The effect of trades' societies as an education in the art of self-government is important'. 'The conduct of employers' associations towards trades' societies has, on too many occasions, been most unfortunate'. 'There will be very little difference of opinion as to the advantage of the trade circulars issued by some of the societies'. The Committee wished 'to record their opinion distinctly, that improvement in the management of trades societies has been most marked and satisfactory'. And the majority drew up a series of nineteen conclusions, of which the only important ones for practical purposes are the following:

11. That the legal difficulties which rich societies experience in finding a profitable investment for their funds often increase greatly the temptation to employ them in strikes.

14. That minor questions connected with trades, which often produce serious irritation, might be advantageously referred to a mixed tribunal of masters and men; but that, in the opinion of a majority of the Committee, it would be over-sanguine to hope for a removal of the more direct and serious causes of strife by such arrangement.

15. That the rate of wages must be settled between the masters and men, and that the intervention of third parties, unless specially invited by both, and possessing in a very high degree the confidence of both, can be of little avail.

16. That the legislature may do much good service to the workmen, by providing an easy and cheap remedy, both in law and equity, to meet the case of disputes between trades societies and the members, especially in respect to the application of benefit funds.

17. That the slightest return to the old policy of prohibiting combinations would be mischievous, and that no legislative measures for preventing strikes and lockouts would be effectual.

The last quoted 'conclusion' is invaluable, as showing the unanimous opinion of so important a Committee, against 'the slightest return' to the old anti-combination policy. On the other hand, conclusion 16 showed a singular ignorance of the feeling of working-men, which, as respects the vast majority, is dead against any interference at law or in equity, with the internal affairs of their trade unions. But even if it were not so, any such remedy would imply as a necessary pre-requisite the giving of a legal constitution to trade unions, and this the majority did not dare to recommend, thereby stultifying itself.

The conclusions of the minority – which I may say were drawn up by myself, and slightly amended by Godfrey Lushington and one or two other friends, are sufficiently brief to be set forth in full:

1. That the simplest and universal function of trades societies is the enabling the workman to maintain himself while casually out of employment, or travelling in search of it.

2. That the organisation requisite for the exercise of this function enables and induces the labourer to attempt to compete with the capitalist for a share in the temporary advantage produced by the increase of demand, or the improvement of methods of production in different trades.

3. In pursuing this latter object, trades societies often create by a strike the want of employment, which it is their first purpose to provide against.

4. That nevertheless the existence of a trade society, however strong, has no necessary connexion with the practice of striking in a trade.

5. That the temptation to strike is often stimulated by the accumulation of funds in the hands of a strong society, and the legal impediments to its safe and profitable investment.

6. That what we have called the universal function of trades societies has the effect of securing to them the support of large numbers of the most prudent and moderate-minded workmen.

7. This support is further accrued, in most cases, by the combination with the purposes of a trade society of those of an ordinary benefit society, and sometimes of other institutions, such as libraries, reading rooms, etc.

8. This combination of purposes often affords the majority the means of involving an unwilling minority in strikes, and all the consequences which may flow from them.

9. The first or universal function of trades societies might (we think) be safely legalised under the Friendly Societies Acts.

10. The second function might (we think) be practically
superseded by habits of frank communication between masters
and workmen on all matters affecting their common interest and
mutual relations.

11. In cases in which masters and workmen are unable or
unwilling to arrange matters of dispute, we think the ultimate
resort to strikes or lockouts might be avoided by the establishment of united associations of capitalists and workmen to the
arbitration of which united associations all disputed questions
might be referred.

Let me say at once that whilst I still hold by the great bulk
of the conclusions of the minority, I should now in great
measure differ from the fifth and eighth. I think 'the
accumulation of funds in the hands of a strong society',
instead of 'often' stimulating the 'temptation to strike', acts in
the great majority of cases as a check upon that temptation.
And I think the combination of benefit and other purposes
with trade purposes instead of 'often' affording to a majority
'the means of involving an unwilling minority in strikes', has
in all cases a precisely opposite effect. Subject to these
reservations, the course of events has substantially justified
the conclusions of the minority and not those of the majority.
The suggestion by the latter of 'an easy and cheap remedy,
both in law and equity, to meet the case of disputes between
trades societies and the members, especially in respect to the
application of benefit funds' is one that shows an entire
misconception of working class feeling, which has always
been dead against any interference of the law with their
internal disputes. On the other hand, the suggestion of the
minority of legalising the primary object of trade unions has
been carried out by law, if not strictly under the Friendly
Societies Act, yet as an additional group of societies in
connexion with the Friendly Societies' registry office. And the
last suggestion of the minority, that of avoiding strikes or
lockouts by the establishment of united associations of
capitalists and workmen, – a proposal drenched with cold
water by the majority (see conclusions 14 and 15) has been
carried out in several trades, and forms the chief object of the
lately formed 'Industrial Union of Employers and Employed'.
But the discussion (Sept. 27th, 1860) at the Glasgow

meeting of the Association ought not to be overlooked, as showing the growth of public opinion on the subject. A future Postmaster-General, Mr. Henry Fawcett, spoke of the 'great advantages which labourers gained by forming themselves into combinations'; considered the report of the Committee 'most valuable, in having cleared up those unjust and untrue assertions which had been cast upon the characters and motives of the delegates of trade unions' . . . Perhaps the most important testimony was that of Sir Archibald Alison who, having had to conduct prosecutions against trade unionists, expressed the opinion that 'trades unions in themselves are not only proper, but are a necessary balance in the fabric of society'. The ablest advocates of the employers' views were Mr. Edmund Potter and Messrs. Ashworth, whom Hughes and I sought to answer, insisting that labour, though a purchasable commodity as maintained by the masters, differed from every other such commodity in having a will of its own. To my observations we appended in the report of the Debate the following conclusions, which I had not time to read:

1. I believe that all trade societies and employers' associations should be compelled to register their rules, and on their being certified not to be illegal, should obtain legal means of protecting their funds, recovering subscriptions, etc.

2. I believe that, with the advance of economic knowledge, the ascertainment of the wages and conditions of work which may be reasonably demanded by the labourer at any given time and place becomes more and more a matter, not of mystery nor of guesswork, but of practical science, capable of determination by competent persons.

3. I believe, therefore, that disputes relating to such matters are capable of solution at the hands of mixed courts of employers and workmen.

4. I believe that the legalisation of trades societies and employers' associations would greatly facilitate the working of such courts.

5. I believe that such courts should act primarily by way of arbitration, [Writing at the present day, I should use the term 'conciliation' instead of 'arbitration'. But in those days there were no legal means of enforcing an award in a trade dispute, so that I was understood at once as distinguishing between voluntary and legal action.] but that means could be devised for making their action compulsory.

6. I believe that strikes and lockouts, or, in other words, private commercial wars, are a remnant of barbarism in the midst of civilisation, and a disgrace to our social state; that the allowal of them is only justifiable so long as no competent tribunals exist for settling trade disputes; that such tribunals (towards which Mr. Mundella's bill would have supplied the first step) must ere long be devised; and that when this takes place, the public will be entitled to insist on measures of penal legislation against strikes and lockouts.

I still (1895) hold by all the above conclusions. I may mention that the number of the *National Review* of the day, then edited by R.H. Hutton, contains an article by me on 'Builders' combinations in London and Paris', contrasting the London builders' strike as reported on by Lefevre and Bennett with the co-operative *Association des maçons* of Paris, then at the height of its prosperity. More important literary deliverances were two articles by me on 'Trade Societies and the Social Science Association' published in *Macmillan's Magazine*, February and March, 1861. The London builders' strike of 1859-60 was followed in 1861-2 by a more partial one of the London stonemasons, of which a very good account by Frederic Harrison was printed in the Report of the Social Science Association for 1862, p. 710. This followed immediately a paper of mine 'On the Investigation of Trade Differences', which I consider the best thing I ever wrote on a social subject . . . I have been surprised at Mr. and Mrs. Webb's neglect of the somewhat abundant material relating to their subject which is contained in the *Transactions* of the Social Science Association, the earlier volumes more especially.

I was by this time in friendly relations with several of the London trade union leaders, and some of the provincial ones. To the Amalgamated Society of Engineers I became virtually Standing Counsel, – had their law cases referred to me, settled a complete amendment of their rules . . .

G. Howell was another trade union leader with whom for many years I was on friendly terms, though I own that of late my opinion of him has become less favourable. His *Conflicts of Capital and Labour*, and *Handbook of the Labour Laws* are valuable works on the subject, especially the former. I should not rate so highly his *Trade Unionism Old and New*, which is disfigured by, I must say, a passionate and unintelligent disparagement of the new unions, and does not take account of the facts which make them what they are. George Potter, on the other hand, I never foregathered with, though he made various advances to me, and I own I was rather pleased than otherwise when after a railway journey we took alone together returning from some festive gathering of trade unionists, he told Lloyd Jones (who then wrote for Potter's *Beehive*) that 'his friend Mr. Ludlow was the closest man he had ever met with'. There was something mysterious and underhand about G. Potter which, coupled with my experience of him at the conference I have mentioned, kept me always on my guard when with him. George Odger is another man who did not suit me. He was indeed perfectly honest, but filled with conceit, and to a great extent, a wind-bag. I never could understand Edmund Maurice's admiration for him, and he seems to me greatly overpraised by Mr. and Mrs. Potter[38] . . . [I may here mention that having been in full contact with Trade Unionism at the time referred to, I should set down Mr. and Mrs. Potter's account of the 'Junta' as largely mythical. There is no doubt on the one hand that Newton, as the ablest Trade Union leader of the day, exercised very great personal influence; nor yet that his firm friendship with W. Allan and the position of the latter as Secretary to the wealthiest Trade Union of the country gave the two together, on most questions, a preponderating position in the trade union world; nor again that the Amalgamated Carpenters and Joiners and the Iron Founders generally followed the lead of the engineers. But I never heard Coulson of the Bricklayers, whom I think I only saw once, spoken of as exercising any particular influence; with George Odger I believe there was frequent friction on the part of the men I have mentioned. And above all, I am quite certain that the great trade unions of Lancashire, Yorkshire, Durham, Northumberland would always have laughed at the idea of accepting the authority of a London 'Junta'. Mr. and Mrs. Potter have probably been led away by some London Trade Unionist, representing the feelings of the London Trades Council in after days, when it sought to dominate the whole movement, but was foiled by the massive strength of the older leaders.]

. . . I need hardly say that the more I saw of Trade Unions the more I realised their importance in the social economy of our country, and there is nothing in my life that I look back to with greater satisfaction than to the delivery of my lectures on the 'Master Engineers and their workmen' as having opened for me a door which can never be closed. I am perfectly certain that no man who does not understand and sympathise with the Trade Union movement can ever understand the working class of our country.

I must now go back to the year 1855, when Trade Unions thought to have obtained protection for their funds under the Friendly Societies Act of that year . . . But eleven years later (1866) the case of 'Hornby v. Close' decided that trade unions were not entitled to take advantage of the clause. This was a terrible blow to the best of the trade unions, and, coinciding as it did with the Sheffield outrages, led to the appointment of the Trade Union Commission, on which T. Hughes and Frederic Harrison were named. [I must own that one of the bitterest disappointments of my life – renewed many years later on the nomination of the Labour Commission – was that of not being appointed on this Commission. I think I may fairly say that I was entitled to be on it. I heard afterwards that the advocates of Trade Unionism were to be limited to two, and if so, I certainly should not have wished to supplant either of the friends who were appointed.] They did excellent service, the Minority Report, signed by them and Lord Lichfield, having served as the real basis for subsequent legislation. A change of Ministry having taken place in 1868, a deputation went up to the Home Secretary urging the need of protection to trade union funds. I was asked to attend on behalf of the Amalgamated Society of Engineers, and suggested as I had already done to my clients, the passing of a temporary Bill. The suggestion was adopted and the Bill passed. It is remarkable as giving by its title (Mr. and Mrs. Sidney Webb have not noticed this), the first legal recognition of Trade Unions: 'An Act to protect the funds of Trades Unions from Embezzlement and Misappropriation'.

With the Trade Union Act, 1871, I had nothing to do. It was drawn by my friend Godfrey Lushington, then Assistant Under Secretary at the Home Office. It granted substantially all that the Trade Unions had asked for. But it was disfigured by a definition clause, no less absurd than mischievous. A trade union was defined as meaning 'such combination, whether temporary or permanent, for regulating the relations between workmen and masters, or between workmen and workmen, or between masters and masters, or for imposing restrictive conditions on the conduct of any trade or business *as would, if this Act had not been passed, have been deemed to have been an unlawful combination, by reason* of some one or more of its purposes being in restraint *of trade.* Provided etc . . . ' This previous illegality was actually made a condition of legalisation. I shall show hereafter, in speaking of my work at the Friendly Societies' Registry Office, how I was able to free Trade Unions from this fetter.[39]

There is however another incident of my connexion with Trade Unions which deserves to be mentioned. Partly through the publication of *The Progress of the Working Class, 1852-1867* by myself and my friend Lloyd Jones (extracts from which had already appeared in *Questions for a Reformed Parliament*), a young German student of political economy, then 23 years old, had been led to give up the views antagonistic to trade unions which then prevailed in Germany alike among Individualists and Socialists . . . He came to England in August, 1868, with Dr. Engel, then Director of the Prussian Statistical Bureau to which the young student was attached, and both called on me at my chambers . . . By means of this visit, Brentano says, he 'was at once carried into the heart of the then Labour movement', so that both the course of his studies was greatly facilitated and his personal position as towards the labour question influenced. He found me in relations with all the prominent personages of the English Labour movement. When Dr. Engel and he went to the North of England the doors of all associations were open to them. And seven months' hard study among the records of the Amalgamated Society of Engineers, to whose leaders I had introduced him, had for their result ((firstly)) the Essay (dedicated to myself) *On the History and Development of Gilds*; ((secondly)) his great work in two volumes, *Die Arbeiter-gilden der Gegenwart*, published in two volumes, 1871-2, which may be said to have completely revolutionised public opinion in

Germany on the subject of trade unions ... [I have always been on such friendly terms with Dr. Brentano, and he has always so generously acknowledged whatever help I have been able to afford him, that I have never been able to understand why he should never have done more than notice by name, as an authority for the facts of a struggle, my lectures on the *Master Engineers and their workmen*. When I gave this to read to another German friend, introduced to me by Brentano himself, the lamented Professor Adolf Held of Bonn, his observation on returning the little book to me was – 'Why, all Brentano's book is there!' This could of course only apply to the second volume and would be an exaggeration even so, but considering the notice taken by Brentano of, and his extracts from, other publications of mine of far less importance, the omission is a singular one.]

((Editor's note – The next chapter of Ludlow's memoirs is omitted. It was entitled 'The Crises of my Life' and tells of the 'seven great crises in my spiritual and moral life'. Six of these – his decision to become an Englishman in 1837, the Martinique earthquake of 1840, the sacrifice of his fortune to his brother-in-law in 1842, his falling in love with his cousin in 1843, the effect on him of the February Revolution of 1848, the turning away of Maurice from the Christian Socialist movement in 1854 – are fully described in the narrative. The seventh concerns a family crisis in which Ludlow and his wife took in the children of another brother-in-law, a clergyman, Granville Forbes, who had become insane, and is without importance for the narrative of Ludlow's public life[40].

It should be added that Ludlow did not live to write, or did not think it worth writing, several chapters of his autobiography which, from various remarks in the ms., we can infer he intended to write. One would have been on his period as Chief Registrar of Friendly Societies (1875-91), another on his activities after his retirement from Public Office, during the period of the Christian Socialist Revival. As Chief Registrar of Friendly Societies, Ludlow had responsibility for overseeing some 25,000 (in 1888) societies, including all Trade Unions, co-operative associations and stores, building societies, besides a multitude of other in some way co-operative organisations, and with the additional role of arbitration in any disputes in 10,000 savings banks and societies. Of this work Ludlow has left no record, save a 25-page chapter of 'Curiosities' which are of no real historical interest. From his letters to Professor Brentano of this period [41] it is clear that Ludlow was not altogether satisfied with his position or with the limitations on his outside activities it imposed, and this may have made him unwilling to write on the period.))

CHAPTER XXV

Couldn't-Have-Beens and Might-Have-Beens

As I look back upon my life I cannot help sometimes speculating upon what would have been possible and impossible to me, had now circumstances, now my own will, been different from what they have been.

In an early part of these reminiscences, I have told how the desire for fame passed away from me, while yet a mere boy. I suppose it never was strong in me, for I must then already have long read Milton, and I remember when first reading 'Lycidas'

> 'Comes the blind Fury with the abhorred shears
> And slits the thin-spun life – but not the praise
> Phoebus replies etc . . . '

the feeling of pain and shame that I felt for such a man as Milton, that he should have consolation in anything so hollow and trumpery as praise even though it were from Jove.

But I was not on that account the less ambitious to do great and good things, though I should have been perfectly satisfied to do them under an iron mask, under a false name, or best of all, as the trusted lieutenant of one greater than myself, merging myself completely in him. This was the relation in which for some short time I really stood towards Mr. Maurice and it was in that respect the happiest time of my life.

The experience of my five years' schooling had shown me,

and I think both my teachers and my school-fellows, that I had abilities beyond the average. To the end of my fourth year, I had shot gradually to the top of my class in almost everything, though I had successful rivals in one or two subjects. And I had learnt moreover, that if there were some subjects in which I could excel virtually without trouble, there were others in which I could only do so by resolute hard work, – that whilst I had facilities for some, I had only capacities for others. My comparative failure in my fifth year (*Rhétorique*) was not disheartening to me, as I had felt it mainly proceeded from the inferior teaching, larger work and want of all emulation during my six months at Tours. And the words of M. Filon, our Professor, when urging me to *redoubler* or spend a second year in the class – '*Vous aurez le prix discours français*', showed me his opinion of my capability, since the winning of that prize in a great Paris college means virtually a career of literary success. For the French world at large my passing for my degree of *Bachelier-ès-lettres* without any special training, immediately after the close of my scholastic year, and at the earliest age when I could have obtained it, only by special permission (which I had vainly solicited the year before), was evidence to the same effect, – but I was by no means satisfied with myself in the matter, having lost my head through fright, and given many answers or none to several questions which I might have answered correctly.

At Mr. Ker's chambers again, though the field was but a small one, I found myself, without any special effort, by mere ordinary hard work such as I was used to, carried into the position of favourite pupil, entrusted with the drawing of everything out of the way, Royal Charters, deeds of settlement, bills for Parliament, or the looking through some unusually big pile of abstracts.

My 6 or 7 months' book-keeping at Martinique for my brother-in-law, beginning some months before I was 20, showed me again that I was capable of doing practical work to earn my livelihood, nay, that my intellectual training placed me on a level with men greatly my senior. Again, on returning from my second visit to the West Indies, I found myself, when little more than of age, the holder with a man of more than twice my years, an experienced French lawyer, of a joint power of attorney entailing the greatest responsibi-lity, – forced to differ from him as to the exercise of our powers, to the extent of refusing to take a step urged not only by him, but by a third person also more than twice as old as myself, whose opinion I had been expressly enjoined to take, – and justified by the result.

In fact, I have no hesitation in saying that at two and twenty I was at least 10 or 12 years older than my age.

And my ambition at this time, I am bound to say, was great, though it was a somewhat singular one. My dream was, that I should get into Parliament, rise to be Prime Minister, pass certain great measures, then resign at once, take Orders, and finish my life as a clergyman at the East End of London. I am not ashamed of the dream. It was a good one in both of its parts. For I am persuaded that neither politics nor the Ministry ought to be professional.

I think that men ought only to go into Parliament with certain definite objects, and when they have seen these carried, with any others which when there they have felt to be necessary, – or conversely, when they have become convinced that such objects were undesirable or impracticable, – ought to withdraw. It is only thus that the work of a Parliament can be really well done, by earnest, whole-hearted men. But when men go into Parliament, as I am afraid 9/10ths do nowadays, not for the sake of doing any specific work, but merely for the sake of being there, with or without blathering, legislation must be in a muddle, and Parliament, as it verily is, a nuisance, so that its breaking up at every vacation is a positive relief to the country. Gladstone would be a far greater man had he kept to his resolution and withdrawn from Parliament after that grand first Midlothian campaign of his, which toppled down the Tory Ministry of the day.

Again, I am convinced that what paralyses the Church of England, and more or less all the Churches, is the clerical profession, ((especially)) embraced in early youth. I don't think any man ought to be ordained to the Ministry unless he has *bona fide* exercised some profession for at least ten years. It is only thus that he can really be in touch with his fellow-men. This is not the place for an essay on Church-government, but I venture to think that there is not one of our Lord's Apostles who would not have lifted up his hands in amazement at the idea of *presbuteroi* of five and twenty, and that scarcely less would be His astonishment at hearing that the 'overlookers', whose duties involved far more physical exertion than those of the resident elders, were almost invariably selected from among the aged ones of that class and were supposed capable of 'overlooking' in their decrepitude. I stick, therefore, as much as ever to my ideal of the prime minister, ripe with all the experience of an active political life, taking up the duties of the Ministry in the Church when he has done that work which he had felt called on to do in the State.

But I have long since known that as far as I was concerned, this was one of the *couldn't-have-beens*. For the Prime Minister who carries the measures he thinks ought to be carried, which he feels himself called upon to carry, must be a first-rate man, and I have long known myself perfectly well to be only good second-rate. Not that a second-rate man does not often attain to be a prime minister, and may not do fairly well as such, but then he only carries the measures which he is allowed to carry, which he may not even think the best. I waive here, it will be seen, the question how a struggling obscure lawyer like myself could ever have reached the highest office (though indeed, had I taken a particular line at a particular moment, I do not think even that would have been impossible to me), because I am convinced that the first-rate man, who is really called of God to rule his fellows, must end by doing so, though he may begin, like Abraham Lincoln, by being a log-splitter.

Let us take a few more of my couldn't-have-beens. To begin with, – I have no manual dexterity, and therefore could never have been a first-rate workman, so that it is fortunate for me that I have had to earn my living with my wits and not with my hands.

I could never have been a great mathematician. I can learn, and by dint of pains, understand mathematics. But I have no mathematical imagination, without which I am firmly persuaded, no mathematician can be great. For there is no greater mistake than to suppose that the exact sciences and imagination have nothing to do with each other. On the contrary, every mathematical problem implies imagination. Now scarcely ever in my life have I been able to solve the simplest problem. I can, as I have said before, understand mathematical reasoning, and when I have gripped hold of it, demonstrate a theorem. But that is all.

I could not have been a great soldier. I believe a great soldier to be a compound of the observer and the mathematician. Now my powers of observation (partly owing no doubt to the nearsightedness which has only passed away in my old age) are limited, and in many respects sluggish. In my volunteer days I took part in various field days and sham fights, and I am bound to say that whilst trying my best, I never understood the reason of any single movement which my corps was called upon to make. Very likely, if I had accepted the command of my corps which I was offered when Webb took it, I might have made a fair second-rate officer. But I felt I had no call that way, and could not waste the time and money required for such a purpose. Moreover, although a soldier's son, I could never have entered either army or navy. For I could not accept for myself the principle of passive obedience. I must be responsible for my acts. I hold to the full with Hosea Biglow that

'Ef you take a sword and dror it,
Aw go stick a feller thru
Gov'munt ain't to answer for it
God'll send the bill to you'.

As for the salve of the 'My country, right or wrong' – a sentiment which would have shut the mouth of every prophet of the Old Testament, – I loathe it. Again, though I am a fairly good sailor, and can enjoy the sea in fine weather, I never had the slightest fancy for the sea as a profession, and would rather be a street-sweeper in London than an Admiral aboard. So that army and navy are for me alike decided couldn't-have-beens.

I do not think I could have been a great doctor – certainly not a great surgeon. I feel quite sure I was not born to cut and slash living flesh, and doubt if I could ever overcome my repugnance to cutting and slashing it when dead. At the same time, I found the practice of mesmerism [See appendix at end of chapter.] (it would be absurd to call it hypnotism when I made it a rule never to send my patients to sleep) deeply interesting, and that the field was one in which my limited powers of observation were fully, and I think intelligently called out. I am not therefore altogether disposed to reckon the practice of curative medicine theoretically among my couldn't-have-beens, though it must have been practically, as at present taught.

I could not have been a great engineer, again for want of sufficient powers of observation, though I look upon the career as a deeply interesting one, and if I had been in the way of taking it up, which I never was, might possibly have won a place among the second-rates.

I might say much the same as to chemistry, in which I took a great interest up to a certain point, though it has now got far far beyond me. But I have never had the feeling of being called to pursue it and devote myself to it. Still, I feel pretty confident that had I had due training, a pretty fair place among second-rate chemists was one of my might-have-beens.

I feel this still more positively about natural history in which I have always continued to take a deep interest. But still, for want of scientific imagination, I could never have been among the great masters, but only among the second-rates.

I am more puzzled as to art and music, but my capacity for either could only have been thoroughly brought out had I been compelled (as all boys should be) to get through the drudgery of both during boyhood, before taste has developed, and eye and ear have become critical. I have always taken great delight in both, and in both I think I might say my taste is a true one. I believe I got on very well on the piano and especially in the science of music during the few months I had (at 15 or 16) lessons from Mme. De la Haye, and later on I certainly greatly enjoyed Mainzer's singing class ... Still, whatever aptitude I may originally have possessed for music, there remains the fact that I have never mastered a single instrument, not even my own voice, so that all that I dare say is that musical competence should rank among my might-have-beens. In art all my training consisted in a certain number of lessons between the ages of 15 and 16, all from the flat, and therefore worth nothing, – some feeble attempts at outline sketching in the West Indies, – and some months of real work under Ruskin at the Working Men's College, when I got to dabble in water colours from life (i.e. still life). Could I have had any such training when a boy, I believe the gift would have developed itself in me, but less congenially than that of music, as I have never felt the need of drawing for myself as I have felt that of setting words to song . . .

On the other hand, eminence in exercises, sports, games, is to be reckoned among my couldn't-have-beens, partly for physical, generally for moral reasons. A weakness in the ankle as a boy prevented me from running, jumping, skating, and by the time (say after 20 years of age) that I grew stronger, I no longer though it worth while to train myself for the purpose. I know what a philistine I declare myself in saying so, but whilst I think it quite right that boys should be trained in athletic exercises, I think there is too much hard work for men still to do in the world to leave them leisure for sports . . .

Could I have been a successful merchant, manufacturer, land or mine-owner, farmer? I believe I had some of the qualities required, – sufficient clearness of head, perseverance, capacity for accounts. I did for several years contemplate the part-owning and managing of a slate quarry. But the current commercial morality, – or immorality, – of making good bargains out of the ignorance or the necessities of others I could not have practised. Could I have succeeded on such terms? I am afraid eminence in such careers would have still for me been among the couldn't-have-beens, though the mere business capacity, as I think will presently be seen, I certainly possessed.

Conversely, I never could have been a successful man-about-town, diner-out, brilliant talker. I have never been able to feel the least relish in the chitter-chatter of drawing-rooms or dining-rooms about Amy, Ethel and Mary, Jack, Tom and Harry, about the last opera, play, concert, race, match, function, scandal, flirtation. I have never what is called flirted, and although in one matter I have been perilously near temptation, I have never cared to be flirted with.

Moreover, from the time I found myself in love I avoided society lest I should be in any wise entangled, and I have continued to do so since my marriage because my wife could but to a very small extent enter into society, and I preferred her company to all other. Dancing I had given up, from distaste of the thing generally (though I am quite sensible of the charm there is in waltzing with a good partner to music) years before, and continued to do so for the same reason. So that, although I am perhaps a not uninteresting talker on my own subjects with those who can enter into them, drawing room and dining room success have always been among my couldn't-have-beens.

I come back now, and with greater assurance of determination, to my might-have-beens. Within the limits of my acquirements, I was a good lawyer, and had I been born twenty years before, I believe I might have attained quite to the front rank among conveyancers. Unfortunately for me, not for the public, I selected conveyancing as my speciality when conveyancing was already doomed, – when the simplification of real property law, on the one hand, by curtailing the length of titles, was to dry up in great measure one great source of the conveyancer's fees, the examination of abstracts, whilst the multiplication of statutory forms, the growing taste for shortening them, the embodying in the law of much that had been matter of habitual private contract, and last not least, the better education of solicitors, were to reduce the drawing of deeds to a minimum. As it was, I believe I was among the last of the pure conveyancers.

Unfortunately for me, I could do nothing else, and I lacked, by my own free will, such training as would have enlarged my knowledge of law, and thereby the sphere of my opinion-giving. Be it observed that I never could, conscientiously, have become either an equity or a common law man. Whatever may be possible to others, I never could have gone into Court to do my best towards proving that black is white without becoming a rascal. I know that it is possible to do so without dishonesty, though I am afraid that in nine cases out of ten a great advocate ends by hardly knowing what he really thinks or feels, but it was not possible to me. [The effect of advocacy on the mind was put to me in a rather original way by T. Hughes. 'When I began', he said, 'I used always to think my client a very ill-used man. Now, I always incline to think him a rogue, till the contrary be absolutely proved'.] Still, it would have been of advantage to me to have spent, say, a twelve-month with an equity man and six months with a special pleader, and so to have rounded off my legal knowledge . . .

Still, without any but a theoretical knowledge of procedure, I was, in my time, a really good jurist, possessing a larger acquaintance with the jurisprudence of foreign countries than most of my contemporaries. Moreover, as I have never been able to endure the divorce of law from morality, and have thus always felt Ulpian and not Paulus to be my real master, I think I was distinctly fitted for the place of legal adviser to the Government, which I failed to obtain in opposition to (Lord) Thring. Thring has no doubt done a great deal to simplify the language of legislation, but looking calmly back, I believe still that I was a better man for the place. Of course, I have long since given up the pursuit of juristic knowledge.

To be a good opinion-giving conveyancer is virtually to be a good judge. In after days, when I was at the Friendly Societies' office, I acquired practical judicial experience through the practice of arbitration, the Registrar of Friendly Societies being the sole arbiter in all Savings banks, and being also liable to be called upon to arbitrate in disputes under the Friendly Societies Acts, the Industrial and Provident Societies Acts, and the Building Societies Acts. I had no doubt the very efficient assistance of my colleague Brabrook, then Assistant Registrar, but without any tinge of vanity I may say that I realised in myself the possession of the judicial faculty, and that had it been my lot to sit on the bench, I should have made a good judge. This then is another of my might-have-beens.

Again, I was a good Secretary to the Friendly Societies' Commission. As to this, I need only refer to the concluding paragraph of the 4th Report of the Commissioners.

I believe I was a good Registrar, and perhaps the conferring on me of the C.B. (though I endeavoured to escape from it) in the Queen's Jubilee Year (1887) is sufficient evidence of the fact.

Nor do I hesitate to say, on the strength of my 17 years' experience at the head of the Friendly Societies Office, that I should have felt myself competent to fill, as Minister, any larger office. I say this, with reference to administrative functions; anything beyond is doubtful. Utterly intolerant of party as I am, it is very questionable whether even if I had had money to spend for the purpose (which I never had), I could ever have got into the House of Commons. If indeed I had done so, I feel confident that I could have made myself listened to – this I have always managed to do, though I have never been, and never cared to be an orator – but it is very doubtful if I could ever have been more than a voice of some individual weight. Cabinet office then, if intellectually a might-have-been, would, I rather think, have always remained practically a couldn't-have-been.

It would have been different had I lived in France. There the ball was literally at my feet. Perfectly master of the language, with a distinguished college record and many friends, I simply *must* have risen to eminence, and my English qualities of perseverance and bottom would have given me a real superiority. On the other hand, it is extremely probable that my career would have been cut short, as I must have resisted to the death the Napoleonic despotism. My political might-have-beens would thus have been unlimited on the other side of the Channel – including an early death.

I believe I write good English and good French. Elisée Reclus once said of me, in a meeting composed solely of Frenchmen, that there were not 300 Frenchmen who wrote as good French as I. This was after I had lived many years in England and had ceased to think in French. But here also, I know that I am only good second-rate, and that my style has no distinctive character. Bits of mine have been printed as parts of the writings of Mr. Maurice, of Kingsley, of T. Hughes, without anyone ever being the wiser . . . I am thus perfectly certain that I possess no originality, and have therefore no claim to a permanent place in the history of English literature. Still, I think that my *Epics of the Middle Ages* is really a good book, and that in my *War in Oude* there are pages not surpassed for vigour of style by any but really great contemporary writers.

Could I have been a novelist? I don't know. My dear old Tours tutor, M. Hatry, prophesied of me quite falsely, from my French narrative exercises, that I should write (according to the then fashion) a two-volume French novel by the time I was twenty. During the Christian Socialist Movement I did project with Hughes a series of historical novels bringing out the socialist or communist element in the past history of England. My period was to be the 17th century, dealing with the tendencies in this direction evolved out of Puritanism, such as the familist movement. I wrote, however, only the concluding scene, which was printed in *Macmillan* under the stupid title given it by Masson of 'The Restoration'. It should have been the most pathetic scene of the book, but I felt that if this was the climax the rest must be dreadfully tame, and gave up the idea altogether. It is not impossible that if I had applied myself to it, I might have made a fair second-rate novelist. I have written dialogues which have occasionally some vivacity; occasionally good bits of narration; now and then good portraits of character (as in my two *Economic Review* papers on 'Some of the Christian Socialists of the middle of the century'). But I have no great liveliness of imagination, and do not on the whole look upon the career of a novelist as a 'might-have-been' for me. Still less that of the playwright, for which I never had the slightest inclination.

As respects poetry, – well, I think I have written a few good bits of verse, and Kingsley, when I asked him to look over a volume which I had prepared for publication, marked several in pencil 'beautiful' and the like. But he very frankly and truly said that as a whole it was second rate, and I thereupon wisely gave up the idea of attempting to publish it. I think indeed the best verse I have written is what he never saw, and that I have written perhaps 3 or 4 things in all which might deserve insertion in some *Golden Treasury* of the future.

Where I believe I have excelled is as a literary critic (both as reviewer and as editor). Meredith Townsend once said that a review by me of Helps's *Realmah* was the best review he had ever read. I could not say as much myself, but I do feel in myself the critical, the judging faculty. Hence, I do not think that – although there are a few feeble papers in it – any three months of a literary paper have surpassed the first three months (barring No. 1) of the *Reader*, during my editorship. Although he retained many of my staff of writers (some refused to write except for me), I think the falling off was immediate as soon as Masson took the editorship in my stead. I think now, as I thought at the time, that Macmillan made a great mistake in not giving me the editorship of *Macmillan's Magazine* when he started it, and handing it to Masson. I believe that I know what is good as literature, and that publishers have been unwise in not securing me as an editor for their periodicals.

I have the capacity to teach, men more especially. I was a good teacher for men and boys in Little Ormond Yard, for men in the Hall of Association, and again at the Wimbledon Lecture Hall, and for men again in my little home Bible class in Warwick Square. With one exception, that of my unfortunate attempt at the Hall of Association to cram into, one lecture, with next to no physical illustrations, the principles of natural history, my lectures have always been successful. I believe therefore I should have made a good Professor, whether of law or jurisprudence, language, history, philosophy, economics or ethics, i.e. of any study really interesting to me. My lectures and addresses, I may observe, have been equally successful whether read or delivered from notes, though I have found from my latest experiment, an address to the ladies of St. Margaret's House, on 'Friendly and other Societies' that with my failing memory the strain of speaking from notes only has become too great for me.

I have occasionally spoken *extempore* with effect. But I am quite conscious of not being an orator.

To sum up: I have been a good conveyancer, jurist, reviewer, lecturer, editor, secretary, official. I believe I might have been a good judge, professor, M.P., Minister (not Prime).

I have no scruple in saying so, because I believe every honest man ought to be able to gauge his own intellectual capacities.

It is for others to judge his moral worth. On this head all I will say is that many good men have thought me good enough to call their friend; that I have seldom lost a friend once made (though in some few cases I have been mistaken in my judgment of others, over-rating them till undeceived); that my second oldest friend died last year (1895) after 59 years of friendship, and that with the oldest of all I have now enjoyed 63 years of such.

APPENDIX

((Ludlow included several supplements to his autobiography on aspects of his life not relevant to his public life (including a 45-page chapter entitled 'Animals whom I have known'). An extract from one of these follows.))

SOME MESMERIC EXPERIENCES
The subject of mesmerism or animal magnetism as it is called (both names are quite unsatisfactory, the second more so) had only come before me in an occasional cursory way, before I knew Charles Mansfield. I remember when we lived in France, a 'somnambuli' being brought to my invalid sister Julia. I was present once with my sister Eliza at a mesmeric seance, where the operator, some German, failing entirely to show anything at all, dismissed his audience with a confession of failure, which he attributed to the presence of some adverse influence in the room ... [As a matter of fact, I had been praying within myself that if the thing were of God it should succeed, but if it were of the devil it should fail. In another world I may know whether my prayer had any effect.]

With Charles Mansfield I found myself in the presence of a man of the acutest powers of observation, and of the most rigidly scientific turn of mind, who had thoroughly studied the subject. I think his conclusions may be summed up as follows: first, that health is communicable, and that the so-called mesmeric passes afford an apt medium for such communications; secondly, that such passes, directed to the head, produce a state of artificial catalepsy accompanied by abnormal phenomena, such as insensibility to pain, rigidity of the muscles, subjection of the movements to the will of the

319

operator and the establishment of peculiar mental relations between the two, and in certain cases an insight into the ordinarily invisible commonly called *clairvoyance* . . .

In my own cases, I freely admit that it was not the love of my fellow creatures generally that suggested my efforts in mesmeric therapeutics. They proceeded simply from my love to the one who has since become my wife. The ills from which she was suffering were such as I was informed yielded freely to mesmeric treatment. If I was still too poor to ask her to be my wife, I might at least pour my strength into her. I talked the idea over with Charles Mansfield, the only friend to whom I ever confided my love . . . and he fully encouraged me. There was then in Weymouth Street, Portland Place, a 'Mesmeric Infirmary'. I called there, saw the Secretary, and told him that I wished to practice mesmerism, gratuitously, and wished to be taught. I bought a small *Mesmerist's Manual*, which I believe I have still somewhere, unless I have lent it, and he took me through the men's portion of the infirmary, enjoining me to keep silence and observe. I remember in particular a bricklayer, who had been suffering from acute rheumatism, whom I spoke to when his 20 minutes of passes were over, and who told me how greatly he had already benefitted at the infirmary. I asked the Secretary what liability the mesmerist incurred of contracting disease. He answered that the infirmary did not receive what are called infectious cases, but that by taking the precaution of washing one's hands immediately after the cessation of the passes, no evil effects were to be apprehended, with this exception, that women mesmerics could not safely be employed in diseases peculiar to women – for these the mesmerisers must be men. I asked him finally, if he had any case which he could assign to me on trial. He referred to a list of applicants for treatment, and selected one in which the applicants stated themselves unable to pay the very stiff fees charged to patients. I undertook the case.

It was that of the epileptic son of a small bootmaker working at home. I used to go there, three mornings a week, before attending at chambers . . . The son was a lawyer's clerk – very clever, they said, in his calling – but his epileptic attacks were constantly making him lose his place, coming upon him sometimes twice a day, and he himself was now out of a place and becoming in a manner desperate. In the course of about six weeks his fits, which had been of daily occurrence before I took him in hand, had diminished to about three a week. Before long he found another place, and kept it till he changed it for a better one, the fits diminishing always in frequency and violence till he ceased to dread them. I was able to mesmerise him at less and less frequent intervals, and at last to discontinue doing so. But in the meanwhile I had found another patient in the same house.

Besides the son, there was also a daughter deformed from infancy in a way I never saw elsewhere. Owing probably to maltreatment at delivery, the backbone towards the bottom took a complete twist, so that one of the *nates* was nearly twice the size of the other, of course making her lame. But in addition to this she had on one side of the neck a large goitre. Whilst I was treating the son the parents took the girl to Ferguson at King's College Hospital . . . After a further examination, he told them that the only remedy was to excise the tumour.

The brother was by this time decidedly improving under my treatment. When only a fortnight remained from the intended operation, I was asked whether mesmerism could do anything for the girl. I said I could not tell in the least, though I felt convinced it would do no harm. Would I try it? I said I would do so if they wished, but they had better go to the Mesmeric Infirmary. No, their son had improved so much under my hands, that they preferred my trying. The girl said the same. By this time I did not mesmerise the son as often as three times a week, and I gave three days to the girl. After a week, it seemed to me that the swelling was diminishing a little. The next week the mother expressed to me the same feeling. The girl herself was persuaded of the fact. She was evidently looking better. At the end of the fortnight they did not take her to King's College Hospital. I continued to mesmerise her, and at last had the satisfaction of seeing the tumour disappear entirely. It was purely a nervous growth, and the lancet would have been her death. This was proved by the fact that, after I had ceased to mesmerise her, the mother told me that when her daughter was worried, the throat showed a tendency to swell, but the girl had ceased to be afraid of it, and would say of the swelling: 'Oh, if it gets bad, Mr. Ludlow will take it away' . . .

Armed with the experience of these two cases, I asked my cousin to let me mesmerise her, but met with such an absolute refusal that I gave up systematic mesmerising. At the same time I had on several occasions a remarkable confirmation of mesmerism as a curative force . . .

EDITOR'S NOTES

1. In the margin beside this page in the ms., Ludlow wrote, 'too egotistic & personal'.
2. A tapster in the Royal Household.
3. The Gurkha War 1814-16.
4. The 1st Burmese War 1823-25.
5. Ludlow uses 'brother-in-law' and 'brother' more or less indiscriminately throughout the narrative.
6. From a winter spent in Tours.
7. Loutherburg anglicised his name to Loutherbourgh on winning great success in England.
8. The word appears clearly as 'boroughrun' in the ms., but this reading must be correct, meaning Alderman.
9. Absolutely, quite simply.
10. Ludlow appended a pamphlet expounding the work of the *Société des Amis des Pauvres* to this chapter; it can be found among the Ludlow papers.
11. Sir Charles Forbes (see chapters VII and VIII).
12. The German Biblical Scholar and radical Higher Critic.
13. These articles, written by Henry Mayhew, were the forerunners of his massive *London Labour and the London Poor* (reprinted by Cass, London 1967), and have been collected and edited by Anne Humphreys in 'Voices of the Poor', Cass, London, 1971.
14. Live and Let Live Health League
    ## THE HEALTH LEAGUE
    ### (Objects)
    For uniting all classes of society in the promotion of the Public Health, and the removal of all causes of disease which unnecessarily abridge man's right to live;
    ### (Means of Action)
    1. By collecting and diffusing information;
    2. By furthering the due execution, and where necessary, the amendment of the law.
    3. By stimulating and assisting all public bodies and private persons in the performance of their respective duties in reference to the Public Health;

EDITOR'S NOTES

In the margin beside this page in the ms., Ludlow wrote, 'too egotistic & personal'.

A tapster in the Royal Household.

The Gurkha War 1814-16.

The 1st Burmese War 1823-25.

Ludlow uses 'brother-in-law' and 'brother' more or less indiscriminately throughout the narrative.

From a winter spent in Tours.

Loutherburg anglicised his name to Loutherbourgh on winning great success in England.

The word appears clearly as 'boroughrun' in the ms., but this reading must be correct, meaning Alderman.

Absolutely, quite simply.

Ludlow appended a pamphlet expounding the work of the *Société des Amis des Pauvres* to this chapter; it can be found among the Ludlow papers.

Sir Charles Forbes (see chapters VII and VIII).

The German Biblical Scholar and radical Higher Critic.

These articles, written by Henry Mayhew, were the forerunners of his massive *London Labour and the London Poor* (reprinted by Cass, London 1967), and have been collected and edited by Anne Humphreys in 'Voices of the Poor', Cass, London, 1971.

Live and Let Live Health League

THE HEALTH LEAGUE
(Objects)

For uniting all classes of society in the promotion of the Public Health, and the removal of all causes of disease which unnecessarily abridge man's right to live;

(Means of Action)

1. By collecting and diffusing information;

2. By furthering the due execution, and where necessary, the amendment of the law.

3. By stimulating and assisting all public bodies and private persons in the performance of their respective duties in reference to the Public Health;

(Machinery)

Through the means of local and district committees.

(Terms of Admission)

All members may become members of the Health League on payment of a yearly subscription of one shilling and one penny, payable in advance . . .

15. See Professor Christensen's presentation of the controversy between Maurice and Ludlow on this point in *Origins and Theory of Christian Socialism 1848-54*, Aarhus 1962.

16. T. L. Jackson, the so-called 'Thieves' Missionary', and a close confidant of Lord Shaftesbury, was a member of the managing committee.

17. Ludlow appended a footnote to this paragraph in the ms., but later crossed it out; part of it seems, nonetheless, worth reproducing. 'Whilst the East End Needlewomen's Workshop was in existence, the managing committee were in the habit of meeting every month at Lord Shaftesbury's. On no single occasion did he shake hands with one of us until the last meeting, when he did so with every one . . .

It is of course impossible to read Lord Shaftesbury's *Life* without feeling how good he was. Yet I believe (and my experience is not a solitary one) that he was essentially an aristocrat, and that, no doubt without being aware of it, some sense of condescension formed an inseparable ingredient in his otherwise boundless charity.

I met him some years after at a Social Science Congress of which he was Chairman. We stood some moments within close hand-shaking distance, looked each other straight in the face; nothing more.'

18. This is of course a surprising statement, and can be traced to a failure of Ludlow to correct his ms. accurately. The original opening sentence began – 'The work which we did in those years may be divided under three heads – literary, Parliamentary, and the more properly social and practical work' – he later altered the opening incompletely.

19. Ludlow here appends a list of some twenty or so pamphlets and tracts published by members of the group between 1850 and 1852.

20. The tractarian organ, not the Liberal newspaper.

21. Ludlow is quite correct in referring to him as Lord Ashley; he succeeded to the Earldom of Shaftesbury in 1851, before he became involved with the East End Needlewomen's Workshop described earlier in this chapter.

22. T. H. S. Sotheron-Estcourt, Tory M. P. for Devizes.

23. Ludlow described in an appendix to this chapter the odd circumstances of his receiving a legacy of 150 guineas as executor of a certain Stevenson estate.

24. The account of this dispute in Christensen, op.cit., pp. 176-219, is extremely full and well-balanced.

25. In Chapter XVI.

26. Ludlow appended the text of this pamphlet to the present chapter.

27. This bitter struggle is fully analysed in Christensen, op.cit., pp. 319-326.

28. In the ms. Ludlow includes at this point a long list of these activities in detail from which one or two have been extracted.
29. Short extracts only are given from the report which Ludlow reproduced.
30. The more or less permanent Executive Committee – see Chapter XIX.
31. The Committee on Teaching and Publication.
32. This final paragraph is partially and somewhat half-heartedly struck out in the ms.
33. For a full account of Furnivall's contribution to, and controversies with, the College, see J. F. C. Harrison, *History of the Working Men's College* (the early chapters).
34. This phrase is crossed out in the ms.
35. In an appendix to this chapter, Ludlow described the career of the Working Gilders' Association, one of the most successful of the associations, lasting from 1858-1891. It was founded by two students of the Working Men's College.
36. The successful playwright and writer (Editor of *Punch* 1874-80).
37. The registrar's office was located in Abingdon Street.
38. Many of Ludlow's references to Trade Union leaders have here been omitted.
39. Either Ludlow never completed this chapter, or it has been lost.
40. See N. C. Masterman's biography: John Malcolm Ludlow (Cambridge 1963), p. 238, for an explanation of these family affairs. N. B., however, Masterman's mistake in writing of the sudden death of 'a clergyman brother of hers ((his wife's)) and his wife'. In fact what seems to have happened is that the wife died (in 1883) and the clergyman brother, Granville Forbes, became more or less intermittently insane. But he certainly did not die, and remained Rector of Broughton, Kettering, until at least 1894 (see Crockford Clerical Directory).
41. See Masterman, op.cit., pp. 227-235.

INDEX

Page references in bold type indicate either an important reference, or one in which the subject is dealt with in some detail.

THE AUTOBIOGRAPHY
OF
JOHN LUDLOW